Edited by
Katja Loos

**Biocatalysis
in Polymer Chemistry**

Further Reading

Fessner, W.-D., Anthonsen, T. (Eds.)

Modern Biocatalysis

Stereoselective and Environmentally Friendly Reactions

2009
ISBN: 978-3-527-32071-4

Matyjaszewski, K., Müller, A. H. E. (Eds.)

Controlled and Living Polymerizations

From Mechanisms to Applications

2009
ISBN: 978-3-527-32492-7

Grogan, G.

Practical Biotransformations

A Beginner's Guide

2009
ISBN: 978-1-4051-7125-0

Dubois, P., Coulembier, O., Raquez, J.-M. (Eds.)

Handbook of Ring-Opening Polymerization

2009
ISBN: 978-3-527-31953-4

Crabtree, R. H. (Ed.)

Handbook of Green Chemistry – Green Catalysis

2009
ISBN: 978-3-527-31577-2

Rothenberg, G.

Catalysis

Concepts and Green Applications

2008
ISBN: 978-3-527-31824-7

Morokuma, K., Musaev, D. (Eds.)

Computational Modeling for Homogeneous and Enzymatic Catalysis

A Knowledge-Base for Designing Efficient Catalysts

2008
ISBN: 978-3-527-31843-8

Edited by Katja Loos

Biocatalysis in Polymer Chemistry

WILEY-VCH Verlag GmbH & Co. KGaA

The Editor

Prof. Katja Loos
University of Groningen
Dept. of Polymer Chemistry
Nijenborgh 4
9747 AG Groningen
The Netherlands

All books published by **Wiley-VCH** are carefully produced. Nevertheless, authors, editors, and publisher do not warrant the information contained in these books, including this book, to be free of errors. Readers are advised to keep in mind that statements, data, illustrations, procedural details or other items may inadvertently be inaccurate.

Library of Congress Card No.: applied for

British Library Cataloguing-in-Publication Data
A catalogue record for this book is available from the British Library.

Bibliographic information published by the Deutsche Nationalbibliothek
The Deutsche Nationalbibliothek lists this publication in the Deutsche Nationalbibliografie; detailed bibliographic data are available on the Internet at http://dnb.d-nb.de.

© 2011 Wiley-VCH Verlag & Co. KGaA, Boschstr. 12, 69469 Weinheim, Germany

All rights reserved (including those of translation into other languages). No part of this book may be reproduced in any form – by photoprinting, microfilm, or any other means – nor transmitted or translated into a machine language without written permission from the publishers. Registered names, trademarks, etc. used in this book, even when not specifically marked as such, are not to be considered unprotected by law.

Composition   Toppan Best-set Premedia Limited, Hong Kong
Printing and Binding   Fabulous Printers Pte. Ltd., Singapore
Cover Design   Adam Design, Weinheim

Printed in Singapore
Printed on acid-free paper

ISBN: 978-3-527-32618-1

Contents

Preface *XIII*
List of Contributors *XIX*
List of Abbreviations *XXIII*

1 Monomers and Macromonomers from Renewable Resources *1*
Alessandro Gandini
1.1 Introduction *1*
1.2 Terpenes *2*
1.3 Rosin *4*
1.4 Sugars *6*
1.5 Glycerol and Monomers Derived Therefrom *8*
1.6 Furans *11*
1.7 Vegetable Oils *16*
1.8 Tannins *21*
1.9 Lignin Fragments *23*
1.10 Suberin Fragments *26*
1.11 Miscellaneous Monomers *28*
1.12 Conclusions *29*
References *29*

2 Enzyme Immobilization on Layered and Nanostructured Materials *35*
Ioannis V. Pavlidis, Aikaterini A. Tzialla, Apostolos Enotiadis, Haralambos Stamatis, and Dimitrios Gournis
2.1 Introduction *35*
2.2 Enzymes Immobilized on Layered Materials *36*
2.2.1 Clays *36*
2.2.1.1 Introduction *36*
2.2.1.2 Enzymes Immobilization on Clays *38*
2.2.2 Other Carbon Layered Materials *43*
2.3 Enzymes Immobilized on Carbon Nanotubes *44*
2.3.1 Introduction *44*
2.3.2 Applications *45*
2.3.3 Immobilization Approaches *46*

Biocatalysis in Polymer Chemistry. Edited by Katja Loos
Copyright © 2011 WILEY-VCH Verlag GmbH & Co. KGaA, Weinheim
ISBN: 978-3-527-32618-1

2.3.4	Structure and Catalytic Behavior of Immobilized Enzymes 50
2.4	Enzymes Immobilized on Nanoparticles 52
2.4.1	Introduction 52
2.4.2	Applications 53
2.4.3	Immobilization Approaches 55
2.4.4	Structure and Catalytic Behavior of Immobilized Enzymes 57
2.5	Conclusions 57
	References 57

3	**Improved Immobilization Supports for *Candida Antarctica* Lipase B** 65
	Paria Saunders and Jesper Brask
3.1	Introduction 65
3.2	Industrial Enzyme Production 66
3.2.1	Fermentation 66
3.2.2	Recovery and Purification 66
3.2.3	Formulation 67
3.3	Lipase for Biocatalysis 67
3.3.1	*Candida Antarctica* Lipase B (CALB) 67
3.4	Immobilization 68
3.4.1	Novozym 435 69
3.4.2	NS81018 71
3.5	CALB- Catalyzed Polymer Synthesis 71
3.5.1	Polymerization 72
3.5.2	Polymer Separation and Purification 72
3.5.3	Characterization and Performance Assays 73
3.5.4	CALB Immobilization 73
3.5.5	Results and Discussion 74
3.5.5.1	Effect of Synthesis Time on Molecular Weight 74
3.5.5.2	Comparison of NS 81018 and Novozym 435 75
3.5.5.3	Determination of Polycaprolactone Molecular Weight by GPC 75
3.5.5.4	Effect of Termination of Reaction 77
3.5.5.5	Effect of Solvent 78
3.5.5.6	Effect of Water 78
3.5.5.7	Effect of Immobilization Support 79
3.6	Conclusions 80
	Acknowledgment 81
	References 81

4	**Enzymatic Polymerization of Polyester** 83
	Nemanja Miletić, Katja Loos, and Richard A. Gross
4.1	Introduction 83
4.2	Synthesis of Polyesters 84
4.3	Enzyme-Catalyzed Polycondensations 85
4.3.1	A-B Type Enzymatic Polyesterfication 86
4.3.2	AA-BB Type Enzymatic Polyesterification 92
4.3.3	Use of Activated Enol Esters for in vitro Polyester Synthesis 97

4.4	Enzyme-Catalyzed Ring-Opening Polymerizations	102
4.4.1	Unsubstituted Lactones	102
4.4.2	Substituted Lactones	109
4.4.3	Cyclic Ester Related Monomers	111
4.5	Enzymatic Ring-Opening Copolymerizations	113
4.6	Combination of Condensation and Ring-Opening Polymerization	121
4.7	Conclusion	122
	References	123

5 Enzyme-Catalyzed Synthesis of Polyamides and Polypeptides 131
H. N. Cheng

5.1	Introduction	131
5.2	Catalysis via Protease	132
5.3	Catalysis via Lipase	134
5.4	Catalysis via Other Enzymes	136
5.5	Comments	137
	References	138

6 Enzymatic Polymerization of Vinyl Polymers 143
Frank Hollmann

6.1	Introduction	143
6.2	General Mechanism and Enzyme Kinetics	143
6.3	Peroxidase-Initiated Polymerizations	146
6.3.1	Mechanism of Peroxidase-Initiated Polymerization	147
6.3.2	Influence of the Single Reaction Parameters	148
6.3.2.1	Enzyme Concentration	148
6.3.2.2	Hydrogen Peroxide Concentration	148
6.3.2.3	Mediator and Mediator Concentration	150
6.3.2.4	Miscellaneous	152
6.3.3	Selected Examples for Peroxidase-Initiated Polymerizations	153
6.4	Laccase-Initiated Polymerization	156
6.5	Miscellaneous Enzyme Systems	159
6.6	The Current State-of-the-Art and Future Developments	160
	References	161

7 Enzymatic Polymerization of Phenolic Monomers 165
Hiroshi Uyama

7.1	Introduction	165
7.2	Peroxidase-Catalyzed Polymerization of Phenolics	165
7.3	Peroxidase-Catalyzed Synthesis of Functional Phenolic Polymers	170
7.4	Laccase-Catalyzed Polymerization of Phenolics	176
7.5	Enzymatic Preparation of Coatings	177
7.6	Enzymatic Oxidative Polymerization of Flavonoids	179
7.7	Concluding Remarks	182
	References	182

8	**Enzymatic Synthesis of Polyaniline and Other Electrically Conductive Polymers** *187*
	Rodolfo Cruz-Silva, Paulina Roman, and Jorge Romero
8.1	Introduction *187*
8.2	PANI Synthesis Using Templates *188*
8.2.1	Polyanion-Assisted Enzymatic Polymerization *188*
8.2.2	Polycation-Assisted Templated Polymerization of Aniline *190*
8.3	Synthesis of PANI in Template-Free, Dispersed and Micellar Media *192*
8.3.1	Template-Free Synthesis of PANI *192*
8.3.2	Synthesis in Dispersed Media *192*
8.3.3	Enzymatic Synthesis of PANI Using Anionic Micelles as Templates *193*
8.4	Biomimetic Synthesis of PANI *194*
8.4.1	Hematin and Iron-Containing Porphyrins *194*
8.4.2	Heme-Containing Proteins *195*
8.5	Synthesis of PANI Using Enzymes Different From HRP *195*
8.5.1	Other Peroxidases *196*
8.5.2	Synthesis of PANI Using Laccase Enzymes *197*
8.5.3	Synthesis of PANI Using Other Enzymes *198*
8.6	PANI Films and Nanowires Prepared with Enzymatically Synthesized PANI *199*
8.6.1	In Situ Enzymatic Polymerization of Aniline *199*
8.6.2	Immobilization of HRP on Surfaces *200*
8.6.2.1	Surface Confinement of the Enzymatic Polymerization *200*
8.6.2.2	Nanowires and Thin Films by Surface-Confined Enzymatic Polymerization *201*
8.6.3	PANI Fibers Made with Enzymatically-Synthesized PANI *202*
8.6.4	Layer-by-Layer and Cast Films of Enzymatically-Synthesized PANI *202*
8.7	Enzymatic and Biocatalytic Synthesis of Other Conductive Polymers *203*
8.7.1	Enzymatic and Biocatalytic Synthesis of Polypyrrole *203*
8.7.2	Enzymatic and Biocatalytic Synthesis of Polythiophenes *205*
8.8	Conclusions *207*
	References *207*
9	**Enzymatic Polymerizations of Polysaccharides** *211*
	Jeroen van der Vlist and Katja Loos
9.1	Introduction *211*
9.2	Glycosyltransferases *213*
9.2.1	Phosphorylase *214*
9.2.1.1	Enzymatic Polymerization of Amylose with Glycogen Phosphorylase *215*
9.2.1.2	Hybrid Structures with Amylose Blocks *220*

9.2.2	Branching Enzyme	224
9.2.3	Sucrase	227
9.2.4	Amylomaltase	228
9.2.5	Hyaluronan Synthase	229
9.3	Glycosidases	231
9.3.1	Cellulase	232
9.3.2	Hyaluronidase	234
9.3.3	Glycosynthases	236
9.4	Conclusion	237
	References	238

10 Polymerases for Biosynthesis of Storage Compounds *247*
Anna Bröker and Alexander Steinbüchel

10.1	Introduction	247
10.2	Polyhydroxyalkanoate Synthases	249
10.2.1	Occurrence of Polyhydroxyalkanoate Synthases	249
10.2.2	Chemical Structures of Polyhydroxyalkanoates and their Variants	250
10.2.3	Reaction Catalyzed by the Key Enzyme	251
10.2.4	Assay of Enzyme Activity	252
10.2.5	Location of Enzyme and Granule Structure	252
10.2.6	Primary Structures of the Enzyme	253
10.2.7	Special Motifs and Essential Residues	254
10.2.8	The Catalytic Mechanism of Polyhydroxyalkanoate Synthases	254
10.2.9	*In Vitro* Synthesis	255
10.2.10	Embedding in General Metabolism	255
10.2.11	Biotechnological Relevance	256
10.3	Cyanophycin Synthetases	257
10.3.1	Occurrence of Cyanophycin Synthetases	257
10.3.2	Chemical Structure of Cyanophycin	258
10.3.3	Variants of Cyanophycin	259
10.3.4	Reaction Catalyzed by the Key Enzyme	260
10.3.5	Assay of Enzyme Activity	260
10.3.6	Location of Enzyme–Granule Structure	261
10.3.7	Kinetic Data of Wild Type Enzyme	261
10.3.8	Primary Structures and Essential Motifs of the Enzyme	262
10.3.9	Catalytic Cycle	263
10.3.10	Mutant Variants of the Enzyme	265
10.3.11	*In Vitro* Synthesis	266
10.3.12	Embedding in General Metabolism	267
10.3.13	Biotechnological Relevance	267
10.4	Conclusions	268
	References	268

11 Chiral Polymers by Lipase Catalysis 277
Anja Palmans and Martijn Veld

11.1 Introduction 277
11.2 Reaction Mechanism and Enantioselectivity of Lipases 278
11.3 Lipase-catalyzed Synthesis and Polymerization of Optically Pure Monomers 280
11.4 Kinetic Resolution Polymerization of Racemic Monomers 284
11.4.1 KRP of Linear Monomers 284
11.4.2 KRP of Substituted Lactones 286
11.5 Dynamic Kinetic Resolution Polymerization of Racemic Monomers 287
11.5.1 Dynamic Kinetic Resolutions in Organic Chemistry 288
11.5.2 Extension of Dynamic Kinetic Resolutions to Polymer Chemistry 289
11.5.3 Dynamic Kinetic Resolution Polymerizations 290
11.5.4 Iterative Tandem Catalysis: Chiral Polymers from Racemic ω-Methylated Lactones 294
11.6 Tuning Polymer Properties with Chirality 296
11.6.1 Chiral Block Copolymers Using Enzymatic Catalysis 296
11.6.2 Enantioselective Acylation and Deacylation on Polymer Backbones 299
11.6.3 Chiral Particles by Combining eROP and Living Free Radical Polymerization 300
11.7 Conclusions and Outlook 301
References 301

12 Enzymes in the Synthesis of Block and Graft Copolymers 305
Steven Howdle and Andreas Heise

12.1 Introduction 305
12.2 Synthetic Strategies for Block Copolymer Synthesis Involving Enzymes 306
12.2.1 Enzymatic Polymerization from Functional Polymers (Macroinitiation) 307
12.2.2 Enzymatic Synthesis of Macroinitiators Followed by Chemical Polymerization 310
12.2.2.1 Dual Initiator Approach 310
12.2.2.2 Modification of Enzymatic Blocks to Form Macroinitiators 316
12.3 Enzymatic Synthesis of Graft Copolymers 319
12.4 Summary and Outlook 320
References 320

13 Biocatalytic Polymerization in Exotic Solvents 323
Kristofer J. Thurecht and Silvia Villarroya

13.1 Supercritical Fluids 324

13.1.1	Lipase-catalyzed Homopolymerizations	*326*
13.1.2	Lipase-catalyzed Depolymerization (Degradation)	*328*
13.1.3	Combination of Polymerization Mechanisms: Polymerization from Bifunctional Initiators	*329*
13.1.4	Free Radical Polymerization Using Enzymatic Initiators	*333*
13.2	Biocatalytic Polymerization in Ionic Liquids	*334*
13.2.1	Free Radical Polymerization	*334*
13.2.2	Lipase-catalyzed Polymerization in Ionic Liquids	*337*
13.3	Enzymatic Polymerization under Biphasic Conditions	*339*
13.3.1	Ionic Liquid-Supported Catalyst	*340*
13.3.2	Biphasic Polymerization of Polyphenols	*342*
13.3.3	Fluorous Biphasic Polymerization	*342*
13.4	Other 'Exotic' Media for Biocatalytic Polymerization	*342*
13.5	Conclusion	*343*
	References	*343*
14	**Molecular Modeling Approach to Enzymatic Polymerization**	*349*
	Gregor Fels and Iris Baum	
14.1	Introduction	*349*
14.2	Enzymatic Polymerization	*352*
14.3	*Candida antarctica* Lipase B – Characterization of a Versatile Biocatalyst	*353*
14.4	Lipase Catalyzed Alcoholysis and Aminolysis of Esters	*354*
14.5	Lipase-Catalyzed Polyester Formation	*357*
14.6	CALB -Catalyzed Polymerization of β-Lactam	*357*
14.7	General Remarks	*367*
	References	*367*
15	**Enzymatic Polymer Modification**	*369*
	Georg M. Guebitz	
15.1	Introduction	*369*
15.2	Enzymatic Polymer Functionalization: From Natural to Synthetic Materials	*369*
15.3	Surface Hydrolysis of Poly(alkyleneterephthalate)s	*370*
15.3 1	Enzymes and Processes	*370*
15.3.2	Mechanistic Aspects	*372*
15.3.3	Surface Analytical Tools	*375*
15.4	Surface Hydrolysis of Polyamides	*376*
15.4.1	Enzymes and Processes	*376*
15.4.2	Mechanistic Aspects	*377*
15.5	Surface Hydrolysis of Polyacrylonitriles	*378*
15.6	Future Developments	*380*
	Acknowledgment	*380*
	References	*381*

16	**Enzymatic Polysaccharide Degradation** *389*
	Maricica Munteanu and Helmut Ritter
16.1	The Features of the Enzymatic Degradation *389*
16.2	Enzymatic Synthesis and Degradation of Cyclodextrin *390*
16.2.1	Cyclodextrins: Structure and Physicochemical Properties *390*
16.2.1.1	The Discovery Period from 1891–1935 *392*
16.2.1.2	The Exploratory Period from 1936–1970 *392*
16.2.1.3	The Utilization Period: from 1970 Onward *392*
16.2.2	Cyclodextrin Synthesis via Enzymatic Degradation of Starch *392*
16.2.2.1	Cyclodextrin Glycosyltransferases: Structure and Catalytic Activity *393*
16.2.2.2	Cyclodextrin Glycosyltransferase: Cyclodextrin-Forming Activity *394*
16.2.2.3	Other Industrial Applications of Cyclodextrin Glycosyltransferase *397*
16.2.3	Cyclodextrin Hydrolysis *398*
16.2.3.1	Acidic Hydrolysis of Cyclodextrin *399*
16.2.3.2	Cyclodextrin Enzymatic Degradation *400*
16.2.3.3	Cyclodextrin Degradation by the Intestinal Flora *404*
16.2.4	Enzymatic Synthesis of Cyclodextrin-Derivatives *405*
16.2.5	Cyclodextrin-Based Enzyme Mimics *405*
16.2.6	Specific-Base-Catalyzed Hydrolysis *406*
16.3	Hyaluronic Acid Enzymatic Degradation *406*
16.3.1	Hyaluronic Acid: Structure, Biological Functions and Clinical Applications *406*
16.3.2	Hyaluronidase: Biological and Clinical Significance *408*
16.4	Alginate Enzymatic Degradation *409*
16.4.1	Alginate as Biocompatible Polysaccharide *409*
16.4.2	Alginate Depolymerization by Alginate Lyases *411*
16.5	Chitin and Chitosan Enzymatic Degradation *411*
16.5.1	Enzymatic Hydrolysis of Chitin *411*
16.5.2	Enzymatic Hydrolysis of Chitosan *413*
16.6	Cellulose Enzymatic Degradation *414*
16.7	Conclusion *415*
	References *415*

Index *421*

Preface

Biocatalytic pathways to polymeric materials are an emerging research area with not only enormous scientific and technological promise, but also a tremendous impact on environmental issues.

Whole cell biocatalysis has been exploited for thousands of years. Historically biotechnology was manifested in skills such as the manufacture of wines, beer, cheese etc., where the techniques were well worked out and reproducible, while the biochemical mechanism was not understood.

While the chemical, economic and social advantages of biocatalysis over traditional chemical approaches were recognized a long time ago, their application to industrial production processes have been largely neglected until recent breakthroughs in modern biotechnology (such as robust protein expression systems, directed evolution etc). Subsequently, in recent years, biotechnology has established itself as an indispensable tool in the synthesis of small molecules in the pharmaceutical sector including antibiotics, recombinant proteins and vaccines and monoclonal antibodies.

Enzymatic polymerizations are a powerful and versatile approach which can compete with chemical and physical techniques to produce known materials such as 'commodity plastics' and also to synthesize novel macromolecules so far not accessible via traditional chemical approaches.

Enzymatic polymerizations can prevent waste generation by using catalytic processes with high stereo- and regio-selectivity; prevent or limit the use of hazardous organic reagents by, for instance, using water as a green solvent; design processes with higher energy efficiency and safer chemistry by conducting reactions at room temperature under ambient atmosphere; and increase atom efficiency by avoiding extensive protection and deprotection steps. Because of this enzymatic polymerizations can provide an essential contribution to achieving industrial sustainability in the future.

In addition, nature achieves complete control over the composition and polydispersities of natural polymers – an achievement lacking in modern polymer synthesis even by using living polymerization techniques. Biotechnology therefore holds tremendous opportunities for realizing unique new functional polymeric materials.

In this first textbook on the topic we aim to give a comprehensive overview on the current status of the field of sustainable, eco-efficient and competitive production of (novel) polymeric materials via enzymatic polymerization. Furthermore an outlook on the future trends in this field is given.

Enzyme Systems Discussed

Enzymes are responsible for almost all biosynthetic processes in living cells. These biosynthetic reactions proceed under mild and neutral conditions at a low temperature and in a quantitative conversion. This, together with the high catalytic activity and selectivity, makes enzymes highly dedicated catalysts. The reaction rates of enzyme catalyzed reactions are typically 10^6 to 10^{12} times greater than the uncatalyzed reactions but can be as high as 10^{17}. In general, the selectivity is higher than conventional catalysts and side products are rarely formed.

According to the first report of the Enzyme Commision from 1961 all enzymes are classified in six enzyme classes, depending on the reaction being catalyzed. Within the scheme of identification each enzyme has an Enzyme Commission number denominated by four numbers after the abbreviation E.C. The first number indicates one of the six possible reaction types that the enzyme can catalyze; the second number defines the chemical structures that are changed in this process; the third defines the properties of the enzyme involved in the catalytic reaction or further characteristics of the catalyzed reaction; the fourth number is a running number.

At present, enzymes from 4 of the 6 E.C. enzyme classes are known to induce or catalyze polymerizations. An overview of the main enzyme and polymer systems discussed in this book is shown in the following table.

Enzyme class	Biochemical function in living systems	Typical enzymes inducing polymerization	Typical polymers	Covered in (parts) in Chapter
I. Oxidoreducates	Oxidation or reduction	Peroxidase Laccase	Polyanilines, Polyphenols, Polystyrene, Polymethyl methacrylate	6, 7, 8, 13, 15
II. Transferases	Transfer of a group from one molecule to another	PHA synthase Hyaluronan synthase Phosphorylase	Polyesters Hyaluronan Amylose	9, 10, 16
III. Hydrolases	Hydrolysis reaction in H_2O	Lipase Cellulase, Hyaluronidase Papain	Polyesters, Cellulose, Glycosaminoglycans (Oligo)peptides	4, 5, 11, 12, 13, 14, 15, 16

Enzyme class	Biochemical function in living systems	Typical enzymes inducing polymerization	Typical polymers	Covered in (parts) in Chapter
IV. Lyases	Nonhydrolytic bond cleavage			
V. Isomerases	Intramoleular rearrangement			
VI. Ligases	Bond formation requiring triphosphate	Cyanophycin synthetase	Cyanophycin	5, 10

Outline of This Book

Biocatalytic approaches in polymer synthesis have to include an optimized combination of biotechnological with classical processes. Therefore, this book starts with a thorough review on the sustainable, 'green' synthesis of monomeric materials (Chapter 1). While few of the monomers presented in this chapter have been used in enzymatic polymerizations so far, the examples given could provide inspiration to use sustainable monomers more often in the future for enzymatic polymerizations and also for classical approaches.

Many of the polymerizations presented in this book proceed in organic solvents. To enhance the stability of enzymes in these solvent systems and to ensure efficient recovery of the biocatalysts the enzymes are commonly immobilized. Chapter 2 reviews some of the new trends of enzyme immobilization on nanoscale materials, while Chapter 3 sheds light on some new approaches to improve the commercial immobilization of Candida antarctica lipase B – the biocatalyst most often employed in enzymatic polymer synthesis.

The most extensively studied enzymatic polymerization system is that of polyesters via polycondensations or ring-opening polymerizations. The state of the art of *in vitro* enzymatic polyester synthesis is reviewed in Chapter 4.

Polyamides are important engineering plastics and excellent fiber materials and their worldwide production amounts to a few million tons annually. Therefore, it is astonishing that not many approaches to synthesize polyamides via enzymatic polymerization have been reported so far. Chapter 5 reviews these approaches and hopefully inspires future research in this direction.

In Chapter 6 the enzymatic polymerization of vinyl monomers is presented. Polymers, such as polystyrene and poly(meth)acrylates can be readily polymerized under catalysis of oxidoreductases like peroxidases, oxidases, etc. In addition oxidoreductases can be used to polymerize phenolic monomers (Chapter 7) and even to synthesize conducting polymers such as polyaniline (Chapter 8).

Well-defined polysaccharides are extremely difficult to synthesize via conventional organic chemistry pathways due to the diverse stereochemistry of the monosaccharide building blocks and the enormous number of intersugar linkages that can be formed. Chapter 9 shows that enzymatic polymerizations are superior alternatives to traditional approaches to synthesize polysaccharides.

The synthesis of bacterial storage compounds is reviewed in Chapter 10, focusing on two systems, namely polyhydroxyalkanoic acids and cyanophycin. Bacterial storage compounds are very interesting biopolymers having attractive material properties, sometimes similar to those of the petrochemical-based polymers.

Chapter 11 draws our attention towards the possibility of synthesizing chiral polymers via biocatalytic pathways. It becomes obvious in that chapter that chiral macromolecules can be achieved by enzymatic polymerizations that would not be synthesizable via traditional methods.

At present not many block copolymer systems using enzymatic polymerizations are reported. Chapter 12 reviews the current status of this field and shows the potential of future research in this direction.

Many enzymatic polymerizations suffer from low solubility of the synthesized polymers limiting the obtained degree of polymerization (e.g. polyamides, cellulose etc.). Chapter 13 illustrates several solutions by reviewing 'exotic' solvents and the possibilities of using them in biocatalysis. Not many reports on using such solvent systems for enzymatic polymerizations have yet been reported but the potential of such solvent systems becomes obvious immediately.

Chapter 14 introduces an interesting way to establish/solve the mechanism of enzymatic polymerizations via computer simulation. This method is quite well-established in other fields of chemistry but has only been used for solving the reaction mechanism of one enzymatic polymerization (the enzymatic ring opening polymerization of β-lactam). The outline of the technique in this chapter proves the power of this method and hopefully inspires future research on other enzymatic polymerization mechanisms.

In Chapters 15 and 16 the modification and degradation of respectively synthetic (e.g. PET, polyamides) and natural polymers (e.g. polysaccharides) are reviewed. It becomes obvious that biocatalytic modifications can offer advantages over chemical modifications therefore building a bridge between 'traditional' polymerization techniques and enzymatic polymerizations.

On most topics described in these chapters an increase in publications in recent years can be observed. This is a very promising trend showing that more and more researchers realize the importance of enzymatic polymerizations. We hope that with this book we can attract more researchers worldwide to this field and thus to tremendously extend the range of polymer classes synthesized by enzymes so far.

Acknowledgement

First of all I would like to acknowledge all authors of this book for their contribution to the book content. Each author is a leading authority in her/his field and generously offered effort and time to make this book a success.

Many thanks go to Iris Baum and Lars Haller for designing the cover and creating the lipase structure shown on the cover. Frank Brouwer is acknowledged for providing the photo on the book cover.

In addition, I would like to thank the Wiley team – especially Heike Nöthe, Elke Maase, Claudia Nussbeck, Hans-Jochen Schmitt, Rebecca Hübner und Mary Korndorffer – for their professional support, assistance and encouragement to make this book a reality.

Katja Loos Groningen, August 2010

Enzymatic Polymerizations

Book Series

Palmans, A., and K. Hult, eds. *Enzymatic Polymerizations*. Advances in Polymer Science Vol. 237. 2010, Springer.

Cheng, H.N., and R.A. Gross, eds. *Green Polymer Chemistry: Biocatalysis and Biomaterials*. ACS Symposium Series. Vol. 1043. 2010, American Chemical Society.

Cheng, H.N., and R.A. Gross, eds. *Polymer Biocatalysis and Biomaterials II*. ACS Symposium Series. Vol. 999. 2008, American Chemical Society.

Kobayashi, S., H. Ritter, and D. Kaplan, eds. *Enzyme-Catalyzed Synthesis of Polymers*. Advances in Polymer Science. Vol. 194. 2006, Springer.

Cheng, H.N., and R.A. Gross, eds. *Polymer Biocatalysis and Biomaterials*. ACS Symposium Series. Vol. 900. 2005, American Chemical Society.

Gross, R.A., and H.N. Cheng, eds. *Biocatalysis in Polymer Science*. ACS Symposium Series. Vol. 840. 2002, American Chemical Society.

Scholz, C., and R.A. Gross, eds. *Polymers from Renewable Resources: Biopolyesters and Biocatalysis*. ACS Symposium Series. Vol. 764. 2000, American Chemical Society.

Gross, R.A., D.L. Kaplan, and G. Swift, eds. *Enzymes in Polymer Synthesis*. ACS Symposium Series. Vol. 684. 1998, American Chemical Society.

Review Articles

Kobayashi, S., Makino, A., *Chemical Reviews* 2009, 109, 5288.

Kobayashi, S., Uyama, H., Kimura, S., *Chemical Reviews* 2001, 101, 3793.

Gross, R.A., Kumar, A., Kalra, B., *Chemical Reviews* 2001, 101, 2097.

Kobayashi, S., *Journal of Polymer Science Part A-Polymer Chemistry* 1999, 37, 3041.

Kobayashi, S., Shoda, S.-i., Uyama, H., in *Advances in Polymer Science*, Vol. 121, 1995, pp. 1.

Biocatalysis

Books

Fessner, W.-D., Anthonsen, T., *Modern Biocatalysis: Stereoselective and Environmentally Friendly Reactions*, Wiley-VCH 2009.

Tao, J., Lin, G.-Q., Liese, A., *Biocatalysis for the Pharmaceutical Industry – Discovery, Development, and Manufacturing*, John Wiley & Sons, 2009.

Grunwald, P., *Biocatalysis: Biochemical Fundamentals and Applications*, Imperial College Press 2009.

Liese, A., Seelbach, K., Wandrey, C., *Industrial Biotransformations*, Wiley-VCH, 2006.

Faber, K., *Biotransformations in Organic Chemistry: A Textbook*, Springer, 2004.

Bommarius, A.S., Riebel, B.R., *Biocatalysis – Fundamentals and Applications*, Wiley-VCH, 2004.

Drauz, K., Waldmann, H., *Enzyme Catalysis in Organic Synthesis: A Comprehensive Handbook*, Wiley-VCH, 2002.

List of Contributors

Iris Baum
University of Paderborn
Department of Chemistry
Warburger Straße 100
33098 Paderborn
Germany

Jesper Brask
Novozymes A/S
Krogshoejvej 36
2880 Bagsvaerd
Denmark

Anna Bröker
Westfälische Wilhelms-Universität Münster
Institut für Molekulare Mikrobiologie und Biotechnologie
Corrensstrasse 3
48149 Münster
Germany

H. N. Cheng
USDA Agricultural Research Service
Southern Regional Research Center
1100 Robert E. Lee Blvd.
New Orleans, LA 70124
USA

Rodolfo Cruz-Silva
Universidad Autonoma del Estado de Morelos
Centro de Investigacion en Ingenieria y Ciencias Aplicadas
Ave. Universidad 1001
Col. Chamilpa
Cuernavaca, Morelos, CP62209
Mexico

Apostolos Enotiadis
University of Ioannina
Department of Materials Science and Engineering
45110 Ioannina
Greece

Gregor Fels
University of Paderborn
Department of Chemistry
Warburger Straße 100
33098 Paderborn
Germany

Alessandro Gandini
University of Aveiro
CICECO and Chemistry Department
3810-193 Aveiro
Portugal

Biocatalysis in Polymer Chemistry. Edited by Katja Loos
Copyright © 2011 WILEY-VCH Verlag GmbH & Co. KGaA, Weinheim
ISBN: 978-3-527-32618-1

Dimitrios Gournis
University of Ioannina
Department of Materials Science and Engineering
45110 Ioannina
Greece

Richard A. Gross
Fruit Research Institute
Kralja Petra I no 9
32000 Čačak
Serbia

Georg M. Guebitz
Graz University of Technology
Department of Environmental Biotechnology
Petersgasse 12
8010 Graz
Austria

Andreas Heise
Dublin City University
School of Chemical Sciences
Glasnevin
Dublin 9
Ireland

Frank Hollmann
Delft University of Technology
Department of Biotechnology
Biocatalysis and Organic Chemistry
Julianalaan 136
2628BL Delft
The Netherlands

Steven Howdle
University of Nottingham
School of Chemistry
University Park
Nottingham NG7 2RD
UK

Katja Loos
University of Groningen
Zernike Institute for Advanced Materials
Department of Polymer Chemistry
Nijenborgh 4
9747 AG Groningen
The Netherlands

Nemanja Miletić
Fruit Research Institute
Kralja Petra I no 9
32000 Čačak
Serbia

Maricica Munteanu
Heinrich-Heine-Universität Düsseldorf
Institute für Organische Chemie und Makromolekulare Chemie
Lehrstuhl II
Universitätsstraße 1
40225 Düsseldorf
Germany

Anja Palmans
Eindhoven University of Technology
Department of Chemical Engineering and Chemistry
Molecular Science and Technology
PO Box 513
5600 MB Eindhoven
The Netherlands

Ioannis V. Pavlidis
University of Ioannina
Department of Biological Applications and Technologies
45110 Ioannina
Greece

Helmut Ritter
Heinrich-Heine-Universität Düsseldorf
Institute für Organische Chemie und
Makromolekulare Chemie
Lehrstuhl II
Universitätsstraße 1
40225 Düsseldorf
Germany

Paulina Roman
Universidad Autonoma del Estado de
Morelos
Centro de Investigacion en Ingenieria
y Ciencias Aplicadas
Ave. Universidad 1001
Col. Chamilpa
Cuernavaca, Morelos, CP62209
Mexico

Jorge Romero
Centro de Investigacion en Quimica
Aplicada
Blvd. Enrique Reyna 120
Col. Los Pinos
Saltillo, Coahuila, CP 25250
Mexico

Paria Saunders
Novozymes North America
Inc.
77 Perry Chapel Church Road
Franklinton, NC 27525
USA

Haralambos Stamatis
University of Ioannina
Department of Biological Applications
and Technologies
45110 Ioannina
Greece

Alexander Steinbüchel
Westfälische Wilhelms-Universität
Münster
Institut für Molekulare Mikrobiologie
und Biotechnologie
Corrensstrasse 3
48149 Münster
Germany

Kristofer J. Thurecht
The University of Queensland
Australian Institute for Bioengineering
and Nanotechnology and Centre for
Advanced Imaging
St Lucia, Queensland, 4072
Australia

Aikaterini A. Tzialla
University of Ioannina
Department of Biological Applications
and Technologies
45110 Ioannina
Greece

Hiroshi Uyama
Osaka University
Graduate School of Engineering
Department of Applied Chemistry
Suita 565-0871
Japan

Martijn Veld
Eindhoven University of Technology
Department of Chemical Engineering
and Chemistry
Molecular Science and Technology
PO Box 513
5600 MB Eindhoven
The Netherlands

Silvia Villarroya
G24 Innovations Limited
Wentloog Environmental Centre
Cardiff CF3 2EE
United Kingdom

Jeroen van der Vlist
University of Groningen
Faculty of Mathematics and
Natural Sciences
Department of Polymer Chemistry
Zernike Institute for
Advanced Materials
Nijenborgh 4
9747 AG Groningen
The Netherlands

List of Abbreviations

3D	three-dimensional
3-MePL	α-methyl-β-propiolactone
3MP	3-mercaptopropionic acid
4MCL	4-methyl caprolactone
4-MeBL	α-methyl-γ-butyrolactone
5-MeVL	α-methyl-δ-valerolactone
6-MeCL	α-methyl-ε-caprolactone
7-MeHL	α-methyl-ζ-heptalactone
8-MeOL	α-methyl-8-octanolide
8-OL	8-octanolide
10-HA	10-hydroxydecanoic acid
11MU	11-mercaptoundecanoic acid
12-MeDDL	α-methyl-dodecanolactone
ABTS	2,2′-azino-bis(3-ethylbenzothiazoline-6-sulfonate) diammonium salt
Acac	acetylacetone
ADM	Archer Daniels Midland
ADP	adenosine diphosphate
AM	amylose
AP	Amylopectin
ATP	adenosine triphosphate
ATRP	atom transfer radical polymerization
BCL	*Burkholderia cepacia* lipase
BG	benzyl glycidate
BHET	bis(2-hydroxyethyl) terephthalate
BMIM BF_4	1-butyl-3-methylimidazolium tetrafluoroborate
BMIM DCA	1-butyl-3-methylimidazolium dicyanamide
BMIM $FeCl_4$	1-butyl-3-methylimidazolium tetrachloroferrate
BMIM NTf_2	1-butyl-3-methylimidazolium bistriflamide
BMIM PF_6	1-butyl-3-methylimidazolium hexafluorophosphate
BMPy BF_4	1-butyl-1-methylpyrrolidinium tetrafluoroborate
BMPy DCA	1-butyl-1-methylpyrrolidinium dicyanamide

Biocatalysis in Polymer Chemistry. Edited by Katja Loos
Copyright © 2011 WILEY-VCH Verlag GmbH & Co. KGaA, Weinheim
ISBN: 978-3-527-32618-1

BOD	bilirubin oxidase
BPAA	Biphenyl acetic acid
BrNP	2-(bromonethyl)naphthalene
BSA	bovine serum albumin
BTMC	5-benzyloxytrimethylene carbonate
Buk	butyrate kinase
CA	*Candida antarctica*
CAL	*Candida antarctica* lipase
CALB	*Candida antarctica* lipase B
CCK	cholecystokinin
CCVD	catalytical chemical vapor deposition
CD	circular dichroism (Chapter 2)
CD	cyclodextrin (Chapter 7, 16)
CDase	cyclomaltodextrinase
CGP	cyanophycin granule polypeptide, cyanophycin
CGTase	cyclodextrin glycosyltransferase
CL	caprolactone
CMC	critical micellar concentration
CMD	cyclomaltodextrinase
CNSL	cashew nut shell liquid
CNT	carbon nanotube
CphA	cyanophycin synthetase
CphB	cyanophycinase
CPO	chloroperoxidase
CRL	*Candida rugosa* lipase
CS	chondroitin sulfate
CSA	camphor sulfonic acid
CSD	Cambridge structural database
CTAB	cetyltrimethylammonium bromide
CVL	*Chromobacterium viscosum* lipase
DA	degree of acetylation
DA	Diels–Alder
DB	degrees of branching
DBSA	dodecylbenzensulfonic acid
DDL	ω-dodecanolactone
DFT	density function theory
DKR	dynamic kinetic resolution
DKRP	dynamic kinetic resolution polymerization
DLLA	D,L-lactide
DMA	dimethyl adipate
DMP	2,4-dimethyl-3-pentanol
DMSO	dimethyl sulfoxide
DM-β-CD	2,6-di-*O*-methyl-β-cyclodextrin
DM-α-CD	2,6-di-*O*-methyl-α-cyclodextrin
DNA	deoxyribonucleic acid

DO	*p*-dioxanone
DODD	dodecyl diphenyloxide disulfonate
DON	1,4-dioxan-2-one
DP	degree of polymerization
DSC	differential scanning calorimetry
DTA	differential thermal analysis
DTC	5,5-dimethyl-trimethylene carbonate
DTNB	5,5'-dithiobis-(2-nitrobenzoic acid)
DWCNT	double-wall carbon nanotube
DXO	1,5-dioxepan-2-one
EACS	enzyme activated chain segment
EAM	enzyme activated monomer
EC	(−)-epicatechin (Chapter 7)
EC	Enzyme Commission (Chapter 9)
ECG	(−)-epicatechin gallate
ECM	extracellular matrix
EDC	1-ethyl-3-(3-dimethylaminopropyl) carbodiimide
EDOT	(3,4-ethylendioxythiophene)
ee	enantiomeric excess
EG	ethylene glycol
EGC	(−)-epigallocatechin
EGCG	(−)-epigallocatechin gallate
EGP	ethyl glucopyranoside
eROP	enzymatic ring-opening polymerization
eROP	enzyme-catalyzed ring-opening polymerization
ESI-MS	electrospray ionization mass spectrometry
F	furfural
FA	furfuryl alcohol
f-CNT	functionalized CNT
FED	flexible electronic device
FOMA	perfluorooctyl methacrylate
FT-IR	Fourier transform infrared
GA	glutaraldehyde
GAG	glycosaminoglycan
GAP	granule-associated protein
GCE	glassy carbon electrode
GlcA	D-glucuronic acid
GMA	glycidyl methacrylate
GME	glycidyl methyl ether
GPC	gel permeation chromatography
GPE	glycidyl phenyl ether
GPEC	gradient polymer elution chromatography
GT	glycosyltransferase
GTFR	glycosyltransferase R
HA	hyaluronan

HAS	hyaluronan synthase
HDL	hexadecanolactone
HEMA	hydroxyethyl methacrylate
HiC	*Humicola insolens* cutinase
HiC-AO	*Humicola insolens* cutinase immobilized on Amberzyme oxiranes
HMF	hydroxymethylfuraldehyde
HMWHA	high-molecular-weight HA
HPLC	high performance liquid chromatography
HRP	horseradish peroxidase
ICP	intrinsically conducting polymer
IL	ionic liquid
IPDM	3(S)-isopropylmorpholine-2,5-dione
IPPL	immobilized porcine pancreas lipase
ITC	iterative tandem catalysis
ITO	indium tin oxide
KRP	kinetic resolution polymerization
L	large
lacOCA	pure lactic acid derived O-carboxy anhydride
LCCC	liquid chromatography under critical condition
LDL	low-density lipoprotein
LLA	lactide
LMS	laccase-mediator-system
LMWC	low-molecular-weight chitosan
LMWHA	low-molecular-weight HA
LS	light scattering
M	medium
MA	malic acid
MALS	multi-angle light scattering
MBC	5-methyl-5-benzyloxycarbonyl-1,3-dioxan-2-one
MBS	maltose-binding site
MCL	4-methyl caprolactone (Chapter 12)
MCL	medium-carbon-chain length (Chapter 10)
MD	molecular dynamics
MDI	methylene-diphenyl diisocyanate
MF	5-methylfurfural
MHET	mono(2-hydroxyethyl) terephthalate
ML	laccase derived from *Myceliophthore*
MM	molecular mechanics
MMA	methyl methacrylate
MML	*Mucor miehei* lipase
MNP	Magnetic nanoparticle
MOHEL	3-methyl-4-oxa-6-hexanolide
MPEG	methoxy-PEG
MRI	magnetic resonance imaging
mRNA	messenger ribonucleic acid

MW	molecular weight
MWCNT	multi-wall carbon nanotube
NAG	*N*-acetylglucosamine
NMP	nitroxide-mediated polymerization (Chapter 11, 12)
NMP	*N*-methyl-2-pyrrolidinone (Chapter 8)
NMR	nuclear magnetic resonance
NP	nanoparticle
NRPS	nonribosomal peptide synthetases
O/W	oil-in-water
OC	2-oxo-12-crown-4-ether
OCP	open circuit potential
OMIM DCA	1-octyl-3-methylimidazolium dicyanamide
PA	phthalic anhydride
PA	polyamide
Pam	peptide amidase
PAMPS	poly(2-acrylamido-3-methyl-1-propanesulfonic acid)
PAN	polyacrylonitrile
PANI	polyaniline
PAT	poly(alkylene terephthalate)
PBS	poly(butylene succinate)
PCL	Laccase derived from *Pycnoporus coccineus*
PCL	polycaprolactone (Chapter 3, 4)
PCL	proceeded using laccase (Chapter 7)
PCL	*Pseudomonas cepacia* lipase (Chapter 11, 12, 13)
PDADMAC	poly(diallyldimethyl ammonium chloride)
PDB	protein database
PDDA	poly(dimethyl diallylammonium chloride)
PDI	polydispersity index
PDL	pentadecalactone
PDMS	poly(dimethylsiloxane)
PEDOT	poly(3,4-ethylendioxythiophene)
PEF	poly(2,5-ethylene furancarboxylate)
PEG	poly(ethylene glycol)
PEGMA	poly(ethylene glycol) methacrylate
PEI	polyethyleneimine
PET	poly(ethylene terephthalate)
PFL	*Pseudomonas fluorescens* lipase
PGA	poly-(γ-glutamate)
PHA	polyhydroxyalkanoic acid (Chapter 10)
PhaC	PHA synthase
PhaG	3-hydroxyacyl-ACP:CoA transferase
PHB	poly(3-hydroxybutyrate) (Chapter 10)
PHB	poly[(*R*)-3-hydroxybutyrate] (Chapter 11, 12)
Pi	inorganic phosphate
pI	isoelectric point

PL	poly-(ε-lysine)
PLA	polylactic acid
PLP	pyridoxal-5′-phosphate
PLU	propyl laurate units
POA	poly(octamethylene adipate)
poly(ε-CL)	poly(ε-caprolactone)
PPG	PEG-poly(propylene glycol)
PPL	porcine pancreatic lipase
PPO	poly(phenylene oxide)
PPP	pentose phosphate pathway
PPy	polypyrrole
PS	Polystyrene
PSorA	poly(sorbitol adipate)
Ptb	phosphotransbutyrylase
PTE	polythioester
PTHF	poly(tetrahydrofuran)
PTP	palm tree peroxidase
PTT	poly(trimethyleneterephthalate)
PVA	poly(vinyl alcohol)
PVP-OH	mono-hydroxyl poly(vinyl pyrolidone)
QCM	quartz crystal microbalance
QM	quantum mechanical
RAFT	reversible addition fragmentation chain-transfer
RI	refractive index
ROP	ring opening polymerization
ROS	reactive oxygen species
RTIL	room temperature ionic liquid
SA	succinic anhydride
SBE	starch branching enzyme
SBP	soybean peroxidase
scCO$_2$	supercritical CO_2
SCL	short-carbon-chain length
SDBS	sodium dodecylbenzensulfonate
SDS	sodium dodecyl sulfate
SEC	size exclusion chromatography
SEM	scanning electron microscopy
SET	single electron transfer
SO	styrene oxide
SP	starch-urea phosphate
SPS	sulfonated polystyrene
SWCNT	single-wall carbon nanotube
TA	terephthalic acid
TGA	thermogravimetric analysis
THF	tetrahydrofuran
TMC	trimethylene carbonate

TMP	trimethylolpropane
TSA	toluene sulfonic acid
TVL	laccase from *Trametes versicolor*
UDL	undecanolactone
UV-Vis	ultraviolet-visible
VOC	volatile organic compound
w/c	water-in-CO_2
WCA	water contact angle
XO	xanthine oxidase
XPS	X-ray photoelectron spectroscopy
XRD	X-ray diffraction
β-PL	β-propiolactone
δ-VL	δ-valerolactone
ε-CL	ε-caprolactone
ωHA	ω-hydroxyalkanoic acids

1
Monomers and Macromonomers from Renewable Resources
Alessandro Gandini

1.1
Introduction

Renewable resources constitute an extremely rich and varied array of molecules and macromolecules incessantly produced by natural biological activities thanks to solar energy. Their exploitation by mankind has always been at the heart of its survival as sources of food, remedies, clothing, shelter, energy, etc., and of its leisure as sources of flowers, dyes, fragrances, and other amenities. In the specific context of materials, good use has always been made of cotton, paper, starch products, wool, silk, gelatin, leather, natural rubber, vegetable oils and terpenes, among others, through progressively more sophisticated and scaled-up technologies.

The meteoric ascension of coal and petroleum chemistry throughout the twentieth century gave rise to the extraordinary surge of a wide variety of original macromolecules derived from the rich diversity of monomers available through these novel synthetic routes. This technical revolution is still very much alive today, but the dwindling of fossil resources and their unpredictable price oscillations, mostly on the increase, is generating a growing concern about finding alternative sources of chemicals, and hence of organic materials, in a similar vein as the pressing need for more ecological and perennial sources of energy. The new paradigm of the *biorefinery* [1] represents the global strategic formulation of such an alternative in both the chemical and the energy fields, with progressive implementations, albeit with different approaches, throughout the planet.

Within the specific context of this chapter, renewable resources represent the obvious answer to the quest for macromolecular materials capable of replacing their fossil-based counterparts [2, 3]. This is not as original as it sounds, because, apart from the role of natural polymers throughout our history evoked above, the very first *synthetic* polymer commodities, developed during the second half of the nineteenth century, namely cellulose esters, vulcanized natural rubber, rosin derivatives, terpene 'resins', were all derived from renewable resources. What is new and particularly promising, has to do with the growing momentum that this trend has been gathering in the last decade, as witnessed by the spectacular increase in the number of publications, reviews, reports, books, scientific symposia and, concurrently, by the correspondingly growing involvement of both the public and the industrial sectors in fostering pure and applied research in this broad field.

The purpose of this chapter is to provide a concise assessment of the state of the art related to the realm of monomers and macromonomers from renewable resources and their polymerization, and to offer some considerations about the prospective medium-term development of its various topics, which are also the section headings. Natural polymers are not covered here, nor are monomers like lactide, which are discussed elsewhere in the book. The reader interested in more comprehensive information on any of these topics, will find it in a recent comprehensive monograph [3].

1.2
Terpenes

The term 'terpene' refers to one of the largest families of naturally-occurring compounds bearing enormous structural diversity, which are secondary metabolites synthesized mainly by plants, but also by a limited number of insects, marine micro-organisms and fungi [4, 5].

Most terpenes share isoprene (2-methyl-1,4-butadiene) as a common carbon skeleton building block and can therefore be classified according to the number of isoprene units. Among the huge variety of structures of terpenes, associated with their different basic skeletons, stereoisomers, and oxygenated derivatives, the only members relevant to the present context are unsaturated hydrocarbon monoterpenes, which bear two such units, *viz.* the general formula $C_{10}H_{16}$, as exemplified in Figure 1.1 for the most representative structures found in turpentine, the volatile fraction of pine resin, which is itself the most representative and viable source of terpenes, whose world yearly production amounts to some 350 000 tons.

Among these molecules, only a few have been the subject of extensive studies related to their polymerization, namely those which can be readily isolated in appreciable amounts from turpentine: α-pinene, β-pinene, limonene and, to a lesser extent, myrcene [5].

Cationic polymerization has been shown to be the most appropriate type of chain reaction for these monomers. Indeed, the very first report of *any* polymerization reaction was published by Bishop Watson in 1798 when he recorded that adding a drop of sulfuric acid to turpentine resulted in the formation of a sticky resin. It was of course much later that the actual study of the cationic polymerization of pinenes was duly carried out, leading to the development of oligomeric adhesive materials still used today, mostly as tackifiers.

The mechanism of β-pinene cationic polymerization is well understood, together with its accompanying side reactions, as shown in Figure 1.2 [5, 6]. As in many other cationic systems, transfer reactions are prominent and hence the

Figure 1.1 Structure of the most common monoterpenes found in turpentine.

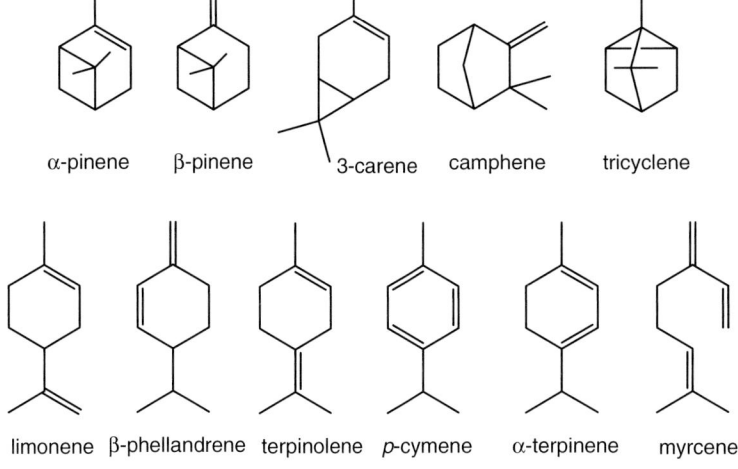

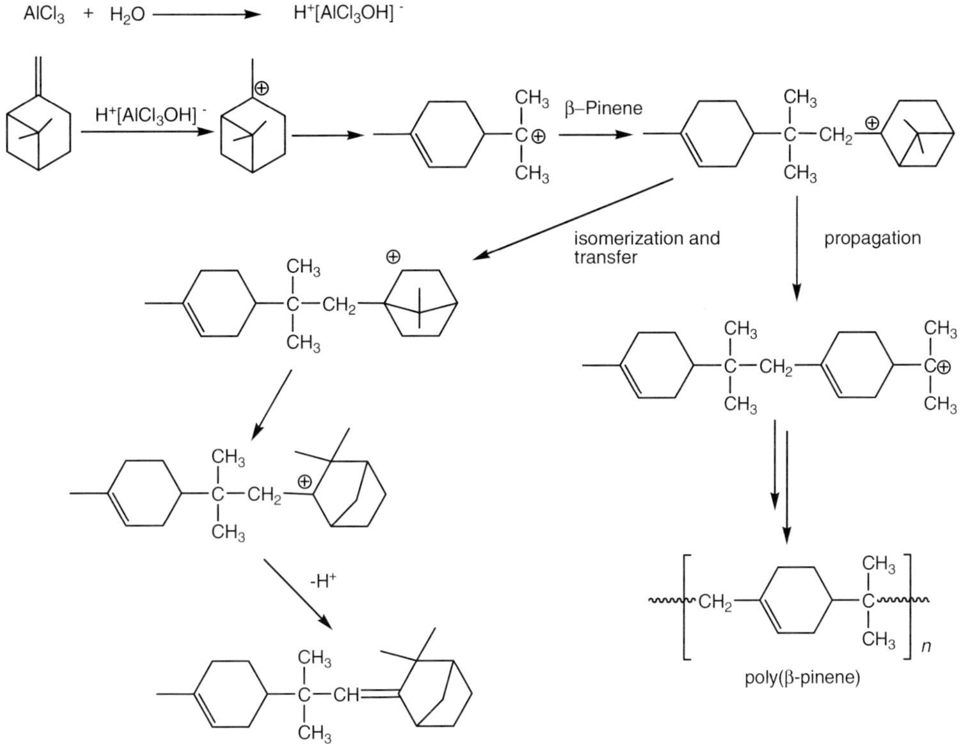

Figure 1.2 The reactions involved in the cationic polymerization of β-pinene.

degree of polymerization (DP) of the ensuing materials tend to be low. The development of cationic initiators capable of providing controlled, or quasi-living, conditions opened a new perspective also for the polymerization and copolymerization of β-pinene, and novel materials were synthesized with higher molecular weight and regular structures [5, 7].

The prevalence of investigations devoted to β-pinene stems from the relative simplicity of its cationic polymerization, compared with the more complex behavior of α-pinene.

The free radical polymerization of pinenes and limonene is of little interest, because of the modest yields and DPs obtained with their homopolymerizations. However, their copolymerization with a variety of conventional monomers has been shown to produce some interesting materials, particularly in the case of controlled reversible addition fragmentation chain-transfer (RAFT) systems involving β-pinene and acrylic comonomers [5].

A recently published original alternative mode of preparing polymeric materials from terpenes [8], makes use of a ring-opening metathesis mechanism in the presence of diclycopentadiene and generates hyperbranched macromolecules bearing a complex structure.

Terpenes epoxides, prepared by the straightforward oxidation of their unsaturation [9], have also been submitted to cationic polymerization [5, 10] and to insertion copolymerization with CO_2 [11], but these studies were not systematic in their approach.

It can be concluded that the advent of both free radical and cationic living polymerizations has brought new life into the area of terpene polymers and copolymers. No study is however known to the author regarding the use of enzymes to induce the polymerization of these natural monomers and such an investigation deserves some attention.

1.3
Rosin

Rosin [12, 13], also known by the name of colophony, is the designation traditionally given to the non-volatile residue obtained after the distillation of volatiles from the resin exuded by many conifer trees, mostly pine. It is therefore the complement of turpentine, which is the major source of terpenes discussed in the preceding section. Rosin played a fundamental role in waterproofing wooden naval vessels in the past and its use declined with the progressive decrease in their construction. Its annual worldwide production is however still more than one million tons and in recent years new uses are being actively sought within the general trend toward the valorization of renewable resources.

All rosins are made up of 90–95% of diterpenic monocarboxylic acids, or 'resin acids', $C_{19}H_{29}COOH$, in different specific molecular architectures. Their most common structures can be subdivided into those bearing two conjugated double bonds, as in Figure 1.3, and those in which the unsaturations are not conjugated, as in Figure 1.4.

The chemical modifications of these molecules which have been thoroughly studied include [12]:

- the oxidation of one unsaturation to give an endoperoxide;
- the aromatization of the rings through dehydrogenation and the subsequent functionalization of the ensuing aromatic moieties;
- the hydrogenation of one or both unsaturations;
- the isomerization relative to the position of the unsaturations;
- the Diels–Alder (DA) reaction with dienophiles;
- the reactions with formaldehyde and phenol;
- the preparation of salts of the carboxylic acid.

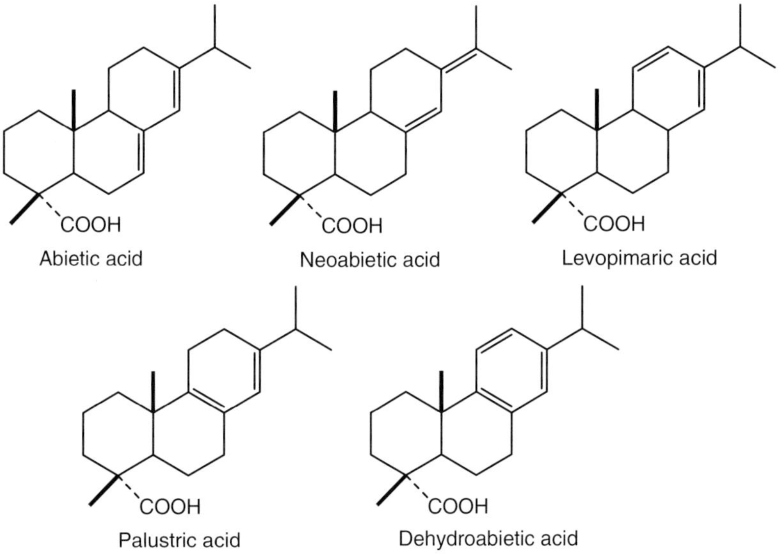

Figure 1.3 Structures of the most common conjugated resin acids.

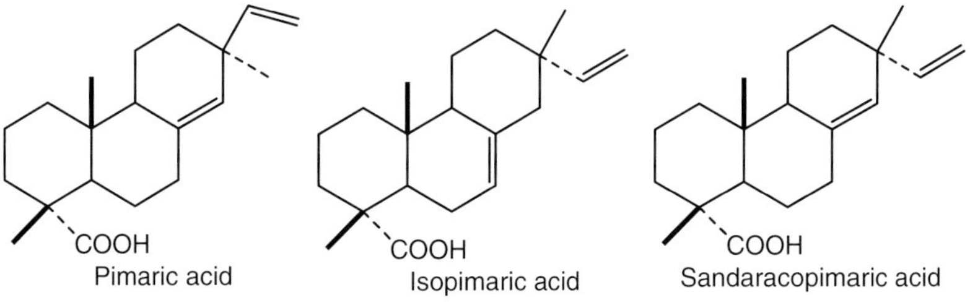

Figure 1.4 Structures of the most common non-conjugated resin acids.

The aspects relevant to the use of rosin as such, or one of the derivatives arising from its appropriate chemical modification as monomer or comonomer [12–14], have to do with the synthesis of a variety of materials based on polycondensations and polyaddition reactions of structures bearing such moieties as primary amines, maleimides, epoxies, alkenyls and, of course, carboxylic acids. These polymers find applications in paper sizing, adhesion and tack, emulsification, coatings, drug delivery and printing inks.

A recent addition to the realm of rosin derivatives used in polymer synthesis dealt with rosin-based acid anhydrides as curing agents for epoxy compositions [15] and showed that their performance was entirely comparable with that of petroleum-based counterparts, with the advantage of a simple process and, of course, their renewable character.

As with terpenes, there is no record of the use of rosin or its derivatives as monomers in enzymatic polymerizations, a fact that should stimulate research aimed at filling this gap.

1.4
Sugars

Carbohydrates constitute a very important renewable source of building blocks for the preparation of a variety of macromolecular materials, which find key applications particularly in the biomedical field, because of their biocompatibility and biodegradability (see also Chapter 9 and 16). The introduction of sugar-based units into the polymer architecture can be achieved via (i) polyaddition reactions involving vinyl-type saccharides; (ii) functionalizations that append the carbohydrate onto a reactive backbone; or (iii) polycondensation reactions of sugar based monomers. Whereas the first two approaches generate macromolecules in which the carbohydrate moieties are in fact side groups to conventional vinyl or acrylic chains, the latter alternative is more interesting, because it gives rise to *real* carbohydrate-based polymers in which the repeating units of the main chain are the sugar derivatives themselves [16]. This type of polymerization is only briefly discussed here, moreover the coverage is limited to chemical catalysis, since the enzymatic approach is dealt with in Chapter 4.

Sugars as such are polyols and hence if linear polymers are sought from them, the number of OH functions, or indeed of functions derived from them, must be reduced to two, either by adequate protection procedures, or by appropriate chemical modifications.

The three anhydroalditol diols shown in Figure 1.5, resulting from the intramolecular dehydration of the corresponding sugars, are among the most extensively studied sugar-based monomers with different polycondensation systems leading to chiral polymers [16, 17]. Isosorbide is readily prepared from starch, isomannide from D-mannose, and both are industrial commodities. Isoidide is prepared from isosorbide by a three-step synthesis, because L-idose is a rare sugar. Given the diol nature of these compounds, it follows that polyesters, polyurethanes and

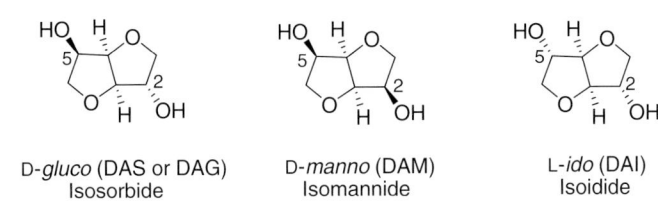

Figure 1.5 Three important anhydroalditol monomers.

Figure 1.6 Terephthalates based on anhydroalditols [18].

5-endo/2-exo, D-*gluco*
5-endo/2-endo, D-*manno*
5-exo/2-exo, L-*ido*

Figure 1.7 Aliphatic polyesters based on anhydroalditols [19].

5-endo/2-exo, D-*gluco*
5-endo/2-endo, D-*manno*
5-exo/2-exo, L-*ido*

m = 2–10

polyethers are among the obvious macromolecular structures investigated, but polycarbonates and polyester-amides have also attracted some interest. All these materials exhibit higher T_g values than counterparts synthesized with standard aliphatic diols, because of the inherent stiffness of the anhydroalditol structure. The macromolecular rigidity can of course be further enhanced by using similarly stiff complementary monomers, such as aromatic diacids, as in the example of Figure 1.6. The use of aliphatic diacids for the preparation of the corresponding polyesters (Figure 1.7) gives rise to biodegradable materials.

Carbohydrate-based polyamides and polyurethanes constitute two major families of polymers and hence the interest in preparing aminosugars and, from them, the corresponding isocyanates. This wide research field has produced very interesting materials including the first chiral nylon-type polyamides [16].

Typical polyamides and polyurethanes prepared from aminosugars, both from anhydroalditols and protected monosaccharides, are shown in Figures 1.8 and 1.9, respectively.

A new family of linear polyurethanes and poly(ester-urethanes), prepared from both aliphatic and aromatic diisocyanates and isorbide [21] or conveniently protected sugar alditols, has recently been reported and the properties of

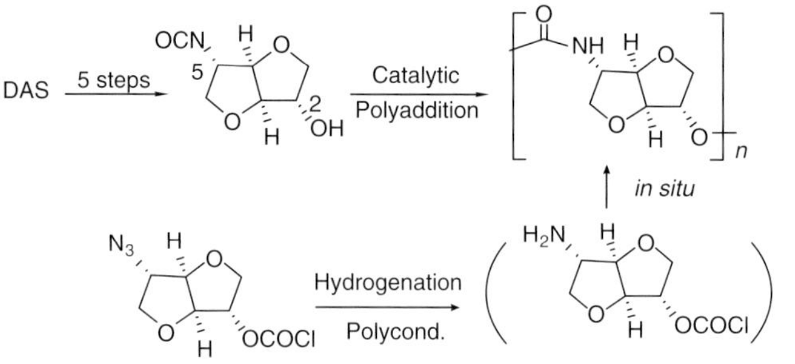

Figure 1.8 Approaches to stiff polyurethanes based on anhydroalditols [20].

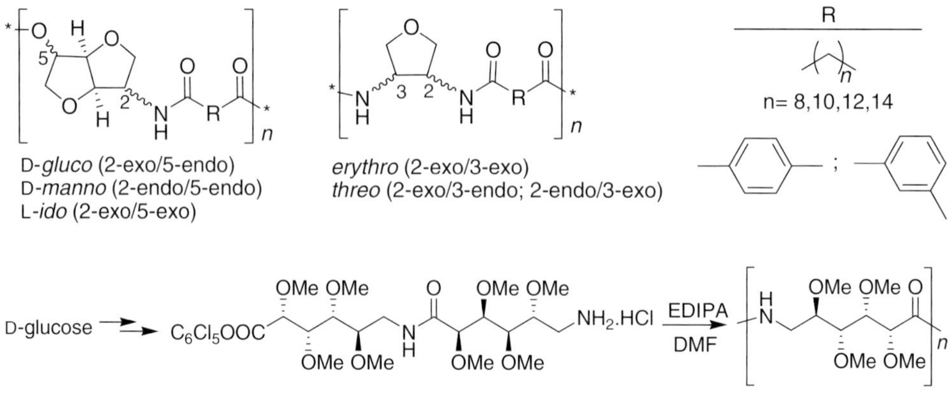

Figure 1.9 Various polyamide structures derived from sugars [16].

the ensuing materials fully characterized [22]. This interesting investigation has been extended to include aliphatic biodegradable polyurethanes bearing L-arabinitol and 2,2'-dithiodiethanol [23]. Furthermore, the preparation of a novel carbohydrate lactone and its ring-opening polymerization were shown to yield a functionalized cyclic aliphatic polyester [24]. A very thorough study of the synthesis of sugar monoisocyanates [25], aimed at preparing ureido-linked disaccharides, should inspire polymer chemists to extend it to diisocyanate homologs.

1.5
Glycerol and Monomers Derived Therefrom

The current boom associated with biodiesel production from vegetable oils has generated a spectacular rise in glycerol availability, with a yearly world production

1.5 Glycerol and Monomers Derived Therefrom

estimated in 2008 at 1.8 MT and a correspondingly low price of ~0.6 Euro per kg. The traditional uses of glycerol in pharmaceuticals, cosmetics and personal care products, alkyd resins, etc., are now being progressively complemented with commodities arising from its chemical conversion, as with the examples detailed in the section below.

The clearest indication of the rapidly rising scientific and technological research on chemicals from glycerol is the number of monographs published on this topic in the last few years [26]. Before discussing these aspects, it is instructive to show that glycerol itself is being investigated as a monomer in novel polymerization systems. There are in fact two distinct areas in this context, namely the synthesis of glycerol oligomers and the preparation of its polymers or copolymers using appropriately modified monomers.

The self-condensation of glycerol catalyzed by both homogeneous and heterogeneous basic catalysis yields a mixture of linear and branched oligomers (DPs typically from 2 to 6), mostly linear with the former medium, while mostly branched with the latter [27]. Figure 1.10 shows a schematic illustration of the linear growth, involving only the primary OH groups, since condensation with the secondary ones generates ramifications. The applications of these oligomers include cosmetics, food additives and lubricants. Carboxylic acid oligoglycerol esters [27, 28] are also gaining relevance as materials in similar domains, as well as as emulsion stabilizers and antistatic or antifogging agents.

The preparation of hydroxyesters or hydroxyacids from glycerol and their polycondensation by transesterification has been the subject of recent studies aimed at preparing biodegradable hyperbranched polycarbonates [29] and polyesters

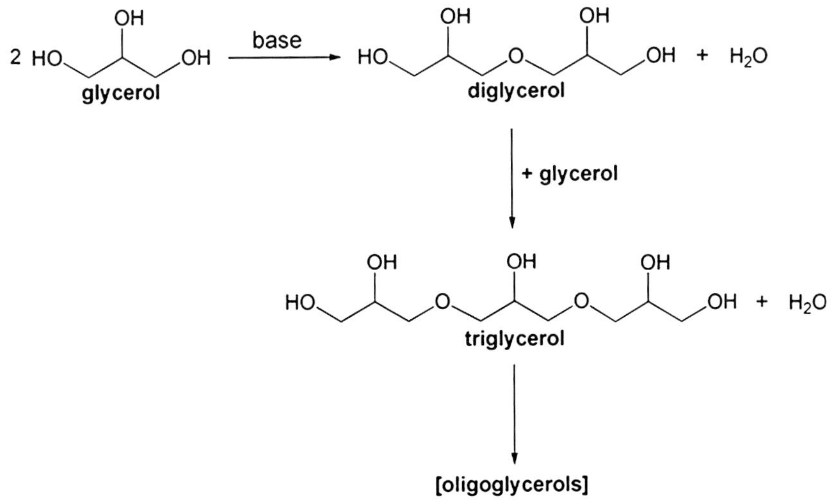

Figure 1.10 Etherification mechanism leading to oligoglycerols.

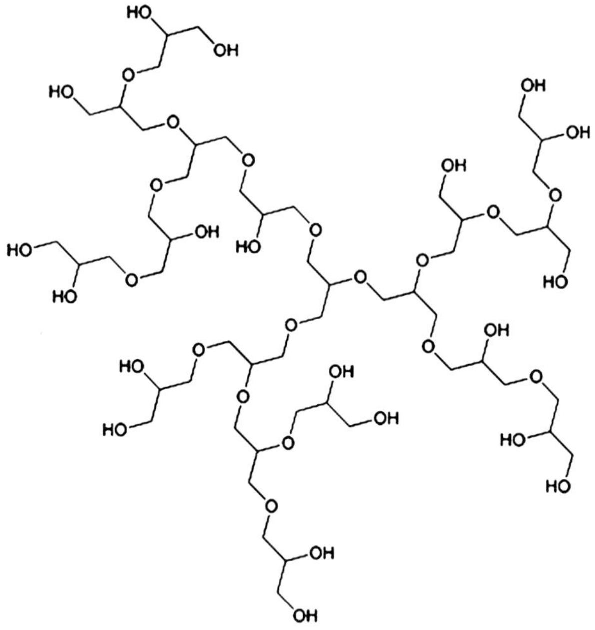

Figure 1.11 A model structure of hyperbranched polyglycerol.

[30, 31]. Although hyperbranched polyglycerols (Figure 1.11) have not thus far been synthesized directly from glycerol, much work has been devoted to their preparation, because of their interesting properties, including biocompatibility, biodegradability, water solubility and aptitude to easy functionalization. Structures of this type, incorporating poly(ethylene oxide) blocks have been reported, using, for example, glycidyl ethers as monomers [32].

Notwithstanding the promising aspect of investigations related to glycerol-based macromolecules, it is fair to state that much more is to be expected from the wide variety of mechanisms and catalysts being studied in order to convert glycerol into a panoply of chemicals [26] and, in particular, of monomers suitable for both step and chain polymerizations. This is extremely important in the present context, because more and more monomers, which were only available through petrochemical routes, are being reported as accessible through glycerol chemistry, notably diols, diacids, hydroxyacids, oxiranes, acrolein, and acrylic acid, among others. The specific instance of ethylene and propylene glycols [33] is particularly relevant, considering the importance of poly(ethylene terephthalate) (PET), the most important polyester on the market and of the novel poly(trimethylene terephthalate) produced by DuPont under the commercial name of Sorona. These considerations extend to polymers fully derived from renewable resources, as in the case of some of the furan polyesters discussed below.

1.6 Furans

Whereas natural monomers like terpenes and sugars constitute building blocks for a limited number of macromolecular structures, associated with their own peculiar chemical features, the realm of furan polymers bears a qualitatively different connotation in that it resembles the context of petrol refinery, that is, it is open to a whole domain of monomers, whose only specificity is the fact that they all incorporate the furan heterocycle in their structure. This state of affairs stems from the fact that, as in petroleum chemistry, saccharide-based renewable resources are used to produce two first-generation furan derivatives, which constitute the substrates capable of being converted into a vast array of monomers and hence a correspondingly large number of macromolecular structures associated with materials possessing different properties and applications [34].

Furfural (F) has been an industrial commodity for over a century and its production has spread throughout the world reaching a yearly production close to 300 000 Tons (at a price of ~0.5 Euros per kg), with some 70% in China. Its precursors range from agriculture to forestry residues, namely the pentoses present in such different by-products of the food industry as corn cobs, oat and rice hulls, sugar cane bagasse, cotton seeds, olive husks and stones, as well as wood chips. This variety of sources explains why any given country possesses a renewable resource which can be turned into F, considering moreover that the technology associated with its production is particularly simple, being based on the acid catalyzed hydrolytic depolymerization of the hemicellulose and the subsequent dehydration of the ensuing aldopentoses [35]. The dominant xylose component is converted into F, whereas rhamnose, present in modest proportions, gives rise to 5-methylfurfural (MF), as shown in Figure 1.12. The two furan compounds can be readily separated by distillation.

Figure 1.12 Mechanism of F (R=H) and MF (R=CH$_3$) formation from aldopentoses.

Most of the furfural is to-day converted into furfuryl alcohol (FA), which is extensively employed as a precursor to a variety of resins for high-tech applications [34], whose spectrum is constantly broadening.

FA

The second furan derivative which can be prepared from the appropriate C6 polysaccharides or sugars is hydroxymethylfuraldehyde (HMF). The mechanism of its formation from hexoses is entirely similar to that of F, but difficulties associated with the recovery of the product have delayed its industrial production, despite, again, the ubiquitous character of its natural precursors. A very substantial effort has been devoted in recent years to investigate and optimize novel processes and/or catalysts [36] and the ensuing results suggest that HMF will be a commercial commodity very soon. Interestingly, because of its relative fragility, some of these approaches consider the *in situ* conversion of HMF into its very stable dialdehyde (FCDA) or diacid (FDCA) derivatives.

HMF   **FCDA**   **FDCA**

This impressive surge of interest in HMF is yet another clear example of the ferment around the search for original alternatives to petrochemistry, based on the exploitation of renewable resources, in the field of chemicals, materials and energy sources.

This said, the ensuing research strategy, whose implementation began some 40 years ago [34] relies on the following working hypothesis: two first-generation furan derivatives, F and HMF, constitute the starting structures from which two sets of monomers can be synthesized, viz. those suitable for chain polymerizations and copolymerizations, and those associated with step-growth mechanisms. Figure 1.13 shows some of the monomers which have been prepared from F and then assessed in terms of their individual aptitude to the appropriate chain polymerization and copolymerization reactions [34]. Figure 1.14 illustrates a different approach for the exploitation of furfural, and in general monosubstituted furans, but in this case with the purpose of synthesizing byfunctional monomers

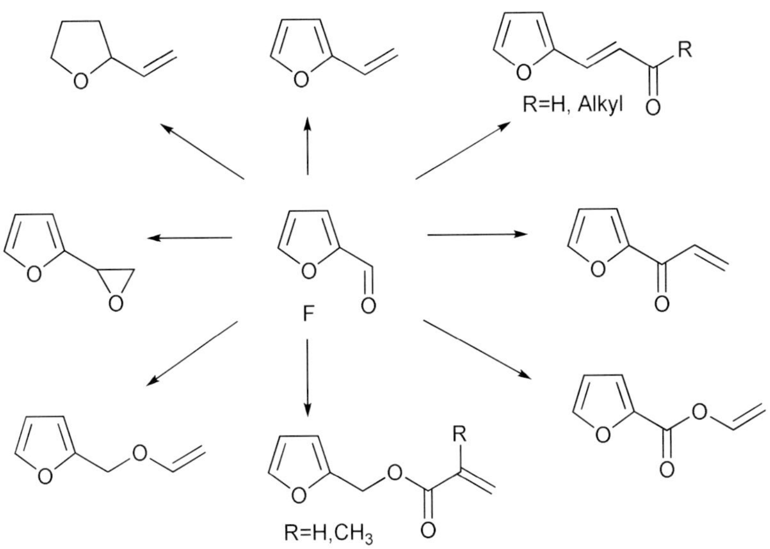

Figure 1.13 A selection of monomers derived from F.

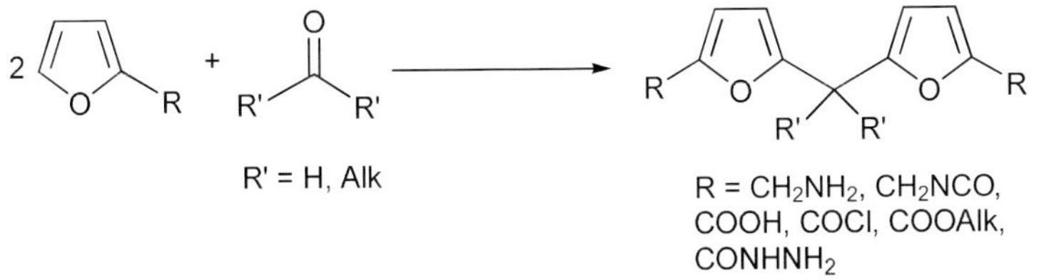

Figure 1.14 Difuran monomers derived from 2-substituted furans (prepared in turn from F).

for polycondensation systems [34]. The ensuing structures incorporate two furan rings, each bearing a reactive group, bridged by a variety of moieties.

Figure 1.15 depicts the type of monomers which have been obtained from HMF, namely, again, structures suitable for step-growth polymerizations [34], but here mostly with a single heterocycle bearing the two reactive moieties.

Only some recent contributions to this large field are discussed here, since thorough reviews are available elsewhere. The most interesting studies relate to two very different approaches to furan polymers, namely the synthesis of novel polyesters and the use of the Diels–Alder reaction to prepare thermoreversible materials [37].

The announced awakening of the 'sleeping giant' HMF has brought furan polycondensates back to the forefront of both fundamental and industrial research. The obvious underlying thought here is the elaboration of macromolecular materials in which aromatic units (fossil resources) are replaced by furan counterparts (renewable resources) and the assessment of the novel structures in terms of their capability of replacing the existing ones advantageously, or at least at par.

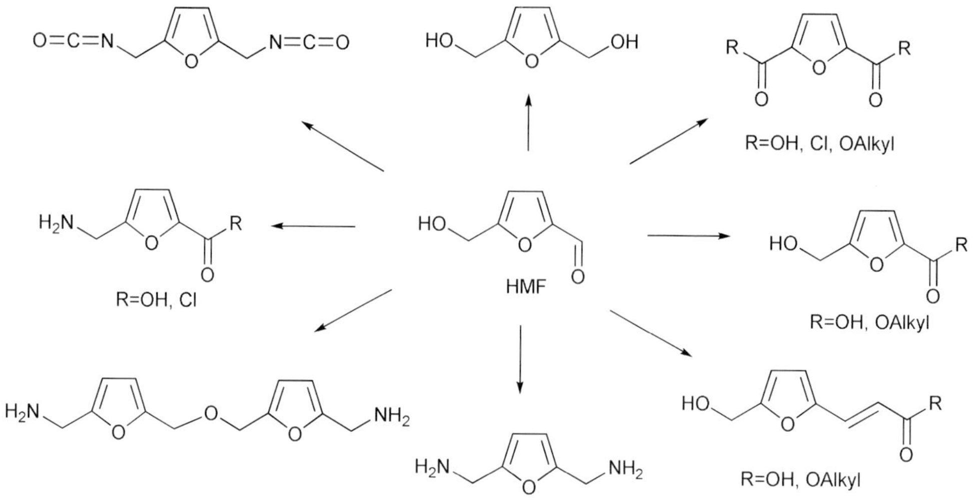

Figure 1.15 A selection of monomers derived from HMF.

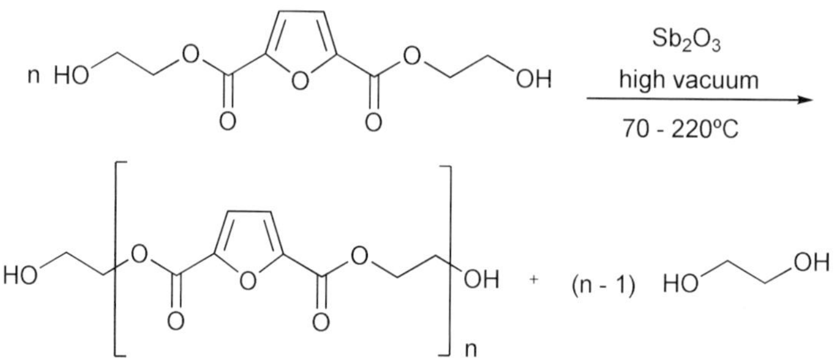

Figure 1.16 Polytransesterification mechanism leading to high-DP PEF.

Polyesters come to mind, since PET and some of its more recent homologs play a very prominent role on the polymer market. FDA was therefore used in conjunction with ethylene glycol to synthesize poly(2,5-ethylene furancarboxylate) (PEF), that is, the furan analog of PET [38]. The use of polytransesterification (Figure 1.16) proved particularly effective in generating high-molecular weight semicrystalline polymers.

Interestingly, the replacement of the aromatic ring by the furan counterpart did not alter in any appreciable fashion such polymer properties as glass transition and melting temperature, the high aptitude to crystallize and the thermal stability [37, 38]. The same considerations apply to the polymer homologs bearing 1,3-propylene glycol units [37]. Work is in progress to extend these comparisons to mechanical and processing properties. Given the recent work mentioned above on the synthesis of glycols from glycerol, these novel furan polyesters can be

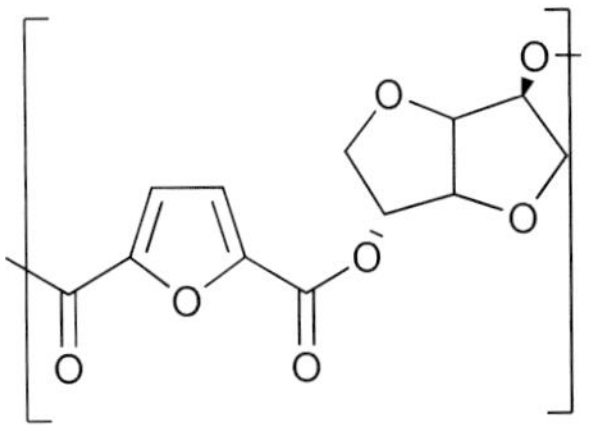

Figure 1.17 Repeat unit of the polyester prepared from FDA and isosorbide.

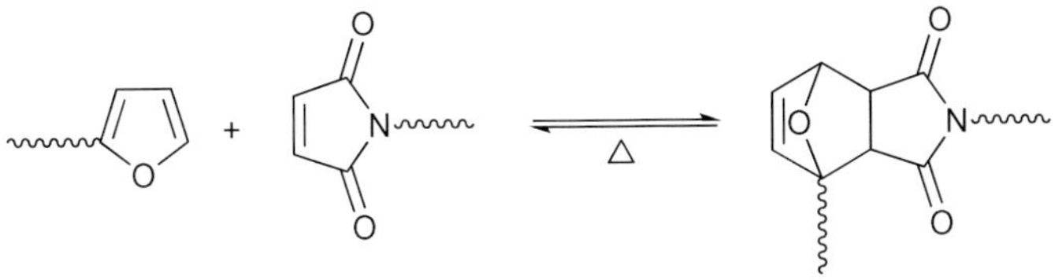

Figure 1.18 The DA equilibrium between growing species bearing respectively furan and maleimide end groups.

considered as being entirely based on renewable resources. The same applies to the combination of FDA and isosorbide (Figure 1.17) which produced, as expected, a much stiffer polyester macromolecule and hence a high T_g [37]. The range of furan polyester structures is currently being widened to include other diols and to study random and block copolymers.

In a very different vein, the purpose of applying the DA reaction to synthesize furan polymers stems from two considerations: (i) the fact that furans are particularly well suited to participate in that coupling interaction, functioning as dienes; and (ii) the thermal reversibility of the DA reaction which opens the way to the preparation of intelligent materials, which are moreover easily recyclable [37, 39, 40]. The most logical complementary function in this context is the maleimide dienophile, because of its excellent reactivity and the fact that multifunctional homologs can be readily prepared. The system therefore can be represented schematically by the reversible interaction shown in Figure 1.18, which depicts any forward/backward DA reaction applied to any polymerization involving furan and maleimide moieties.

Among the different applications of this reaction, the systems which have received most attention are: (i) linear polycondensations involving difuran and bismaleimide monomers (AA + BB) and, more recently, monomers incorporating both moieties (AB); (ii) non-linear polycondensations with at least one monomer with a functionality higher than two; polymer crosslinking, for example, a polymer bearing pendant furan heterocycles reacting with a bismaleimide, or vice-versa. Of course in all these instances, the growth or crosslinking processes are thermoreversible, with temperatures above 100 °C shifting the equilibrium in favor of the reagents, that is, the opening of the adduct, accompanied by the corresponding depolymerization, or decrosslinking. Given the clean-cut nature of

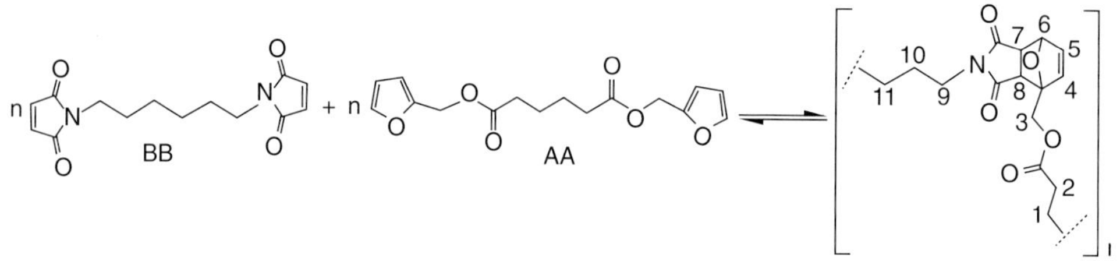

Figure 1.19 Reversible DA polycondensation between complementary bifunctional monomers [40].

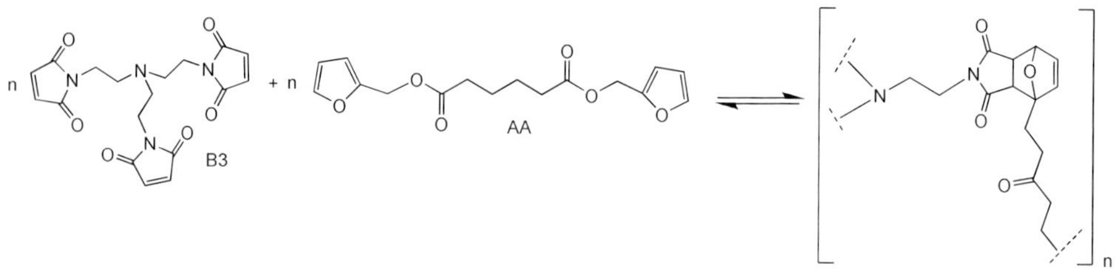

Figure 1.20 Reversible DA non-linear polycondensation between a tris-maleimide and a difuran monomer [37].

the DA reaction, several cycles have been applied to these various systems with reproducible forward and backward features.

A recent example of a thorough study of a linear polymerization using UV and NMR spectroscopy included various monomer combinations [40], like that shown in Figure 1.19.

Similarly, Figure 1.20 shows a typical monomer combination leading to branched and ultimately crosslinked polymers, which are thermally reversible all the way to the initial monomer mixture [37]. Finally, Figure 1.21 shows the reversible crosslinking of a linear furan copolymer by an oligomeric bismaleimide [41].

The present choice of systems based on furan monomers is not exhaustive of course, but hopefully sufficient to illustrate the enormous potential of this area of polymer science, which does justice to the plea for more macromolecular materials from renewable resources.

1.7
Vegetable Oils

The exploitation of vegetable oils, triglycerides of fatty acids, as a source of materials is as old as the inception of civilization, first for coating applications and soap manufacture, then for progressively wider and more sophisticated applications including inks, plasticizers, alkyd resins, agrochemicals, etc. This qualitative and

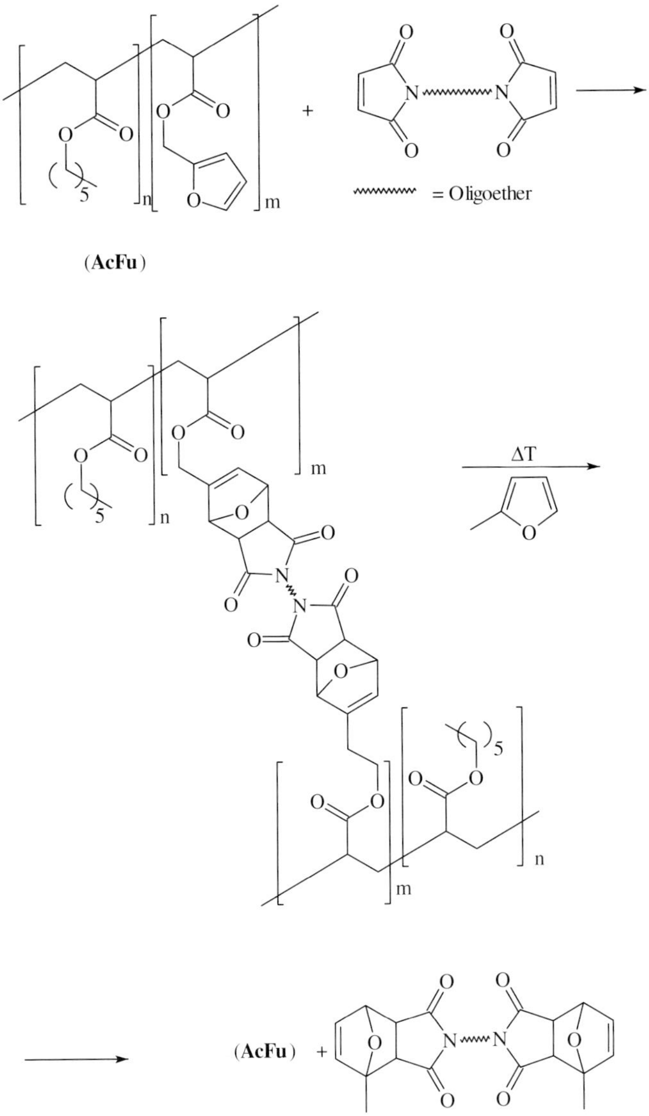

Figure 1.21 Reversible DA crosslinking reaction of a linear furan copolymer [41].

quantitative increase in the realm of materials has been complemented in recent years by the use of vegetable oils for the production of biodiesel. This section highlights briefly recent progress in the use of vegetable oils as macromonomers for the elaboration of novel macromolecular materials, following more thorough reviews published in the last few years [42].

The structure of these natural products can be schematically depicted by the simple generic triglyceride formula given in Figure 1.22, where R_1, R_2 and

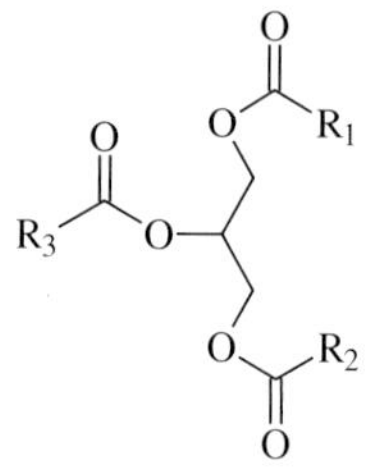

Figure 1.22 Generic structure of a triglyceride.

R_3 represent fatty acid chains, viz. linear aliphatic structures which can vary in length (from C12 to C22) in the number of C=C unsaturations (from zero to 6) and in the possible presence of other moieties, like OH or epoxy groups. Figure 1.23 portrays some of the most characteristic structures of fatty acids present in triglycerides. A given triglyceride often incorporates identical groups R, but a given oil is always made up of a mixture of triglycerides (typically 3 to 5), frequently with one or two predominant structures making up 70–80% of the composition [42]. The degree of unsaturation of a vegetable oil is perhaps the most relevant parameter in terms of such criteria as siccativity in coatings, lacqers, paints and inks, that is, the drying of the oil by atmospheric oxidopolymerization through a well-established mechanism [42], but also as a source of other types of polymerization and of chemical modification.

The annual global production of vegetable oils is expected to reach some 150 Mt in 2010, of which about 20% is devoted to industrial applications (materials and energy), compared with ~75 and 5% for food and feed uses. Within the combined contexts of this huge amount of renewable resources and of the growing emphasis on their exploitation as alternative precursors to polymers instead of fossil counterparts, it is not surprising to witness a burgeoning research effort in the area, which appears to be set to grow in the near future.

Among the most interesting chemical modifications of oils or fatty acids, epoxidation occupies a privileged position for the number and variety of studies and applications it has spurred [43]. This reaction involves the transformation of C=C alkenyl moieties along the fatty acid chain into oxirane groups using a variety of oxidation systems, for example, that shown in Figure 1.24, although terminal epoxy functions have also been appended to saturated triglycerides [43]. These epoxidized macromonomers have found numerous applications in thermosetting resins using diamines and anhydrides as hardeners, including reinforced nano-hybrid materials and flame retardancy [43].

The other mode of activation of the oxirane moieties is through cationic polymerization, which will generate a network for polyfunctional epoxidized glycerides and thermoplastic materials with fatty acids bearing a single epoxy group [43].

Apart from these direct uses as macromonomers, epoxydized vegetable oils have also been employed as precursors to other polymerizable structures, follow-

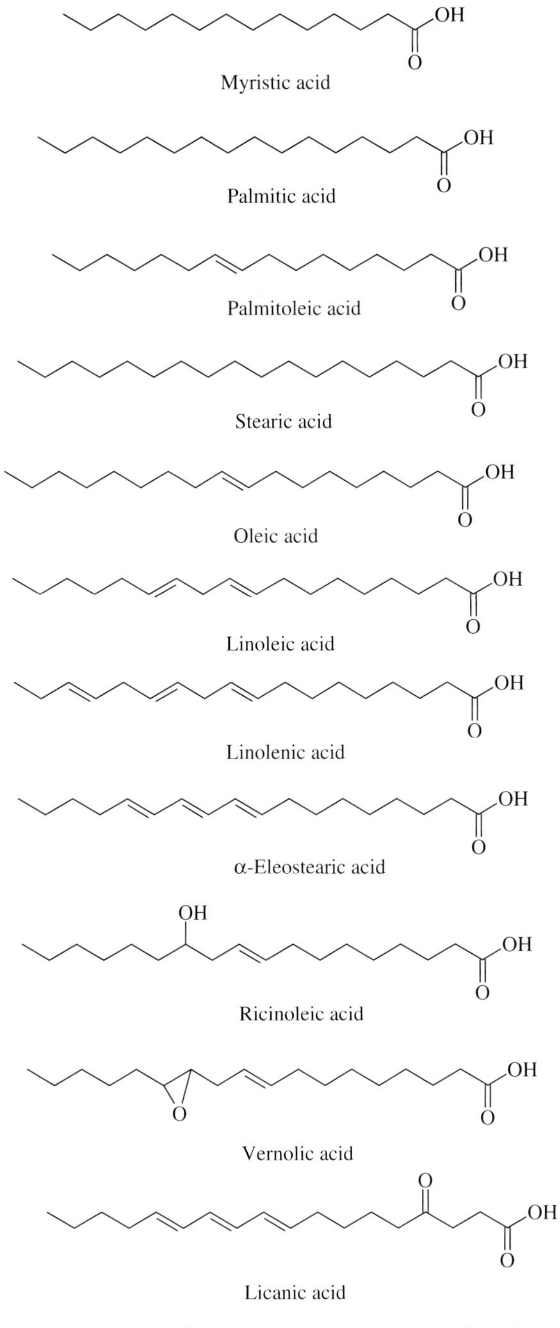

Figure 1.23 Typical fatty acids borne by vegetable oil triglycerides.

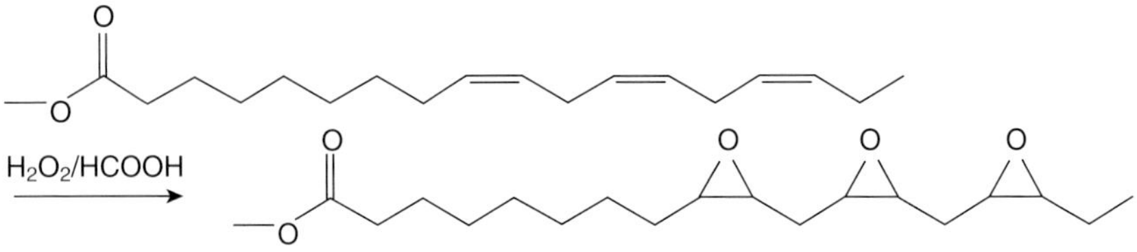

Figure 1.24 The epoxydation of methyl linolenate by performic acid.

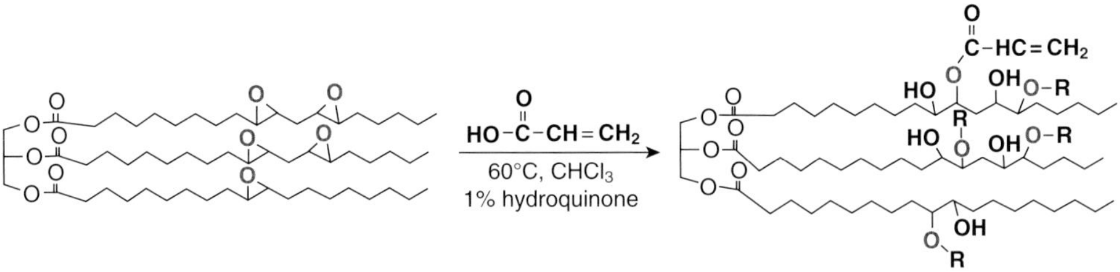

Figure 1.25 Hydroxylated and acrylated derivative of epoxidized soy oil.

ing appropriate chemical modifications, namely the formation of polyols for the synthesis of polyurethanes and the insertion of acrylic functions (Figure 1.25) for photopolymerizable structures [42, 43].

Other modifications of vegetable oils in polymer chemistry include the introduction of alkenyl functions, the study of novel polyesters and polyethers and the synthesis of semi-interpenetrating networks based on castor oil (the triglyceride of ricinoleic acid) [42], and also the production of sebacic acid and 10-undecenoic acid from castor oil [44]. Additionally, the recent application of metathesis reactions to unsaturated fatty acids has opened a novel avenue of exploitation leading to a variety of interesting monomers and polymers, including aliphatic polyesters and polyamides previously derived from petrochemical sources [42, 45].

The vitality of this field is further demonstrated by a continuous search for original ways of exploiting vegetable oils. Examples include self-healing elastomers in which fatty acids play a central role [46]; the esterification of cellulose with fatty acids to give thermoplastic materials [47]; the synthesis of a saturated aliphatic diisocyanate from oleic acid and its subsequent use in the preparation of fully biobased polyurethanes in conjunction with canola oil-derived polyols [48]; and novel elastomers from the concurrent cationic and ring-opening metathesis polymerization of a modified linseed oil [49].

Regrettably, all this *chemical* ferment has not yet found a *biochemical* counterpart in the sense that no study on the use of biocatalytic systems has been published in the context of the use of vegetable oils or their fatty acids as sources of polymeric materials.

1.8
Tannins

A large variety of trees and shrubs contain tannins, a term used loosely to define two broad classes of natural phenolic compounds, viz. condensed (or polyflavonoid) tannins and hydrolyzable tannins. Whereas the former are oligomeric in nature, the latter are essentially non-polymeric. The most important varieties of these compounds include the bark of oak and black wattle as well as of pines and firs, and the woods of chestnut, mangrove, and sumach, among others [50]. After millennia of empirical use and a century of intensive industrial exploitation of tannins, particularly in leather tanning, their use began to dwindle after the Second World War up to the 1970s, when new applications were actively sought. This trend saw a considerable boost with the beginning of the third millennium within the more general context of the rational exploitation of renewable resources associated with the paradigm of the biorefinery.

Condensed tannins constitute more than 90% of the total world production of commercial tannins which amounts to about 200 kT per year. Figure 1.26 depicts two typical flavonoid monomer units of condensed tannins, whose average molecular weights vary between 1000 and 4000, depending on the species involved.

Although hydrolyzable tannins are generally considered as being mixtures of phenols, they must be viewed as more complex substances made up of simple structures as those shown in Figure 1.27, together with higher oligomers, some of which contain carbohydrates covalently linked with phenols [50]. Given the relative paucity and poor reactivity of this family of tannins, only flavonoids will be discussed further here.

Within the realm of polymer science and technology, the major application of tannins is in adhesives, mostly for wood, because of their aptitude to react with formaldehyde, although considerable research has been devoted to reduce the use of the latter by optimizing the conditions favoring tannin self-condensation and/or reactions with other aldehydes, as reviewed recently by Pizzi [50], the most authoritative expert in the field.

A major development in the use of tannins in macromolecular materials has taken place in the last few years and again this original contribution comes from Pizzi's laboratory. Condensed tannins were found to crosslink in polycondensation reactions with furfuryl alcohol and small amounts of formaldehyde, thus

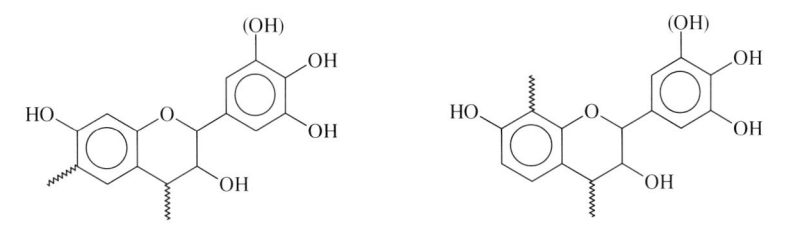

Figure 1.26 The most common monomer units in condensed tannins (polyflavonoids).

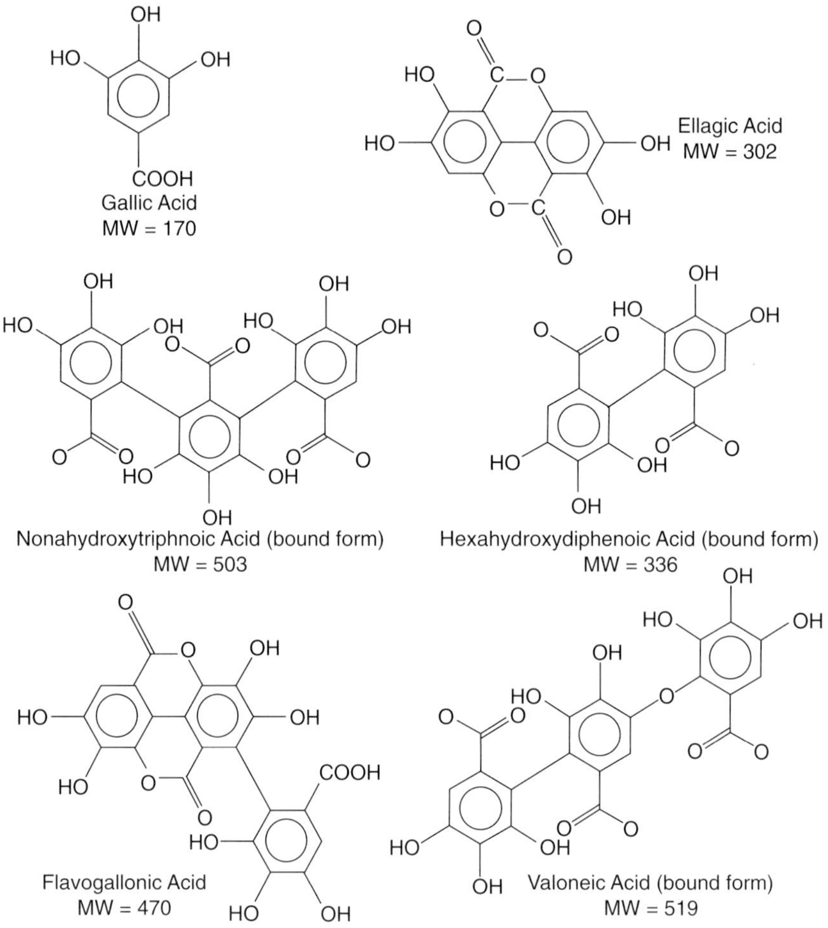

Figure 1.27 Low molecular weight components of hydrolysable tannins.

making these compositions almost totally based on renewable resources. When these reactions were conducted in the presence of a volatile additive, foaming took place giving rise to rigid cellular materials [51]. These foams displayed remarkable properties in terms of insulation, fire resistance, chemical inertness and metal ion sequestration [52].

Further work on these remarkable materials involved their carbonization and the thorough characterization of the ensuing carbon foams [53], as well as their subsequent chemical activation, which produced a dramatic increase in surface area, reaching $1800 \, m^2 \, g^{-1}$ when $ZnCl_2$ was used as promoter [54].

In conclusion, it is heartening to witness the lively rebirth of tannins as precursors of promising materials. As with other natural monomers and macromonomers, it appears that the possible interest of polymer bioprocesses could also be assessed on tannins, a type of study that has not been published thus far.

1.9
Lignin Fragments

Lignin is a fundamental component of plants, where it plays the fundamental role of amorphous matrix around the cellulose fibers and the hemicelluloses in the cell wall to form a supramolecular lignin–carbohydrate structure. Lignin is a phenolic–aliphatic polymer, whose biosynthesis is well documented and whose structure varies as a function of the vegetable species [55]. Figure 1.28 illustrates the typical building blocks found in lignin and the fact that this complex crosslinked macromolecule is also linked to the hemicellulose through occasional covalent bridges.

All chemical pulping processes and, today, all biorefineries applied to woods and annual plants are associated with the chemical splicing of lignin to produce fragments with molecular weights which vary from 1000 to 40 000 and possess specific structural features [56]. Notwithstanding this very wide range of features, all these lignin fragments are characterized by a common structural peculiarity, which is particularly relevant to the polymer chemist, namely the ubiquitous presence of both aliphatic and phenolic hydroxyl groups, albeit in different frequencies and proportions.

The exploitation of lignin fragments in macromolecular materials can be divided into three different approaches: (i) as additives in physical blends [57]; (ii) as macromonomers in polymer synthesis; and (iii) as a source of phenolic monomers [58]. Only the latter two aspects are briefly discussed here.

The chemical participation of lignin macromonomers in polymerization or copolymerization reactions has been focussed mostly on the reactivity of both types of OH groups, and hence in the synthesis of polyesters, polyurethanes and polyethers, although some research has also dealt with their intervention through the unsubstituted aromatic sites in different formaldehyde-based resins in partial replacement of phenol [58, 59].

The use of the fragments as such has given contrasting results regarding their reactivity, because of the different topochemical situations in terms of the steric availability of the hydroxyl groups. For obvious reasons of simplicity and economy, the possibility of calling upon lignin macromonomers as produced in industrial operations is nevertheless much more attractive than having to modify them chemically to enhance their reactivity. For this reason, the low molecular weight fragments obtained from organosolv processes are much more interesting lignin macromonomers, owing to their modest steric crowding.

The most important problem associated with lignin macromonomers, as such or after appropriate chemical modifications (mostly aimed at making the OH groups more available), is however the reproducibility of their characteristics, since variations in structural features and molecular weight or molecular weight distribution inevitably induce corresponding variations in the properties of the final materials. A novel biorefinery organosolv process applied to annual plants [60], appears to provide a much better control over the properties of the ensuing low-molecular weight lignin.

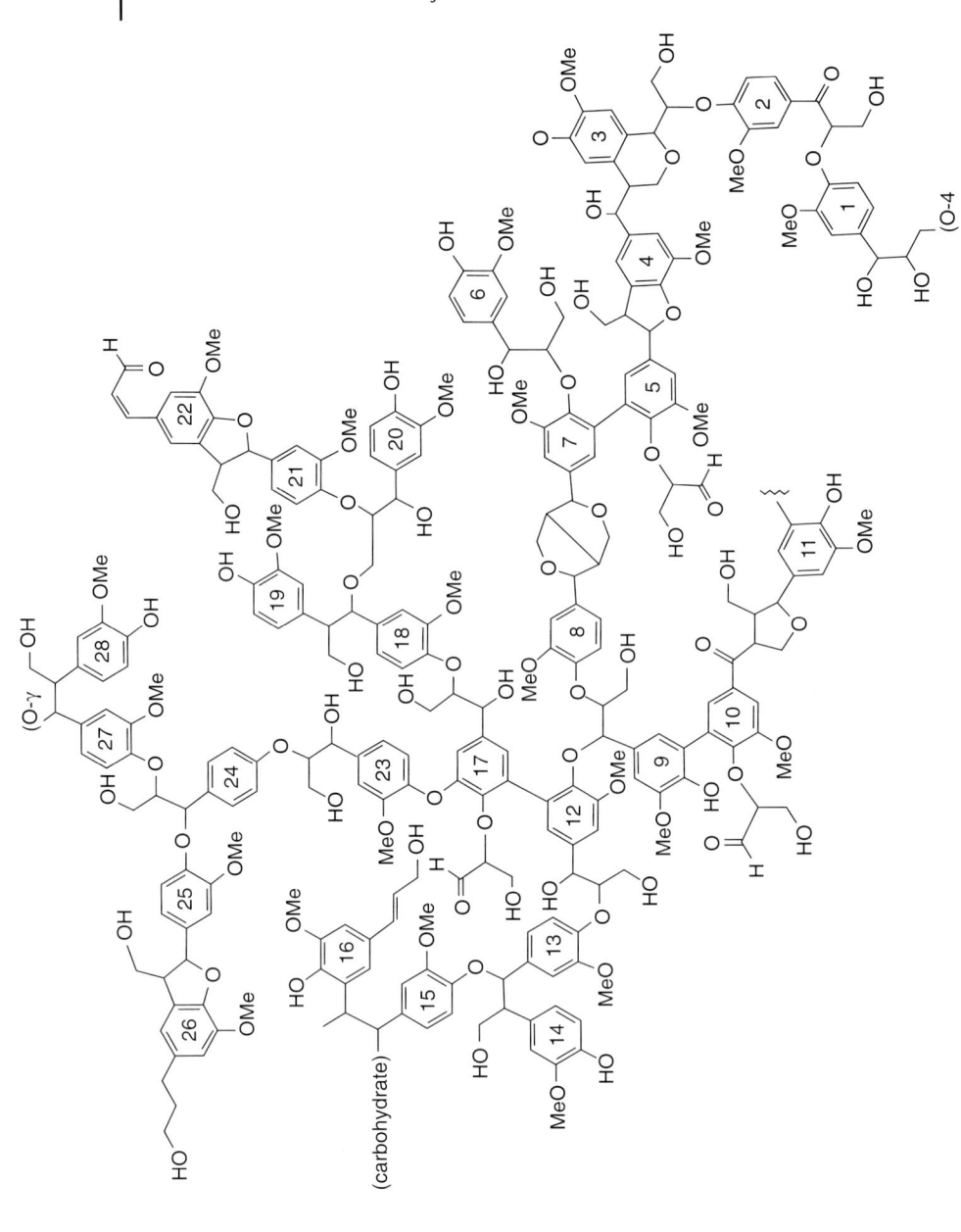

24 | *1 Monomers and Macromonomers from Renewable Resources*

Figure 1.28 An illustration of the typical building blocks in the structure of lignin.

1.9 Lignin Fragments

An altogether different strategy of valorization of lignin fragments is their radical conversion to liquid polyols through an oxypropylation reaction induced by the activation of the lignin hydroxyl groups with a Lewis or Bronsted base, followed by the grafting-from oligomerization of propylene oxide initiated by the ensuing oxianions [61, 62]. This process possesses a very wide applicability to OH-bearing substrates, many of them, like sugarbeet pulp, otherwise intractable, with the additional advantage of having a green connotation associated with the fact it does not require a solvent, nor any separation or purification treatment at the end of the reaction.

In essence, its double role is, on the one hand, to bring the OH groups out of the compact macromolecular assembly of the substrate, thus making them more prone to react and, on the other, to enhance the chain flexibility and mobility by appending low T_g grafts, hence turning solid residues into viscous liquids. Figure 1.29 provides a schematic view of this chemical transformation and emphasizes that the grafting reaction is always accompanied by some propylene oxide homopolymerization to give macrodiols, which are themselves potential monomers [61].

Numerous biomass by-products of little use other than combustion, have been valorized in this way, namely poor quality cork powder, olive stones, sugarbeet pulp, chitin and chitosan residues and lignins from different pulping technologies, among others [61, 62]. The ensuing lignin polyols were found to be very appropriate macromonomers for the synthesis of rigid polyurethane foams.

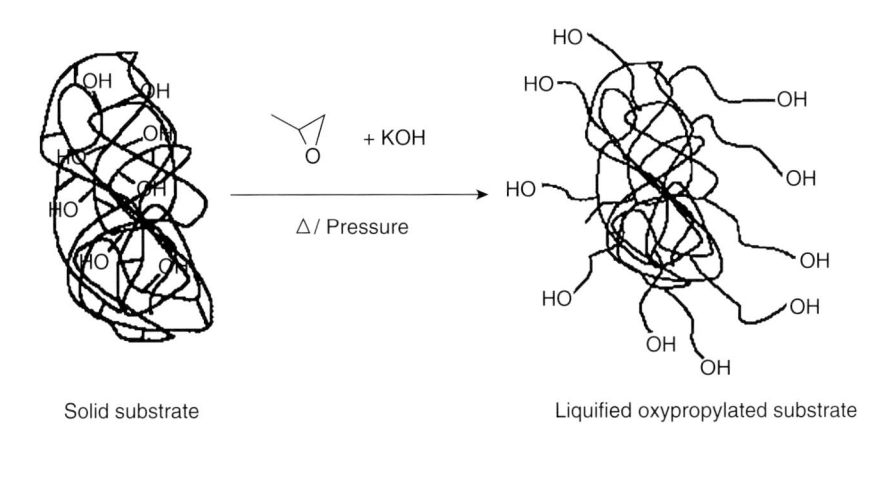

Figure 1.29 Schematic representation of the oxypropylation of solid biomass residues.

The other strategy of lignin exploitation for polymer synthesis revolves around the 'total' splicing of the fragments down to phenolic units [58, 63], which could be used, as such, or after appropriate chemical modifications, as monomers for either step-growth or chain polymerizations. Two major problems are implicit in this approach, namely the optimization of the depolymerization process and, more importantly, the efficient separation of the different ensuing building blocks. This situation has been tackled successfully in a somewhat different context, viz. the production of vanillin from kraft lignin within a biorefinery operation which made use of the remaining lignin as a source of polyurethanes [64]. Similar efforts should be devoted to integrated processes in which some specific monomer structures would be privileged in the lignin depolymerization, leaving the rest for other uses. Biochemical treatments would certainly contribute to the positive outcome of this type of study.

Recent additions to the search of lignin-based polymers include the use of lignin oligomers and vegetable oils to elaborate paper hydrophobic coatings [65] and the preparation of epoxy networks derived from lignosulfonate and glycerol [66].

1.10
Suberin Fragments

Suberin is a natural highly hydrophobic polyester almost ubiquitous in the vegetable realm, where it plays the important role of a protecting barrier between the plant and the environment. The amounts of suberin needed for this protection are very modest and in fact only two trees produce barks which are sufficiently rich in suberin to justify its exploitation, namely *Quecus suber* (cork), a Mediterranean species, and *Betula pendula* (birch), one of the most important hardwood species in Northern Europe. Most of the work on the extraction, hydrolysis and fragment characterization of suberin components has been carried out on cork, which contains between 40 to 60% of this polymer [67]. The suberin structure is made up of crosslinked macromolecules comprising mostly aliphatic polyesters and to a lesser extent a lignin-like network. From a polymer chemist's standpoint, the aliphatic portion constitutes by far the most interesting source of monomers, as clearly suggested by the most representative families of compounds isolated from suberin hydrolytic depolymerization shown in Figure 1.30. The alternative depolymerization by methanolysis yields the same structures with methyl ester moieties instead of COOH groups.

These monomer mixtures have been fully characterized [67, 68] and thereafter submitted to polycondensation reactions with diisocyanates to produce polyurethanes [67] and on their own to prepare polyesters [69]. The preliminary results of the latter study, which called upon both chemical and enzymatic catalyses, showed that the ensuing aliphatic polyesters were interesting materials with

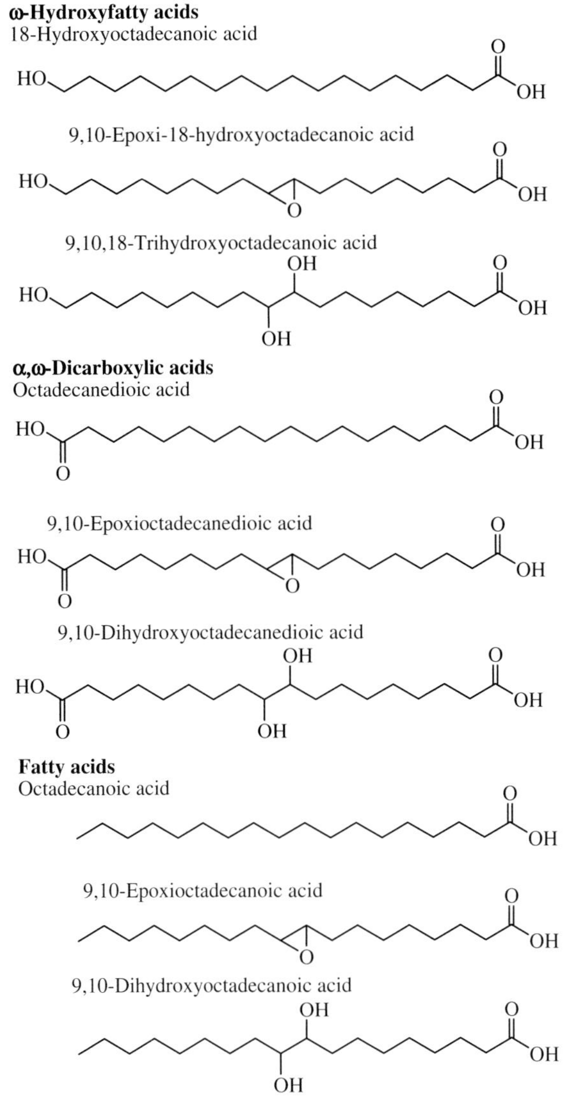

ω-Hydroxyfatty acids
18-Hydroxyoctadecanoic acid

9,10-Epoxi-18-hydroxyoctadecanoic acid

9,10,18-Trihydroxyoctadecanoic acid

α,ω-Dicarboxylic acids
Octadecanedioic acid

9,10-Epoxioctadecanedioic acid

9,10-Dihydroxyoctadecanedioic acid

Fatty acids
Octadecanoic acid

9,10-Epoxioctadecanoic acid

9,10-Dihydroxyoctadecanoic acid

Figure 1.30 The most representative monomers isolated by hydrolyzing suberin.

a marked hydrophobic character. This work is still in progress and the relative merits of the two types of activations are being assessed.

It is important to emphasize that the exploitation of suberin monomers is not done to the detriment of the *noble* parts of cork, since these are extremely useful manifestations of renewable resources. What is used instead are the rejects of cork processing, like powder from stopper manufacturing and irregular morphologies.

1.11
Miscellaneous Monomers

This final section is not intended to cover exhaustively the numerous scattered mentions of the use of different natural monomers, or monomers prepared from natural compounds, in polymerization reactions, mostly because this odd catalog would not reflect the actual potential of these often isolated investigations. The choice was instead placed on a couple of monomers which have received serious attention and produced interesting materials.

Citric acid (Figure 1.31) is a low cost commodity produced industrially via fermentation with an annual production of about 7×10^5 tons. Its use in polymer science was concentrated mostly on its role as a chemical modifier and/or plasticizer of starch, but in more recent years citric acid has been employed as a monomer in the synthesis of polyesters, from crosslinked materials for controlled drug release [70], to oligomers with diol mixtures [71], but particularly of biodegradable polyesters for tissue engineering applications [72]. A different approach was recently proposed [73], in which citric acid was introduced in a polyesterification system with the purpose of enhancing its functionality, following a thorough study of model reactions. The specific role of citric acid was to introduce COOH-terminated branching moieties in an isosorbide-based polyester and thereafter favor its crosslinking with various curing agents in view of preparing high-tech coatings [73].

Tartaric acid (Figure 1.32) is an equally widespread and cheap natural product and its use as a monomer has involved a number of groups, particularly that of Muñoz-Guerra. Among the recent studies, the most interesting deal with the synthesis of biodegradable polycarbonates with anhydroalditols [74], optically active hydrophilic aliphatic polyamides [75] capable of associating themselves to produce supramolecular stereocomplexes [76], and polyurethanes with different diisocyanates bearing COOH side groups, which displayed a unique proneness to degrade with water upon incubation under physiological conditions [77].

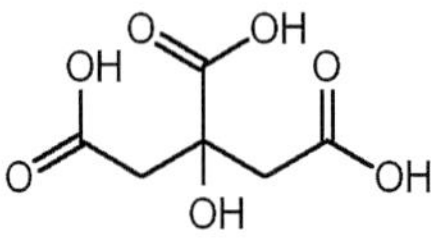

Figure 1.31 Citric acid.

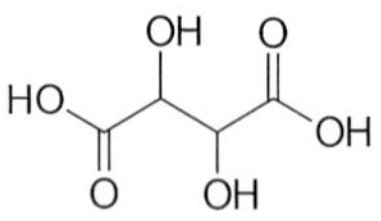

Figure 1.32 Tartaric acid.

Succinic acid, readily available from the fermentation of glucose, is a very prolific molecule that can be usefully exploited in the synthesis of a large spectrum of interesting compounds, including, of course, monomeric structures [78]. Its potential in macromolecular chemistry is awaiting to be intensively exploited.

1.12
Conclusions

Despite the condensed format of this overview, I hope that it manages to convince readers that the interest in monomers from renewable resources is not a passing whim of some polymer chemists led astray by fashionable trends, but, instead, a very sound strategy that should help to shape the future of polymer science and technology. Although obviously not all these studies will reach viable practical realizations in terms of novel macromolecular materials, it is indispensable to build a rich database in preparation for the progressive dwindling of fossil resources, to enable decisions and choices to be made with competence and experience.

References

1 Kamm, B., Gruber, P.R., and Kamm, M. (2006) *Biorefineries-Industrial Processes and Products*, vols. 1 and 2, Wiley-VCH Verlag GmbH, Weinheim, Germany; Amidon, T.E., Wood, C.D., Shupe, A.M., Wang, Y., Graves, M., and Liu, S. (2008) *J. Biobased Mater. Bioen.*, **2**, 100; Campbell, G.M., Vadlani, P., and Azapagic, A. (2009) *Chem. Eng. Res., Special Issue on Biorefinery Innovations*, **87** (9), 1101–1348.

2 Gandini, A. (2008) *Macromolecules*, **41**, 9491; Coates, G.W., and Hillmyer, M.A. (2009) A Virtual Issue of Macromolecules: 'Polymers from Renewable Resources', *Macromolecules*, **42** (21), 7987–7989, doi: 10.1021/ma902107; Various, (2008) *Polym. Rev.*, **48** (1). A special issue devoted to polymers from renewable resources; Eichhorn, S.J., and Gandini, A. (2010) Materials from Renewable Resources, a Special Issue of *MRS Bull.*, **35**, 187–225.

3 M.N. Belgacem, and A. Gandini (eds) (2008) *Monomers, Polymers and Composites from Renewable Resources*, Elsevier, Amsterdam.

4 Erman, W.F. (1985) *Chemistry of Monoterpenes*, Marcel Dekker, New York.

5 Silvestre, A.J.D., and Gandini, A. (2008) Terpenes: major sources, properties and applications, in *Monomers, Polymers and Composites from Renewable Resources* (eds M.N. Belgacem and A. Gandini), Elsevier, Amsterdam, Ch 2.

6 Roberts, W.J., and Day, A.R. (1950) *J. Am. Chem. Soc.*, **72**, 1226.

7 Lu, J., Kamigaito, M., Sawamoto, M., Higashimura, T., and Deng, Y.X. (1997) *Macromolecules*, **30**, 27.

8 Mathers, R.T., Damodaran, K., Rendos, M.G., and Lavrich, M.S. (2009) *Macromolecules*, **42**, 1512.

9 Neuenschwander, U., Guignard, F., and Hermans, I. (2010) *ChemSusChem*, **3**, 75.

10 Crivello, J.V., Conlon, D.A., and Lee, J.L. (1985) *Polym. Bull.*, **14**, 279;

Crivello, J.V. (2009) *J. Polym. Sci. Part A Polym. Chem.* **47**, 866 and 1825.

11 Byrne, C.M., Allen, S.D., Lobkovsky, E.B., and Coates, G.W. (2004) *J. Am. Chem. Soc.*, **126**, 11404.

12 Silvestre, A.J.D., and Gandini, A. (2008) Rosin: major sources, properties and applications, in *Monomers, Polymers and Composites from Renewable Resources* (eds M.N. Belgacem and A. Gandini), Elsevier, Amsterdam, Ch 4.

13 Maiti, S., Ray, S.S., and Kundu, A.K. (1989) *Prog. Polym. Sci.*, **14**, 297.

14 Wang, H., Liu, B., Liu, X., Zhang, J., and Xian, M. (2008) *Green Chem.*, **10**, 1190.

15 Liu, X., Xin, W., and Zhang, J. (2009) *Green Chem.*, **11**, 1018; Wang, H., Liu, X., Liu, B., Zhang, J., and Xian, M. (2009) *Polym. Int.*, **58**, 1435; Liu, X., Xin, W., and Zhang, J. (2010) *Bioresour. Technol.*, **101**, 2520.

16 Galbis, J.A., and García-Martín, M.G. (2008) Sugars as monomers, in *Monomers, Polymers and Composites from Renewable Resources* (eds M.N. Belgacem and A. Gandini), Elsevier, Amsterdam, Ch 5.

17 Kricheldorf, H.R. (1997) *J. Macromol. Sci., Rev. Macromol. Chem. Phys.*, **C37**, 599.

18 Storbeck, R., Rehahn, M., and Ballauff, M. (1993) *Makromol. Chem.*, **194**, 53.

19 Okada, M., Okada, Y., Tao, A., and Aoi, K. (1995) *J. Polym. Sci. Part A: Polym. Chem.*, **33**, 2813; Okada, M., Okada, Y., Tao, A., and Aoi, K. (1996) *J. App. Polym. Sci.*, **62**, 2257.

20 Bachmann, F., Reimer, J., Ruppenstein, M., and Thiem, J. (1998) *Macromol. Rapid Comunn.*, **19**, 21.

21 Lee, C.-H., Tagaki, H., Okamoto, H., Kato, M., and Usuki, A. (2009) *J. Polym. Sci. Part A Polym. Chem.*, **47**, 6025; Marín, R., and Muñoz-Guerra, S. (2009) *J. Appl. Polym. Sci.*, **114**, 3723.

22 Marín, R., de Paz, M.V., Ittobane, N., Galbis, J.A., and Muñoz-Guerra, S. (2009) *Macromol. Chem. Phys.*, **210**, 486.

23 Violante de Paz, M., Zamora, F., Begines, B., Ferris, C., and Galbis, J.A. (2010) *Biomacromolecules*, **11**, 269.

24 Tang, M., White, A.J.P., Stevens, M.M., and Williams, C.K. (2009) *Chem. Commun.*, 941.

25 Ávalos, M., Babiano, R., Cintas, P., Hursthouse, M.B., Jiménez, J.L., Light, M.E., Palacios, J.C., and Pérez, E.M.S. (2006) *Eur. J. Org. Chem.*, 3, 657.

26 Johnson, D.T., and Taconi, K.A. (2007) *Environ. Prog.*, **26**, 338; Behr, A., Eilting, J., Irawadi, K., Leschinski, J., and Lindner, F. (2008) *Green Chem.*, **10**, 13; Pagliaro, M., and Rossi, M. (2008) *The Future of Glycerol*, RSC Publishing, Cambridge; Zhou, C.-H., Beltramini, J.N., Fan, Y.-X., and Lu, G.Q. (2008) *Chem. Soc. Rev*, **37**, 527; Zheng, Y., Chen, X., and Shen, Y. (2008) *Chem. Rev.*, **108**, 5253; Prati, L., Spontoni, P., and Gaiassi, A. (2009) *Top. Catal.*, **52**, 288.

27 Barrault, J., Jerome, F., and Pouilloux, Y. (2005) *Lipid Technol.*, **17**, 131.

28 Ding, Z.Y., Hao, A.Y., and Wang, Z.N. (2007) *Fuel*, **86**, 597.

29 Parzuchowski, P.G., Jaroch, M., Trynowski, M., and Rokicki, G. (2008) *Macromolecules*, **41**, 3859.

30 Parzuchowski, P.G., Grabowska, M., Jaroch, M., and Kusznerczuk, M. (2009) *J. Polym. Sci. Part A Polym. Chem.*, **47**, 3860.

31 Zhao, X., Liu, L., Dai, H., Ma, C., Tan, X., and Yu, R. (2009) *J. Appl. Polym. Sci.*, **113**, 3376.

32 Wurm, F., Nieberle, J., and Frey, H. (2008) *Macromolecules*, **41**, 1184 and 1909.

33 Yin, A.-Y., Guo, X.-Y., Dai, W.-L., and Fan, K.-N. (2009) *Green Chem.*, **11**, 1514.

34 Gandini, A., and Belgacem, M.N. (2008) Furan derivatives and furan chemistry at the service of macromolecular materials, in *Monomers, Polymers and Composites from Renewable Resources* (eds M.N. Belgacem and A. Gandini), Elsevier, Amsterdam, Ch 6; Gandini, A., and Belgacem, M.N. (1997) *Progr. Polym. Sci.*, **22**, 1203; Moreau, C., Gandini, A., and Belgacem, M.N. (2004) *Topics Catal.*, **27**, 11; Gandini, A. (2010) *Polymer Chem.*, **1**, 245–251.

35 Theander, O., and Nelson, D.A. (1988) *Adv. Carbohydr. Chem. Biochem.*, **46**, 273; Zeitsch, K.J. (2000) *The Chemistry and Technology of Furfural and Its Many By-Products*, Elsevier, Amsterdam.

36 (a) Carlini, C., Patrono, P., Galletti, A.M.R., Sbrana, G., and Zima, V. (2004) *Appl. Catal. A Gen.*, **275**, 111; (b) Bicker, M., Kaiser, D., Ott, L., and Vogel, H. (2005) *J. Supercrit. Fluids*, **36**, 118; (c) Román-Leshkov, Y., Chheda, J.N., and Dumesic, J.A. (2006) *Science*, **312**, 1933; (d) Asghari, F.S., and Yoshida, H. (2006) *Carbohydr. Res.*, **341**, 2379; Asghari, F.S., and Yoshida, H. (2006) *Ind. Eng. Chem. Res.*, **45**, 2163; (e) Zhao, H., Holladay, J.E., Brown, H., and Zhang, C. (2007) *Science*, **316**, 1597; (f) Amarasekara, A.S., Williams, L.D., and Ebede, C.C. (2008) *Carbohydr. Res.*, **343**, 3021; (g) Hu, S., Zhang, Z., Zhou, Y., Han, B., Fan, H., Li, W., Song, J., and Xie, Y. (2008) *Green Chem.*, **10**, 1280; (h) Yong, G., Zhang, Y., and Ying, J.Y. (2008) *Angew. Chem. Int. Ed.*, **47**, 9345; (i) Qi, X., Watanabe, M., Aida, T.M., and Smith, R.L. Jr. (2008) *Green Chem.*, **10**, 799; *Ind. Eng. Chem. Res.*, 2008, **47**, 9234; (j) Hu, S., Zhang, Z., Zhou, Y., Song, J., Fan, H., and Han, B. (2009) *Green Chem.*, **11**, 873; (k) Román-Leshkov, Y., and Dumesic, J.A. (2009) *Top. Catal.*, **52**, 297; (l) Chan, J.Y.G., and Zhang, Y. (2009) *ChemSusChem*, **2**, 731; (m) Li, C., Zhang, Z., Bao, Z., and Zhao, K. (2009) *Tetrahedron Lett.*, **50**, 5403; (n) Qi, X., Watanabe, M., Aida, T.M., and Smith, R.L. Jr. (2009) *Green Chem.*, **11**, 1327; (o) Gorbanev, Y.Y., Klitgaard, S.K., Woodley, J.M., Christensen, C.H., and Riisager, A. (2009) *ChemSusChem*, **2**, 672; (p) Mascal, M., and Nikitin, E.B. (2009) *ChemSusChem*, **2**, 859; (q) Verevkin, S.P., Emel'yanenko, V.N., Stepurko, E.N., Ralys, R.V., and Zaitsau, D.H. (2009) *Ind. Eng. Chem. Res.*, **48**, 10087–10093. doi: 10.1021/ie901012g; (r) Binder, J.B., and Raines, R.T. (2009) *J. Am. Chem. Soc.*, **131**, 1979; (s) Boisen, A., Christensen, T.B., Fu, W., Gorbanev, Y.Y., Hansen, T.S., Jensen, J.S., Klitgaard, S.K., Pedersen, S., Riisager, A., Ståhlberg, T., and Woodley, J.M. (2009) *Chem. Eng. Res. Des.*, **87**, 1318; (t) Qi, X., Watanabe, M., Aida, T.A., and Smith, R.L. Jr. (2009) *ChemSusChem*, **2**, 944; (u) Hu, S., Zhang, Z., Song, J., Zhou, Y., and Han, B. (2009) *Green Chem.*, **11**, 1746; (v) Hansen, T.S., Woodley, J.W., and Riisager, A. (2009) *Carbohydr. Res.*, **344**, 2568; (w) Zhang, Z., and Zhao, Z.K. (2009) *Biores. Technol.*, **101**, 1111; (x) Ilgen, F., Ott, D., Kralisch, D., Palmberger, A., and König, B. (2009) *Green Chem.*, **11**, 1948; (y) Casanova, O., Iborra, S., and Corma, A. (2009) *ChemSusChem*, **2**, 1138; (z) Stålberg, T., Sørensen, M.G., and Riisager, A. (2010) *Green Chem.*, **12**, 321; (aa) Zhang, Z., and Zhao, Z.K. (2010) *Biores. Technol.*, **101**, 1111.

37 Gandini, A., Coelho, D., Gomes, M., Reis, B., and Silvestre, A. (2009) *J. Mater. Chem.*, **19**, 8656; Gomes, M. (2009) Síntese de poliésteres a partir do ácido 2,5-furanodicarboxílico. Thesis, University of Aveiro, Portugal.

38 Gandini, A., Silvestre, A.J.D., Pascoal Neto, C., Sousa, A.F., and Gomes, M. (2009) *J. Polym. Sci. Part A Polym. Chem.*, **47**, 295.

39 Gandini, A., and Belgacem, M.N. (2007) *ACS Symp. Ser.*, **954**, 280.

40 Gandini, A., Coelho, D., and Silvestre, A.J.D. (2008) *Eur. Polym. J.*, **44**, 4029; Gandini, A., Coelho, D., and Silvestre, A.J.D. (2010) *J. Polym. Sci. Part A: Polym. Chem.*, **48**, 2053–2056.

41 Gheneim, R., Pérez-Berumen, C., and Gandini, A. (2002) *Macromolecules*, **35**, 7246.

42 Meier, M.A.R., Metzger, J.O., and Schubert, U.S. (2007) *Chem. Soc. Rev.*, **36**, 1788; Belgacem, N.M., and Gandini, A. (2008) Materials from vegetable oils: major sources, properties and applications, in *Monomers, Polymers and Composites from Renewable Resources* (eds M.N. Belgacem and A. Gandini), Elsevier, Amsterdam, Ch 3; Sharma, V., and Kundu, P.P. (2008) *Prog. Poly. Sci.*, **33**, 1199; Petrovič, Z.S. (2008) *Polym. Rev.*, **48**, 109; Lu, Y., and Larock, R.C. (2009) *ChemSusChem*, **2**, 136; Metzger,

J.O. (2009) *Eur. J. Lipid Sci. Technol.*, **111**, 865; Galià, M., Montero de Espinosa, L., Ronda, J.C., Lligadas, G., and Cádiz, V. (2010) *Eur. J. Lipid Sci. Technol.*, **112**, 87–96.

43 Gandini, A. (2009) Epoxy polymers based on renewable resources, in *Epoxy Polymers: New Materials and Innovation* (eds J.P. Pascault and R.J.J. Williamd), Wiley-VCH Verlag GmbH, Weinheim, Germany, Ch 4; Liu, Z., Doll, K.M., and Holser, R.A. (2009) *Green Chem.*, **11**, 1774.

44 Ogunniyi, D.S. (2006) *Bioresour. Technol.*, **97**, 1086.

45 Meier, M.A.R. (2009) *Macromol. Chem. Phys.*, **210**, 1073.

46 Montarnal, D., Cordier, P., Soulié-Ziakovic, C., Tournilhac, F., and Leibler, L. (2008) *J. Polym. Sci. Part A Polym. Chem.*, **46**, 7925.

47 Crépy, L., Chaveriat, L., Banoub, J., Martin, P., and Joly, N. (2009) *ChemSusChem*, **2**, 1.

48 Hojabri, L., Kong, X., and Narine, S.S. (2009) *Biomacromolecules*, **10**, 884.

49 Jeong, W., Mauldin, T.C., Larock, R.C., and Kessler, M.R. (2009) *Macromol. Mater. Eng.*, **294**, 756.

50 Pizzi, A. (2008) Tannins: major sources, properties and applications, in *Monomers, Polymers and Composites from Renewable Resources* (eds M.N. Belgacem and A. Gandini), Elsevier, Amsterdam, Ch 8.

51 Pizzi, A., Tondi, G., Pasch, H., and Celzard, A. (2008) *J. Appl. Polym. Sci.*, **110**, 1451; Tondi, G., and Pizzi, A. (2009) *Ind. Crops Prod.*, **29**, 356.

52 Tondi, G., Pizzi, A., and Olives, R. (2008) *Maderas Cien. Tecnol.*, **10**, 219; Tondi, G., Zhao, W., Pizzi, A., Du, G., Fierro, V., and Celzard, A. (2009) *Bioresour. Technol.*, **100**, 5162; Tondi, G., Oo, C.W., Pizzi, A., Trosa, A., and Thevenon, M.F. (2009) *Ind. Crops Prod.*, **29**, 336.

53 Tondi, G., Pizzi, A., Pasch, H., and Celzard, A. (2008) *Polym. Degrad. Stab.*, **93**, 968; Tondi, G., Pizzi, A., Masson, E., and Celzard, A. (2008) *Polym. Degrad. Stab.*, **93**, 1539; Tondi, G., Fierro, V., Pizzi, A., and Celzard, A. (2009) *Carbon*, **47**, 1480.

54 Tondi, G., Blacher, S., Léonard, A., Pizzi, A., Fierro, V., Leban, J.-M., and Celzard, A. (2009) *Microsc. Microanal.*, **15**, 395.

55 Gellerstedt, G., and Henriksson, G. (2008) Lignins: major sources, structure and properties, in *Monomers, Polymers and Composites from Renewable Resources* (eds M.N. Belgacem and A. Gandini), Elsevier, Amsterdam, Ch 9.

56 Lora, J. (2008) Industrial commercial lignins: sources, properties and applications, in *Monomers, Polymers and Composites from Renewable Resources* (eds M.N. Belgacem and A. Gandini), Elsevier, Amsterdam, Ch 10.

57 Banu, D., El-Aghoury, A., and Feldman, D. (2006) *J. Appl. Polym. Sci.*, **101**, 2732.

58 Gandini, A., and Belgacem, M.N. (2008) Lignins as components of macromolecular materials, in *Monomers, Polymers and Composites from Renewable Resources* (eds M.N. Belgacem and A. Gandini), Elsevier, Amsterdam, Ch 11.

59 Wang, M., Leitch, M., and Xu, C. (2009) *Eur. Polym. J.*, **45**, 3380.

60 Banoub, J.H., Benjelloun-Mlayah, B., Ziarelli, F., Joly, N., and Delmas, M. (2007) *Rapid Commun. Mass Spectrom.*, **21**, 2867; Delmas, M. (2008) *Chem. Eng. Technol.*, **31**, 792.

61 Gandini, A., and Belgacem, M.N. (2008) Partial or total oxypropylation of natural polymers and the use of the ensuing materials as composites or polyol macromonomers, in *Monomers, Polymers and Composites from Renewable Resources* (eds M.N. Belgacem and A. Gandini), Elsevier, Amsterdam, Ch 12.

62 Cateto, C.A., Barreiro, M.F., Rodrigues, A.E., and Belgacem, M.N. (2009) *Ind. Eng. Chem. Res.*, **48**, 2583.

63 Kleinert, M., and Barth, T. (2008) *Chem. Eng. Technol.*, **31**, 736.

64 Borges da Silva, E.A., Zabkova, M., Araújo, J.D., Cateto, C.A., Barreiro, M.F., Belgacem, M.N., and Rodrigues, A.E. (2009) *Chem. Eng. Res. Des.*, **87**, 1276.

65 Antonsson, S., Henriksson, G., and Lindröm, M.E. (2008) *Ind. Crops Prod.*, **27**, 98.

References

66 Ismail, T.N.M.T., Hassan, H.A., Hirose, S., Taguchi, Y., Hatakeyama, T., and Hatakeyama, H. (2010) *Polym. Intern.*, **59**, 181.

67 Silvestre, A.J.D., Pascoal Neto, C., and Gandini, A. (2008) Cork and suberins: major sources, properties and applications, in *Monomers, Polymers and Composites from Renewable Resources* (eds M.N. Belgacem and A. Gandini), Elsevier, Amsterdam, Ch 14; Gandini, A., Pacoal Neto, C., and Silvestre, A.J.D. (2006) *Prog. Polym. Sci.*, **31**, 878.

68 Silvestre, A.J.D., Pascoal Neto, C., Gandini, A., and Sousa, A.F. (2009) *Appl. Spectrosc.*, **63**, 873.

69 Silvestre, A.J.D., Pascoal Neto, C., Gandini, A., and Sousa, A.F. (2008) *ChemSusChem*, **1**, 1020.

70 Pramanik, D., Ray, T.T., and Bakr, M.A. (1996) *J. Polym. Mater*, **13**, 173.

71 Barroso-Bujans, F., Martínez, R., and Ortiz, P. (2003) *J. Appl. Polym. Sci.*, **88**, 302.

72 Tsutsumi, N., Oya, M., and Sakai, W. (2004) *Macromolecules*, **37**, 5971; Yang, J., Webb, A.R., and Ameer, G.A. (2004) *Adv. Mater.*, **16**, 511; Ding, T., Liu, Q., Shi, R., Tian, M., Yang, J., and Zhang, L. (2006) *Polym. Degrad. Stab.*, **91**, 733; Doll, K.M., Shogren, R.L., Willett, J.L., and Swift, G. (2006) *J. Polym. Sci. Part A Polym. Chem.*, **44**, 4259; Yang, J., W1.ebb, A.R., Pickerill, S.J., Hageman, G., and Ameer, G.A. (2006) *Biomaterials*, **27**, 1889.

73 Noordover, B.A.J., Duchateau, R., van Benthem, R.A.T.M., Ming, W., and Koning, C.E. (2007) *Biomacromolecules*, **8**, 3860.

74 Yokoe, M., Aoi, K., and Okada, M. (2005) *J. Polym. Sci. Part A Polym. Chem.*, **43**, 3909.

75 Iribarren, I., Alemán, C., Regaño, C., Martínez de Ilarduya, A., Bou, J.J., and Muñoz-Guerra, S. (1996) *Macromolecules*, **29**, 8413.

76 Marín, R., Martínez de Ilarduya, A., Romero, P., Sarasua, J.R., Meaurio, E., Zuza, E., and Muñoz-Guerra, S. (2008) *Macromolecules*, **41**, 3734.

77 Marín, R., Martínez de Ilarduya, A., and Muñoz-Guerra, S. (2009) *J. Polym. Sci. Part A Polym. Chem.*, **47**, 2391.

78 Delhomme, C., Weuster-Botz, D., and Kühn, F.E. (2009) *Green Chem.*, **11**, 13.

2
Enzyme Immobilization on Layered and Nanostructured Materials

Ioannis V. Pavlidis, Aikaterini A. Tzialla, Apostolos Enotiadis, Haralambos Stamatis, and Dimitrios Gournis

2.1
Introduction

Enzymes are versatile biocatalysts bearing some excellent properties (such as high activity and chemo-, regio-, and stereospecificity) which render them able to perform various reactions under mild conditions. The use of immobilized enzymes instead of fully or partially soluble preparations presents many advantages, such as enhanced stability, repeated or continuous use, easy separation from the reaction mixture, possible modulation of the catalytic properties, prevention of protein contamination in the product, and simple performance and design of the bio-reactor. Enzymes have been immobilized on different supports using methods such as adsorption, deposition, precipitation, gel entrapment, covalent binding or cross-linking [1, 2]. Among various supports used, nanostructured composite materials are now the focus of intense fundamental and applied research in a number of areas, including biocatalysis.

Recent scientific advances in nanotechnology have allowed the design and synthesis of a variety of nanostructured materials, including nanoporous media, nanofibers, carbon nanotubes and nanoparticles, with large active surface area and desirable pore sizes [3, 4]. Proteins, enzymes included, are nanometer-scaled molecules, which benefit from the high surface area of these nanostructures [5, 6]. This large surface area results in increased enzyme loading, and therefore high apparent enzyme activity per unit mass or volume, compared with that provided by conventional supports. Nanostructured materials have manifested great efficiency in the manipulation of the nanoscale environment of the enzyme and thus to its catalytic behavior and stability [6]. Such a characteristic can be of significant scientific and industrial interest, since stabilization of enzymes is an important factor in large-scale applications of immobilized enzyme systems in the carbohydrate, food and pharmaceutical industries. The unique properties of nanomaterials as immobilization supports for biomolecules, together with other desirable properties, such as conductivity and magnetism, offer particularly exciting opportunities in molecular imaging, therapy, and delivery of biomolecules, as well as in construction of biosensors and biomedical devices [7–9].

Biocatalysis in Polymer Chemistry. Edited by Katja Loos
Copyright © 2011 WILEY-VCH Verlag GmbH & Co. KGaA, Weinheim
ISBN: 978-3-527-32618-1

The immobilization of enzymes onto nanoparticles, such as surface-modified polymeric nanogranules and magnetic particles, was first reported in the late 1980s [10, 11]. Up to date, materials of various composition and structure have been used as supports for enzyme immobilization. Nanoparticles made of silica, magnetite and gold comprise the first group of nanomaterials employed for biocatalysis, while enzymes were immobilized onto these materials using conventional approaches, such as simple adsorption and covalent attachment. Since then, various nanomaterials, such as functionalized single-walled or multi-walled carbon nanotubes, nanoporous media with pores 2–50 nm, electrospun polymer nanofibers, nanoporous silica and organically modified nanoclays, have been used as immobilization supports [6, 12–14].

In this chapter we will focus on the most recent developments in the use of nanostructured materials such as nanoclays, carbon nanotubes and magnetic nanoparticles for enzyme immobilization and stabilization, together with their potential applications in various fields, such as development of biosensors and biofuel cells, biocatalytic processes, enzyme purification/separation, intracellular protein transportation etc.

2.2
Enzymes Immobilized on Layered Materials

2.2.1
Clays

2.2.1.1 Introduction
Production of earthware is probably one of the oldest manufacturing techniques known to mankind. Yet the main materials, clays, are here to stay in the twenty-first century to develop into fundamental and inexpensive technological applications. Smectite clays are layered minerals, consisting of nanometer-sized aluminosilicate nanoplatelets, with a unique combination of swelling, intercalation and ion exchange properties that make them valuable nanostructures in diverse fields [15–18]. Their structure (Figure 2.1) consists of essentially four layers: (a) a gallery in which the intercalated species reside together with molecular H_2O; (b) a tetrahedral sheet consisting of SiO_4 tetrahedra joined at the corners to form a hexagonal arrangement; (c) an octahedral sheet composed of edge-sharing $AlO_4(OH)_2$ octahedra also in a hexagonal arrangement; and (d) another tetrahedral sheet symmetrically disposed to the first with respect to the octahedral sheet. The basal oxygen sheets are arranged in a Kagomé lattice whose hexagonal 'pockets' form triangular lattice gallery sites.

Such clays have a cation exchange capacity which depends upon substitution of lower valent atoms, for example, Mg^{2+} for Al^{3+} in the octahedral sheet and Al^{3+} for Si^{4+} in the tetrahedral sites. As a consequence, the sheets have a fixed negative charge and neutrality is provided, for example, by hydrated cations that are present in the galleries and on the outer surfaces. The intercalation process in

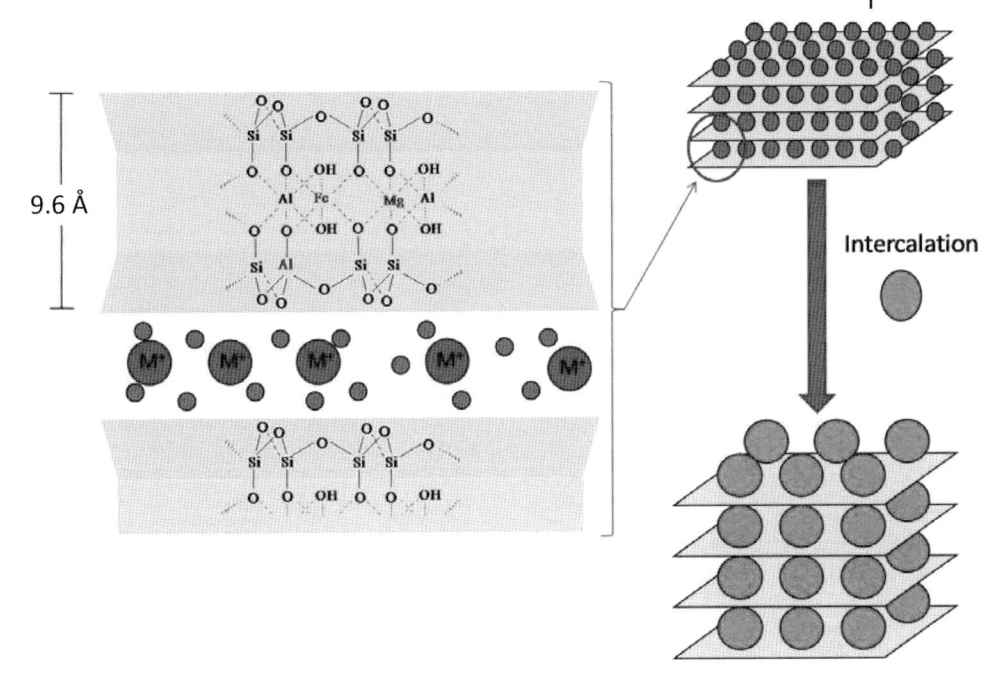

Figure 2.1 Structure of smectite clays.

these systems is equivalent to ion exchange and, differently from intercalation compounds of graphite, it does not necessarily involve charge transfer between the guest and host molecules. The charge on the sheets affects many fundamental properties of the clays, including cation exchange capacity, cation fixation, swelling ability, water holding and specific surface area. These materials have the natural ability to adsorb organic or inorganic guest cationic species (or neutral molecules) from solutions. This cation 'storage', which gives unique properties to clay minerals, makes them such important components in various industrial and technological applications. Clays are widely used as catalysts [19], sorbents [20], templates [21] in organic synthesis or as building blocks for composite materials [22]. Clays are used as nanofillers of engineering polymers resulting in nanocomposites with enhanced thermal stability, mechanical, and barrier properties. The nature of the microenvironment between the aluminosilicate sheets regulates the topology of the intercalated molecules and affects possible supramolecular rearrangements or reactions, such as self-assembling processes that are usually not easily controlled in the solution phase [23]. Smectite clays can be used as matrices for immobilization of various biomolecules [18, 24–27] and for other medical applications [28, 29]. The properties of the parent nanoclay can be tailored using simple chemical methods such as acid activation, pillaring and intercalation with organic or inorganic molecules. Moreover, the hydrophobicity of the parent clays can be increased through treatment with an organic surfactant.

As a result, the presence of the surfactant expands the interlayer gallery and lowers the clay surface energy rendering the nanoclay compatible with organic solvents.

2.2.1.2 Enzymes Immobilization on Clays

The swelling ability of these naturally occurring materials provides unusual properties and appreciable surface area for adsorption of organic molecules [15, 30], as was mentioned in Section 2.2.1.1. For the last 25 years different research groups have studied the immobilization of different biomolecules, including enzymes, proteins, nucleic acids [31] as well as cells [32–35] in layered materials. More specifically, various enzymes have been immobilized on clays, including hydrolases and oxidoreductases (see Table 2.1).

The immobilization of enzymes on clays usually takes place through adsorption, based on electrostatic interactions between the external charges on protein and support [56–58]. The adsorption of a protein to a surface is a complex process in which the structural stability of the protein, the surface properties of the sorbent, the ionic strength and the pH of the solution have an influence on the affinity of the protein for a given surface [59, 60]. The affinity of proteins for various types of interface originates from the flexibility of the enzyme molecule and from the reactivity of side chain groups on the surface of the protein molecules. These properties give rise to a large variety of interactions with layered materials. It is well established that for most proteins the maximum adsorption occurs near their isoelectric point (pI) [61]. Generally, strong electrostatic interactions between a highly charged protein and the clay can lead to enzyme deactivation. Near their pI, protein molecules are adsorbed with minor structural changes and thus enzymes preserve their catalytic activity [61]. It must be noted, that the formation of an electrostatic complex is usually a reversible process depending on pH and ionic strength of the solution.

When clay minerals are used as supports for enzymes, different types of binding mechanisms are possible, including ion exchange, van der Waals interactions, hydrogen bonding, and ion-dipole interactions with metal exchanged ions on the clay surfaces. Concerning the site of immobilization, it is proposed that the enzyme molecules are anchored on the external surface, the edges of the clay sheets, or intercalated within the interlayer space of the clay [24, 56, 57]. Taking into consideration that the enzyme molecule in most cases is bigger than the clay galleries, the molecule is not expected to intercalate between the clay nanosheets [56]. It was proposed that some side chains of amino acid residues of the enzyme take part in intercalation, while the polypeptide backbone is localized outside the pores. More specifically, the adsorption of different enzymes as α-amylase [25], glucoamylase [26], invertase [62], β-glucosidase [58] and *Candida antarctica* lipase B (CALB) (see also Chapter 3) [56] over synthetic or natural clays is in accordance with this theory.

Recently, the functionalization of clays has been used to increase the immobilization efficiency of enzymes and other biomolecules [63]. The clay surface may be modified through reactions with various functional groups, such as thiol,

Table 2.1 Examples of immobilized enzymes on clays and possible applications.

Enzyme	Clay	Application	Reference
Glucose oxidase	Modified Y zeolite matrix	Amperometric glucose biosensor	[36]
Glucose oxidase	Kaolinite (KGa-1), illite, bentonite, nontronite(SWa-1, ferruginous smectite), montmorillonite (SWy-1), and vermiculite (VTx-1),	Amperometric glucose biosensor	[37]
Glucose oxidase	Laponite	Amperometric glucose biosensor	[38]
Glucose oxidase	Anionic clay; layered double hydroxides [Zn3-Al–Cl]	Amperometric detection of glucose	[39]
Glucose oxidase Polyphenol oxidase from mushroom	Organoclay: natural Cameroonian smectites grafted with either aminopropyl or trimethylpropylammonium groups.	Modified electrodes	[40]
Glucose oxidase	Mixture of chitosan-laponite	Amperometric glucose biosensor	[41]
α-Chymotrypsin	HY, NH4Y, NaY, HNH4DAY, HDAY and MCM-41 zeolites	Peptide synthesis	[42]
α-Chymotrypsin	Y zeolites (HY, NH4Y, NaY) and mesoporous dealuminized zeolites	Peptide synthesis	[43]
Phosphohydrolase	soil	Hydrolysis of organic phosphorus	[44]
Urease	Laponite	Conductimetric detection of urea	[45]
Urease	Laponite	Urea biosensor	[46]
Lactate dehydrogenase, alcohol dehydrogenase	Laponite gel-methylene blue polymer	Biosensor	[47]
Polyphenol oxidase	Laponite	Amperometric biosensor	[48]
Trametes versicolor laccase	Kaolinite	Oxidation of anthracene	[49]
Trametes versicolor laccase	Calcareous clay soil	Atrazine bioremediation	[50]
Pectinlyase from *Aspergillus niger*	Betonite	Food technology	[51]
Bacterial hydrogenase	Montmorillonite nanoparticles	Modified electrodes	[52]

Table 2.1 Continued

Enzyme	Clay	Application	Reference
Pseudomonas sp. and *Candida rugosa* lipases	Clay	Synthesis of lipids from tricaprylin and trilinolein	[53]
Pseudomonas cepacia lipase	Phyllosilicate sol–gel matrix	Biocatalysis	[54]
Pseudomonas cepacia and *Rhizopus oryzae* lipase	Phyllosilicate sol-gel matrix	Esterification of glycerol	[55]
Lipase B from *Candida antarctica*	Smectite group (Laponite, SWy-2 and Kunipia) and organically modified derivatives	Oxidation of terpenes	[56, 57]

amine, or long carbon chains. Towards to this direction, palygorskites were modified either by acid treatment, or reacting the surface silanol groups present with 3-aminopropyltriethoxysilane, or treating with a quaternary ammonium compound (octodecyl trimethyl ammonium chloride), to produce derivatives with suitable functional groups for further use as supports for enzyme immobilization [64]. Also smectite clays such as laponite, kunipia and SWy-2 were modified by treating with a cationic surfactant (octadecyl trimethyl ammonium bromide) [65] for the immobilization of CALB [56, 57]. A schematic illustration of immobilized CALB on organo-modified clays is presented in Figure 2.2.

Enzymes immobilized on clays have been applied in various industrial processes including food technology [51], peptide synthesis [42, 43], pharmaceuticals [66], biosensors [39, 41, 46, 48], construction of modified electrodes [40, 52] as well as in bioremediation processes [50]. Table 2.1 presents some examples concerning the use of immobilized enzymes on clays.

A combination of techniques, such as powder X-ray diffraction (XRD) [56, 58], thermogravimetric analysis (TGA) [57], differential thermal analysis (DTA) [57], X-ray photoelectron spectroscopy (XPS) [56, 58], scanning electron microscopy (SEM) [26, 57], Fourier transform infrared (FT-IR) spectroscopy [57, 58] and BET N_2 adsorption measurements [67], was used for structural characterization of the enzyme–clay conjugates.

More specifically, XRD measurements provide a powerful tool to understand the changes in the interior of the clay microenvironment and thus to evaluate the different types of nanocomposites formed. In most cases, as enzyme loading increases, the intensity of the reflection increases, confirming enhanced intercalation of enzyme or a higher degree of ordering [25, 58], while other studies showed that the enzyme loading leads to exfoliated nanocomposite structures [56, 57]. In the case of organo-modified clays, the interaction between the clay and

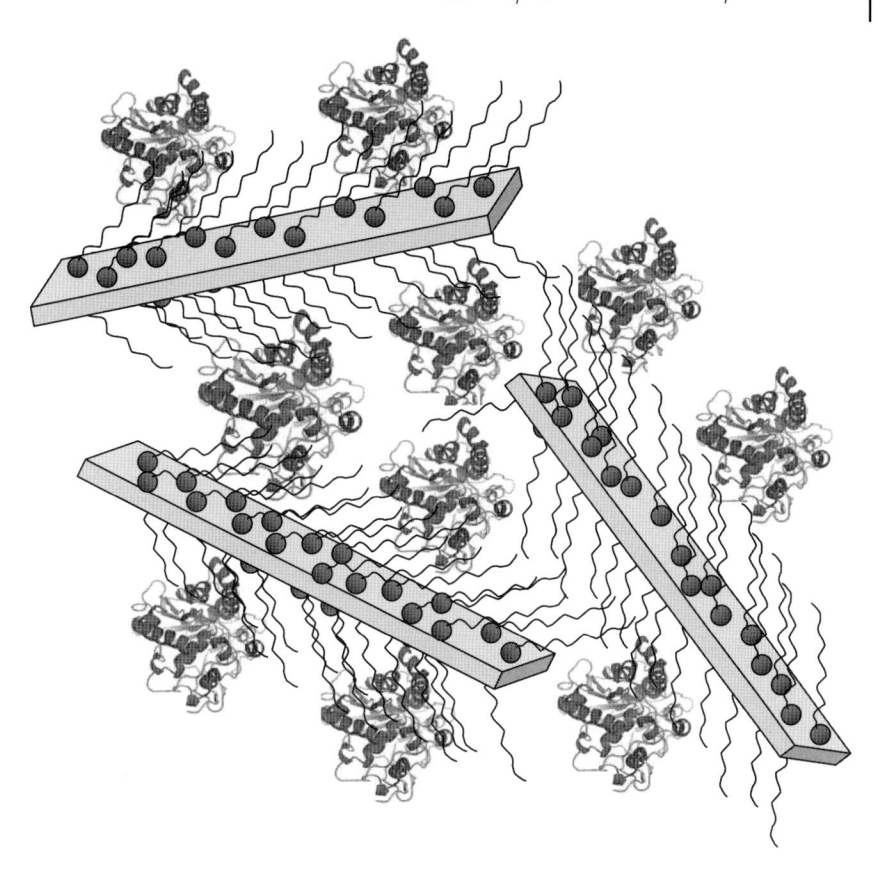

Figure 2.2 Schematic illustration of exfoliated organo-modified clays with immobilized enzyme molecules.

the surfactant led to a shift of the d_{001} diffraction peak of the organo-modified clay toward lower 2θ values, implying the expansion of the interlayer space due to the alkylammonium intercalation [56].

In many cases, the enzyme–clay systems were studied using XPS, a surface chemical analysis technique that can be used to investigate how enzyme binds to the siloxane interface and generally gives quantitative information on the elemental composition of hybrid systems. While many chemically distinct carbon environments exist in enzymes, XPS may distinguish them. For example, XPS can distinguish between aliphatic and carbonyl carbons but not between different types of carbonyls present in the molecule. Recent studies with XPS confirmed the adsorption of β-glucosidase on laponite and kunipia [62] as well as the adsorption of CALB on parent and organo-modified clays [56].

FT-IR is another tool to confirm the successful immobilization of an enzyme on a solid support [57, 58, 68]. More specifically, the binding can be confirmed by the FT-IR spectra of the immobilized enzyme onto various modified and unmodified clays, in which are presented all the characteristic bands of clays and protein [58, 64, 68].

Scanning electron microscopy (SEM) has been successfully applied to study the morphology of enzyme-containing carriers. The surface features of immobilized CALB on parent and organo-modified smectite clays [57], glucose oxidase on anionic clays [69], α-amylase, glucoamylase onto acid activated montmorillonite (K-10) [25], amylase on K-10 montmorillonite [67] have been studied by this technique. In most cases the immobilization of enzymes on the solid matrix changes the morphology of the support.

Information about the secondary structures (α-helices, β-sheets, random coil) can be useful for understanding conformation changes of proteins upon the immobilization process. More specifically, circular dichroism (CD) [70] and FT-IR spectroscopy [56, 58, 61, 71–73] have been applied to study the structural characteristics of various proteins adsorbed on mineral surfaces. Kondo and coworkers [70] have studied the modification in α-helix content of proteins adsorbed on ultrafine silica particles with CD and found a decrease upon immobilization. Circular dichroism is not usually used because this technique is applicable only for the study of enzymes immobilized on nano-sized mineral particles due to problems arising from light scattering effects. On the other hand, infrared spectroscopy does not suffer from light scattering perturbations and has thus been used for the study of the conformation of proteins when they are immobilized on solid supports [57, 58].

Generally, in recent studies on the adsorption of bovine serum albumin (BSA) on montmorillonite, a decrease in the α-helix content and an increase in intermolecular β-sheet content have been observed for the BSA by FT-IR analysis [72, 74]. Similar structural changes were observed in the case of adsorption of β-glucosidase on laponite and kunipia in which the immobilized enzyme retains an ordered structure, which is different from its native conformation. More specifically, the native structure of the enzyme was modified upon immobilization on clays by a substantial decrease of the α-helix and a lower percentage of the β-sheet form. On the other hand the percentage of β-turns increased, but the random coil did not [58]. In other studies, the immobilization of CALB in smectite clays led to major changes in the secondary structure of CALB in comparison to its structure in water [56]. In this study, the organo-modification of the clays does not have the same impact on the structure of the enzyme with the parent clay. More specifically, the increased α-helix content of the enzyme, which was observed for the organo-modified clays, could be attributed to the increased hydrophobicity of these materials compared with the parent ones. Very recently it was reported that an increase of hydrophobicity of the immobilization matrix results to an increase in the α-helix content and thus the fraction of properly-folded functional enzyme, enhancing the biocatalytic activity [75]. Finally, other studies revealed that the adsorption of α-chymotrypsin on montmorillonite has only a very small effect on the secondary structure of this protein [71].

2.2.2
Other Carbon Layered Materials

Apart from classical carbon layered materials, such as graphite and graphite oxide, a wide variety of new carbon nanostructures such as fullerenes, nanotubes, nanohorns, nanobuds, nano-onions and graphenes, have emerged as new and fascinating forms of carbon whose chemical and physical properties are currently being unraveled [76]. Of these forms, graphene, a single layer of sp^2-hybridized carbon atoms found in graphite, is well known for its outstanding electronic (mainly), mechanical and structural properties and has attracted interest worldwide for its possible applications in various fields [77]. Because of their unique physicochemical characteristics, such as surface-to-volume ratio, high catalytic activity and efficient electron transfer capabilities, carbon layered materials have been reported to be used for the immobilization of different enzymes on graphite for the construction of various biosensors [79]. Only a few reported studies are currently published about the enzyme immobilization onto graphene, in which the beneficial effect of graphene instead of graphite is emphasized [80]. Generally, the application of graphene in the area of biosensors is a large field and shows an immense potential to be explored. Table 2.2 shows the main applications of enzymes immobilized on graphite or graphene which are used as transducing elements in various biosensors.

Table 2.2 Applications of enzymes immobilized on graphite (modified or not) or graphene.

Enzyme	Support	Application	Reference
Trametes versicolor laccase	Graphite	Biosensors for phenolic compounds	[81]
Trametes hirsuta laccase	Graphite	Electrode for direct electrocatalytic reduction of O_2 to H_2O	[82]
Cerrena unicolor laccase	Graphite	Electrode for direct electrocatalytic reduction of O_2 to H_2O	[78]
Trametes versicolor laccase	Carbon nanotubes-ionic liquid gel on graphite	Electrode for direct electrocatalytic reduction of O_2 to H_2O	[83]
Glucose oxidase	Graphite	Biofuel cells	[84]
Glucose oxidase	Modified graphite	Amperometric glucose biosensor	[85]
Glucose oxidase	Graphite nanoplatelets decorated with Pt and Pd nanoparticles	Amperometric glucose biosensor	[79]

Table 2.2 Continued

Enzyme	Support	Application	Reference
Glucose oxidase	Polyvinyl pyrrolidone-protected graphene	Amperometric glucose biosensor	[86]
Glucose oxidase	Graphene oxide modified glassy carbon	Amperometric glucose biosensor	[80]
Glucose oxidase	Graphene layered carbon nanofibers	Amperometric glucose biosensor	[87]
Tyrosinase	Graphite modified with carbon nanotubes	Quantification of methimazole in pharmaceuticals formulations	[88]
Tyrosinase	Graphite–Teflon composite matrix	Amperometric biosensor	[89]
Urease	Graphite	Amperometric urea biosensor	[90]
Lactate oxidase, horseradish peroxidase	Graphite–Teflon	Amperometric lactate biosensor	[91]

2.3
Enzymes Immobilized on Carbon Nanotubes

2.3.1
Introduction

Carbon nanotubes (CNTs) are a new class of nanomaterials that have attracted considerable attention due to their electronic, mechanical, optical and magnetic properties [92]. In recent years, many efforts have led to the development of versatile chemical modification methodologies, in order to solve the insolubility obstacle, targeting CNT derivatives with even more attractive features [93]. Towards this end, a wide range of derivatives have been prepared and fully characterized that exhibit promising applications in nanoelectronics [94], biology [95, 96], catalysis [97], or for the synthesis of composite materials [98]. CNTs can be synthesized as multi-wall carbon nanotubes (MWCNTs), double-wall carbon nanotubes (DWCNTs) and single-wall carbon nanotubes (SWCNTs). Nowadays three main techniques are widely used to synthesize CNTs: laser ablation, arc discharge and catalytical chemical vapor deposition (CCVD). Each method has its own advantages and limitations, with the CCVD method being an economic technique for synthesizing CNTs at low temperature and ambient pressure using hydrocarbon gases (methane, ethane, and acetylene), over catalytically active metallic centers (commonly transitional metal nanoparticles) embedded in solid supports zeolites, mesoporous silica, aluminosilicate layers, alumina and graphite [99]. Two main paths are usually followed for the functionalization of carbon nanotubes: attachment of organic moieties either to carboxylic groups that are formed by oxidation of CNTs with strong acids or direct bonding to the surface double bonds [93]. Highly soluble and purified nanotubes are obtained by 1,3 dipolar cycloaddition and also by attachment of aryl groups through electrochemical reduction of aryl diazonium salts. Fluorination, addition of carbenes and nitrenes, electrophiles or peroxy radicals were found to be successful reactions for sidewall covalent functionalization of CNTs [93]. Another type of nanotubes used for enzyme immobilization is the CNx nanotubes. Nitrogen atoms doped into the surface of CNTs produced a high density of defective sites on the graphene sheets and a C–N microenvironment, favorable for the immobilization of proteins [100].

Carbon nanotubes are receiving a great deal of attention as alternative matrices for enzyme immobilization and to improve stability and sensitivity of biosensors. These interesting nanomaterials provide high surface area for higher enzyme loading, reduced diffusion limitations and a biocompatible microenvironment helping enzyme to retain its catalytic properties. Apart from proteins, other biological molecules can also be immobilized on carbon nanotubes, such as nucleic acids, antigens, peptides and drugs, leading to a hybrid biomaterial. The type of the molecule immobilized on the carbon nanotubes lead to different applications, which renders these novel biomaterials one of the most versatile type of materials [101, 102]. For efficient use in biomedical and biocatalytic applications, the nanomaterials should combine the biocompatibility and the capability of interacting with the biomacromolecules.

2.3.2
Applications

Carbon nanotubes, due to their unique properties, have numerous applications, especially when conjugated with biomacromolecules. As far as the immobilization of proteins concerns, the high specific surface area of these nanomaterials facilitates the immobilization of more protein molecules on the carrier material, which is accompanied by an increased specific enzyme activity.

CNTs present good electrical communication, which renders feasible the electron transfer from protein to the electrode. For this reason many laboratories have turned their scientific interests in the fabrication of CNT-modified electrodes onto which enzymes or nucleic acids are immobilized. As it can be seen from Table 2.3, most of the works in the field of CNT-protein conjugates are about the development of new biosensors. CNT-biosensors have shown efficient electrical communications and promising sensitivities required for applications as antigen recognition, enzyme-catalyzed reactions and deoxyribonucleic acid (DNA) hybridizations [124]. The presence of CNTs facilitates the transportation of the signal from the enzyme to the electrode. The use of CNT-modified electrodes permits good communication with redox proteins when the redox active center is close to the surface of CNTs [125].

The development and application of biosensors has been discussed in several reviews [125, 126]. For instance, fabrication of cholesterol biosensors relies on the immobilization of cholesterol oxidase and cholesterol esterase onto CNTs, while horseradish peroxidase and flavocytochrome P450scc are also used for the same reason [125, 126]. Beyond the biosensing field, carbon nanotubes are also used as carriers for peptide, nucleic acid and drug delivery, due to their intrinsic property to cross cell membranes [102, 124]. The fact that the functionalized CNTs (f-CNTs) are not immunogenic and low-toxic opens the pathway for more research in the field of CNT-abetted drug delivery [102].

Another promising application of enzymes immobilized onto carbon nanotubes is the development of biofuel cells. Some works on this field are cited in Table 2.3. Enzyme-based biofuel cells are a promising technology that uses natural molecules for the generation of electricity. Enzymes or whole cells could be used to produce energy. Their practical application has been hampered so far by the low stability of the enzymes and the low power density of the cell. An increase in enzyme loading is anticipated to lead to increased power density. The use of conductive nanomaterials, such as CNTs, might contribute to an increased power density of biofuel cells by facilitating the charge transport between the enzyme and the electrode, while stabilization of proteins when immobilized onto these materials has been reported [108].

2.3.3
Immobilization Approaches

Proteins can interact with nanotubes with multiple types of interactions. A coarse categorization of the types of interactions observed between enzyme molecules and CNTs should separate the covalent and the non-covalent bonding of the enzymes on the CNTs surface. Non-covalent bonding is an easy way to immobilize protein on the surface of an immobilization matrix, though is less controllable than a covalent bonding. Moreover, the covalent conjugation may deactivate the protein, either disrupt the π-networks on the CNT surfaces and may diminish their mechanical and electronic properties [127]. In most cases, the interaction between protein and carbon nanotubes is attributed to hydrophobic and electrostatic interactions [95, 104]. The fact that pristine CNTs are highly hydrophobic led to the assumption that the non-specific absorption of proteins can be attributed to hydrophobic interactions, through interactions with the side chains of hydrophobic amino acids [127]. As far as the electrostatic interactions concerns, the π electrons on the surface of the CNTs interact with the π electrons of the aromatic ring of amino acids (e.g., phenylalanine and tryptophan). The number of hydrophobic or aromatic residues of a protein cannot help predicting the binding ability of the protein, due to the protein folding. Some amino acid residues are buried inside the 'core' of the protein and thus cannot contribute to the interactions between the protein and the carbon nanotubes [128]. Azamian and

Table 2.3 Applications of carbon nanotube–protein bioconjugates.

Protein	Carbon nanotube nanomaterials	Application	Reference
α-Chymothrypsin	SWCNTs/poly(methyl methacrylate) film	Antifouling surface materials	[103]
β-Glucosidase	MWCNT-COOH	Biocatalysis	[104]
Candida rugosa lipase	MWCNTs or MWCNTs-COOH	Biocatalysis	[105]
Horseradish peroxidase	SWCNT/chitosan modified glassy carbon electrode (GCE)	Bioelectrochemical sensor	[106]
Trametes hirsuta laccase	MWCNT modified GCE	Biofuel cells	[107]
Horseradish peroxidase, myoglobin, cytochrome c	SWCNT/ionic liquid modified GCE	Biofuel cells, Biosensors	[108]
Cytochrome c	SWCNT modified GCE	Biofuel cells, biosensors	[109]
Cerrena unicolor laccase	CNT modified boron-doped diamond electrode	Biofuel cells, biosensors	[110]
Xylanase	MWCNTs	Enzyme refolding	[111]
Alcohol dehydrogenase	SWCNT/ poly(dimethyl diallylammonium chloride) (PDDA) modified GCE	Ethanol biosensor	[112]
Glucose oxidase	SWCNT/silica modified GCE	Glucose biosensor	[113]
Glucose oxidase horseradish peroxidase	MWCNT-toluidine blue/ nafion modified GCE	Glucose biosensor	[114]
Cytochrome c	MWCNT/chitosan/ionic liquid modified GCE	H_2O_2 detector	[115]
Horseradish peroxidase myoglobin, hemoglobin	SWCNT/ CTAB film on saturated calomel electrode	H_2O_2 detector	[116]
Horseradish peroxidase	MWCNT/chitosan/sol-gel modified GCE	H_2O_2 detector	[117]
Cytochrreptavidin, protein A, bovine serum albumin	SWCNT-COOH	Intracellular protein transportation	[118]
Cytochrome c	MWCNT/poly(amidoamine)/ chitosan modified GCE	Nitrite biosensor	[119]

Table 2.3 Continued

Protein	Carbon nanotube nanomaterials	Application	Reference
Bovine odorant-binding protein	CNTs	Odorant biosensor	[120]
Trametes versicolor laccase	MWCNT/chitosan	Oxygen biosensor, biofuel cells	[121]
Cytochome c, hemoglobin	MWCNTs	Protein purification	[122]
RNA-ase A	Carboxylated MWCNTs or SWCNTs	RNA extraction	[123]

coworkers underlined these difficulties and proved that the adsorption of proteins on SWCNTs is insensitive to the protein pI, thus inconsistent with an electrostatic interaction mechanism [129]. In this way they proved that the interactions between the carbon nanotubes and the proteins are complicated; not only the electrostatic interactions are responsible for the non-specific binding, but also other interaction types contribute to this interaction between the CNTs an the proteins. Apart from electrostatic and hydrophobic interactions, hydrogen bonds and non-specific adsorption can also play a role for enzyme adsorption onto carbon nanotubes.

The non-specific adsorption of proteins on carbon nanotubes is an interesting phenomenon but represents a relatively less controllable mode of protein-CNT interaction. Moreover, in non-covalent immobilization process, the immobilized protein is in equilibrium between the surface of the carbon nanotubes and the solution and can therefore be gradually detached from the nano-material surface, a phenomenon called 'protein leakage' [127]. To prevent the leaching of enzymes, covalent bonds have been used to attach the enzyme molecules to the nanostructured materials, which lead to more robust and predictable conjugation. Experimental evidences prove that proteins can be immobilized either in their hollow cavity or on the surface of carbon nanotubes [130].

A common method for covalent immobilization of enzymes and proteins onto the nanomaterials is the use of molecules which act as 'bridge' between the material and the protein. One such molecule widely used is 1-ethyl-3-(3-dimethylaminopropyl) carbodiimide (EDC), which can lead proteins to covalently attach on carboxylated CNTs [127]. There is an ongoing discussion whether this coupling agent is efficient or not. Azamian and coworkers have immobilized proteins on SWCNTs in the presence and absence of the coupling reagent EDC and followed by prolonged incubations of the product material in high salt or surfactant levels. In both cases, the immobilization efficiency was at comparable levels. The authors claimed that this is consistent with adsorption being predominantly non-covalent. It was not possible to significantly alleviate this physical immobilization by either washing the modified tubes or by carrying out the immobilization in the presence of surfactant [129]. On the other hand, Lin and coworkers compared ferritin–SWCNT conjugation in water under both non-specific adsorption and covalent functionalization with another carbodiimide coupling agent [131]. An important difference was the fact that the non-covalently adsorbed ferritin species were more removable in vigorous dialysis, while the use of poly(ethylene glycol) (PEG) prevented the non-specific binding of ferritin on SWCNTs but not on the functionalized-SWCNTs, on which the protein was immobilized covalently. Chen and coworkers reported the use of 1-pyrenebutanoic acid succinimide ester as a linker to immobilize proteins on SWCNT surfaces [132]. The assumption was that the pyrene moiety would be $\pi-\pi$ stacking with the graphitic nanotube sidewall so that the succinimidyl ester group at the other end of the bifunctional molecule would be available for reaction with primary and secondary amines in the proteins.

Whatever the coupling agent is, the control of non-specific protein absorption is important to the use of nanomaterials in specific protein binding. There are plenty of molecules used for protection of various surfaces from proteins with mechanisms as steric repulsion, hydration and solvent structuring. For example, the modification of CNTs with the absorption of biotinylated Tween 20 allowed streptavidin recognition by the specific biotin-streptavidin interaction, but provided resistance towards other protein absorption [133].

Polyethylene glycol (PEG) is a widely used molecule for the prevention of non-specific binding [95, 131], although it does not absorb well on the surface of CNTs, so various techniques have been developed [124]:

1) Surfactant molecules (as Triton X, Tween 20 or Pluronic P103) are used to cover the surface of the CNTs (with hydrophobic interactions) and subsequently facilitate PEG absorption.

2) Surfactant molecules functionalized with PEG units provide sufficient protein resistance.

3) Direct functionalization of CNTs with PEG moieties.

In order to increase the immobilization efficiency and specificity, among with biocompatibility and dispersability, carbon nanotubes can be covered by polymers. Nanomaterials can be coated with several materials including a broad band of molecules such as organic stabilizers, inorganic molecules and functionalization with targeting ligands. Polymeric coating materials can be classified into synthetic and natural polymers. Polymers based on poly(amidoamine) [119], poly(dimethyldiallylammonium chloride) [112, 134], poly(methyl methacrylate) [103], poly(ethylene glycol) [95], sol-gels [113, 117] etc. are typical examples of synthetic polymeric systems. Natural polymer systems include use of dextran, chitosan etc. [106, 115, 117, 119, 121, 135]. Various surfactants, as cetyltrimethylammonium bromide (CTAB) [116], Triton X-100 [95] or Tween 20 [133], are also usually used.

Another technique to increase the dispersability and the biocompatibility of the carbon nanotubes is the insertion of terminal groups on the surface of CNTs with a procedure called 'functionalization' [104, 105, 118, 123, 136]. The creation of free carboxylic acid and amino acid moieties on the surface are the most commonly used functionalization for enzyme immobilization and stabilization [104, 105, 118, 123, 136]. Apart from these two functional groups, other groups are also used, for various purposes. For example, Jeykumari and Narayanan functionalized MWCNTs with toluidine blue, in order to prevent the leakage of the redox mediator from their bienzymic biosensor [114].

2.3.4
Structure and Catalytic Behavior of Immobilized Enzymes

The immobilization procedure has been developed aiming mainly to two targets; first, to render feasible the recovery of the biocatalyst (and in some cases also the co-enzyme used) and second, to enhance the enzyme stability, especially when used in low-water media. Poor biocatalytic efficiency of immobilized enzymes, however, is a main drawback that hinders the large-scale application. The nonporous materials, to which enzymes are attached onto their surfaces, are subject to minimum diffusion limitation. Additionally, 'nano'-sized materials provide large surface area per weight, which allows high enzyme loading, up to 1.3 protein to CNT weight ratio [13].

The immobilization of enzymes onto CNTs leads in most cases in the stabilization of the immobilized enzyme [108, 135]. The stability was further enhanced when covalent bond between the enzyme and the nanomaterials were used [13, 119, 137]. The covalent bond prevents the enzyme leakage during repeating used of the enzyme-nanomaterial conjugates. When the enzymes are non-covalently attached onto the nanomaterials, the enzyme molecules are in equilibrium between the solid state (the microenvironment of the nanomaterial) and the bulk liquid phase, so that in every use some molecules are detached from the nanomaterial.

Over the last years some research groups investigate also other aspects of the enzyme immobilization onto nanomaterials, such as the activity-structure relationship of the immobilized enzymes. In many cases, the higher stability is attributed in a more rigid structure that the enzyme adopts, while altered substrate specificity is observed in some cases after immobilization, which results from the structural changes happening in the region of the active site of the enzyme. Several spectroscopic techniques such as CD, fluorescence, FT-IR and ultraviolet-visible (UV-Vis) spectroscopy are applied in this context in order to monitor possible structural changes upon immobilization. Table 2.4 presents the results of some recent works on this field. It is observed that the microenvironment provided from nano-structured materials can affect the enzyme structure and catalytic behavior. For instance, α-chymotrypsin unfold when immobilized on the surface of SWCNTs, whereas soybean peroxidase retained higher fraction of its 3D structure and its catalytic activity [138]. The type of the enzyme and the

Table 2.4 Structural studies of proteins immobilized onto carbon nanotubes.

Protein	Nanomaterials	Spectroscopic Techniques	Results	Reference
Myoglobin	Bmim[BF$_4$]-SWCNTs	CD UV-Vis	No structural changes observed upon immobilization	[108]
Albumin ovalbumin carbonic anhydrase hemoglobin hexokinase	MWCNTs MWCNT-COOH MWCNT-tyrosine MWCNT-isobutane amine	CD Fluorescence	The functionalized MWCNTs selectively induced protein secondary structure changes. Structural changes depend on the enzyme used and the functional group and the concentration of MWCNTs.	[136]
Albumin lysozyme	SWCNTs	CD Vis-Near IR	The proteins are partially unfolded upon immobilization.	[128]
α-Chymotrypsin soybean peroxidase	SWCNTs	FT-IR	Both enzymes undergo structural changes upon adsorption.	[138]
Horseradish peroxidase subtilisin Carlsberg lysozyme	SWCNTs	CD Fluorescence	Enzymes retain a high fraction of their native structure when attached to SWCNTs.	[13]
RNAase	SWCNT-COOH MWCNT-COOH	CD	Conformational changes where observed upon immobilization.	[123]
Ferricytochrome c	SWCNTs	CD UV-Vis	The presence of CNTs decreased the mobility of the protein and induced a more tight conformation.	[139]
Cytochrome c	MWCNT/poly (amidoamine)/chitosan	UV-Vis	No structural changes observed upon immobilization	[119]
Amyloglucosidase	SWCNTs, MWCNTs (in combination with coupling agents)	CD	Enzyme retains its structure when physically adsorbed, while covalent linkage leads to significant loss of structure	[140]

nanotubes, the functional groups on the surface of the CNT and the immobilization procedure are crucial factors which affect the structure of the enzyme upon immobilization [136].

2.4
Enzymes Immobilized on Nanoparticles

2.4.1
Introduction

Nanomaterials and especially metal nanoparticles (NPs) [141–143] have emerged as a new class of compounds that are particularly interesting for materials science due to their unique electronic, optical, magnetic and catalytic properties. These features differ importantly from those of the bulk materials [144, 145] and depend on the size and shape of the NP [146–148]. Current issues in NP research focus: (i) on the synthesis of NPs such as noble metals (Au, Ag, Pt) [149], transition metals (Fe, Co, Ni) and alloys (FePt, CoPt) [150, 151], ceramics (metals oxides, metal chalcogenides) [152] and semiconductor NPs [146]; (ii) on studies of their properties [153], and (iii) on their applications in several areas such as chemistry [154], physics [155], material sciences [156, 157], biology and medicine [158, 159]. Of particular interest are new synthetic strategies to control the shape and the dimensionality of the NPs [159]. A large number of methods have been developed for synthesizing nano-crystalline solids to control size, dispersity and morphology such as ball milling [160] and laser ablation [161], gas-phase condensation [162], sputtering [163], and flame hydrolysis [164]. Thin film deposition and annealing methods can also be used to form nanoparticles supported on a substrate [165]. Chemical methods are also widely used for synthesizing nanocrystalline solids [149]. Passing the transmission of the macro to nanoscale the properties of many nanoparticles change upon dimensional confinement including magnetism [166], catalysis [167], and mechanical properties [168]. Importantly, the unique properties of nanoscale solids can be used for a variety of applications, including information storage [169], medical and biological imaging [170], and catalysis for energy [171] (Figure 2.3).

Nanoparticles provide functional surface area to increase enzyme loading and a biocompatible microenvironment helping enzyme to retain its catalytic properties, while the diffusion limitations are minimized due to the nano-scale dimensions. Nanoparticles made of magnetite and gold comprise the first group of nanomaterials employed for enzyme immobilization [10, 172]. Magnetic nanoparticles (MNPs) are commonly used for enzyme immobilization, because of their unique properties. The first work on the field of protein immobilization on magnetic nanoparticles was published in 1987 [10]. Glucose oxidase and uricase were covalently bonded to magnetic nanoparticles isolated from magnetotactic bacteria. Two of the most interesting preparation techniques of magnetic nanoparticles are the water-in-oil microemulsions (which are widely used in enzyme

2.4 Enzymes Immobilized on Nanoparticles | 53

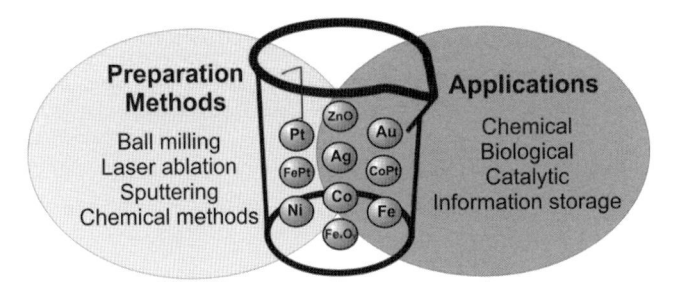

Figure 2.3 Preparation methods and applications of various nanoparticles.

biocatalysis) and the fabrication of magnetic NPs inside cells, in organelles which are called magnetosomes.

2.4.2
Applications

Nanoparticles, as it is discussed before, are a very versatile group of nanomaterials. The different types of materials possess different properties, which, in combination with the different types of biomacromolecules immobilized onto their surface, can lead to a variety of applications. As discussed in Section 2.3 on carbon nanotubes, proteins, nucleic acids, antigens, peptides and drugs, can interact and be immobilized onto nanoparticles.

Gold nanoparticles are widely used for protein immobilization, due to the intrinsic property of the gold to interact with proteins. The protein molecules can be directly attached on the gold nanoparticles surface, which renders gold nanoparticles one of the most biocompatible nanomaterials. Numerous works have been published on this field the last years [173–175].

Magnetic NPs, on the other hand, present better dispersability, while they are easily purified and separated, because of their magnetic properties. The dispersability of the nanoparticles also enables smooth contact with the substrate minimizing the mass-transfer limitation phenomenon. Enzymes immobilized on magnetic supports generally have higher thermal and storage stability which facilitates repeated use [176]. Taking advantage of the magnetic properties of magnetic nanoparticles, numerous applications have arisen. In Table 2.5 some of the most interesting works concerning the immobilization of proteins in various NPs are presented.

The most interesting applications of metallic or magnetic NPs, either used as conjugates with biological molecules or alone, are listed below [187]:

1) **Cellular labeling/cell separation** Cell labeling with MNPs is a method for *in vivo* cell separation, as the labeled cells can be detected by magnetic resonance imaging (MRI) [187, 188].

Table 2.5 Applications of nanoparticle–protein bioconjugates.

Protein	Nanoparticles	Possible application	Reference
Alcohol dehydrogenase	Sol-gel/Ru(bpy)$_3^{2+}$/Gold NPs	Electrochemiluminescence biosensor	[175]
Candida rugosa lipase	Fe$_3$O$_4$-chitosan NPs	Enzyme immobilization	[177]
Candida rugosa lipase	Acetylated γ-Fe$_2$O$_3$ MNPs	Enzyme immobilization	[178]
Candida rugosa lipase	Ionic liquid functionalized magnetic silica NPs	Enzyme immobilization	[179]
Cholesterol oxidase	Fe$_3$O$_4$ MNPs	Enzyme immobilization, stabilization	[180]
α-chymotrypsin	Amine functionalized polyacrylamide coated-Fe$_3$O$_4$ NPs	Enzyme immobilization, stabilization	[181]
Peroxidase	PEG hydrogel microparticles containing enzyme-linked MNPs	H$_2$O$_2$ biosensor	[182]
Horseradish peroxidase	Magnetic dextran microspheres	H$_2$O$_2$ biosensor	[183]
CD34$^+$ monoclonal antibodies	Silica coated MNPs	Immunomagnetic separation of cells	[184]
Streptococcus protein G	Gold NPs	Immunosensors	[174]
Trypsin	Amine functionalized MNPs	Protein digestion	[185]
Candida rugosa lipase	Lauric acid stabilized MNPs	Resolution of ($\pm$) menthol	[186]

2) **Tissue repair** The MNPs can be coupled to cells and used to target these cells at the desired site in the body. In this context, various proteins, growth factors etc., could be bound to these NPs that might be delivered at the damaged tissue, where it would play a role in tissue development.

3) **Drug delivery** Site-specific delivery of drugs. Ideally, these NPs could bear a pharmaceutical drug that could be driven to the target organ and be released there.

Table 2.6 Applications of carbon nanotube–nanoparticle–protein bioconjugates.

Protein	Nanomaterials	Possible application	Reference
Tyrosinase	Fe_3O_4 MNP/MWCNT modified GCE	Coliform detector	[189]
Glucose oxidase	MWCNT/ZnO NP/PDDA modified pyrolytic graphite electrode	Glucose biosensor	[134]
Glucose oxidase	CNT-modified titanium nanotube covered by Pt NP	Glucose biosensor	[190]
Glucose oxidase, Horseradish peroxidase	Silver NP/CNT/Chitosan modified indium tin oxide electrode	Glucose biosensor	[135]
Cytochrome c	Gold NP/ionic liquid/ MWCNT modified GCE	H_2O_2 biosensor	[173]

4) **Hyperthermia** Magnetic particles embedded around a tumor site and placed within an oscillating magnetic field will heat up. Cancer cells are destroyed at lower temperatures than the normal cells.

5) **Magnetofection** MNPs associated with vector DNA are transfected into cells by the influence of an external magnetic field.

6) **Protein and nucleic acid purification** [176, 187].

7) **Enzyme (and nucleic acid) immobilization** This application of NPs is the one that is discussed in the present chapter. Various types of enzymes can be immobilized onto NPs [176]. Functionalization of NPs with co-factors (as NAD^+) facilitates the biocatalysis from enzymes that need the presence of a cofactor (as lactate dehydrogenase or glucose oxidase) [188].

In some cases, the properties of NPs are combined with these of the carbon nanotubes, in order to develop new hybrid biomaterials (see Table 2.6), especially for the biosensing field.

2.4.3
Immobilization Approaches

The interaction of nanoparticles with the proteins is governed from the same type of interactions described for carbon nanotubes. Since NPs carry charges, they can electrostatically adsorb biomolecules with different charges, which depend on the pH that the immobilization takes place and the pI of the protein [3, 191]. Moreover, hydrophobic interactions, hydrogen bonds and non-specific absorption can play a role for enzyme non-covalent adsorption onto the surface of nanoparticles.

Most of the works with magnetic nanoparticles are depended on the covalent immobilization of proteins on the surface of these nanomaterials. Gold nanoparticles, as discussed earlier, can be used as nanotemplates for protein immobilization, through covalent bonds formed between the gold atoms and the amine groups and cysteine residues of proteins. Various proteins such as alcohol dehydrogenase [175], *Streptococcus* protein G [174], horseradish peroxidase [192], tyrosinase [193] and hemoglobin [194] were immobilized onto gold nanoparticles. A rapid method for the covalent immobilization of proteins onto NPs is based on glutaraldehyde-induced crosslinking. Glutaraldehyde is a widely used molecule in order to create covalent bond between two surfaces that both have free amine moieties [177, 178]. The first citation of protein immobilization onto NPs used this method of covalent bonding [10].

Another immobilization technique proposed is 'nanoentrapment' into NPs. In this method, a water-in-oil microemulsion system is used for the fabrication of NPs and for the dispersion of enzyme. This procedure leads to the creation of discrete NPs through polymerization in the water phase or on the interface, in which the enzyme is dispersed [195, 196]. One of the challenges of this approach is the difficulty in controlling the size of reverse micelles, as well as the number of enzyme molecules within each reverse micelle, which will directly affect the final properties of enzyme-entrapped nanoparticles [6].

Recently, a 'ship-in-a-bottle' approach was employed [197]. This approach improved both enzyme loading and enzyme activity by effectively preventing the leaching of enzyme. The first step involves the adsorption of enzymes into nanoporous material, which results in a high degree of enzyme loading. The second step is glutaraldehyde treatment, which covalently links the enzyme molecules, resulting in crosslinked enzyme aggregates at the nanometer-scale within the nanopores.

In order to increase the biocompatibility and the immobilization efficiency, nanomaterials can be covered by polymers or functionalized. Nanomaterials can be coated with several materials including a broad band of molecules such as organic stabilizers, inorganic molecules and targeting ligands. Polymeric coating materials can be classified into synthetic and natural polymers. Polymers based on polyacrylamide [181], poly(ethylene glycol) [182], sol–gels [175, 184] etc. are typical examples of synthetic polymeric systems. Various natural polymer systems was used including dextran, chitosan etc. [135, 177, 183].

Terminal groups can be inserted on the surface of NPs with a procedure called 'functionalization', in order to increase their dispersability and their biocompatibility [181, 185]. The creation of free carboxylic acid and amino acid moieties on the surface are the most commonly used functionalization procedure for enzyme immobilization and stabilization [181, 185]. In the same concept, Jiang and his coworkers functionalized silica coated-NPs with imidazolium-based ionic liquids [179]. The immobilization of ionic liquid facilitates the reuse of these solvents, while it provides a microenvironment in which the enzyme exhibits high activity and stability.

2.4.4
Structure and Catalytic Behavior of Immobilized Enzymes

One of the main advantages of the nanomaterials as immobilization matrices is the high active surface area per weight provided to the protein which facilitates the interaction and the immobilization. The 'nano'-sized order of these materials allows high enzyme loading up to 40% wt. [179]. Moreover, the enzyme molecules immobilized on the surface of nanoparticles are subject to minimum diffusion limitation.

A stabilization effect is observed when the enzymes are immobilized onto magnetic nanoparticles [135, 175, 179]. As discuss also in the carbon nanotubes section, the stability can be further enhanced when covalent bond between the enzyme and the nanoparticles were used [177, 178, 197]. Moreover, the enzyme and the nanoparticles used, the functional groups doped on the surface of nanoparticles and the immobilization procedure are crucial factors which affect the activity and the structure of the enzyme upon immobilization [179, 181, 198].

2.5
Conclusions

Nanostructured materials, such as nano-clays, carbon nanotubes and metallic or magnetic nanoparticles, are versatile materials that possess unique mechanical, optical and electrical properties. Various biomolecules (proteins, enzymes or genes) can interact and be immobilized onto these nanostructured materials, leading to a wide field of applications, including preparation of biosensors, biofuel cells and efficient biocatalysts, as well as intracellular protein and DNA transportation. A better fundamental understanding of the interaction of biomolecules with nanostructured materials and the development of efficient methods for the immobilization of biomolecules will aid the preparation of innovative functional nanocomposites, biomaterials and nanobiocatalytic systems, which might lead to numerous applications in the nanobiotechnology and enzyme technology fields.

References

1 Bornscheuer, U.T. (2003) *Angew. Chem. Int. Ed.*, **42**, 3336–3337.
2 Mateo, C., Palomo, J.M., Fernandez-Lorente, G., Guisan, J.M., and Fernandez-Lafuente, R. (2007) *Enzyme Microb. Technol.*, **40**, 1451–1463.
3 Kim, J., Grate, J.W., and Wang, P. (2006) *Chem. Eng. Sci.*, **61**, 1017–1026.
4 Kohli, P., and Martin, C.R. (2005) *Curr. Pharm. Biotechnol.*, **6**, 35–47.
5 Mann, S. (2008) *Angew. Chem. Int. Ed.*, **47**, 5306–5320.
6 Kim, J., Grate, J.W., and Wang, P. (2008) *Trends Biotechnol.*, **26**, 639–646.
7 Wang, P. (2006) *Curr. Opin. Biotechnol.*, **17**, 574–579.

8. Willner, I., Basnar, B., and Willner, B. (2007) *FEBS J*, **274**, 302–309.
9. Vamvakaki, V., and Chaniotakis, N.A. (2007) *Biosens. Bioelectron.*, **22**, 2650–2655.
10. Matsunaga, T., and Kamiya, S. (1987) *Appl. Microbiol. Biotechnol.*, **26**, 328–332.
11. Khmelnitsky, Y.L., Neverova, I.N., Momtcheva, R., Yaropolov, A.I., Belova, A.B., Levashov, A.V., and Martinek, K. (1989) *Biotechnol. Tech.*, **3**, 275–280.
12. Asuri, P., Bale, S.S., Karajanagi, S.S., and Kane, R.S. (2006) *Curr. Opin. Biotechnol.*, **17**, 562–568.
13. Asuri, P., Bale, S.S., Pangule, R.C., Shah, D.A., Kane, R.S., and Dordick, J.S. (2007) *Langmuir*, **23**, 12318–12321.
14. Das, D., and Das, P.K. (2009) *Langmuir*, **25**, 4421–4428.
15. Pinnavaia, T.J. (1983) *Science*, **220**, 365–371.
16. Konta, J. (1995) *Appl. Clay Sci.*, **10**, 275–335.
17. Lagaly, G. (1986) *Solid State Ionics.*, **22**, 43–51.
18. Newman, A.C.D. (1987) *Chemistry of Clays and Clay Minerals, Mineralogical Society Monograph, No. 6*, Longman, London.
19. Ballantine, J.A. (1986) *NATO-ASI Ser. Ser. C*, **165**, 197.
20. Madsen, F.T. (1998) *Clay Miner.*, **33**, 109–129.
21. Georgakilas, V., Gournis, D., and Petridis, D. (2001) *Angew. Chem. Int. Ed.*, **40**, 4286–4288.
22. Theng, B.K.G. (1974) *The Chemistry of Clay Organic Reactions*, Adam Hilger, London.
23. Gil, A., Gandia, L.M., and Vicente, M.A. (2000) *Catal. Rev. Sci. Eng.*, **42**, 145–212.
24. Secundo, F., Miehe-Brendle, J., Chelaru, C., Ferrandi, E.E., and Dumitriu, E. (2008) *Microporous Mesoporous Mater.*, **109**, 350–361.
25. Gopinath, S., and Sugunan, S. (2007) *Appl. Clay Sci.*, **35**, 67–75.
26. Sanjay, G., and Sugunan, S. (2007) *J. Porous Mater.*, **14**, 127–136.
27. de Fuentes, I.E., Viseras, C.A., Ubiali, D., Terreni, M., and Alcantara, A.R. (2001) *J. Mol. Catal. B Enzym.*, **11**, 657–663.
28. Dong, Y., and Feng, S.-S. (2005) *Biomaterials*, **26**, 6068–6076.
29. Lin, F.-H., Chen, C.-H., Cheng, W.T.K., and Kuo, T.-F. (2006) *Biomaterials*, **27**, 3333–3338.
30. Gournis, D., Jankovic, L., Maccallini, E., Benne, D., Rudolf, P., Colomer, J.F., Sooambar, C., Georgakilas, V., Prato, M., Fanti, M., Zerbetto, F., Sarova, G.H., and Guldi, D.M. (2006) *J. Am. Chem. Soc.*, **128**, 6154–6163.
31. De Cristofaro, A., and Violante, A. (2001) *Appl. Clay Sci.*, **19**, 59–67.
32. Massalha, N., Basheer, S., and Sabbah, I. (2007) *Ind. Eng. Chem. Res.*, **46**, 6820–6824.
33. Jézéquel, K., and Lebeau, T. (2008) *Bioresour. Technol.*, **99**, 690–698.
34. Akar, S.T., Akar, T., Kaynak, Z., Anilan, B., Cabuk, A., Tabak, Ö., Demir, T.A., and Gedikbey, T. (2009) *Hydrometallurgy*, **97**, 98–104.
35. Chen, X.C., Hu, S.P., Shen, C.F., Dou, C.M., Shi, J.Y., and Chen, Y.X. (2009) *Bioresour. Technol.*, **100**, 330–337.
36. Liu, B., Hu, R., and Deng, J. (1997) *Anal. Chem.*, **69**, 2343–2348.
37. Zen, J.-M., Lo, C.-W., and Chen, P.-J. (1997) *Anal. Chem.*, **69**, 1669–1673.
38. Poyard, S., Jaffrezic-Renault, N., Martelet, C., Cosnier, S., and Labbe, P. (1998) *Anal. Chim. Acta*, **364**, 165–172.
39. Shan, D., Yao, W., and Xue, H. (2006) *Electroanalysis*, **18**, 1485–1491.
40. Kemmegne Mbouguen, J., Ngameni, E., and Walcarius, A. (2006) *Anal. Chim. Acta*, **578**, 145–155.
41. Shi, Q., Li, Q., Shan, D., Fan, Q., and Xue, H. (2008) *Mat. Sci. Eng. C Biomimetic Supramol. Syst.*, **28**, 1372–1375.
42. Ye, Y.-H., Xing, G.-W., Tian, G.-L., and Li, C.-X. (2000) in *Peptides Biology and Chemistry; Chinese peptide Symposia 1998*, Springer Netherlands, pp. 21–25.
43. Xing, G.W., Li, X.W., Tian, G.-L., and Ye, Y.-H. (2000) *Tetrahedron*, **56**, 3517–3522.
44. Quiquampoix, H., and Mousain, D. (2005) *Enzymatic Hydrolysis of Organic Phosphorus*, CAB International, Wallingford.

45 Senillou, A., Jaffrezic, N., Martelet, C., and Cosnier, S. (1999) *Anal. Chim. Acta*, **401**, 117–124.

46 de Melo, J.V., Cosnier, S., Mousty, C., Martelet, C., and Jaffrezic-Renault, N. (2002) *Anal. Chem.*, **74**, 4037–4043.

47 Cosnier, S.L.L., and Le Lous, K. (1996) *J Electroanalytical. Chem.*, **406**, 243–246.

48 Shan, D., Mousty, C., Cosnier, S., and Mu, S. (2003) *Electroanalysis*, **15**, 1506–1512.

49 Dodor, E., Hwang, H.-M., and Ekunwe, S.I.N. (2004) *Enzyme Microb. Technol.*, **35**, 210–217.

50 Bastos, A.C., and Magan, N. (2009) *Int. Biodeterior. Biodegradation*, **63**, 389–394.

51 Spagna, G., Pifferi, P.G., and Gilioli, E. (1995) *Enzyme Microb. Technol.*, **17**, 729–738.

52 Lojou, E., and Bianco, P. (2006) *Electroanalysis*, **18**, 2426–2434.

53 Lee, K.-T., and Akoh, C. (1998) *Biotechnol. Tech.*, **12**, 381–384.

54 Hsu, A.-F., Foglia, T.A., and Shen, S. (2000) *Biotechnol. Appl. Biochem.*, **31**, 179–183.

55 Hsu, A.-F., Jones, K., and Foglia, T.A. (2002) *Biotechnol. Lett.*, **24**, 1161–1165.

56 Tzialla, A.A., Pavlidis, I.V., Felicissimo, M.P., Rudolf, R., Gournis, D., and Stamatis, H. (2010) *Bioresour. Technol.*, **101**, 1587–1594.

57 Tzialla, A.A., Kalogeris, E., Enotiadis, A., Taha, A.A., Gournis, D., and Stamatis, H. (2009) *Mat. Sci. Eng. B*, **165** (3), 173–177.

58 Serefoglou, E., Litina, K., Gournis, D., Kalogeris, E., Tzialla, A.A., Pavlidis, I.V., Stamatis, H., Maccallini, E., Lubomska, M., and Rudolf, P. (2008) *Chem. Mater.*, **20**, 4106–4115.

59 Haynes, C.A., and Norde, W. (1994) *Colloids Surf. B Biointerfaces*, **2**, 517–566.

60 Alkan, M., Demirbaa, Ö., Doana, M., and Arslan, O. (2006) *Microporous Mesoporous Mater.*, **96**, 331–340.

61 Quiquampoix, H., Staunton, S., Baron, M.H., and Ratcliffe, R.G. (1993) *Colloids Surf. A Physicochemical. Eng. Asp.*, **75**, 85–93.

62 Gougeon, R.D., Soulard, M., Reinholdt, M., Miehe-Brendle, J., Chezeau, J.M., Le Dred, R., Marchal, R., and Jeandet, P. (2003) *Eur. J. Inorg. Chem.*, (7), 1366–1372.

63 de Paiva, L.B., Morales, A.R., and Valenzuela Diaz, F.R. (2008) *Appl. Clay Sci.*, **42**, (1–2), 8–24.

64 Huang, J., Liu, Y., and Wang, X. (2008) *J. Mol. Catal. B Enzym.*, **55**, 49–54.

65 Litina, K., Miriouni, A., Gournis, D., Karakassides, M.A., Georgiou, N., Klontzas, E., Ntoukas, E., and Avgeropoulos, A. (2006) *Eur. Polym. J.*, **42**, 2098–2107.

66 Choi, W.G., Lee, S.B., and Ryu, D.D.Y. (2004) *Biotechnol. Bioeng.*, **23**, 361–371.

67 Sanjay, G., and Sugunan, S. (2005) *Clay Miner.*, **40**, 499–510.

68 Cârjă, G., Răçtoi, S., Ciobanu, G., and Balasanian, I. (2009) *J. Environ. Eng. Manag.*, **8**, 55–58.

69 Guadagnini, L., Ballarin, B., Mignani, A., Scavetta, E., and Tonelli, D. (2007) *Sens. Actuators. B Chem.*, **126**, 492–498.

70 Kondo, A., Oku, S., and Higashitani, K. (1991) *J. Colloid Interface Sci.*, **143**, 214–221.

71 Baron, M.H., Revault, M., Servagent-Noinville, S., Abadie, J., and Quiquampoix, H. (1999) *J. Colloid Interface Sci.*, **214**, 319–332.

72 Servagent-Noinville, S., Revault, M., Quiquampoix, H., and Baron, M.H. (2000) *J. Colloid Interface Sci.*, **221**, 273–283.

73 Noinville, S., Revault, M., Quiquampoix, H., and Baron, M.H. (2004) *J. Colloid Interface Sci.*, **273**, 414–425.

74 Quiquampoix, H., Servagent-Noinville, S., and Baron, M.H. (2002) *Enzyme Adsorption on Soil Mineral Surfaces and Consequences for the Catalytic Activity*, Marcel Dekker, New York.

75 Menaa, B., Herrero, M., Rives, V., Lavrenko, M., and Eggers, D.K. (2008) *Biomaterials*, **29**, 2710–2718.

76 Delgado, J.L., Herranz, M.A., and Martin, N. (2008) *J Mater. Chem.*, **18**, 1417–1426.

77 Geim, A.K., (2009) *Science*, **324**, 1530–1534.
78 Shleev, S., Klis, M., Wang, Y., Rogalski, J., Bilewicz, R., and Gorton, L. (2007) *Electroanalysis*, **19**, 1039–1047.
79 Lu, J., Do, I., Drzal, L.T., Worden, R.M., and Lee, I. (2008) *ACS Nano*, **2**, 1825–1832.
80 Zhou, M., Zhai, Y., and Dong, S. (2009) *Anal. Chem.*, **81**, 5603–5613.
81 Portaccio, M., Di Martino, S., Maiuri, P., Durante, D., De Luca, P., Lepore, M., Bencivenga, U., Rossi, S., De Maio, A., and Mita, D.G. (2006) *J. Mol. Catal. B Enzym.*, **41**, 97–102.
82 Vaz-Dominguez, C., Campuzano, S., Rüdiger, O., Pita, M., Gorbacheva, M., Shleev, S., Fernandez, V.M., and De Lacey, A.L. (2008) *Biosens. Bioelectron.*, **24**, 531–537.
83 Liu, Y., Huang, L., and Dong, S. (2007) *Biosens. Bioelectron.*, **23**, 35–41.
84 Boland, S., Jenkins, P., Kavanagh, P., and Leech, D. (2009) *J Electroanalytical Chem.*, **626**, 111–115.
85 Horozova, E., Dodevska, T., and Dimcheva, N. (2009) *Bioelectrochemistry*, **74**, 260–264.
86 Shan, C., Yang, H., Song, J., Han, D., Ivaska, A., and Niu, L. (2009) *Anal. Chem.*, **81**, 2378–2382.
87 Stavyiannoudaki, V., Vamvakaki, V., and Chaniotakis, N. (2009) *Anal. Bioanal. Chem.*, **395**, 429–435.
88 Martinez, N.A., Messina, G.A., Bertolino, F.A., Salinas, E., and Raba, J. (2008) *Sens. Actuators B Chem.*, **133**, 256–262.
89 Carralero, V., Mena, M.L., Gonzalez-Cortés, A., Yáñez-Sedeño, P., and Pingarrón, J.M. (2006) *Biosens. Bioelectron.*, **22**, 730–736.
90 Pizzariello, A., Stredanský, M., Stredanská, S., and Miertus, S. (2001) *Talanta*, **54**, 763–772.
91 Herrero, M., Requena, T., Reviejo, A.J., and Pingarrón, J.M. (2004) *Eur. Food Res. Technol.*, **219**, 557–560.
92 Wildoer, J.W.G., Venema, L.C., Rinzler, A.G., Smalley, R.E., and Dekker, C. (1998) *Nature*, **391**, 59–62.
93 Tasis, D., Tagmatarchis, N., Bianco, A., and Prato, M. (2006) *Chem. Rev.*, **106**, 1105–1136.
94 Thomas, K.M. (2009) *Dalton Trans.*, (9), 1487–1505.
95 Shim, M., Kam, N.W.S., Chen, R.J., Li, Y., and Dai, H. (2002) *Nano Lett.*, **2**, 285–288.
96 Lueking, A.D., Yang, R.T., Rodriguez, N.M., and Baker, R.T.K. (2004) *Langmuir*, **20**, 714–721.
97 Planeix, J.M., Coustel, N., Coq, B., Brotons, V., Kumbhar, P.S., Dutartre, R., Geneste, P., Bernier, P., and Ajayan, P.M. (1994) *J. Am. Chem. Soc.*, **116**, 7935–7936.
98 Mitchell, C.A., Bahr, J.L., Arepalli, S., Tour, J.M., and Krishnamoorti, R. (2002) *Macromolecules*, **35**, 8825–8830.
99 Rao, C.N.R., Sood, A.K., Subrahmanyam, K.S., and Govindaraj, A. (2009) *Angew. Chem. Int. Ed.*, **48**, 7752–7777.
100 Jia, N., Liu, L., Zhou, Q., Wang, L., Yan, M., and Jiang, Z. (2005) *Electrochim. Acta*, **51**, 611–618.
101 Lin, Y., Taylor, S., Li, H., Fernando, K.A.S., Qu, L., Wang, W., Gu, L., Zhou, B., and Sun, Y.P. (2004) *J. Mater. Chem.*, **14**, 527–541.
102 Bianco, A., Kostarelos, K., and Prato, M. (2005) *Curr. Opin. Chem. Biol.*, **9**, 674–679.
103 Rege, K., Raravikar, N.R., Kim, D.Y., Schadler, L.S., Ajayan, P.M., and Dordick, J.S. (2003) *Nano Lett.*, **3**, 829–832.
104 Gómez, J.M., Romero, M.D., and Fernández, T.M. (2005) *Catal. Lett.*, **101**, 275–278.
105 Shi, Q., Yang, D., Su, Y., Li, J., Jiang, Z., Jiang, Y., and Yuan, W. (2007) *J. Nanoparticle Res.*, **9**, 1205–1210.
106 Jiang, H., Du, C., Zou, Z., Li, X., Akins, D.L., and Yang, H. (2009) *J. Solid State Electrochem.*, **13**, 791–798.
107 Smolander, M., Boer, H., Valkiainen, M., Roozeman, R., Bergelin, M., Eriksson, J.E., Zhang, X.C., Koivula, A., and Viikari, L. (2008) *Enzyme Microb. Technol.*, **43**, 93–102.
108 Du, P., Liu, S., Wu, P., and Cai, C. (2007) *Electrochim. Acta*, **52**, 6534–6547.
109 Yin, Y., Wu, P., Lü, Y., Du, P., Shi, Y., and Cai, C. (2007) *J. Solid State Electrochem.*, **11**, 390–397.

110 Stolarczyk, K., Nazaruk, E., Rogalski, J., and Bilewicz, R. (2008) *Electrochim. Acta*, **53**, 3983–3990.
111 Shah, S., and Gupta, M.N. (2008) *Biochim. Biophys. Acta Proteins. Proteomics.*, **1784**, 363–367.
112 Liu, S., and Cai, C. (2007) *J. Electroanalytical Chem.*, **602**, 103–114.
113 Ivnitski, D., Artyushkova, K., Rincón, R.A., Atanassov, P., Luckarift, H.R., and Johnson, G.R. (2008) *Small*, **4**, 357–364.
114 Jeykumari, D.R.S., and Narayanan, S.S. (2009) *Carbon*, **47**, 957–966.
115 Zhang, Y., and Zheng, J. (2008) *Electrochim. Acta*, **54**, 749–754.
116 Wang, S., Xie, F., and Liu, G. (2009) *Talanta*, **77**, 1343–1350.
117 Kang, X., Wang, J., Tang, Z., Wu, H., and Lin, Y. (2009) *Talanta*, **78**, 120–125.
118 Kam, N.W.S., and Dai, H. (2005) *J Am. Chem. Soc.*, **127**, 6021–6026.
119 Chen, Q., Ai, S., Zhu, X., Yin, H., Ma, Q., and Qiu, Y. (2009) *Biosens. Bioelectron.*, **24**, 2991–2996.
120 Ramoni, R., Staiano, M., Bellucci, S., Grycznyski, I., Grycznyski, Z., Crescenzo, R., Iozzino, L., Bharill, S., Conti, V., Grolli, S., and D'Auria, S. (2008) *J Phys. Condens. Matter.*, **20**, 474201.
121 Liu, Y., Qu, X., Guo, H., Chen, H., Liu, B., and Dong, S. (2006) *Biosens. Bioelectron.*, **21**, 2195–2201.
122 Du, Z., Yu, Y.L., Chen, X.W., and Wang, J.H. (2007) *Chem. A Eur. J.*, **13**, 9679–9685.
123 Yi, C., Fong, C.C., Zhang, Q., Lee, S.T., and Yang, M. (2008) *Nanotechnology*, **19**, 095102.
124 Foldvari, M., and Bagonluri, M. (2008) *Nanomedicine Nanotechnol. Biol. Med.*, **4**, 183–200.
125 Aguí, L., Yáñez-Sedeño, P., and Pingarrón, J.M. (2008) *Anal. Chim. Acta*, **622**, 11–47.
126 Arya, S.K., Datta, M., and Malhotra, B.D. (2008) *Biosens. Bioelectron.*, **23**, 1083–1100.
127 Gao, Y., and Kyratzis, I. (2008) *Bioconjug. Chem.*, **19**, 1945–1950.
128 Matsuura, K., Saito, T., Okazaki, T., Ohshima, S., Yumura, M., and Iijima, S. (2006) *Chem. Phys. Lett.*, **429**, 497–502.
129 Azamian, B.R., Davis, J.J., Coleman, K.S., Bagshaw, C.B., and Green, M.L.H. (2002) *J. Am. Chem. Soc.*, **124**, 12664–12665.
130 Kang, Y., Liu, Y.C., Wang, Q., Shen, J.W., Wu, T., and Guan, W.J. (2009) *Biomaterials*, **30**, 2807–2815.
131 Lin, Y., Allard, L.F., and Sun, Y.P. (2004) *J. Phys. Chem. B*, **108**, 3760–3764.
132 Chen, R.J., Zhang, Y., Wang, D., and Dai, H. (2001) *J. Am. Chem. Soc.*, **123**, 3838–3839.
133 Chen, R.J., Bangsaruntip, S., Drouvalakis, K.A., Wong Shi Kam, N., Shim, M., Li, Y., Kim, W., Utz, P.J., and Dai, H. (2003) *Proc. Natl. Acad. Sci. U. S. A.*, **100**, 4984–4989.
134 Wang, Y.T., Yu, L., Zhu, Z.Q., Zhang, J., Zhu, J.Z., and Fan, C.h. (2009) *Sens. Actuators B Chem.*, **136**, 332–337.
135 Lin, J., He, C., Zhao, Y., and Zhang, S. (2009) *Sens. Actuators B Chem.*, **137**, 768–773.
136 Mu, Q., Liu, W., Xing, Y., Zhou, H., Li, Z., Zhang, Y., Ji, L., Wang, F., Si, Z., Zhang, B., and Yan, B. (2008) *J. Phys. Chem. C*, **112**, 3300–3307.
137 Li, J., Wang, Y.B., Qiu, J.D., Sun, D.C., and Xia, X.H. (2005) *Anal. Bioanal. Chem.*, **383**, 918–922.
138 Karajanagi, S.S., Vertegel, A.A., Kane, R.S., and Dordick, J.S. (2004) *Langmuir*, **20**, 11594–11599.
139 Jiang, X., Qu, X., Zhang, L., Zhang, Z., Jiang, J., Wang, E., and Dong, S. (2004) *Biophys. Chem.*, **110**, 203–211.
140 Cang-Rong, J.T., and Pastorin, G. (2009) *Nanotechnology*, **20**, 255102.
141 Cozzoli, P.D., Pellegrino, T., and Manna, L. (2006) *Chem. Soc. Rev.*, **35**, 1195–1208.
142 Eychmüller, A. (2004) *Nanoparticles: From Theory to Application*, G. Schmid (ed.) Wiley-VCH Verlag GmbH, Weinheim.
143 Georgakilas, V., Gournis, D., Tzitzios, V., Pasquato, L., Guldi, D.M., and Prato, M. (2007) *J. Mater. Chem.*, **17**, 2679–2694.

144 Alivisatos, A.P. (1996) *Science*, **271**, 933–937.

145 Hodes, G. (2007) *Adv. Mater.*, **19**, 639–655.

146 Yin, Y., and Alivisatos, A.P. (2005) *Nature*, **437**, 664–670.

147 Burda, C., Chen, X., Narayanan, R., and El-Sayed, M.A. (2005) *Chem. Rev.*, **105**, 1025–1102.

148 El-Sayed, M.A. (2004) *Acc. Chem. Res.*, **37**, 326–333.

149 Cushing, B.L., Kolesnichenko, V.L., and O'Connor, C.J. (2004) *Chem. Rev.*, **104**, 3893–3946.

150 Euliss, L.E., DuPont, J.A., Gratton, S., and DeSimone, J. (2006) *Chem. Soc. Rev.*, **35**, 1095–1104.

151 Yoshida, J., Saruwatari, K., Kameda, J., Sato, H., Yamagishi, A., Sun, L.S., Corriea, M., and Villemure, G. (2006) *Langmuir*, **22**, 9591–9597.

152 Cheng, J., Zhang, X., Luo, Z., Liu, F., Ye, Y., Yin, W., Liu, W., and Han, Y. (2006) *Mater. Chem. Phys.*, **95**, 5–11.

153 Hu, M., Chen, J., Li, Z.Y., Au, L., Hartland, G.V., Li, X., Marquez, M., and Xia, Y. (2006) *Chem. Soc. Rev.*, **35**, 1084–1094.

154 Kiely, C.J., Fink, J., Brust, M., Bethell, D., and Schiffrin, D.J. (1998) *Nature*, **396**, 444–446.

155 Andres, R.P., Bein, T., Dorogi, M., Feng, S., Henderson, J.I., Kubiak, C.P., Mahoney, W., Osifchin, R.G., and Reifenberger, R. (1996) *Science*, **272**, 1323–1325.

156 Shevchenko, E.V., Talapin, D.V., Kotov, N.A., O'Brien, S., and Murray, C.B. (2006) *Nature*, **439**, 55–59.

157 Chen, J.Y., Kutana, A., Collier, C.P., and Giapis, K.P. (2005) *Science*, **310**, 1480–1483.

158 Rosi, N.L., Giljohann, D.A., Thaxton, C.S., Lytton-Jean, A.K.R., Han, M.S., and Mirkin, C.A. (2006) *Science*, **312**, 1027–1030.

159 Fang, X., and Zhang, L. (2006) *J. Mater. Sci. Technol.*, **22**, 721–736.

160 Allemand, P.M., Khemani, K.C., Koch, A., Wudl, F., Holczer, K., Donovan, S., Gruner, G., and Thompson, J.D. (1991) *Science*, **253**, 301–303.

161 Caruso, T., Agostino, R.G., Bongiorno, G., Barborini, E., Piseri, P., Milani, P., Lenardi, C., La Rosa, S., and Bertolo, M. (2004) *Appl. Phys. Lett.*, **84**, 3412–3414.

162 Haubold, T., Bohn, R., Birringer, R., and Gleiter, H. (1992) *Mat. Sci. Eng. A*, **153**, 679–683.

163 Pivin, J.C., Garcia, M.A., Llopis, J., and Hofmeister, H. (2002) *Nucl. Instrum. Methods. Phys. Res. B*, **191**, 794–799.

164 Pratsinis, S.E. (1998) *Prog. Energy Combustion Sci.*, **24**, 197–219.

165 Nie, J.C., Yamasaki, H., and Mawatari, Y. (2004) *Phys. Rev. B Condens Matter. Mater. Phys.*, **70**, 1–11.

166 Vassiliou, J.K., Mehrotra, V., Russell, M.W., Giannelis, E.P., McMichael, R.D., Shull, R.D., and Ziolo, R.F. (1993) *J Appl. Phys.*, **73**, 5109–5116.

167 Astruc, D., Lu, F., and Aranzaes, J.R. (2005) *Angew. Chem. Int. Ed.*, **44**, 7852–7872.

168 Wu, B., Heidelberg, A., and Boland, J.J. (2005) *Nat. Mater.*, **4**, 525–529.

169 Li, F., Son, D.I., Ham, J.H., Kim, B.J., Jung, J.H., and Kim, T.W. (2007) *Appl. Phys. Lett.*, **91**, 162109.

170 Lee, H., Mi, K.Y., Park, S., Moon, S., Jung, J.M., Yong, Y.J., Kang, H.W., and Jon, S. (2007) *J Am. Chem. Soc.*, **129**, 12739–12745.

171 Yao, N., Lordi, V., Ma, S.X.C., Dujardin, E., Krishnan, A., Treacy, M.M.J., and Ebbesen, T.W. (1998) *J. Mater. Res.*, **13**, 2432–2437.

172 Crumbliss, A.L., Perine, S.C., Stonehuerner, J., Tubergen, K.R., Zhao, J., and Henkens, R.W. (1992) *Biotechnol. Bioeng.*, **40**, 483–490.

173 Xiang, C., Zou, Y., Sun, L.X., and Xu, F. (2008) *Electrochem. Commun.*, **10**, 38–41.

174 Jeong, M.L., Hyun, K.P., Jung, Y., Jin, K.K., Sun, O.J., and Bong, H.C. (2007) *Anal. Chem.*, **79**, 2680–2687.

175 Deng, L., Zhang, L., Shang, L., Guo, S., Wen, D., Wang, F., and Dong, S. (2009) *Biosens. Bioelectron.*, **24**, 2273–2276.

176 Horák, D., Babi, M., Macková, H., and Beneš, M.J. (2007) *J. Sep. Sci.*, **30**, 1751–1772.

177 Wu, Y., Wang, Y., Luo, G., and Dai, Y. (2009) *Bioresour. Technol.*, **100**, 3459–3464.
178 Dyal, A., Loos, K., Noto, M., Chang, S.W., Spagnoli, C., Shafi, K.V.P.M., Ulman, A., Cowman, M., and Gross, R.A. (2003) *J. Am. Chem. Soc.*, **125**, 1684–1685.
179 Jiang, Y., Guo, C., Xia, H., Mahmood, I., Liu, C., and Liu, H. (2009) *J Mol. Catal. B Enzym.*, **58**, 103–109.
180 Kouassi, G.K., Irudayaraj, J., and McCarty, G. (2005) *J. Nanobiotechnology*, **3**, 1.
181 Hong, J., Gong, P., Xu, D., Dong, L., and Yao, S. (2007) *J. Biotechnol.*, **128**, 597–605.
182 Park, S., Lee, Y., Kim, D.N., Park, S., Jang, E., and Koh, W.G. (2009) *Reactive Funct. Polym.*, **69**, 293–299.
183 Zhang, H.L., Lai, G.S., Han, D.Y., and Yu, A.M. (2008) *Anal. Bioanal. Chem.*, **390**, 971–977.
184 Liang, X., Xu, K., Xu, J., Chen, W., Shen, H., and Liu, J. (2009) *J. Magnetism Magn. Mater.*, **321**, 1885–1888.
185 Lin, S., Yun, D., Qi, D., Deng, C., Li, Y., and Zhang, X. (2008) *J. Proteome. Res.*, **7**, 1297–1307.
186 Bai, S., Guo, Z., Liu, W., and Sun, Y. (2006) *Food Chem.*, **96**, 1–7.
187 Gupta, A.K., and Gupta, M. (2005) *Biomaterials*, **26**, 3995–4021.
188 Katz, E., Willner, I., and Wang, J. (2004) *Electroanalysis*, **16**, 19–44.
189 Cheng, Y., Liu, Y., Huang, J., Li, K., Xian, Y., Zhang, W., and Jin, L. (2009) *Electrochim. Acta*, **54**, 2588–2594.
190 Pang, X., He, D., Luo, S., and Cai, Q. (2009) *Sens. Actuators B Chem.*, **137**, 134–138.
191 Shang, W., Nuffer, J.H., Dordick, J.S., and Siegel, R.W. (2007) *Nano Lett.*, **7**, 1991–1995.
192 Jia, J., Wang, B., Wu, A., Cheng, G., Li, Z., and Dong, S. (2002) *Anal. Chem.*, **74**, 2217–2223.
193 Liu, Z.M., Wang, H., Yang, Y., Yang, H.F., Hu, S.Q., Shen, G.L., and Yu, R.Q. (2004) *Anal. Lett.*, **37**, 1079–1091.
194 Gu, H.Y., Yu, A.M., and Chen, H.Y. (2001) *J Electroanalytical Chem.*, **516**, 119–126.
195 Daubresse, C., Grandfils, C., Jerome, R., and Teyssie, P. (1994) *J. Colloid Interface Sci.*, **168**, 222–229.
196 Munshi, N., De Tapas, K., and Maitra, A. (1997) *J. Colloid Interface Sci.*, **190**, 387–391.
197 Lee, J., Kim, J., Kim, J., Jia, H., Kim, M.I., Kwak, J.H., Jin, S., Dohnalkova, A., Park, H.C., Chang, H.N., Wang, P., Grate, J.W., and Hyeon, T. (2005) *Small*, **1**, 744–753.
198 Jia, H., Zhu, G., and Wang, P. (2003) *Biotechnol. Bioeng.*, **84**, 406–414.

3
Improved Immobilization Supports for *Candida Antarctica* Lipase B

Paria Saunders and Jesper Brask

3.1
Introduction

In the past few decades, many investigations have been performed on the use of immobilized enzymes as catalysts for organic synthesis [1–4]. Among these efforts, those that use immobilized enzymes as catalysts for polymerization reactions are prominent. Enzyme-catalyzed polymerization not only is environmentally desirable because it replaces toxic catalysts, but also offers a number of technical and economical benefits. Among these benefits are (i) synthesis of new types of monomers, oligomers, and polymers that are not possible with chemical routes; (ii) eliminating steps of synthesis that require protection and deprotection of functional groups; (iii) producing products with high stereo-, and regio-selectivity; (iv) performing synthesis at milder temperatures (60–110 °C); and (v) synthesis of biodegradable or recyclable polymers [2].

Among the enzymes used successfully for polymer synthesis, *Candida antarctica* lipase B (CALB) is by far the most well-known enzyme in the literature (see for instance Chapters 4, 5, 11 and 12). Novozym® 435 is a commercially available heterogeneous biocatalyst that consists of *Candida antarctica* lipase B (CALB) physically immobilized within a macroporous resin of poly(methyl methacrylate). Novozym 435 is capable of catalyzing a range of organic chemical reactions, including polyester synthesis by ring-opening polymerization of various lactones and polycondensation reactions of diacid/diol substrates [5, 6] (see Chapter 4). However, in spite of the many advantages that Novozym 435 offers in the production of polymers, there are two general concerns related to this product. The first concern is the cost: the high cost of the product is at least partly attributed to the immobilization support material. The other concern with the use of Novozym 435 is enzyme leaching from the carrier. This leads to a polymer product that is contaminated with the enzyme, as well as limiting the number of cycles the immobilized enzyme can be used before it requires replacing, and therefore additional costs.

In this chapter we share some of our efforts to find a new immobilized CALB product that is suitable for polymerization reactions. The new immobilized CALB was required to have: (i) a more economic support material for CALB to reduce the overall cost of production, and (ii) another mode of attachment of the enzyme to the carrier to prevent the product from leaching while maintaining the activity of the enzyme. A new experimental product (NS81018) has been developed to overcome these issues. The performance of (NS81018) compared with Novozym 435 in catalyzing polyester synthesis was investigated using the synthesis of polycaprolactone (PCL) by ring-opening polymerization (see Chapter 4) as an example.

3.2
Industrial Enzyme Production

Most industrial enzymes are produced by micro-organisms via submerged fermentation. Most fermentations are aerobic. The production process of industrial enzymes can be divided into three steps: fermentation, recovery, and formulation.

3.2.1
Fermentation

Micro-organisms called production strains are designed and constructed for a specific enzyme production under certain conditions. The production strain is selected or modified so that it can produce a large amount of desired enzyme. Industrial enzyme fermentations may use bacteria, such as *Bacillus*, fungi such as *Aspergillus*, or yeast species such as *Saccharomyces cerevisiae* as production strains. To scale up a fermentation, the production strain is typically first cultivated in a small flask containing nutrients and agar. The cells are then transferred to a seed fermenter containing sterilized raw materials and water, called the medium. The medium contains the nutrients for the micro-organism, usually a carbohydrate source, a protein source, and a source of vitamins and trace minerals. The seed fermentation allows the cells to reproduce, and adapt to the environment and nutrients for main fermentation; the size of the seed fermenter is usually 4–12% of the main fermenter. Once the cells are transferred to the main fermenter, the conditions of temperature, pH, and dissolved oxygen for aerobic fermentation are carefully controlled to be optimal for production of enzymes. Fermentation processes can be batch, fed-batch or continuous. When the main fermentation is complete, the mixture of cells, nutrients, and enzymes is transferred to the recovery step, to recover the enzymes from the broth. Reported yields of enzymes vary from $1\,kg\,m^{-3}$ to $50\,kg\,m^{-3}$ broth [7].

3.2.2
Recovery and Purification

In the downstream processing or recovery step, fermentation biomass is separated from the enzyme. Since most industrial enzymes are extracellular enzymes, whole cells and other particulate matter are removed in the recovery step via centrifugation or filtration. Following enzyme separation, the enzyme is concentrated by means of semi-permeable membranes or evaporation. The biomass is inactivated and recycled for example, as fertilizer in farms.

3.2.3
Formulation

The recovered broth has to be formulated into a final product to comply with requirements appropriate to its final application. In formulation, whether for a solid or liquid product, different issues are addressed. Maintaining the activity of enzyme from the time of manufacture to the time of application through storage is one of the main factors tested during formulation. Other stability issues such as microbial stability and physical stability are important as well. Depending on the final application of the enzyme, the physical appearance can be customized (e.g., colored). A specific example of formulation is immobilization (see also Chapter 2).

3.3
Lipase for Biocatalysis

Lipases are a special class of esterases that hydrolyze fatty acid esters like triglycerides at lipid/water interfaces. In most lipases, a part of the enzyme molecule covers the active site with a short amphiphilic α–helix, called the lid. The side of the α-helical lid facing the catalytic site, as well as the protein chains surrounding the catalytic site are mostly composed of hydrophobic side chains. The lid in its closed conformation (i.e., in the absence of an interphase or organic solvent) prevents access of the substrate to the catalytic triad. Opening of the lid twists and exposes a large hydrophobic surface, and the previously exposed hydrophilic domain becomes buried inside the protein.

All lipases show structural and functional similarities, regardless of the organism from which they were isolated. Hence, they all have a α-β-hydrolase fold structure with a catalytic triad. However, small variations in the substrate binding site may have a strong effect on the catalytic properties and the stability of enzyme [1].

3.3.1
Candida Antarctica Lipase B (CALB)

The basidomycetous yeast *Candida antarctica* produces two different lipases, A and B. This yeast was originally isolated in Antarctica with the aim of finding enzymes with extreme properties. Both the lipase A and lipase B have been purified and characterized. The two lipases are very different. The lipase B is a little less thermostable and is less active toward large triglycerides but very active towards a broad range of other esters, as well as amides, thioesters, etc.. The lipase B of CALB has a molecular weight of 33 kD with 317 amino acid residues and a pI of 6.0. The active site of CALB consists of a Ser105-His224-Asp187 catalytic triad [8, 9, 10]. (see also Chapter 14 for detailed information on the structure and reaction mechanism of CALB).

CALB is industrially produced by submerged fermentation of a genetically modified *Aspergillus* micro-organism. The active site in CALB is deep and narrow, which is believed to be the reason for the substrate specificity and pronounced stereo-specificity of this lipase [10]. The role and even presence of a lid in CALB has been much debated, but it is now believed that CALB in fact has a lid like other lipases [11].

3.4
Immobilization

Immobilization is a technique to physically confine or localize an enzyme with retention of its activity (see also Chapter 2). In immobilization, the enzyme is localized on a matrix or carrier through different modes of attachment. The main purposes of enzyme immobilization include: (i) enzyme stabilization; (ii) improvement of enzyme performance by increasing the contact area of enzyme and substrate; and (iii) re-use or continuous use of enzymes in several reactions or over an extended time [12, 13].

Enzymes have been bound to different types of matrices or carriers such as porous membranes or beads. The immobilization support needs to have several physical and mechanical characteristics. Among these are high mechanical strength, large surface area, controlled swelling behavior and an ideal hydrophobic/ hydrophilic balance. There are several modes of attachments and techniques to load the enzyme in/on the matrix. Physical adsorption, covalent attachment, ionic linkage, and entrapment are the general categories. For lipases, due to their high surface hydrophobicity, physical adsorption of enzymes on suitably hydrophobic supports has been the more popular strategy over covalent conjugation methods. Novozym 435 is an example of physical adsorption of enzyme on porous beads. However, as mentioned earlier, this mode of attachment is reversible and can lead to separation of the enzyme from carrier or leakage.

One strategy to minimize enzyme leakage is crosslinking. Despite reports that larger polyaldehydes, for example, periodate-oxidized polysaccharides, result in higher activity of the crosslinking enzymes [14], glutaraldehyde (GA, pentanedial) continues to be the crosslinking agent of choice [15]. Applications involve direct crosslinking of enzyme in solution, CLEC/CLEA type of immobilizations [16–18], crosslinking of enzyme adsorbed on a carrier, as well as being a bifunctional linker between a carrier and enzyme [15]. GA is known to react with amines (primarily from lysine residues) but the chemistry involved is surprisingly complex [19]. Hence crosslinking procedures have developed largely through empirical methods. Historically, it has been proposed that the crosslinking ability was due to Schiff base formation. Due to the lability of such linkages some protocols suggest reducing the putative Schiff bases with sodium borohydride or other reducing agents. There is, however, little evidence that such reduction has any effect. In fact it is well known that GA immobilized proteins are stable and the linkages cannot be hydrolyzed. Further, GA has unique characteristics that render it a far more effective crosslinking agent than other dialdehydes.

Unfortunately, increasing crosslinking also leads to lower enzyme activity. This could be due to the crosslinking agent reacting with amino acids that are essential for the catalytic activity, or that the overall geometry of the enzyme is distorted so that it becomes inactive, or it could simply be a matter of reduced steric access of substrate to the enzyme. One approach to make crosslinking effective while maintaining good enzyme activity could be through addition of a polyamine such as polyethyleneimine (PEI). The many amine nucleophiles will react with GA, limiting the GA reaction directly on the enzyme. The crosslinked PEI should create a three-dimensional network retaining the enzyme [15].

3.4.1
Novozym 435

Figure 3.1 is a scanning electron microscopy (SEM) photograph of Novozym 435 before and after immobilization of CALB on the matrix (Lewatit). It is obvious that after immobilization, the enzyme has been adsorbed on the surface of the matrix and the surface has been saturated. This observation confirms the results of synchrotron infrared microspectroscopy performed at amide band wavelength on Novozym 435 (Figure 3.2) [5, 6]. The researchers measured the intensity of the amide band across the cross-section of a Novozym 435 bead and attributed the presence of amide groups to the location of the enzyme immobilized on the bead. They showed that distribution of CALB on the bead is not homogenous and it mostly saturates the surface of the beads and hardly enters the center. The CALB enzyme is a globular protein with dimensions of $30\text{Å} \times 40\text{Å} \times 50\text{Å}$ [10], whereas

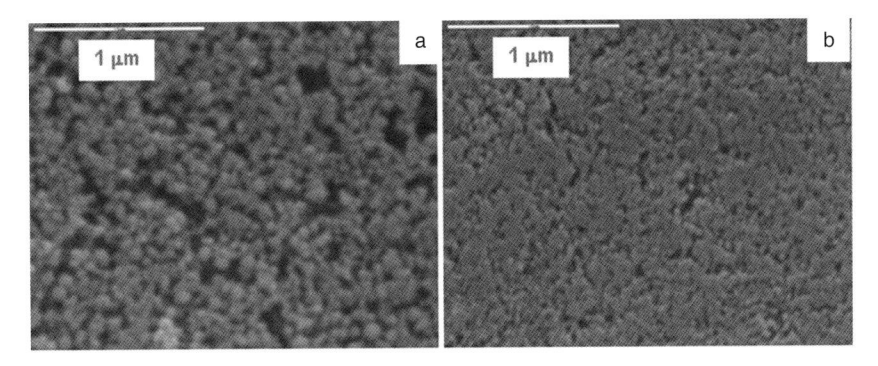

Figure 3.1 Scanning electron microscopy (SEM) photograph of Novozym 435 (a) before and (b) after immobilization of CALB on the matrix (Lewatit).

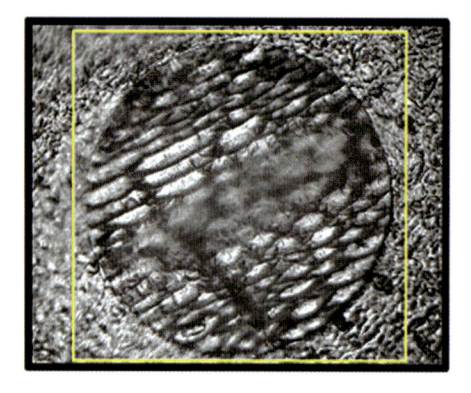

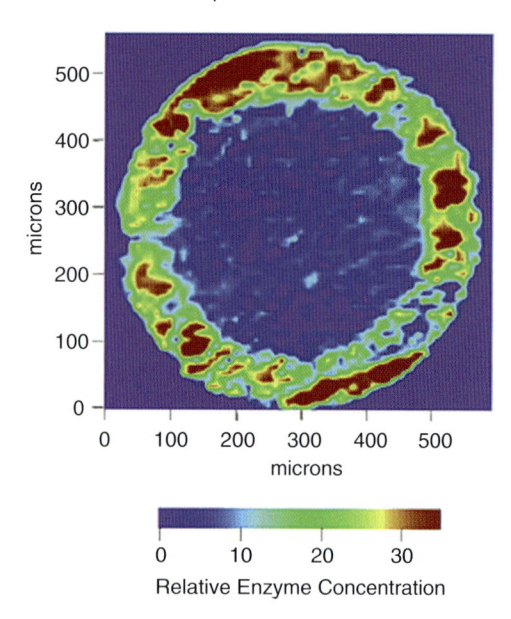

Figure 3.2 The visible light and synchrotron infrared images of protein distribution of CalB on Lewatit in Novozym 435. Reproduced with permission from [6]. (Copyright © (2003) American Chemical Society).

the average pore size of Lewatit is 140–170 Å. So with enough time it would be expected that the enzymes enter the pores of the beads. This suggests that the hydrophobic/hydrophilic balance of the matrix is very crucial on the amount of protein adsorbed on the beads.

In Novozym 435, the enzyme is not covalently linked to the carrier, but merely adsorbed. Therefore, as mentioned, enzyme leaching is a known issue with this product. The problem has been addressed by several authors [20, 21]. Even though the enzyme performance has been satisfactory, there has been a need specifically for a non-leaching immobilized formulation. It has been shown that several physical and chemical treatments such as coating the immobilized particles can enhance the stability of Novozym 435 [20–22].

In line with this, in order to reduce the enzyme leakage using a rather quick and easy procedure, we tested the addition of GA as crosslinking agent to Novozym 435. The crosslinking was performed by treating the immobilized enzyme with 2.3% aqueous GA at pH 7.0 for 4 h. The material was then washed with water and ethanol and dried in vacuum. It is known that CALB can be effectively leached from Novozym 435 with dimethyl sulfoxide (DMSO). Hence, to quantify leaching, Novozym 435 and the crosslinked Novozym 435 were incubated with DMSO at 37 °C for 30 min. The amount of leached enzyme was quantified with BCA coloring using a standard curve for CALB. It was found that the amount of leaching for Novozym 435 was 119 mg g^{-1} whereas the leaching for crosslinked CALB measured to be 21 mg g^{-1}. Hence, crosslinking resulted in a 5–6 times lower level of leaching.

The activity of both Novozym 435 and crosslinked Novozym 435 was measured using the standard PLU-assay. The PLU (propyl laurate units) assay is a method to determine the activity of immobilized lipases by ester synthesis. One PLU corresponds to 1 µmol of propyl laurate formed from condensation of lauric acid and 1-propanol per gram of enzyme per minute under specified conditions. The PLU value is determined by measuring the area under the peak of formed propyl laurate versus a standard sample of propyl laurate with known concentration using a gas chromatograph with an FID detector. Crosslinked Novozym 435 showed an activity of 4800 PLU g^{-1}, whereas the activity of Novozym 435 before crosslinking was measured to be 12 200 PLU g^{-1}.

3.4.2
NS81018

To address the issues of cost and leakage, NS81018 was developed as an experimental product. NS81018 is an immobilized preparation of CALB on silica particles where the enzyme has been crosslinked on the carrier via PEI and GA [23]. It has been shown that such a preparation significantly reduces the leaching of enzyme from carrier. When the enzyme was physically immobilized on the carrier without crosslinking, all the enzyme leached from the carrier in the leakage test with DMSO. However, in all preparations using PEI and GA crosslinking, the enzyme leaching was reduced down to 4–13 mg g^{-1}, where the enzyme loading on the carrier varied from 30–59 mg g^{-1} [23]. PLU-activity of NS81018 was found to be lower than Novozym 435, that is, approximately 5400 PLU g^{-1} for NS81018 vs approx. 12 000 PLU g^{-1} for Novozym 435, partly due to lower enzyme loading in NS81018. However, based on the application a higher dose of NS81018 may be used to reach the same activity.

3.5
CALB- Catalyzed Polymer Synthesis

Immobilized CALB has frequently been applied in the literature as a catalyst for polymerization of aliphatic polyesters, polycarbonates, polyurethanes and their copolymers. In the present work on CALB catalyzed polymerization, the ring opening of ε-caprolactone to polycaprolactone was selected as the model polymerization reaction (Figure 3.3). This model reaction has been well established in the literature [24–27] as an example of a polymerization reaction that can be successfully catalyzed by immobilized lipases (see also Chapter 4). Polymer synthesis and characterization was performed in four steps: (i) polymerization; (ii) separation; (iii) purification; and (iv) characterization.

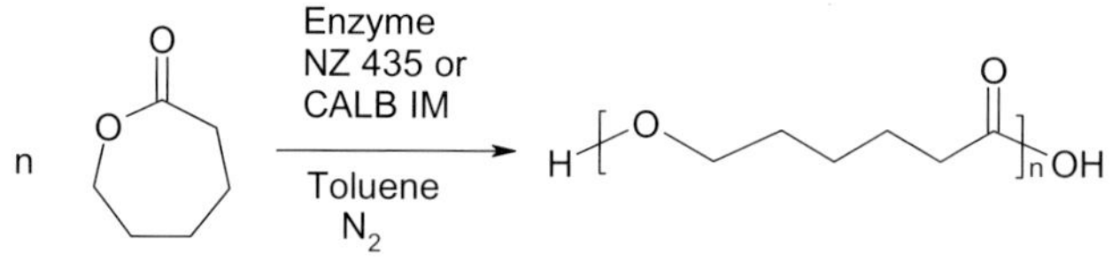

Figure 3.3 Ring opening polymerization of ε-caprolactone to polycaprolactone (PCL).

3.5.1
Polymerization

Polycaprolactone synthesis was performed in two scales: preparative, and screening scale. In preparative scale, synthesis was performed in a three-neck flask with an overhead mechanical stirrer and a temperature controller in presence of an anhydrous solvent. Toluene was selected as the best solvent for PCL synthesis [24]. In screening scale, synthesis was performed in smaller scale inside glass vials in a J-Kem parallel synthesizer with either 6-vial or 12-vial reactor.

In both synthesis procedures, the immobilized enzyme was added to the reactor followed by the solvent and heated to the desired temperature (70 °C). Unless otherwise mentioned, the immobilized enzyme was dried in a vacuum oven for 24 hours at 30–35 °C prior to use. The monomer was added afterwards with a syringe through a rubber septum. In preparative scale, mechanical overhead stirrer was introduced to the flask and a condenser was added to the side neck of the flask. Nitrogen was introduced to the flask with a slight positive pressure. In case of screening scale, stirring was provided by orbital shaking action of the parallel synthesizer and condensation was possible by passing cold water around the top half of reaction vials. Generally, solvent (toluene) to monomer (ε-caprolactone) ratio was 5:1 v/v and monomer (ε-caprolactone) to catalyst (immobilized CALB) ratio was 10:1 (v/w) unless otherwise stated. All reactions were performed at 70 °C for a few hours. Aliquots of about 0.5–1 ml were taken every hour during the reaction through the rubber septum using a glass syringe and immediately filtered by a 0.45 μm PTFE syringe filter for molecular weight determination.

3.5.2
Polymer Separation and Purification

Generally after completion of the solution polymerization, tetrahydrofuran (THF) or chloroform was added to the reaction mixture to dilute the rather viscous solution. Some researchers believe that this step denatures the enzyme and terminate the polymerization reaction, however it has been shown in earlier literature that polymerization can occur in a solvent such as THF. Our work also confirms that this does not lead to inactivation of the enzyme. The immobilized enzyme was then removed from the solution by vacuum filtering the sample through a coarse-filter fritted funnel. Tetrahydrofuran was removed by rotary evaporation at 40 °C.

Polymer was precipitated out of the viscous remainder solution by adding a cold polar solvent such as methanol while stirring vigorously. The precipitated polymer was separated from the solution by vacuum filtration and was dried in a vacuum oven.

3.5.3
Characterization and Performance Assays

Gel permeation chromatography (GPC) was selected as the main tool for monitoring the progress of polymerization by measuring the polymer's molecular weight average and distribution. In this work, the GPC was equipped with refractive index (RI) detector and multi-angle light scattering detector (MALS) to quantify the molecular weight of the polymer.

An Agilent 2010 series high performance liquid chromatography (HPLC) equipped with a chilled auto-sampler, an RI detector and a Wyatt's Mini Dawn tri-angle light scattering detector was used in GPC mode. Agilent Chemstation software with GPC module was used for controlling the GPC and ASTRA 4.0 software was used with MALS detector for molecular weight determination. Three Waters Styragel high resolution columns (HR 0.5, HR 2 and HR 4E) were used in series for separation and HPLC grade THF was the solvent and the mobile phase. Column and detector temperatures were set at 35 °C for all GPC determinations. All samples were filtered via a 0.45 µm PTFE filter syringe before injections and the sample size for all injections were 300 µL. Pump flow was set to 1 ml min^{-1} for all sample collections.

Calibration of the MALS detector was performed with a polystyrene standard with $M_w = 17\,500$, $M_w/M_n = 1.02$ and $d_n/d_c = 0.185$. The RI peak and light scattering (LS) detector peak for polystyrene standard were aligned within the ASTRA software. A sample of commercial PCL purchased from Sigma-Aldrich with $M_w \sim 14\,000$ was diluted in THF and injected to GPC for confirmation of molecular weight value by MALS detector. It was confirmed that GPC-MALS is capable of accurate determination of absolute molecular weight, even for lower molecular weight polymers of about 14 000.

A series of polystyrene standards were purchased from Waters Corporation and American Polymer Standards Corporation for creating a calibration curve and determination of molecular weight by elution time based on polystyrene standards number. The calibration curve was created using Chemstation GPC module software. The molecular weight determined by this method was used for comparison with literature data, since polystyrene standards are commonly used for reference.

3.5.4
CALB Immobilization

CALB was immobilized on different hydrophobic support materials by physical adsorption. Table 3.1 shows the physical properties of the support materials.

Table 3.1 Physical properties of the support materials available for immobilization by physical adsorption.

No.	Carrier	Material	Particle Size (µm)	Pore size or void volume	Bulk Density (Kg/m³)
1	Lewatit	Poly methyl methacrylate copolymer	315–1000	14–17 nm	650–800
2	Silica	Synthetic porous silica	250–375	28 nm 74%	2000
3	Accurel powder XP100	Polypropylene	1500	2000–10000 nm 70 ± 5%	<1500
4	Accurel pellet MP100	Polypropylene	3000×3000	800–2500 nm 78 ± 2%	115–145

The support materials were vacuum dried for 24 hours. CALB enzyme solution was diluted in water. To the support material enough diluted enzyme was added to reach 55 kLU g^{-1} (activity units per g carrier). The buffer used for immobilization was a 1.25 M potassium phosphate buffer pH = 8. The slurry was stirred for 20 hours using a mechanical laboratory overhead stirrer. The immobilization liquor was filter out using vacuum and the particles were rinsed two times with water. The immobilized enzyme was finally vacuum dried at 45 °C for 30 hours. The immobilized enzymes were used to synthesize PCL. The molecular weight of the synthesized polymer and PLU g^{-1} values were determining factors in selecting the best support material.

3.5.5
Results and Discussion

3.5.5.1 Effect of Synthesis Time on Molecular Weight

The molecular weight of PCL synthesized under nitrogen using Novozym 435 (PLU = 11 800) as catalyst was studied. Figure 3.4 shows the increase in the absolute molecular weight of the polymer. The fluctuation could reflect an artifact from the sample preparation. Figure 3.5a shows an overlay of the RI chromatograms of ε-caprolactone and PCL after 1 hour synthesis. This chromatogram shows that within only 1 hour of synthesis, most of the monomer is consumed leaving a small shoulder to the left of the toluene peak. A large broad peak with higher molecular weight is generated which is related to PCL. Figure 3.5b shows an overlay of the RI chromatograms of 1-, 3- and 6-hour samples. As the polymerization time increases, the polymer elutes earlier due to increased molecular weight. The monomer is completely consumed after 3 hours.

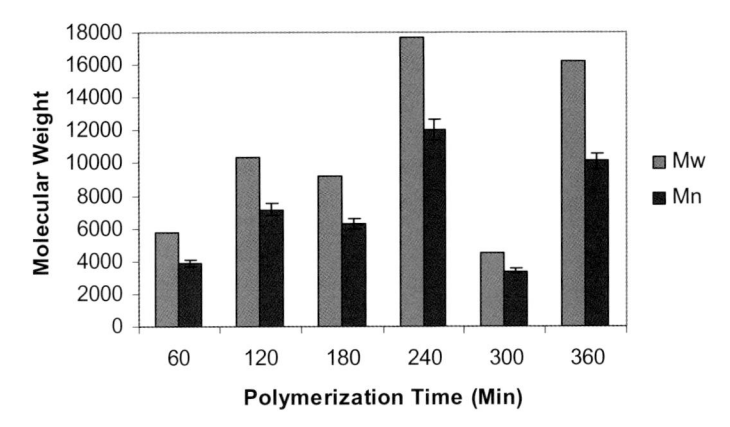

Figure 3.4 Molecular weight of PCL during solution polymerization at 70 °C using 10% Novozym 435 as catalyst.

3.5.5.2 Comparison of NS 81018 and Novozym 435

In this study, PCL synthesis was also performed with NS81018 using a similar procedure and sampling method as for Novozym 435. Since NS81018 (PLU = 5400) is not as active as Novozym 435 (PLU = 11 800), the amount of NS81018 used in this reaction was on equivalent PLU g^{-1} basis. Quantitatively speaking, instead of 1% Novozym 435, 2.2% NS81018 was used. Figure 3.6 shows the molecular weights of the samples taken during synthesis and the final purified polymer using Novozym 435 and NS81018. Even though an increase in the molecular weight with synthesis time is observed using NS81018, generally the molecular weights are lower in comparison with Novozym 435. The final purified sample shows a sudden increase in molecular weight. This could be due to fractionation of the product during filtration, that is, only the fraction of polymer with higher molecular weight was recovered.

3.5.5.3 Determination of Polycaprolactone Molecular Weight by GPC

The absolute molecular weight determination on PCL samples synthesized in-house was measured using a MALS detector. This method was confirmed earlier by determination of molecular weight of a commercial PCL sample. However, the results of this method of measurement can be somewhat different than the relative molecular weights based on polystyrene standards, which appears to be the method of choice in most publications. To reassure that the molecular weight results obtained in this study are accurate, a sample of chemically synthesized commercial PCL and a sample of PCL synthesized using Novozym 435 were sent to two different laboratories for molecular weight determination. All labs used THF as the mobile phase for GPC. The results are reported in Table 3.2. One lab used polylactic acid (PLA) standards for calibration. Since in GPC, separation is based on molecular size and hydrodynamic volume of the polymer in solution,

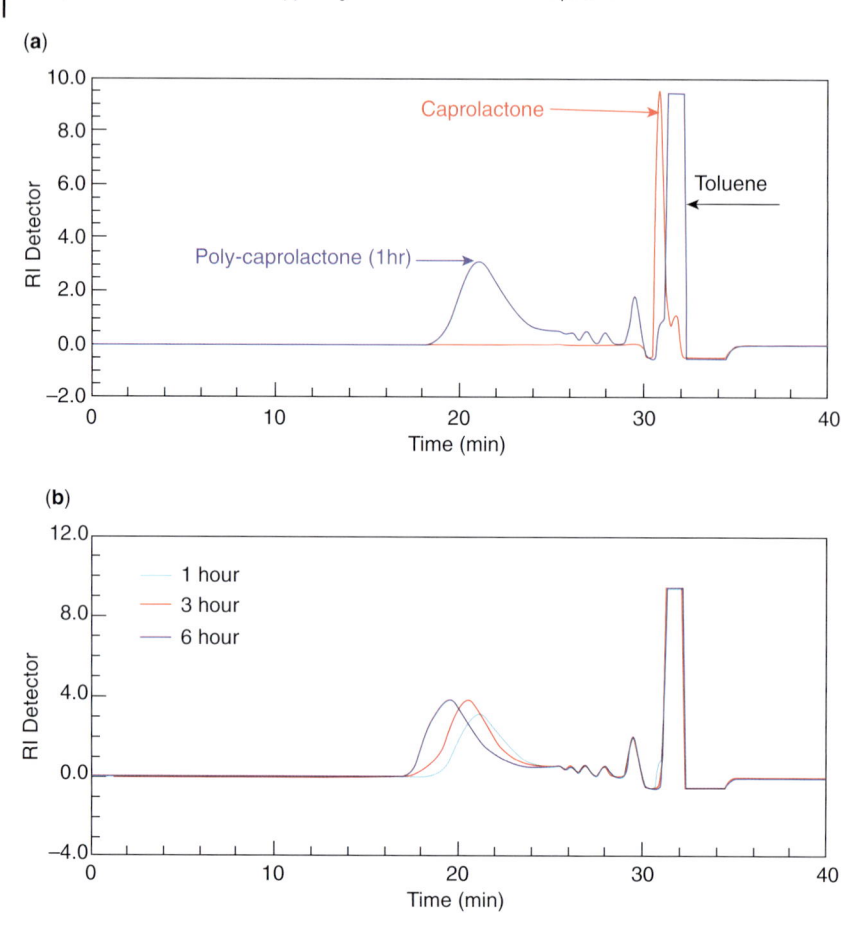

Figure 3.5 (a) Overlay of the GPC-RI chromatograms of ε-caprolactone and PCL after 1 hour synthesis (b) Overlay of the GPC-RI chromatograms of PCL after 1-, 3- and 6-hour synthesis.

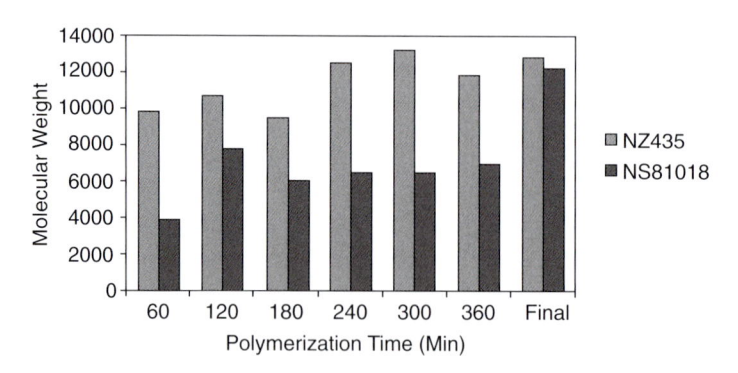

Figure 3.6 Molecular weight of PCL during solution polymerization at 70 °C using immobilized CALB.

Table 3.2 Molecular weights of two polycaprolactone samples determined using different methods at different laboratories.

Method	PCL Samples					
	(commercial)			(synthesis)		
	M_w	M_n	PDI	M_w	M_n	PDI
GPC-MALS	12 750	9 556	1.334	12 870	11 240	1.146
GPC-PS Lab 1	25 406	17 064	1.489	21 031	15 745	1.336
GPC-PS Lab 2	21 197	14 301	1.482	18 485	13 516	1.368
GPC-PLA	12 900	4 500	2.87	10 500	4 400	2.39

using standards that have similar molecular structure to the sample is crucial to get molecular weight results that are realistic and close to the actual values. PLA is a good standard choice for determination of the molecular weight of PCL since both PLA and PCL are aliphatic polyester. As shown in Table 3.2, the GPC-PLA M_w results are very close to GPC-MALS results and this validates our method of determination of molecular weight. Polystyrene (PS), even though a standard choice, could give unrealistic results when used for PCL molecular weight determination. This explains the large difference between the GPC-PS and GPC-MALS results. This study shows the sensitivity of molecular weight numbers to its method of determination, which should be emphasized when reporting molecular weights.

3.5.5.4 Effect of Termination of Reaction

The effect of termination of enzymatic PCL synthesis was studied. These studies were performed in parallel in the parallel synthesizer. In the first trial, the termination method was the addition of 10:1 (v/v) THF to monomer ratio. According to literature [5], THF or chloroform should be able to denature the enzyme and as a result the reaction will be terminated. All the polymer solutions remained in the added solvent for two weeks before filtration and analysis. In the second trial, all the conditions remained the same and the reaction was terminated by THF, but the enzyme was filtered out of the solution immediately and the sample was analyzed soon after. The molecular weight results of this study (Figure 3.7) shows that very low molecular weight products are achieved when the polymer solution stays in THF or chloroform and these solvents do not completely denature the enzyme. The enzyme, however, is active and seems to reverse the reaction toward depolymerization. It is important to note that at higher enzyme doses the depolymerization was more intense, that is, the lowest molecular weights were achieved. This could also be due to the water content of enzyme in the solution. When molecular sieves were added to the solution to remove the excess water, the depolymerization reaction was not observed and the molecular weight of the polymer solution did not change even after 20 days.

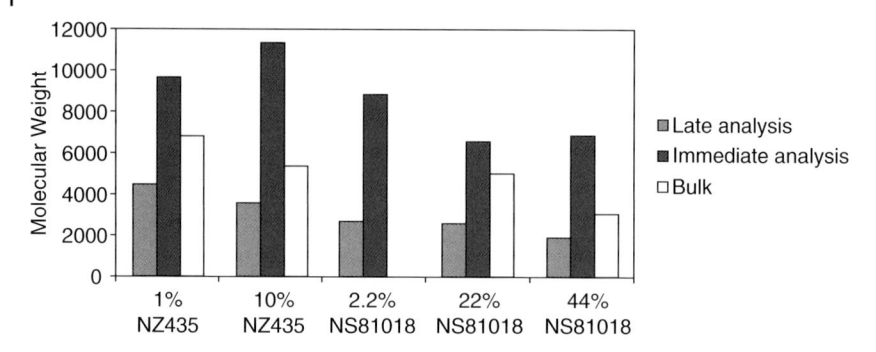

Figure 3.7 Effect of different synthesis parameters on the molecular weight of PCL.

Generally, the molecular weights of the polymers synthesized in the parallel synthesizer were lower than preparative scale; this is due to two reasons: having air as the headspace instead of the nitrogen and using the immobilized enzymes as is without vacuum drying.

3.5.5.5 Effect of Solvent
Polycaprolactone polymerization was also performed solvent-free (bulk) in the parallel synthesizer, under similar conditions. In the absence of solvent (toluene), the reaction resulted in a viscous polymer solution that could not be stirred well by shaking. As a result, polymerization leads to low molecular weight polymers (Figure 3.7).

3.5.5.6 Effect of Water
The effect of vacuum drying of the immobilized enzyme prior to polymerization was studied. In trial 1, Novozym 435 and NS81018 were dried in vacuum oven at 40 °C and 254 mm Hg for 24 hours prior to use in PCL synthesis. In trial 2, both immobilized enzymes were used without any pre-conditioning. The molecular weight results are shown in Figure 3.8. Results show that pre-drying of immobilized enzyme increases the molecular weight of the resulting polymer. This effect is more pronounced in case of NS81018, since this immobilization carrier is silica, a hygroscopic material. The effect of water on the activity of biocatalysts has been studied in the literature. Even though some uncertainties still exist, it has been found that a controlled water content is necessary for achieving repeatable yields and molecular weights.

The presence of a small amount of water for synthesis in non-aqueous media is essential for the enzyme activity. The enzyme-bound water acts as a lubricant and gives the enzyme molecule enough flexibility that is necessary for conformational changes for the enzyme's catalytic activity. Water in enzymatic ring opening polymerizations also acts as initiator for the reaction. Higher water concentration results in fast monomer conversion and low molecular weight. This is due to the high number of polymer chains initially starting to grow when water concentra-

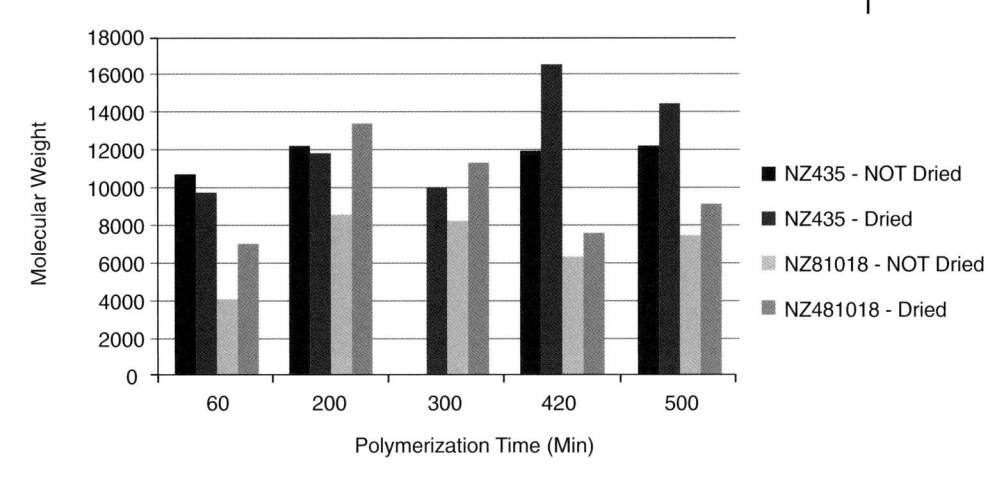

Figure 3.8 Effect of pre-drying of Novozym 435 and NS81018 as catalysts on molecular weight of PCL.

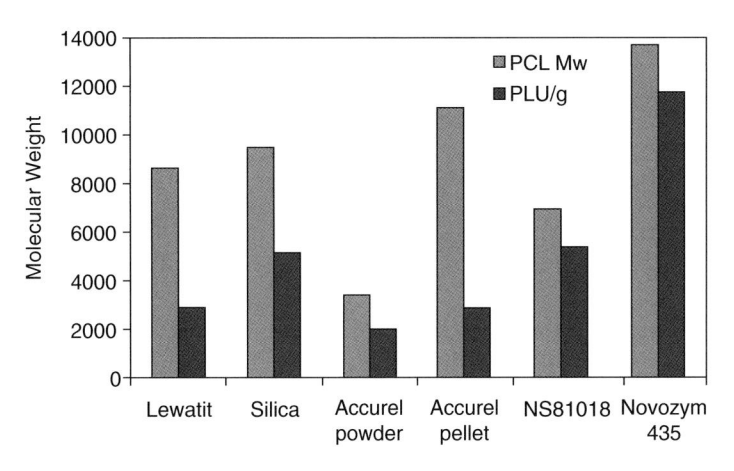

Figure 3.9 The molecular weight of PCL polymerized using four in-house immobilized enzymes.

tion is high. As a result, all the monomer will be consumed to generate lots of low DP chains. Whereas at low concentrations of water a lower number of chains will be initiated, which can grow to result in high DP chains [27, 28].

3.5.5.7 Effect of Immobilization Support

The molecular weight of PCL polymerized using four lab-scale immobilized enzymes are shown in Figure 3.9. The PLU values were also measured. The results are compared with Novozym 435 and NS81018. Both PLU values and the molecular weights show that none of the lab-scale immobilized samples could reach the activity of Novozym 435. Even when Lewatit was used as a support,

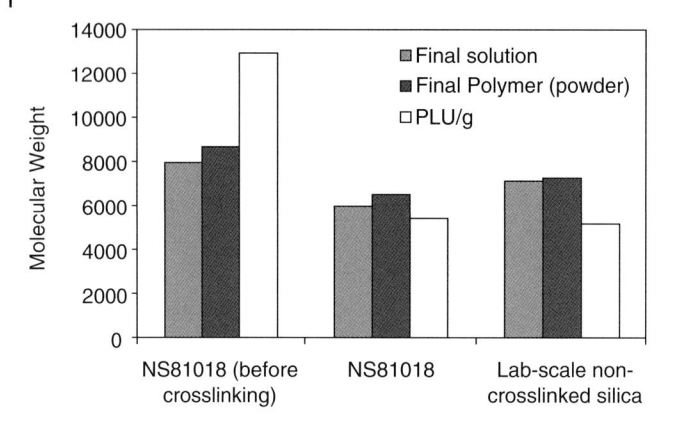

Figure 3.10 Comparison of PLU and molecular weights of PCL synthesized using NS81018 before crosslinking, NS81018 (crosslinked), and lab-scale non-crosslinked silica.

similar activity was not observed, probably due to the method of immobilization. Accurel pellets surprisingly showed relatively high molecular weights. This could be due to particle size, pore sizes or void volume of this product. The PLU values however correlate poorly with polymerization results, suggesting that even though the PLU assay is based on esterification, it might not be very suitable for polymerization studies.

Figure 3.10 shows the molecular weight results of PCL polymerization using CALB immobilized on silica. The performance of NS81018 (crosslinked CALB on silica), NS81018 before crosslinking, and lab-scale immobilized silica is compared. As expected, the non-crosslinked NS81018 had a higher activity than the crosslinked one.

3.6
Conclusions

Enzymatic processes in fields such as food and leather manufacturing have ancient roots, yet application of enzymes for organic synthesis is still a relatively young and a much developing discipline [29]. In a few cases the conversion is straight forward and commercially available enzymes can be applied like any other standard shelf chemical. Specifically for industrial processes, however, where productivity and hence economy are key issues, introducing enzymatic catalysis often requires extensive optimization of the enzyme as well as the process in which the enzyme is applied. In this chapter, we sought to illustrate the development work involved in introducing immobilized enzymes to the polymer industry. Even though enzymatic polymerizations are well-described in the academic literature, only few large-scale industrial processes rely on enzymatic catalysis. In this chapter, we chose to rely on the well-known CALB enzyme and focus our efforts on the formulation, specifically on a low-cost, low-leakage immobilization technology. These efforts were partly successful. It is evident from the trials with the PCL model polymerization reaction that performance of the experimental NS81018 is inferior to that of Novozym 435. However, if introduced to the market, cost of NS81018 would also be much lower than Novozym 435, likely by a factor of 5–10. Therefore, it is possible that for some applications, the lower performance is more than compensated for by the cost-differences, allowing for a lower productivity target that is, less polymer product needs to be produced per kg of enzyme for the process to be economically feasible. In conclusion, the current study has shown that even though challenges for immobilized CALB for polymer synthesis are left, there is potential, need and interest for introduction of other immobilized enzymes for synthesis such as NS81018.

Acknowledgment

The authors thank Joe Jump, Morten Christensen and Poul Poulsen for leading the research, Kåre Jørgensen and Lars Pedersen for NS81018, Sonja Salmon for support, and Richard Gross at Polytechnic University for insights and advice during the course of study.

References

1 Schmid, A., and Dordick, J.S. (2001) Lipases: interfacial enzymes with attractive applications. *Nature*, **409**, 258–268.
2 Gross, R.A., Kumar, A., and Kalra, B. (2001) Polymer synthesis by in vitro enzyme catalysis. *Chem. Rev.*, **101**, 2097–2124.
3 Kobayashi, S., Uyama, H., and Kimura, S. (2001) Enzymatic Polymerization. *Chem. Rev.*, **101**, 3793–3818.
4 Kobayashi, S. (1999) Enzymatic polymerization: a new method of polymer synthesis. *J. Polym. Sci. [A1]*, **37**, 3041–3056.
5 Nakaoki, T., Kalra, B., Kumar, A., Gross, R.A., Kirk, O., and Christensen, M. (2002) Candida antarctica lipase B catalyzed polymerization of lactones: Effect of immobilization matrix on polymerization kinetics and molecular weight. *Abstracts of Papers of the Am. Chem. Soc.*, **224**, U473.
6 Mei, Y., Miller, L., Gao, W., and Gross, R.A. (2003) Imaging the distribution and secondary structure of immobilized enzymes using infrared microspectroscopy, *Biomacromolecules*, **4**(1), 70–74.
7 Kirk, O., Damhus, T., Borchert, T.V., Fuglsang, C.C., Olsen, H.S., Hansen, T.T., Lund, H., Schiff, H.E., and Nielsen, L.K. (2004) Enzyme applications, Industrial in *Kirk-Othmer Encyclopedia of Chemical Technology*, (**10**), 248–317, John Wiley & Sons, Inc., New York, USA.
8 Anderson, E.M., Karin, M., and Kirk, O. (1998) One biocatalyst–Many applications: The use of Candida antarctica B-lipase in organic synthesis. *Biocatalysis and Biotransformation*, **16**, (3) 181–204.
9 Petkar, M., Lali, A., Caimi, P., and Daminati, M. (2006) Immobilization of lipases for non-aqueous synthesis. *Journal of Molecular Catalysis B-Enzymatic*, **39** (1–4), 83–90.

10 Uppenberg, J., et al. (1994) Sequence, crystal-structure determination and refinement of 2 crystal forms of lipase-B from Candida-Antarctica. *Structure*, **2.4**, 293–308.

11 Skjøt, M., De Maria, L., Chatterjee, R., Svendsen, A., Patkar, S.A., Østergaard, P.R., and Brask, J. (2009) Understanding the plasticity of the alpha/beta-hydrolase fold: lid swapping on the Candida antarctica lipase B results in chimeras with interesting biocatalytic properties. *ChemBioChem.*, **10**, 520–527.

12 Brena, B.M., and Batista-Viera, F. (2006) Immobilization of Enzymes: a literature survey in *Immobilization of Enzymes and Cells* (ed. J.M. Guisan) 2nd edn, Springer Verlag, New York, pp. 15–30.

13 Pederson, S., and Christensen, M.W. (1994) Immobilized biocatalysts, in *Applied Biocatalysis* (ed. A.J.J. Straathof and P. Aldercreutz), Harwood Academic Publishers, Amsterdam, pp. 213–228.

14 Schoevaart, R., Siebum, A., van Rantwijk, F., Sheldon, R., and Kieboom, T. (2005) Glutaraldehyde cross-link analogues from carbohydrates. *Starch/Stärke*, **57**, 161–165.

15 Betancor, L., Lopez-Gallego, F., Alonso-Morales, N., Dellamora, G., Mateo, C., Fernandez-Lafuente, R., and Guisan, J.M. (2006) Glutaraldehyde in protein immobilization: a versatile reagent in *Immobilization of Enzymes and Cells* (ed. J.M. Guisan), 2nd edn, Springer Verlag, New York, pp. 57–64.

16 Govardhan, C.P. (1999) Crosslinking of enzymes for improved stability and performance. *Curr. Opin. Biotechnol.*, **10**, 331–335.

17 Cao, L., van Rantwijk, F., and Sheldon, R.A. (2000) Cross-linked enzyme aggregates: A simple and effective method for the immobilization of penicillin acylase. *Org. Lett.*, **2**, 1361–1364.

18 Mateo, C., et al. (2004) A new, mild crosslinking methodology to prepare crosslinked enzyme aggregates. *Biotechnol. Bioeng.*, **86**, 273–276.

19 Migneault, I., Dartiguenave, C., Bertrand, M.J., and Waldron, K.C. (2004) Glutaraldehyde: behaviour in aqueous solution, reaction with protein with proteins, and application to enzyme crosslinking. *BioTechniques*, **37**, 790–802.

20 Wiemann, L.O., et al. (2009) enzyme stabilization by deposition of silicone coatings. *Org. Process Res. Dev.*, **13.3**, 617–620.

21 Chen, B., Hu, J., Miller, E.M., Xie, W., Cai, M., and Gross, R.A. (2008) Candida antarctica lipase B chemically immobilized on epoxy-activated micro- and nanobeads: Catalysts for polyester synthesis. *Biomacromolecules*, **9 (2)**, 463–471.

22 Cabrera, Z., Fernandez-Lorente, G., Fernandez-Laufente, R., Palomo, J.M., and Guisan, J.M. (2009) Enhancement of Novozym 435 catalytic properties by physical or chemical modification. *Process Biochem.*, **44**, 226–231.

23 Mazeaud, I., and Poulsen P.B.R. (2007) PCT application WO2007/036235.

24 Kumar, A., and Gross, R.A. (2000) Candida antarctica lipase B catalyzed polycaprolactone synthesis: effects of organic media and temperature. *Biomacromolecules*, **1**, 133–138.

25 Deng, F., and Gross, R.A. (1999) Ring-opening bulk polymerization of epsilon-caprolactone and trimethylene carbonate catalyzed by lipase Novozym 435. *Int. J. Biol. Macromol.*, **25**, 153–159.

26 Mei, Y., Kumar, A., and Gross, R. (2003) Kinetics and mechanism of Candida antarctica Lipase B catalyzed solution polymerization of epsilon-caprolactone. *Macromolecules*, **36.15**, 5530–5536.

27 Mei, Y., Kumar, A., and Gross, R.A. (2002) Probing water-temperature relationships for lipase-catalyzed lactone ring-opening polymerizations. *Macromolecules*, **35.14**, 5444–5448.

28 Kvittingen, L. (1994) Some Aspects of Biocatalysis in Organic-Solvents. *Tetrahedron*, **50 (28)**, 8253–8274.

29 Bornscheuer, U.T., and Buchholz, K. (2005) *Eng. Life. Sci.*, **5**, 309–323.

4
Enzymatic Polymerization of Polyester

Nemanja Miletić, Katja Loos, and Richard A. Gross

4.1
Introduction

Polyesters are in widespread use in our modern life, ranging from bottles for carbonated soft drinks and water, fibers for shirts and other apparel, to the base for photographic film and recording tape. Household tradenames, such as Dacron®, Fortrel®, Terylene®, Mylar®, etc. demonstrate the ubiquitous nature of polyesters. In addition, of the biodegradable polymers employed in medical applications, polyesters are most often used.

In the past, the term polyester referred to polymers derived essentially from diols and dicarboxylic acids. Earliest reports of polyester resins of this type include those from Berzelius [1], who documented resins from tartaric acid and glycerol, Berthelot [2], who produced a resin from glycerol and camphoric acid, and Van Bemmelen [3], who synthesized glycerides of succinic acid and citric acid. Back in 1901, Watson Smith had already described the reaction product of glycerol and phthalic anhydride [4]. In 1924, Kienle and Hovey began to study the kinetics of polyesterification reactions between glycerol and phthalic anhydride [5]. Carothers' pioneering studies were based on aliphatic polyesters and culminated in laying the foundations for condensation and step-growth polymerization [6–9]. Since then, many research groups have investigated this group of polymers, broadening fundamental studies and working towards developing commercial products.

In recent years, environmental concerns have led to a renewed interest in biodegradable polyesters as an alternative to commodity plastics. Since ester linkages are frequently encountered in nature it is reasonable to assume that at least a subset of the polyester family will be environmentally degradable. Random and block copolymers as well as blends have been investigated with regard to controlling the lifetime of biodegradable polymers as well as improving their mechanical properties. Environmental pollution caused by production and disposal of petrochemical-derived plastics have led to pursuit of alternative approaches using environmentally benign processes to synthesize plastics that are engineered to degrade-on-demand.

Biocatalysis in Polymer Chemistry. Edited by Katja Loos
Copyright © 2011 WILEY-VCH Verlag GmbH & Co. KGaA, Weinheim
ISBN: 978-3-527-32618-1

Enzymatic polymerizations are a promising strategy under study by many groups throughout the world to develop environmental friendly processes for polyester synthesis.

Okumara et al. [10] were the first to attempt the enzyme-catalyzed synthesis of oligoesters from a reaction between dicarboxylic acids and diols. Gutman et al. [11] reported the first study on polyester synthesis by enzyme-catalyzed polymerization of A-B type monomers. Two independent groups in 1993 [12, 13] were first to report enzyme-catalyzed ring-opening polymerization (ROP). Their studies focused on 7- and 6-membered unsubstituted cyclic esters, ε-caprolactone (ε-CL) and δ-valerolactone (δ-VL), respectively.

A variety of *in vitro* polyester synthesis reactions have been developed in the last couple of decades and a couple of excellent reviews on this topic have been published [14–24].

In the present chapter, the current status of enzymatic polyester synthesis is described. For information on the enzymatic synthesis of chiral polyesters and polyester block copolymers using enzymatic polymerizations please refer to Chapters 11 and 12 respectively.

4.2
Synthesis of Polyesters

In nature, various macromolecules are constantly being produced in living organisms for their normal metabolic needs. These macromolecules, such as polysaccharides, polynucleotides (DNA and RNA), proteins, or polyesters, are essential to organism survival. Their synthesis generally involves *in vivo* enzyme-catalyzed chain-growth polymerization reactions of activated monomers, which are generally formed within the cells by complex metabolic processes. Please refer to Chapter 10 for an review on important bacterial storage compounds including polyesters.

Among enzymes, lipases proved to be the most efficient for the *in vitro* polyester synthesis. Lipases or triacylglycerol acylhydrolases are water-soluble enzymes that catalyze the hydrolysis of ester bonds in water-insoluble, lipid substrates, and therefore comprise a subclass of the esterases.

Lipases are ubiquitous enzymes of considerable physiological significance and perform crucial roles in the digestion, transport and processing of dietary lipids in most of living organisms. Thus, lipases can be found in diverse sources, such as plants, animals, and micro-organisms. More abundantly, they are found in bacteria, fungi and yeasts.

Lipases catalyze the hydrolysis of relatively long chain triglycerides (with acyl chain lengths of over ten carbon atoms) to the corresponding diacylglyceride, monoacylglyceride, glycerol and fatty acids. Since the water-insoluble lipid interferes with the water-soluble lipase, digestion of these triglycerides takes place at the water–oil interface. On the other hand, it is well known that the reaction is reversible and lipases can catalyze ester synthesis and transesterification in a

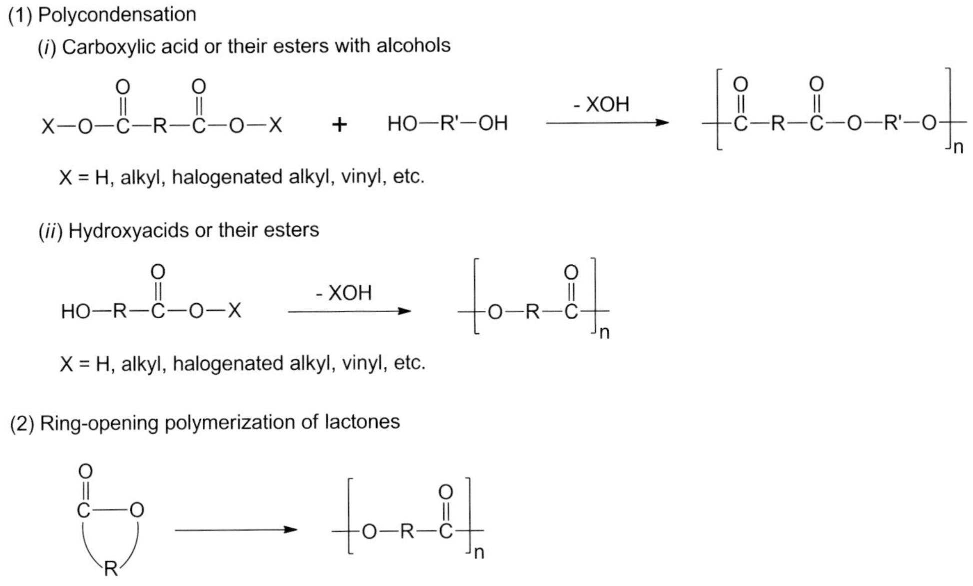

(1) Polycondensation

(i) Carboxylic acid or their esters with alcohols

$$\text{X-O-}\overset{O}{\underset{\|}{C}}\text{-R-}\overset{O}{\underset{\|}{C}}\text{-O-X} + \text{HO-R'-OH} \xrightarrow{-\text{XOH}} {\left[\overset{O}{\underset{\|}{C}}\text{-R-}\overset{O}{\underset{\|}{C}}\text{-O-R'-O}\right]}_n$$

X = H, alkyl, halogenated alkyl, vinyl, etc.

(ii) Hydroxyacids or their esters

$$\text{HO-R-}\overset{O}{\underset{\|}{C}}\text{-O-X} \xrightarrow{-\text{XOH}} {\left[\text{O-R-}\overset{O}{\underset{\|}{C}}\right]}_n$$

X = H, alkyl, halogenated alkyl, vinyl, etc.

(2) Ring-opening polymerization of lactones

$$\overset{O}{\underset{\|}{C}}\text{-O (ring with R)} \longrightarrow {\left[\text{O-R-}\overset{O}{\underset{\|}{C}}\right]}_n$$

Scheme 4.1 Two basic modes of enzyme-catalyzed polyester synthesis.

reaction containing low water concentrations opening up the possibility to synthesize polyester.

Although there are notable exceptions as given below, the most common lipase-catalyst used for polyester synthesis is *Candida antarctica* lipase B (CALB) (please refer to Chapter 14 for more information on the structure and reaction mechanisms of CALB). The immobilized CALB catalyst that has been primarily used is Novozym® 435, manufactured by Novozymes (Bagsvaerd, Denmark). Novozym 435 consists of CALB physically adsorbed within the macroporous resin Lewatit VPOC 1600 (poly[methyl methacrylate-*co*-butyl methacrylate], supplied by Bayer) (please refer to Chapter 3 for more information on Novozym 435).

The *in vitro* polyester synthesis can proceed via two major polymerization modes (see Scheme 4.1):

1) polycondensation between a carboxyl group and an alcohol group (following route (*i*) or route (*ii*)), and
2) ring-opening polymerization (ROP).

4.3
Enzyme-Catalyzed Polycondensations

The enzymatic condensation reaction to form an ester using enzymes is composed of four modes of elemental reactions: (i) dehydration; (ii) alcoholysis;

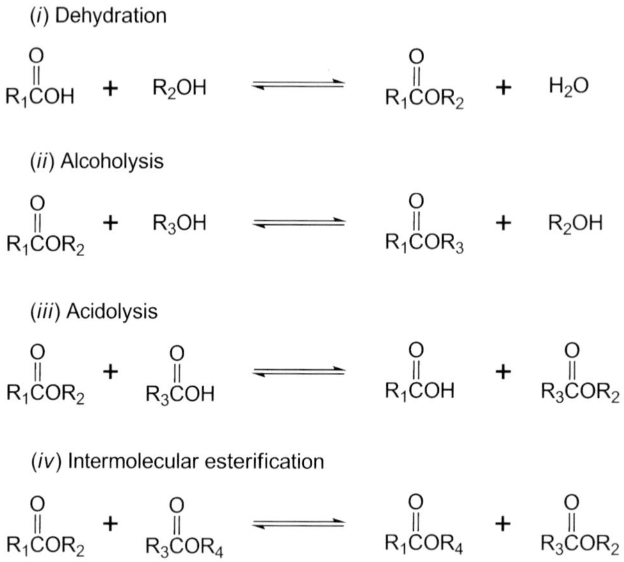

(i) Dehydration

$$R_1COOH + R_2OH \rightleftharpoons R_1COOR_2 + H_2O$$

(ii) Alcoholysis

$$R_1COOR_2 + R_3OH \rightleftharpoons R_1COOR_3 + R_2OH$$

(iii) Acidolysis

$$R_1COOR_2 + R_3COOH \rightleftharpoons R_1COOH + R_3COOR_2$$

(iv) Intermolecular esterification

$$R_1COOR_2 + R_3COOR_4 \rightleftharpoons R_1COOR_4 + R_3COOR_2$$

Scheme 4.2 Four basic modes of elemental reactions of enzyme-catalyzed condensations.

(iii) acidolysis; and (iv) intermolecular esterification (Scheme 4.2). The reactions are all reversible; therefore, to shift the reaction equilibrium to the product side, the by-products, like water or alcohol, are normally removed from the reaction mixture. The lipase-catalyzed polyester synthesis via polycondensation (condensation polymerization) uses the reaction of all four modes, which is the reverse direction of the inherent lipase catalysis (hydrolysis). For detailed information on the mechanism of lipase-catalyzed ester bond formation, readers can refer to Chapter 14 in this book.

4.3.1
A-B Type Enzymatic Polyesterfication

Polyesters can be obtained starting from hydroxyacids or, more generally, A-B type monomers, where the groups A and B can react with other B and A groups, respectively. Condensations of the A-B type generate a leaving group that, in most of the cases, must be efficiently removed in order to obtain high molecular weight polyesters. High purity monomers of A-B type can be used directly for form high molecular weight polyesters, whereas, A-A and B-B type monomers (see Section 4.2.1.2) must be of high purity but also require that they are mixed in precisely equimolar quantities in order to obtain high molecular weight polymers.

Reported hydroxyacids that are self-condensable by enzyme catalysis include: 6-hydroxyhexanoic acid [25], 10-hydroxydecanoic acid [26], 5-hydroxyhexanoic acid [27], 5-hydroxydodecanoic acid [27], 11-hydroxydecanoic acid [28],

12-hydroxydodecanoic acid [29], 15-hydroxypentadecanoic acid [27], 16-hydroxyhexadecanoic acid [25], 18-hydroxyoctadecanoic acid [30], ricinoleic acid [31, 32], *cis*-9,10-epoxy-18-hydroxyoctadecanoic acid [30], cholic acid [33] (see also Table 4.1). Also, alkyl esters of these compounds have been used as monomers for lipase-catalyzed polycondensation reactions. Examples include ethyl esters of 3- and 4-hydroxybutyric acid [27], methyl ricinoleate [32] and isopropyl aleuriteate [46].

In the following we report some of the examples of A-B type enzymatic polyesterifications. The reader should refer to Table 4.1 for further interesting examples.

Methyl ε-hydroxyhexanoate was chosen as a model monomer for the first investigation to determine how important reaction parameters that include enzyme origin, solvent, concentration and reaction time influence its self-condensation polymerization [12]. The degree of polymerization (DP) of the polyester formed followed a S-shaped behavior with solvent $\log P$ ($-0.5 < \log P < 5$) – with an increase in DP around $\log P \sim 2.5$. Decreasing values of DP in good solvents for polyesters were attributed to the rapid removal of product oligomers from the enzyme surface, resulting in reduced substrate concentration near the enzyme.

A time course study of 11-hydroxydecanoic acid polymerization catalyzed by *Candida cylindracea* lipase was reported by O'Hagan and Zaidi [28]. The authors revealed that oligomers are formed relatively rapidly and then later condense to generate higher molecular weight polyesters. After 7 days, they reported formation of a polyester with molecular weights up to $\overline{M_w} = 35000$.

Polyester synthesis activity of *Humicola insolens* cutinase (HiC) immobilized on Amberzyme oxiranes (HiC-AO) was systematically studied by Feder and Gross using ω-hydroxyalkanoic acids (ωHA) with 6, 10, 12 and 16 carbons [39]. Variation of substrate chain lengths showed that immobilized HiC has higher chain length selectivity than Novozym 435, as Novozym 435 was able to polymerize 10-hydroxydecanoic acid, 12-hydroxydodecanoic acid and 16-hydroxyhexadecanoic acid while HiC was just active on ωHAs with 12 and 16 carbons. In other words, Novozym 435 is more promiscuous remaining active on a broader set of substrates relative to immobilized HiC. Therefore, cutinases might be interesting alternatives to lipases for enzymatic polyester synthesis.

The enzymatic polymerization of some rather unconventional hydroxyacids was also reported.

For instance an epoxy-functionalized polyester from the suberin monomer *cis*-9,10-epoxy-18-hydroxyoctadecanoic acid (see also Chapter 1) was synthesized by Olsson *et al.* [30]. The lipase-catalyzed polymerization was performed in toluene in the presence of 4 Å molecular sieves for 68 h and high molecular weight of epoxy-functionalized polyester was obtained ($\overline{M_w} = 20000$; $\overline{M_w}/\overline{M_n} = 2.2$).

Novozym 435-catalyzed self condensation of isopropyl aleuriteate [46] at 90 °C in toluene and 2,4-dimethyl-3-pentanol as cosolvent gave the corresponding polyester in 43% yield ($\overline{M_n} = 5600$). Subsequently, isopropyl aleuriteate was copolymerized with ε-CL and random copolymers were obtained in around 70% yield with $\overline{M_n}$ values up to 10600.

Table 4.1 Enzyme-catalyzed polyester condensation polymerizations.

Enzyme	Monomer	Reference
Aspergillus niger lipase A	1,13-tridecandedoic acid with 1,3-propane diol	[34]
	bis(2-chloroethyl)(+−)2,5-bromoadpate with 1,6-hexanediol	[35]
Candida antarctica lipase B	11-mercaptoundecanoic acid	[36]
	1,18-cis-9,10-epoxyoctadecanedioic acid with 1,8-octanediol	[37]
	1,18-cis-9-octadecenedioic acid with 1,16-hexadecanediol	[37]
	1,18-cis-9-octadecenedioic acid with 1,3-propanediol	[37]
	1,18-cis-9-octadecenedioic acid with 1,8-octanediol	[37]
	1,18-octadecanedioic acid with 1,8-octanediol	[37]
	1,22-cis-9-docosenedioic acid with 1,8-octanediol	[37]
	1,2-benzenedimethanol 4,4-isopropylidenebis[2-(2,6-dibromophenoxy)ethanol] bisphenol A	[38]
	1,3-propanediol divinyl adipate with 1,3-benzenedimethanol	[38]
	1,3-propanediol divinyl adipate with 1,4-benzenedimethanol	[38]
	1,3-propanediol divinyl adipate with 2,6-pyridinedimethanol	[38]
	1,3-propanediol divinyl carbonate with 1,3-propane diol	[10]
	1,4 butane diol divinyl carbonate with glycerol	[38]
	1,6 hexane diol divinyl carbonate with 1,2,4 butane triol	[38]
	10-hydroxydecanoic acid	[25, 39]
	12-hydroxydodecanoic acid	[25, 39]
	16-hydroxyhexadecanoic acid	[25, 39]
	6-hydroxyhexanoic acid	[25]
	adipic acid with 1,4-butanediol	[38, 40]
	adipic acid with 1,8-octanediol	[39]
	adipic acid with glycerol	[41]
	adipic acid with sorbitol	[41]
	azelaic acid with 1,8-octanediol	[39]
	brassylic acid with 1,8-octanediol	[39]
	cholic acid	[33]

Table 4.1 Continued

Enzyme	Monomer	Reference
	cis-9,10-epoxy-18-hydroxyoctadecanoic acid	[30]
	diethyl octane-1,8-dicarboxylate and 1,4-butanediol	[42]
	divinyl adipate with 2,2,3,3,4,4-hexafluoro-1,5-pentanediol	[43]
	divinyl adipate with 2,2,3,3-tetrafluoro-1,4-butanediol	[43]
	divinyl adipate with 3,3,4,4,5,5,6,6-octafluorooctan-1,8-diol	[43]
	divinyl carbonate with 1,10-decanediol	[38]
	divinyl carbonate with 1,12-dodecanediol	[38]
	divinyl carbonate with 1,2,4-butanetriol	[38]
	divinyl carbonate with 1,3-benzenedimethanol	[38]
	divinyl carbonate with 1,3-propanediol	[38]
	divinyl carbonate with 1,4-benzenedimethanol	[38]
	divinyl carbonate with 1,9-nonanediol	[38]
	divinyl carbonate with 2,6-pyridinedimetanol	[38]
	divinyl isophthalate with 1,6-hexanediol divinyl terephthalate divinyl *p*-phenylene diacetate	[44]
	divinyl sebacate with glycerol and the unsaturated fatty acids	[45]
	divinyl sebacate with *p*-xylene glycol	[44]
	isopropyl aleuriteate	[46]
	octanediol adipate with glycerol adipate	[47]
	octanediol adipate with sorbitol adipate	[47]
	poly(octamethylene adipate) with poly(sorbitol adipate)	[47]
	sebacic acid with 1,4-butanediol	[48]
	sebacic acid with 1,6-hexanediol	[39]
	sebacic acid with 1,8-octanediol	[39]
	suberic acid with 1,8-octanediol	[39]
	succinic acid with 1,4-butanediol	[49]
	terephthalic acid/isophthalic acid with 1,4-butanediol/1,6-hexanediol	[50]
Candida cylindracea lipase	11-hydroxyundecanoic acid	[28]
	10-hydroxyundecanoic acid	[26]
	sebacic acid with 1,8-OL	[34]

Table 4.1 Continued

Enzyme	Monomer	Reference
Candida rugosa lipase	bis(2,2,2-trifluoroethyl) sebacate with 1,4-butanediol	[51]
Humicola insolens cutinase (HiC)	12-hydroxydodecanoic acid	[39]
	16-hydroxyhexadecanoic acid	[39]
	sebacic acid with 1,8-octanediol	[39]
	sebacic acid with 1,6-hexanediol	[39]
	azelaic acid with 1,8-octanediol	[39]
	brassylic acid with 1,8-octanediol	[39]
Klebsiella oxytota (Lipase K)	sebacic acid with 1,8-OL	[34]
Mucor meihei	sebacic acid with 1,4-butanediol	[51]
	diethyl sebacate with 1,4-butanediol	[51]
	bis(2,2,2-trifluoroethyl) sebacate with 1,4-butanediol	[51]
	bis(2,2,2-trifluoroethyl) sebacate with 1,4- butanediol	[35, 51]
	bis(2,2,2-trifluoroethyl) sebacate with 1,2-ethanediol	[51]
	bis(2,2,2-trifluoroethyl) sebacate with 1,3-propanediol	[51]
	bis(2,2,2-trifluoroethyl) sebacate with 1,5-pentanediol	[51]
	bis(2,2,2-trifluoroethyl) sebacate with 1,6-hexanediol	[51]
Porcine pancreatic lipase	sebacic acid + 1,8-OL	[34]
	bis(2,2,2 trifluoroethyl) sebacate with 1,4-butanediol	[52]
	bis(2,2,2-trichloroethyl) trans-3-hexanedioate (racemic mixture) with 1,4-butanediol	[53]
	methyl-5-hydroxypentanoate	[12]
	methyl-6-hydroxyhexanoate	[12]
Pseudomonas aeruginosa lipase	sebacic acid with 1,8-OL	[34]
Pseudomonas cepacia lipase	methyl ricinoleate	[32]
	sebacic acid with 1,8-OL	[34]
Pseudomonas fluorescens lipase	sebacic acid with 1,8-OL	[34]
	divinyl adipate with 1,4-butanediol	[54]
	bis(2,2,2-trifluoroethyl) sebacate with 1,4-butanediol	[51]

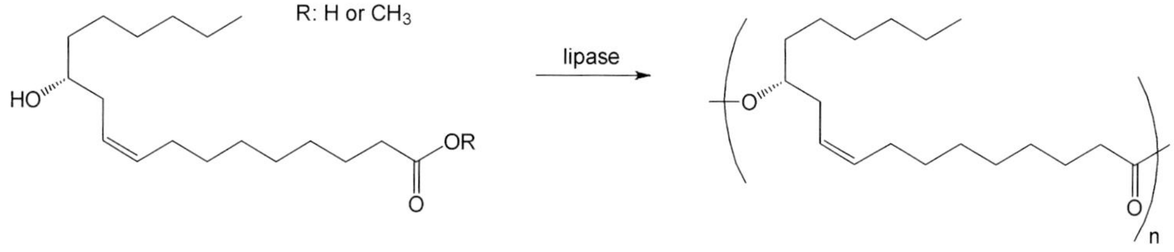

Scheme 4.3 Lipase-catalyzed preparation of polyricinoleate.

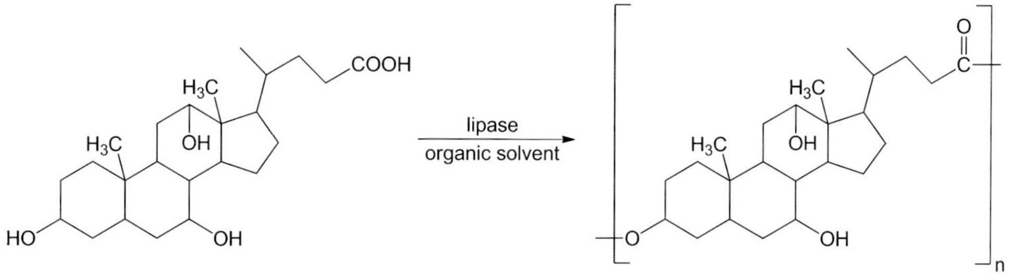

Scheme 4.4 Self-condensation of cholic acid catalyzed by *Candida antarctica* lipase B.

The enzymatic polymerization of methyl ricinoleate was performed using an immobilized lipase from *Pseudomonas cepacia* as catalyst. Reactions were conducted in bulk, with molecular sieves, at 80 °C, for 7 days to give poly(ricinoleic acid) with $\overline{M_w} > 1 \times 10^5$ (Scheme 4.3) [32]. This result is generally uncharacteristic of other reports on related monomers given that lipase-catalyzed esterification of secondary hydroxyls proceeds slowly (see below) and ricinoleic acid purity to achieve such molecular weights must be very high.

Ritter et al. reported the formation of oligomers from cholic acid by self-condensation reaction catalyzed by CALB (Scheme 4.4) [33].

There is a natural interest in enzymatic routes to polymers having thioester [−S−C(=O)−] links since many properties (e.g., higher melting temperature, greater heat stability, and lower solubility in various organic solvents) of these materials are superior to those polymers prepared with ester links. Kato et al. [36] were able to show that the CALB catalyzed direct polycondensation of 11-mercaptoundecanoic acid proceeds readily. An aliphatic poly(11-mercaptoundecanoate) was prepared in bulk for 48 h at 110 °C in the presence of 4 Å molecular sieves as a water absorbent with a $\overline{M_w}$ of 3.4×10^4 in high yield. This is surprising given that thiols are generally considered to react much slower than hydroxyl groups using lipase catalysis. In fact, chemoselective reactions in which hydroxyl groups have been reacted in favor of thiol groups are also known in the literature. Furthermore, 110 °C is considered high to sustain the activity of lipase catalysts over such long reaction periods. Poly(11-mercaptoundecanoate) can be degraded by lipase in dilute *n*-nonane solution forming cyclic

11-mercaptoundecanoate oligomers. The cyclic oligomer could be readily repolymerized by lipase to produce a relatively high molecular weight poly(11-mercaptoundecanoate) in high yield.

A compilation of enzyme-catalyzed self-condensation polymerizations that have thus far appeared in the literature is given in Table 4.1.

4.3.2
AA-BB Type Enzymatic Polyesterification

Okumara and co-workers [10] were the first to attempt the lipase-catalyzed synthesis of oligoesters from reactions between diols (AA) and dicarboxylic acids (BB). They showed that 'trimer', 'pentamer', and 'heptamer' consisting of AA-BB-AA, AA-BB-AA-BB-AA, and AA-BB-AA-BB-AA-BB-AA, respectively, were formed. Klibanov et al. [121] used the stereoselectivity of lipases to prepare enantioenriched oligoesters. The reactions were conducted using racemic diester and an achiral diol or, conversely, a racemic diol and an achiral diester as monomers (see also Chapter 11).

Polymerizations of adipic acid and 1,4-butanediol using Novozym 435 was studied by Binns et al. [40]. They reported that, under solvent-free conditions, the mixture was heated at 40 °C for 4 h, followed by heating at 60 °C for 10 h under pressure. The polymerization proceeds by a step growth mechanism to give a homogeneous reaction medium. The gel permeation chromatography (GPC) of reaction products formed after 4 and 14 h showed very different product distributions. The former showed a discrete array of predominantly hydroxy-terminated oligomers and the latter showed that polyesters had formed with weight average molecular weight of around 2200 and polydispersity of 1.5.

Linko et al. [35] systematically varied the chain length of dicarboxylic acid [C-4, C-6, C-8, C-10, and C-12] and diol [C-2, C-3, C-4, C-5, and C-6] monomers used for enzymatic polycondensation polymerizations. Of the lipases and solvents screened, the *Mucor miehei* lipase and diphenyl ether, respectively, were found to be preferred. Furthermore, product polyester molecular weight increased as the substrate concentration increased to about 0.83 M. The reaction of adipic acid with different diols showed the following trend with respect to polymer DP: 1,6-hexanediol > 1,4-butanediol > 1,5-decanediol > 1,3-decanediol > 1,2-butanediol. Similarly, the reaction of 1,6-hexanediol with different acids showed the following trend toward polymer DP: adipic acid > sebacic acid > octanedioic acid > dodecanoic acid > succinic acid. *Mucor miehei* catalyzed the condensation polymerization of adipic acid and hexanediol in diphenyl ether at 37 °C for 7 days under reduced pressure (0.15 mmHg) to give poly(hexenyl adipate) with $\overline{M_w} = 77\,400$, PDI 4.4. Transesterification between diethyl carbonate and a diol to produce polycarbonates proceeded via two stages; the first to yield oligomers and the second to give higher molecular weight polymers [55].

Lipase-catalyzed synthesis of poly(1,4-butyl sebacate) from reactions of 1,4-butanediol and sebacic acid or activated derivatives of sebacic acid were studied by Linko et al. [51]. Reactions between 1,4-butanediol with sebacic acid,

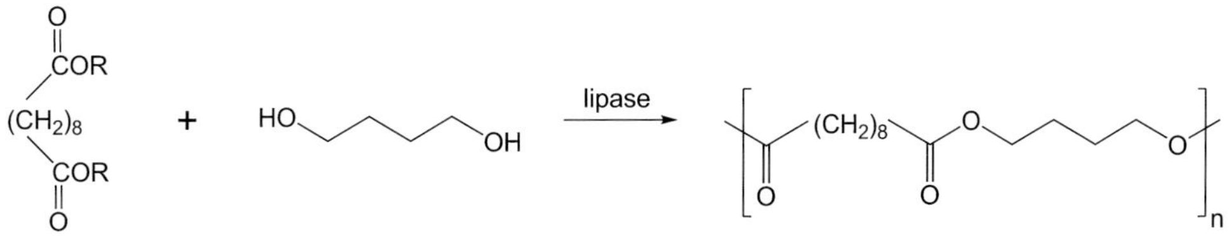

Scheme 4.5 Lipase-catalyzed condensation polymerization of sebacic acid ester with butanediol.

diethyl sebacate, or bis(2,2,2-trifluoroethyl) sebacate) were performed in veratrole or diphenyl ether using the lipase from *Mucor miehei* (36.5 wt %). Influence on poly(1,4-butyl sebacate) molecular weight as a function of removing the condensation by-product, solvent character, and substrate structure was assessed. When vacuum was used to remove water formed during the polymerization, sebacic acid was directly polymerized with 1,4-butanediol in diphenyl ether to give a product with $\overline{M_w} = 42\,000$ in 7 days at 37 °C (Scheme 4.5).

The same research group using the lipase from *Mucor miehei* in diphenyl ether (11.1% w/v) studied copolymerizations of an aromatic diacid (terephthalic or isophthalic) and an aliphatic diol (1,4-butane- or 1,6-hexanediol) [50]. Even at temperatures up to 70 °C, polymerizations of these aromatic diacids were unsuccessful. However, using Novozym 435 as catalyst, polymerizations of aromatic diacids were accomplished, with yield ranging 85–93%. For example, while the Novozym 435-catalyzed reaction of isophthalic acid with butanediol yielded oligomers, a similar reaction between the C-6 diol and isopthalic acid at 70 °C yielded a polymer with $\overline{M_w} = 55\,000$.

Kobayashi and co-workers [48] studied the potential of carrying out condensation reactions in solventless or bulk reactions. They reported the preparation of aliphatic polyesters with $\overline{M_w} > 10\,000$ by reacting sebacic acid with 1,4-butanediol in a solvent-free system, under reduced pressure, using CALB as catalyst.

The phase separation of reactants hindered attempts to carry out lipase-catalyzed synthesis of poly(butylene succinate) (PBS) from succinic acid and 1,4-butanediol via dehydration. Therefore, in order to obtain a monophasic reaction mixture, dimethyl succinate was used in place of succinic acid. [49] The reaction mixture remained monophasic during the reaction course, and after 21 h at 95 °C PBS with $\overline{M_n}$ of 38 000 was obtained.

Polycondensation of diethyl 1,8-octanoic diacid and 1,4-butanediol at room temperature and at 60 °C was carried out in 1-butyl-3-methylimidazolium hexafluorophosphate ionic liquid (see also Chapter 13 for enzymatic polymerizations in unconventional solvents), using lipase PS–C as catalyst [42]. The highest molecular weight polymer ($\overline{M_n} = 4300$; $\overline{M_w} = 5400$) was obtained at 60 °C for 7 days.

The polymerization of substrates varying in α,ω-n-alkane diol and α,ω-n-alkane diacid chain length by *Humicola insolens* cutinase immobilized on Amberzyme oxiranes (HiC-AO) and Novozym 435 was studied [39]. HIC-AO

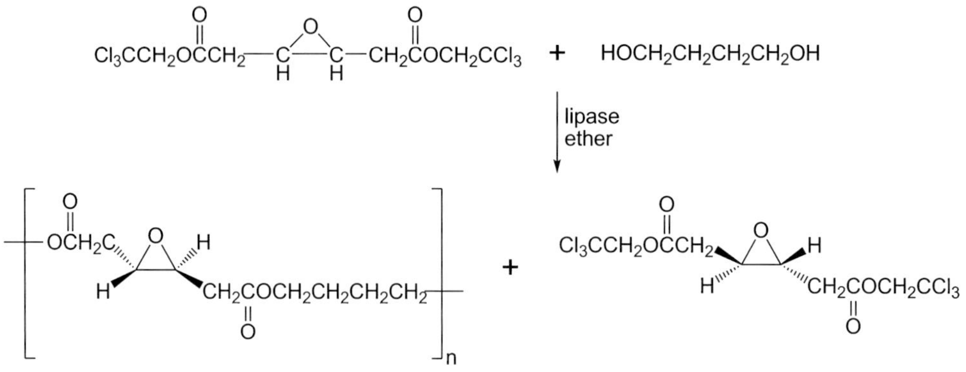

4 Enzymatic Polymerization of Polyester

Scheme 4.6 Porcine pancreatic lipase-catalyzed synthesis of enantio-enriched polyester with epoxy groups in the main chain.

showed a higher chain length selectivity than Novozym 435 (see also Section 4.2.1.1). Novozym 435 was able to polymerize 1,8-octanediol with diacids with chain lengths of 6, 8, 9, 10, and 13 carbons while HIC-AO just polymerized diacids with chain lengths of 9, 10, and 13. Analog the authors could show that HIC-AO was just able to polymerize sebacic acid with diols with chain length of 6 and 8 carbons while Novozym 435 could polymerize diols with chain length of 3, 4, 5, 6 and 8 with sebacic acid.

Wallace and Morrow used halogenated alcohols, such as 2,2,2-trichloroethyl, to activate the acyl donor and thereby improve the polymerization kinetics [53, 56]. They also removed by-products periodically during reactions to further shift the equilibrium toward chain growth instead of chain degradation. They copolymerized bis(2,2,2-trichloroethyl) *trans*-3,4-epoxyadipate and 1,4-butanediol using porcine pancreatic lipase as the catalyst. After 5 days, an enantioenriched polyester with $\overline{M_w} = 7900$ g mol^{-1} and an optical purity in excess of 95% was formed (Scheme 4.6).

The synthesis of aliphatic poly(carbonate-*co*-ester)s with about 1:1 molar ratio of the ester-to-carbonate repeat units was reported by CALB-catalyzed transesterification among diethyl carbonate, a diester, and a diol. Molecular weight $\overline{M_w}$ values reached 59 000 at a reaction temperature of 90 °C. A carbonate-ester transesterification reaction between poly(butylene carbonate) and poly(butylene succinate) was also catalyzed by CALB at 95 °C to result in a block copolymer [57].

Linear unsaturated and epoxidized polyesters via enzymatic polymerization were reported as well [58]. For this long-chain symmetrically unsaturated α,ω-dicarboxylic acid dimethyl esters (C18, C20, C26) were synthesized using metathesis techniques from 9-decanoic, 10-undecanoic, and 13-tetradecanoic acid methyl esters, respectively. The dicarboxylic acid dimethyl esters were epoxidized via chemoenzymatic oxidation with hydrogen peroxide/methyl acetate and Novozym

435 as catalyst. Polycondensation of these dimethyl esters with diols, catalyzed by Novozym 435, resulted in linear unsaturated and epoxidized polyesters. When using propane-1,3-diol, product molecular weights ranged from 1950–3300. Increasing the diol chain length to butane-1,4-diol resulted in polyesters with correspondingly higher molecular weights (7900–11 600).

Hilker et al. [59] studied the Novozym 435-catalyzed copolymerization of racemic α,α′-dimethyl-1,4-benzenedimethanol with secondary hydroxyl groups with dimethyl adipate. Due to CALB enantioselectivity, hydroxyl groups at (R) stereocenters preferably reacted to form ester bonds with liberation of methanol. The reactivity ratio was estimated as (R)/(S) = ≈ 1×10^6. In situ racemization of monomer stereocenters from (S) to (R) by ruthenium catalysis allowed the polymerization to proceed and reach high functional group conversions. Readers should also refer to Chapter 11 for more information on chiral discriminations by lipases.

Enzyme regioselectivity also enables the conversion of multifunctional monomers (functionality ≥3) to linear or nearly linear homo- and copolymers. In 1991, Dordick and co-workers [60] reported that, by using the protease Proleather, condensation polymerizations (45 °C, 5 days) performed in pyridine between sucrose and bis(2,2,2-trifluoroethyl) sebacate proceed with high regioselectivity giving sucrose oligoesters (DP 11) in 20% yield (see also Chapter 1). This inspired subsequent work by others that demonstrated such copolymerizations with polar multifunctional polyols could be performed under bulk reaction conditions without activation of carboxylic acids (see below).

To avoid the use of deactivating polar–aprotic solvents, polyols were combined with monomers to form monophasic liquids at temperatures sufficiently low to maintain immobilized CALB activity (≤95 °C). For example, Novozym 435-catalyzed bulk polycondensations were carried out at 70 °C under vacuum (40–60 mmHg) using adipic acid (A), 1,8-octanediol (O), and glycerol (G) (see also Chapter 1) as comonomers (monomer feed ratio, A:O:G, 1.0:0.8:0.2 mol/mol) [61] (Scheme 4.7). Initially, the reaction media was a two-phase liquid but within 60 min became monophasic with suspended Novozym 435. Products at 45 min and 2 h had little or no unreacted monomers, a $\overline{M_n}$ of 2250 and 2700, respectively, and a $\overline{M_w}/\overline{M_n}$ of 1.2 and 1.6, respectively. Extension of polycondensations from 2 to 6 and 18 h resulted in: (i) substantial increases in $\overline{M_n}$ and broadening of the molecular weight distribution. Furthermore, CALB's regioselectivity circumvented branching (i.e., gave linear polymers) during chain formation for polymerizations up to 18 h. However, as the reaction time was extended towards 42 h, products formed became increasingly branched as reactions moved from kinetic to thermodynamic control. Thus, at 42 h, a hyperbranched polymer with 19 mol-% dendritic glycerol repeat units was obtained in 90% yield with $\overline{M_w}$ and $\overline{M_w}/\overline{M_n}$ of 75 600 and 3.1, respectively (by SEC-MALLS). Even with branching, the product remained soluble in many organic media.

In another study by Kulshresta et al. [62], Novozym 435-catalyzed terpolymerizations of trimethylolpropane (TMP), 1,8-octanediol, and adipic acid were performed in bulk, at 70 °C, for 42 h, under vacuum (40 to 60 mmHg). Variation of

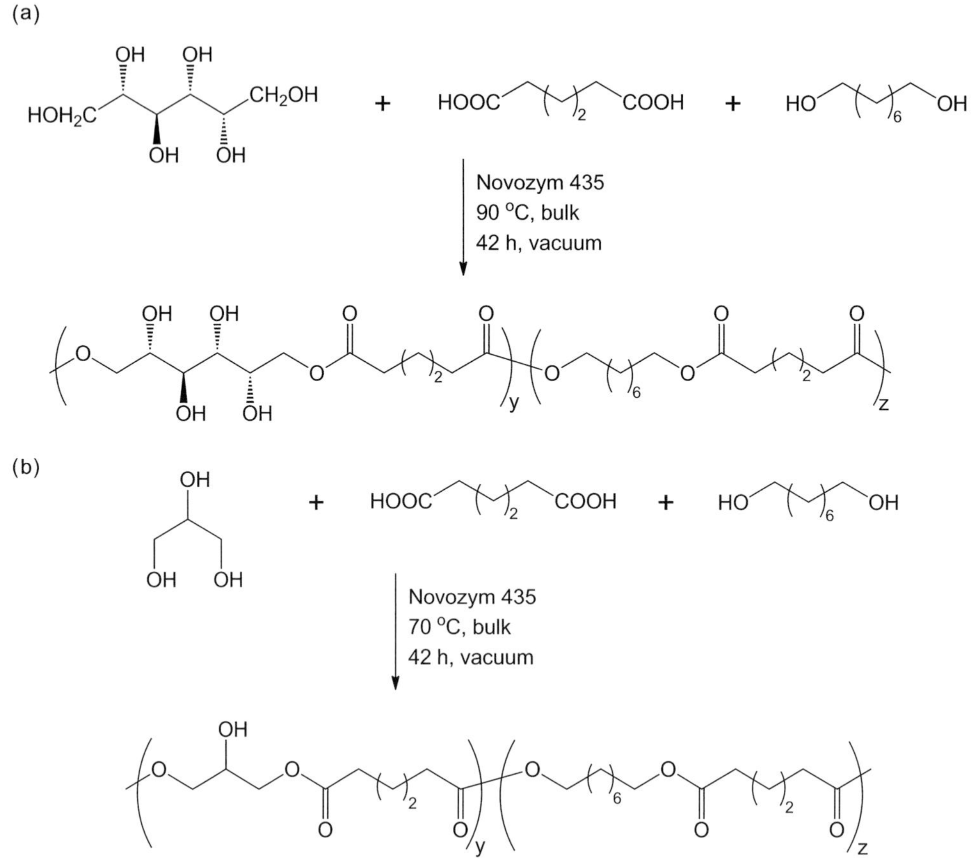

Scheme 4.7 Lipase-catalyzed polymerization of (a) sorbitol and (b) glycerol to form terpolyesters [41].

TMP in the monomer feed gave copolymers with degrees of branching (DB) from 20 to 67%. In one example, a hyperbranched copolyester with 53 mol % TMP-adipate units was formed in 80% yield, $\overline{M_w}$ 14 100, $\overline{M_w}/\overline{M_n}$ 5.3, and degree of branching 36%. As above, steric constraints imposed by CALB result in the formation of soluble branched polyesters. Chemical polymerizations with multifunctional monomers such as glycerol or TMP are plagued with formation of insoluble gels when reaction conditions are not strictly controlled.

Hu et al. [63] used the differential selectivities of CALB with various alditols to 'tune' polyol–polyester branching and, therefore, polymer properties (e.g., viscosity). Thus, CALB-catalyzed bulk terpolymerizations of adipic acid, 1,8-octanediol and a series of alditols (erythritol, xylitol, ribitol, D-glucitol, D-mannitol, and D-galactitol) was studied. Surprisingly, all substrates polymerized forming polyol–polyesters with $\overline{M_w}$ values ranging from 11 K (galactitol) to 73 K (D-mannitol). There was no correlation between sugar reactivity and its chain length. Comparison of exponent *a* values from slopes of log[η] vs log $\overline{M_w}$ showed that copolymers from D-mannitol had the largest degree of branching and, therefore, the greatest propensity for combined reactivity at both primary and secondary hydroxyl groups. Explanations for this difference in reactivity between sugars were proposed by the authors although it was acknowledged that additional experiments with an expanded set of alditol substrates will be needed to reach definitive conclusions.

Malic acid (MA) is a natural AB_2 monomer used to prepare functional polyesters. By chemically-catalyzed ROP, (R,S)-β-benzyl malolactonate has been homo- and copolymerized to prepare malic acid-containing materials [64]. However, protection–deprotection steps involved are tedious. Enzyme regioselectivity offers the potential to develop simple and direct routes to prepare malic acid copolymers. Li *et al.* [64] investigated Novozym 435-catalyzed copolymerization of adipic acid, 1,8-octanediol and L-MA. Reactions were conducted with 20 %-by-wt (relative to monomer) Novozym 435 for 48 h in bulk under reduced pressure (20–40 mm Hg). By using 20 mol% L-MA in the monomer feed at 80 °C a copolyester was formed in 91% yield with $\overline{M_w}$ and $\overline{M_w}/\overline{M_n}$ of 7400 and 1.8, respectively. Most importantly, NMR studies revealed that Novozym 435 was strictly selective for esterification of L-MA carboxylic groups leaving hydroxyl pendant groups unchanged.

In addition copolymers of octanediol adipate and sorbitol adipate, P(OA-*co*-SA), copolymers of octanediol adipate and glycerol adipate, P(OA-*co*-GA), poly(octamethylene adipate), (POA), and poly(sorbitol adipate), (PSorA), were synthesized using Novozym 435 as catalyst [47].

Recently, whole-cell biotransformations (see also Chapter 10) catalyzed by *C. tropicalis* ATCC20962 and other engineered yeast strains were used to synthesize biobased ω-carboxy fatty acid monomers (see also Chapter 1) [37]. For example, bioconversions of oleic, erucic and epoxy stearic acids by *C. tropicalis* ATCC20962 in shake flasks gave 1,18-*cis*-9-octadecenedioic, 1,22-*cis*-9-docosenedioic, and 1,18-*cis*-9,10-epoxy-octadecanedioic with volumetric yields of 17.3, 14.2, and 19.1 g l^{-1}, respectively. Polycondensations with diols were performed in bulk as well as in diphenyl ether (Scheme 4.8). Linear unsaturated and epoxidized polyesters with high $\overline{M_w}$ values (25 000 to 57 000 g mol^{-1}) and low melting points (23–40 °C) were synthesized. In contrast, when saturated polyesters were synthesized by polymerization of ω-carboxyl stearic acid and 1,8-octanediol, the corresponding polymeric materials melted at relatively higher temperature (77–88 °C). Increasing the chain length of diols resulted in higher molecular weights and melting points of unsaturated polyesters from ω-carboxyl OA.

A compilation of enzyme-catalyzed AA-BB type condensation polymerizations that have thus far appeared in the literature is given in Table 4.1.

4.3.3
Use of Activated Enol Esters for in vitro Polyester Synthesis

Much of the earlier work carried out on lipase-catalyzed condensation polymerizations focused on the use of activated diacids such as enol esters (see also Table 4.1). This was due to the belief that such activation was necessary to achieve

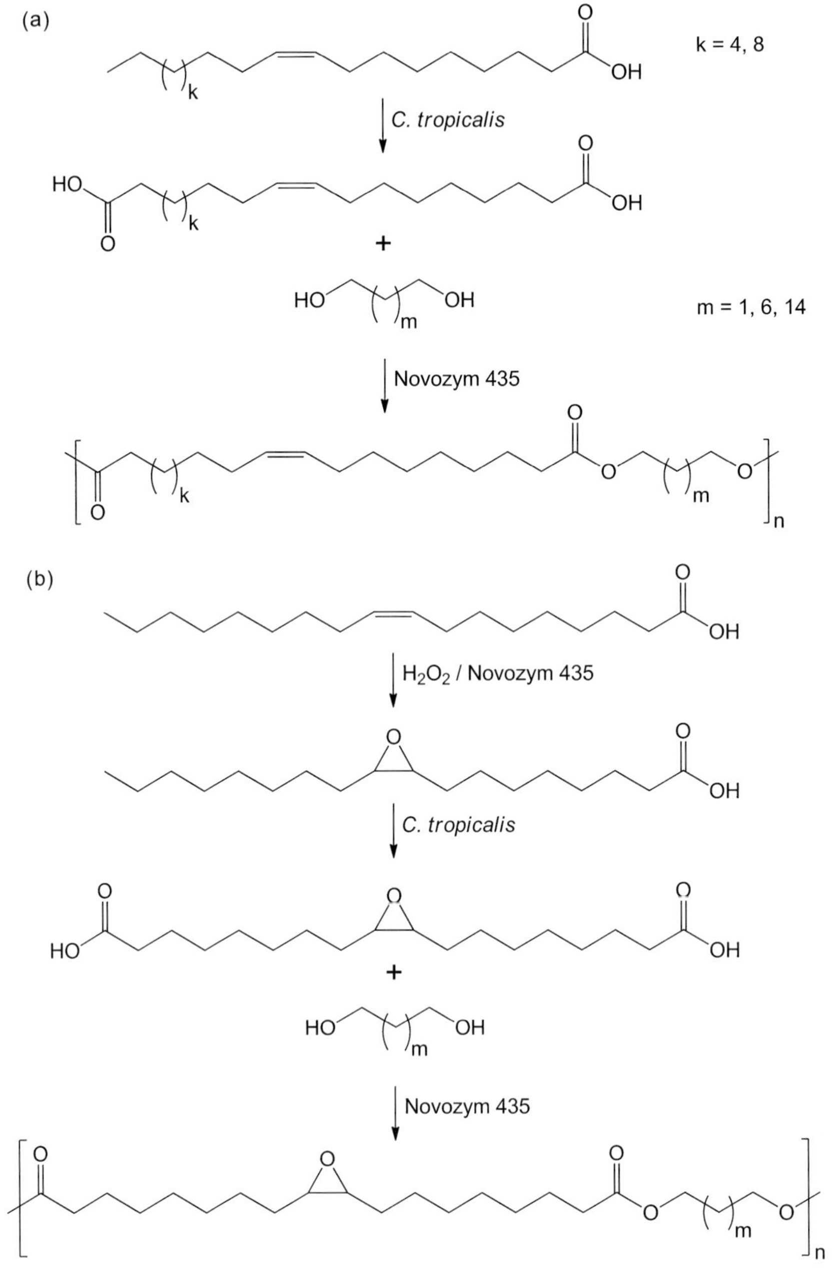

Scheme 4.8 Lipase-catalyzed polycondensation of unsaturated (a) and epoxidized (b) dicarboxylic acids with diols.

suitably high polymer molecular weights. Enol or vinyl esters both accelerate the rate of acyl transfer and shift the reaction equilibrium towards polymer synthesis since the by-product rapidly tautomarizes to a ketone or aldehyde. Acyl transfer using enol esters has been shown to be about 10 times slower than hydrolysis and about 10 to 100 times faster than acyl transfer using other activated esters [65]. For example, enzymatic hydrolysis of non-activated esters such as ethyl esters reacts at rates about 10^{-3} to 10^{-4} times slower than the corresponding enol esters. Uyama et al. used Pseudomonas fluorescens lipase to investigate polymerizations of divinyl adipate with different diols (ethylene glycol, 1,4-butanediol, 1,6-hexanediol, 1,10-decanediol) [54]. These polymerizations were performed in isopropyl ether for 48 h at 45 °C. The polymerization of divinyladipate and 1,4-butanediol gave poly(butyl adipate) in 50% yield with $\overline{M_w}$ 6700 and $\overline{M_w}/\overline{M_n}$ 1.9. The Pseudomonas fluorescens-catalyzed polymerization of divinyl adipate with ethylene glycol, 1,6-hexane diol or 1,10-decanediol as glycols resulted in polyesters of lower $\overline{M_w}$ values (2000, 5900 and 2700, respectively) than the polyester obtained from the polymerization of divinyladipate and butanediol. Uyama et al. [52] extended the above concept to find conditions suitable for enzymatic polymerization of divinyl isophathate, terephthalate, p-phenylene diacetate, and sebacate diesters. Lipases evaluated included those from Candida antarctica (CA), Candida cylinderacea, Mucor meihei, Pseudomonas cepacia, Pseudomonas fluorescens and porcine pancreas. Breadth of reaction conditions explored included temperatures from 45 to 75 °C, solvents of varying polarity (heptane, acetonitrile, cyclohexane, isooctane, tetrahydrofuran, and toluene) and chain length of α,ω-alkylene glycols (Scheme 4.9). Of the lipases studied, lipase CALB gave polyesters of highest molecular weights. Also, non-polar solvents such as heptane and cyclohexane were preferred. Furthermore, the maximum yields and product molecular weights were obtained at 60 °C. For example, the lipase CALB-catalyzed polymerization of divinyl isophthalate and 1,6-hexanediol in heptane at 60 °C resulted in polyester formation in 74% yield with $\overline{M_n}$ and $\overline{M_w}/\overline{M_n}$ of 5500 and 1.6, respectively in 48 h.

In addition the Novozym 435-catalyzed bulk polymerization of divinyl adipate and 1,4-butanediol was reported [44]. The highest $\overline{M_w} = 23\,236$ in 98.3% yield was obtained after 72 h polymerization at 50 °C. It was found that the product molecular weight was decreased when the reaction was conducted without taking proper precautions to exclude water in reactions that can hydrolyze reactive divinyl ester groups. They also found excellent agreement between their experimental data and that predicted by a mathematical model [66].

Russell et al. [38] also studied Novozym 435-catalyzed A-A/B-B type condensation polymerizations to prepare aromatic polyesters and polycarbonates. Polymerizations between divinylesters or dicarbonates with aromatic diols, conducted for 24 h in bulk catalyzed by Novozym 435 (10 wt %) at preferably 70 °C, gave low molecular weight polycarbonates and polyesters. The aromatic diols included 1,2-benzenedimethanol, 1,3-benzenedimethanol, 1,4-benzenedimethanol, 2,6-pyridinedimethanol or 4,4-isopropylidenebis(2-(2,6-dibromophenoxy)ethanol) and bisphenol-A. The $\overline{M_w}$ of polycarbonates and polyesters did not exceed 5200 and 3500 (yields <35%), respectively. When various isomers of benzenedimethanol

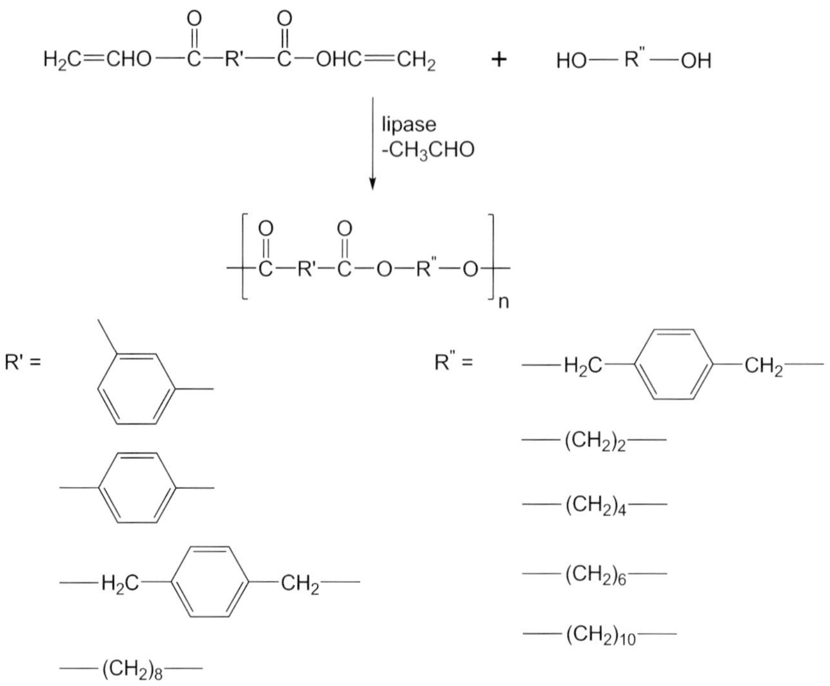

Scheme 4.9 Lipase-catalyzed condensation polymerization of various divinyl esters with diols of varying length.

were used, Novozym 435 exhibited regioselectivity, as *p*-benzenedimethanol reacted to a greater extent than the corresponding *m*- or *o*-isomers. The regioselectivity of lipases can thereby be exploited to preferentially polymerize selected isomers from complex mixtures (see also Chapter 11).

By using unsaturated fatty acids as substrates and enzyme-catalysis, Uyama *et al.* [45] prepared polyesters with epoxy-containing pendant groups. One route to these polyesters was to first copolymerize divinyl sebacate, glycerol, and unsaturated fatty acids followed by epoxidation of unsaturated groups in side chains. An alternative route was to first epoxidize unsaturated fatty acids using hydrogen peroxide in the presence of a lipase catalyst, and subsequently the epoxidized fatty acids were polymerized with divinyl sebacate and glycerol (Scheme 4.10).

Russell *et al.* [43] studied lipase-catalyzed polymerizations of activated diesters and fluorinated diols. The effects of reaction time, continuous enzyme addition, enzyme concentration, and diol chain length were studied to determine factors that might limit chain growth. Potential limiting factors considered were enzyme inactivation, enzyme specificity, reaction thermodynamics, hydrolysis of activated esters and polymer precipitation. The polymer molecular weight at 50 °C steadily increased and then leveled off after 30 h at $\overline{M_w} \sim 1773$.

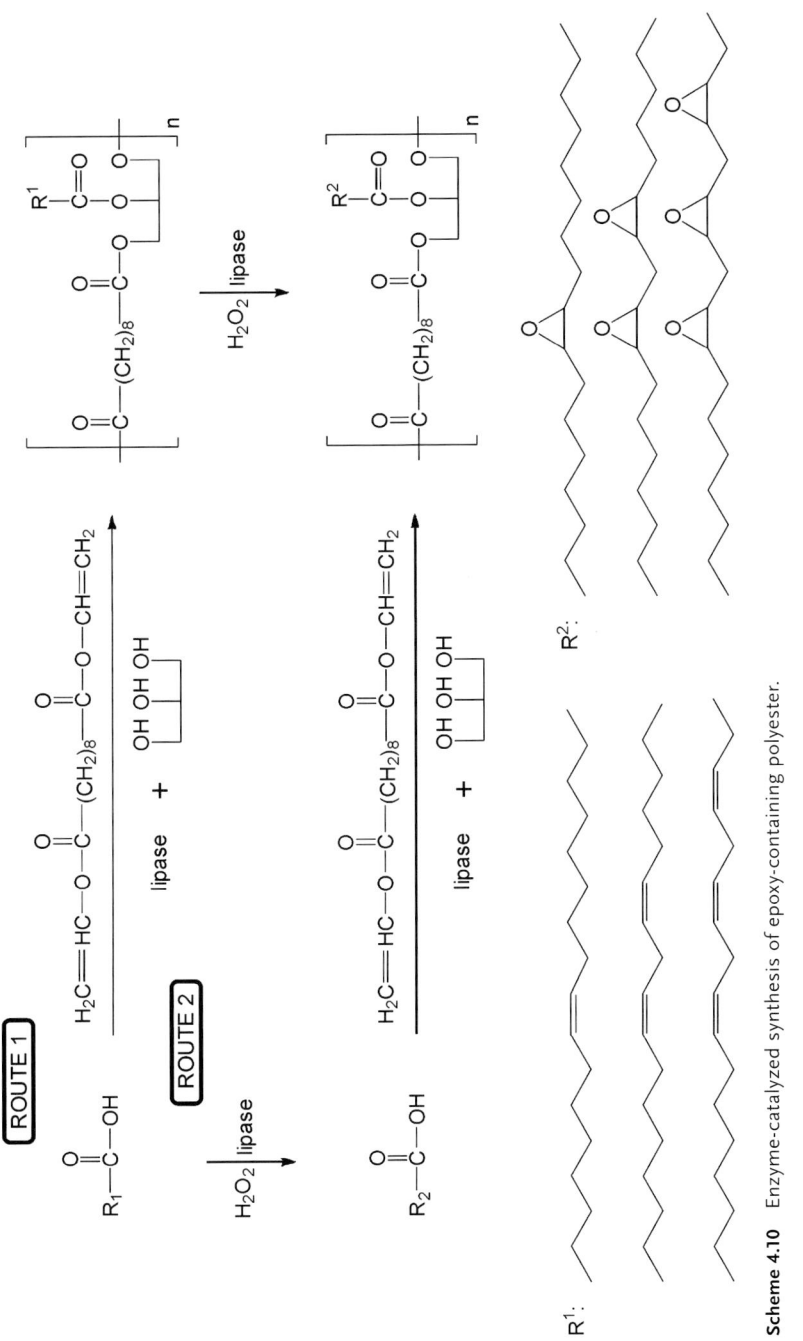

Scheme 4.10 Enzyme-catalyzed synthesis of epoxy-containing polyester.

Enzyme specificity towards shorter chain fluorinated diols appeared to be a prominent factor that limited chain growth. An increase in the product molecular weight resulted when the fluorinated diol contained an additional CH_2 spacer between CF_2 and hydroxyl groups. For example, no polymer was produced for reactions between divinyl adipate and either $HOCH_2(CF_2)_7CH_2OH$ and $HOCH_2(CF_2)_{10}CH_2OH$. Fluorinated oligomers of $\overline{M_w} \sim 1000$ were produced for copolymerizations between divinyl adipate and either $HOCH_2(CF_2)_3CH_2OH$ or $HOCH_2(CF_2)_2CH_2OH$. However, by incorporation of an additional methylene spacer between the fluorinated segment and -CH_2OH, for example, by using $HOCH_2CH_2(CF_2)_4CH_2CH_2OH$ as the diol, chains of relatively high molecular weight ($\overline{M_w} = 8094$ under solvent-free conditions) were obtained [43].

Similar to enol esters, oximes can also be used as irreversible acyl transfer agents for lipase catalysis. Thus, instead of a di-enol ester, Athavale et al. [67] polymerized diols with bis(2,3-butane dione monoxime) alkanedioate using Lipozyme IM-20. The results obtained by activation with enol-esters and their corresponding oximes were comparable. No attempts were made to analyze the end-group of the polyester.

It is important to note that processes using activated monomers such as those above would add substantial cost to corresponding products and, therefore, can only be justified for academic research or to prepare high-value specialty materials.

4.4
Enzyme-Catalyzed Ring-Opening Polymerizations

4.4.1
Unsubstituted Lactones

In contrast to condensation polymerizations, ring-opening polymerizations of lactones and carbonates do not generate a leaving group during the course of the reaction (Scheme 4.11) [68–70]. This characteristic alleviates concerns that the leaving group, if not efficiently removed, might limit monomer conversion or polymer molecular weight [71].

In lipase-catalyzed ROP, it is generally accepted that the lactone activation proceeds via the formation of an acyl-enzyme intermediate by reaction of the serine residue with the lactone rendering the carbonyl prone to nucleophilic attack (Scheme 4.12) [68, 69, 72]. See also Chapter 14 for more details on the reaction

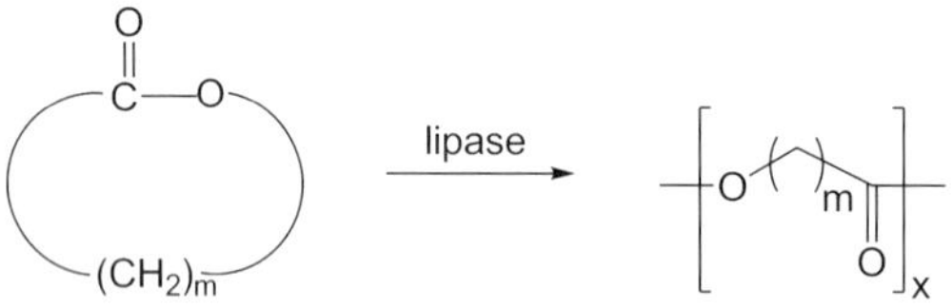

Scheme 4.11 Ring-opening polymerization of unsubstituted lactones.

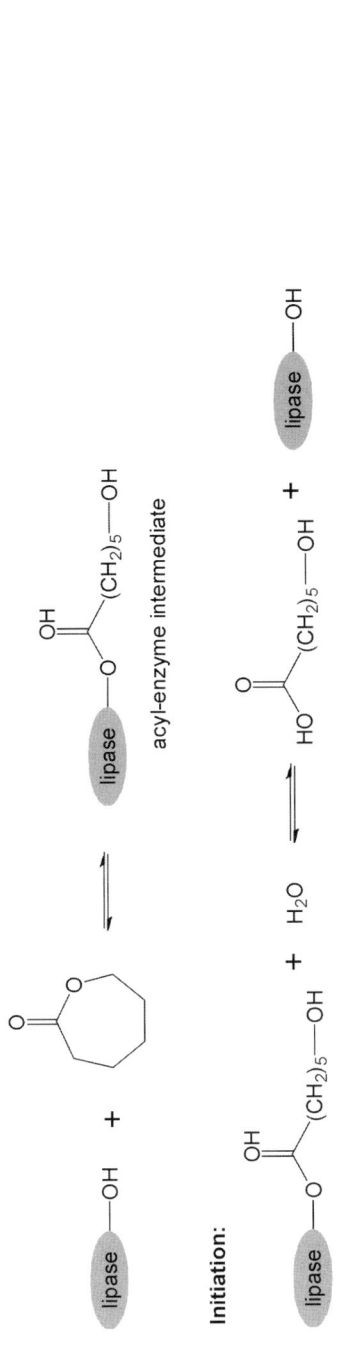

Scheme 4.12 Mechanism for lipase-catalyzed ROP of lactones.

mechanism of lipases. Initiation of the polymerization occurs by deacylation of the acyl-enzyme intermediate by an appropriate nucleophile (water, alcohol, etc.) to produce the corresponding ω-hydroxycarboxylic acid/ester. Propagation occurs by deacylation of the acyl-enzyme intermediate by the terminal hydroxyl group of the growing polymer chain to produce a one unit elongated polymer chain. Careful mechanistic investigations revealed that, the formation of the acyl-enzyme intermediate is the rate-determining step in enzymatic ROP [68, 69].

The first enzyme-catalyzed ROP was found for ε-CL and δ-VL by two independent groups in 1993 [12, 13]. ROP of ε-CL was performed in bulk at 75 °C for 10 days using *Pseudomonas fluorescens* as a catalyst, and poly(ε-CL) was formed in 92% yield with molecular weight $\overline{M_n}$ of 7700 and $\overline{M_w}/\overline{M_n} = 2.4$. Similarly, poly(δ-VL) was obtained at 60 °C with $\overline{M_n}$ of 1900 and $\overline{M_w}/\overline{M_n} = 3.0$. Further investigation of obtained polyesters (terminal structure of a carboxylic acid group at one end and a hydroxyl group at the other) revealed that ROP was initiated by water.

Since then, research has focused on variables that include enzyme used, substrate selectivity and reaction conditions. Examples of published work looking at one or more of these variable includes the following studies with different monomer substrates: ω-dodecanolactone (DDL) [34, 72, 73], ω-pentadecanolactone (PDL) [73–77], β-propiolactone [78, 79], 8-octanolide (8-OL) [80], undecanolactone (UDL) [34, 72, 73, 75, 81], hexadecanolactone (HDL) [72], γ-butyrolactone [78], and others (see also Table 4.2).

Table 4.2 Lipase-catalyzed ring-opening polymerizations.

Enzyme	Monomer	Reference
Aspergillus niger lipase A	ε-CL	[34]
	DDL	[34]
	PDL	[76]
Candida antartica lipase B	5-methyl-5-benzyloxycarbonyl-1,3-dioxan-2-one	[89]
	ε-CL	[83, 90]
	8-OL	[34, 80]
	1,4-dioxane-2-one	[91]
	1,5-dioxepan-2-one	[92]
	2-methylene-4-oxa-12-dodecanolide	[93, 94]
	2-methylene-8-phenyl-4-oxa-8-aza-11-undecanolide	[94]
	TMC	[95]
	PDL	[76]
	5-methyl-5-benzyloxycarbonyl-1,3-dioxan-2-one	[89]
	α-Me-γ-VL	[96]
	α-Me-ε-CL	[96]
	4-methyl-ε-caprolactone	[97]
	4-ethyl-ε-caprolactone	[97]
	4-propyl-ε-caprolactone	[97]

Table 4.2 *Continued*

Enzyme	Monomer	Reference
	α-methyl-β-propiolactone	[98]
	α-methyl-γ-butyrolactone	[98]
	α-methyl-δ-valerolactone	[98]
	α-methyl-ε-caprolactone	[98]
	α-methyl-ζ-heptalactone	[98]
	α-methyl-8-octanolide	[98]
	α-methyl-dodecanolactone	[98]
	monomer from L-tartaric acid	[99]
Candida cylinderacea lipase	8-OL	[80]
	TMC	[95, 100]
	α-Me-β-PL	[101]
	ε-CL	[72, 86]
	δ-VL	[34]
	PDL	[34, 72, 73]
	β-PL	[79]
	DDL	[72, 72]
	UDL	[34, 72]
Candida rugosa lipase	5methyl-5-benzyloxycarbonyl-1,3-dioxan-2-one	[89]
	PDL	[102]
Humicola insolens cutinase (HiC)	ε-CL	[87]
	PDL	[87]
Mucor Javanicus lipase	β-BL (R,RS)	[73]
	TMC	[76]
Mucor meihei	PDL	[73]
	TMC	[95, 100]
	5methyl-5-benzyloxycarbonyl-1,3-dioxan-2-one	[89]
Pencillium rorueforti lipase	PDL	[73]
Porcine pancreatic lipase	DDL	[34]
	ε-CL	[34, 67, 69, 72, 86]
	γ-VL	[34, 86]
	TMC	[95, 100]
	5-methyl-5-benzyloxycarbonyl-1,3-dioxan-2-one	[89]
	β-BL	[78]
	α-Me-β-PL	[101]
	PDL	[73]
	3(S)-isopropylmorpholine-2,5-dione	[103]
Pseudomonas aeruginosa lipase	ε-CL	[34]
	DDL	[34]
	S-MOHEL	[34]
	8-OL	[80]

Table 4.2 Continued

Enzyme	Monomer	Reference
Pseudomonas cepacia lipase	δ-DDL	[34]
	β-BL	[78]
	ε-CL	[34, 72]
	TMC	[95, 100]
	PDL	[73, 76]
	5-methyl-5-benzyloxycarbonyl-1,3-dioxan-2-one	[89]
	8-OL	[34, 80]
	DDL	[34]
	MOHELs	[34]
	HDL	[72]
	α-Me-β-PL	[71]
Pseudomonas fluorescens lipase	ε-CL	[29, 34, 72, 73, 86]
	δ-VL	[34, 86]
	S-MOHEL	[34]
	UDL	[34, 73]
	DDL	[34, 72, 73]
	PDL	[34, 72, 73]
	HDL	[72]
	8-OL	[34, 80]
	TMC	[95]
Rhizopus delemer lipase	ε-CL	[87]
	PDL	[73]
Rhizopus japonicus lipase	ε-CL	[72, 86]
	γ-VL	[86]
	PDL	[73]
HE	PDL	[73]
PD	PDL	[73]
PR	PDL	[73]
CR	PDL	[73]
Pseudomonas sp. lipase	ε-CL, β-BL, γ-BL, δ-DCL, δ-DDL, PDL	[27]
	ethyl 4-hydroxybutyrate	[27]
	ethyl-6-hydroxyhexanoate	[27]
	ethyl-3-hydroxybutyrate	[27]
	ethyl 5-hydroxyhexanoate	[27]
	ethyl 5-hydroxylaurate	[27]
	ethyl 15-hydroxypentadecanoate	[27]

The largest linear aliphatic unsubstituted lactone monomer thus far studied for enzymatic ROP is HDL (17-membered) [72]. ROP of HDL was performed in bulk, using various lipases, at 60 and 75 °C for 120 h, giving rise to poly(HDL). Using *Pseudomonas cepacia* lipase as catalyst resulted in a polyester with $\overline{M_n}$ reaching to 5800 ($\overline{M_w}/\overline{M_n}$ = 2.0) in quantitative yields.

However, for model studies, ε-CL has been the most commonly selected of the lactone monomers [12, 13, 23, 34, 68–70, 72, 73, 77, 82–86]. General difficulties included low product molecular weights and slow polymerization kinetics. In seeking to overcome these, it was found, for instance, that polymerization rate, molecular weight and polydispersity of poly(ε-CL) can be controlled by different combinations of reaction parameters (temperature, solvent, monomer concentration, enzyme concentration, water content, etc.) [70].

Apart from CALB other enzymes were shown to be able to successfully polymerize lactones. For instance Lipase PS-30, immobilized on Celite, was used as catalyst to study PDL-ROP under bulk reaction conditions. Poly(PDL) with $\overline{M_n} = 62\,000$ and PDI 1.9 was reported [76]. Gross and coworkers could show that *Humicola insolens* cutinase (HiC) showed a high catalytic activity for enzymatic ROP of ε-CL and PDL [87]. Poly(ε-CL) with $\overline{M_n} = 16\,000$ ($\overline{M_w}/\overline{M_n} = 3.1$), in >99% yields was produced in bulk (70 °C, 24 h) with 0.1% w/w immobilized HiC. Furthermore, using immobilized HiC in toluene (70 °C, 24 h), PDL was converted to poly(PDL) (99% yield) with $\overline{M_n} = 44\,600$ and $\overline{M_w}/\overline{M_n} = 1.7$.

Kobayashi *et al.* systematically investigated enzyme-catalyzed ROP of δ-VL, ε-CL, UDL, DDL and PDL (6-, 7-, 12-, 13- and 16-membered lactones) [16]. Catalytic activities of lipases of different origin (*Aspergillus niger* lipase A, *Candida cylindracea* lipase, *Candida rugosa* lipase, *Rhizopus delmar*, *Rhizopus javanicus*, *Pseudomonas fluorescens*, phospholipase, porcine pancreas lipase, *Penicillium roqueforti* lipase, *Rhizopus japanicus* lipase and hog liver) for ROP of lactone monomers were screened and selected results from this work are listed in Table 4.3. While these results provide a quantitative understanding of relative enzyme activities, it should be noted that enzyme catalysts were in different forms (powders, immobilized on solid supports of varying types), purities and have

Table 4.3 Enzyme screening in the lipase-catalyzed ROP of ε-CL, δ-VL and PDL. Data reported are from reference [16].

Enzyme	Monomer	Conversion	M_n	M_w / M_n
Candida cylindracea lipase	ε-CL	75	3300	2.5
Pseudomonas fluorescens	ε-CL	85	7000	2.2
Porcine pancreas lipase	ε-CL	69	2500	1.9
Pseudomonas fluorescens	δ-VL	95	1900	3.0
Aspergillus niger lipase A	PDL	16	2800	1.7
Candida cylindracea lipase	PDL	54	5800	2.5
Candida rugosa lipase	PDL	21	2500	1.6
Penicillium roqueforti lipase	PDL	12	3500	1.4
Pseudomonas fluorescens	PDL	97	2800	2.2
Pseudomonas cepacia lipase	PDL	90	2400	2.6
Rhizopus japanicus lipase	PDL	<5		
Porcine pancreas lipase	PDL	27	1800	1.7
Hog liver	PDL	<5		
None	PDL	0		

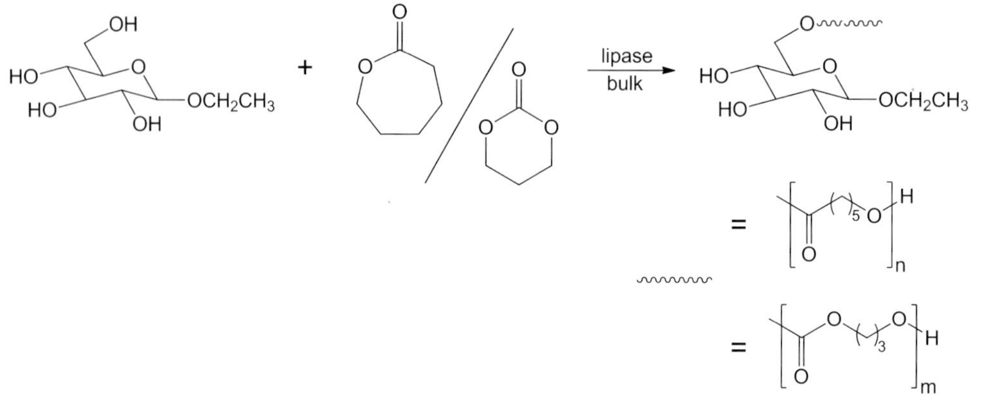

Scheme 4.13 EGP-initiated polymerizations of trimethylene carbonate (TMC) and ε-CL.

different water contents. Furthermore, precipitation of products leads to fractionation and relatively higher molecular weights than are actually formed. For all these reasons it is important for readers to use this information only as a quantitative guide.

In addition lipase-catalyzed ROP of lactones was successfully used to synthesize macromers by using hydroxyl moieties of carbohydrates as sites for initiation [68, 69, 88]. Specifically, ethylglucopyranoside (EGP) was used as a multifunctional initiator and ε-CL/trimethylene carbonate (TMC) as monomers for lipase-catalyzed ROPs. Initiation of ROP occurred selectively from the 6-hydroxyl position forming macromers with a carbohydrate head group with three remaining hydroxyl groups that remained available for other enzymatic or chemical transformations (Scheme 4.13).

Furthermore, Kobayashi and co-workers prepared macromers based on polyesters with methacryloyl end groups, using lipases from different origin [72]. This was accomplished by the polymerization of DDL in the presence of ethylene glycol methacrylate and vinyl methacrylate. The acryl-enzyme intermediate, formed by reaction of the lipase and the vinyl ester, reacted to terminate propagating chains.

Similarly, a telechelic polymer bearing carboxylic acid groups at both chain ends was formed by carrying out the lipase-catalyzed polymerization of DDL in the presence of divinyl sebacate [72]. In this case, divinyl sebacate functioned as a coupling agent creating poly(DDL) chains with hydroxyl groups at both termini.

Hult and co-workers performed a very tedious study on the synthesis of end-functionalized PCL macromers using Novozym 435 as catalyst [83]. Enzyme-catalyzed ROP of ε-CL was performed with addition of potential chain initiators [e.g., 9-decenol, 2-(3-hydroxyphenyl)-ethanol, 2-(4-hydroxyphenyl)ethanol, and cinnamyl alcohol] to the reactions. Alternatively, acids and esters containing

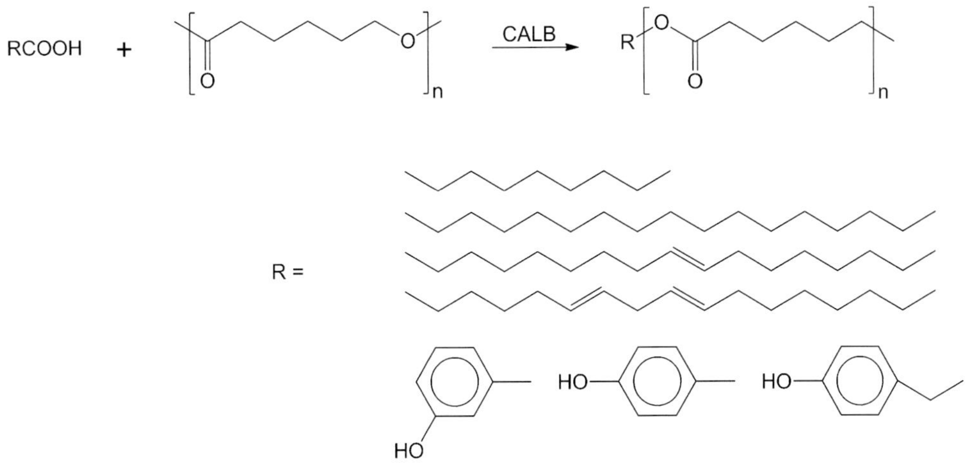

4.4 Enzyme-Catalyzed Ring-Opening Polymerizations

Scheme 4.14 Enzymatic synthesis of end-functionalized poly(caprolactone) monomers using carboxylic acid as chain-terminating group.

the target end-functionality (such as octadecanoic acid, oleic acid, linoleic acid, 2-(3-hydroxyphenyl)-acetic acid, 2-(4-hydroxyphenyl)acetic acid, and 3-(4-hydroxyphenyl)propanoic) acid) were added to prepolymerized ε-CL resulting in acid-terminated PCL (see Scheme 4.14) [83]. In an effort to simultaneously control both the hydroxyl and the carboxyl end groups of macromers initiation and termination was combined either by using a di-functionalized ester or by subsequent addition of initiator and terminator.

A compilation of the unsubstituted lactone polymers, the enzymes used, and the corresponding citation(s) is given in Table 4.2.

4.4.2
Substituted Lactones

Various substituted lactones were used for enzymatic polyester synthesis via ROP: (±)-α-methyl-β-propiolactone [101], β-methyl-β-propiolactone [78], α-decenyl-β-propiolactone [27], α-dodecenyl-β-propiolactone [46], benzyl-β-D,L-malonolactonate [104], α-methyl-ε-caprolactone [96], α-methyl-δ-valerolactone [96], 1,4-dioxane-2-one [91], and others (see also Table 4.2).

Van Buijtenen and coworkers [98] demonstrated Novozym 435-catalyzed ring-opening of a range of α–methylated lactones (α-methyl-β-propiolactone (3-MePL; 4-membered), α-methyl-γ-butyrolactone (4-MeBL; 5-membered), α-methyl-δ-valerolactone (5-MeVL; 6-membered), α-methyl-ε-caprolactone (6-MeCL; 7-membered), α-methyl-ζ-heptalactone (7-MeHL; 8-membered), α-methyl-8-octanolide (8-MeOL; 9-membered), α-methyl-dodecanolactone (12-MeDDL; 13-membered)), in toluene at 70 °C. Ring-opening of small lactones was found to be S-selective (3-MePL and 6-MeCL) or nonselective (5-MeVL). On the other hand, ring-opening of the larger lactones was found to be R-selective with very high enantioselectivity values. The authors reason that differential behaviors as a function of ring-size is due to the corresponding preferences of lactones of varying size to adopt cisoid or transoid conformations at their ester bond (Scheme 4.15). Namely, ester bonds of small lactones with ring sizes of 4 to 8 are more stable in their cisoid conformation and, therefore, are S-selective. In contrast, ester bonds of larger lactones with ring sizes of 9 and 13 are more stable in their transoid conformation and favor R-selectivity (this assumption was supported by molecular modeling studies). For intermediate ring sizes (7-MeHL and 8-MeOL), the significant presence of cisoid conformers does not appear to affect the enantioselectivity of corresponding ROPs. Poly-(R)-7-MeHL, poly-(R)-8-MeOL, and poly-(R)-12-MeDDL were all obtained with good molecular weights (between 14 200 and 16 700, >99% yields) and quite high ee (>99%). (see also Chapter 11)

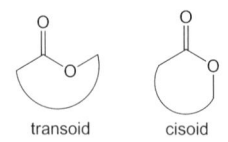

Scheme 4.15 Cisoid and transoid conformations of lactone ester bonds.

ROP of substituted 4-membered β-propiolactones, (β-PL), were reported using lipase-catalysis in bulk. α-methyl-β-PL gave a polymer with an analogous structure to poly(lactic acid) (PLA). *Pseudomonas fluorescent* lipase-catalyzed ROP of α-methyl-β-PL in toluene was found to be selective for (S)- α-methyl-β-PL giving (S) enriched poly(α-methyl-β-PL) with $\overline{M}_n$ ranging from 2000 to 2900 [101].

Peeters et al. [97] performed ROP of 4-substituted ε-CL employing Novozym 435 as the biocatalyst. The focus of their work was to establish the relationship between polymerization rate and substituent size (Scheme 4.16). The polymerization rate decreased by a factor of 2 by substitution at the 4-position of H with CH_3. Furthermore, 4-Et-CL and 4-Pr-CL polymerizes 5 and 70 times slower, respectively, than 4-Me-CL. Moreover, decrease in the polymerization rate is accompanied by a large decrease in enantioselectivity: while the E-ratio of 4-MeCL polymerization is 16.9, the E-ratios of 4-EtCL and 4-PrCL are 7.1 and 2.0, respec-

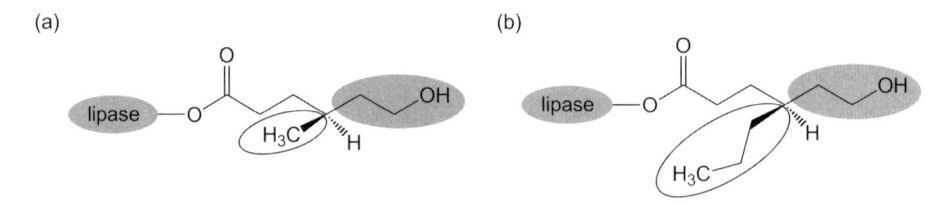

Scheme 4.16 Schematic representation of the acyl-enzyme intermediate of (a) 4-Me-CL and (b) 4-Pr-CL.

tively. In contrast, the rate of hydrolysis is only slightly affected by substituent size. Obtained results indicate that chirality of the propagating alcohol chain end is important in the catalytic cycle and that, in contrast to unsubstituted lactones, the rate-determining step is not necessarily formation of the acyl-enzyme intermediate, but, more likely, is the deacylation of the acyl-enzyme intermediate by the propagation alcohol chain end.

A compilation of polymers synthesized from substituted lactones monomers, the enzymes used, and the corresponding citation(s) is given in Table 4.2.

4.4.3
Cyclic Ester Related Monomers

In addition to (substituted) lactones various cyclic esters related monomers were polymerized via enzyme-catalyzed ROP.

For instance, in the last decade synthesis of poly(ester-*alt*-ether) was intensively studied. A common enzyme used in these syntheses is CALB. Polymerization of 1,5-dioxepan-2-one (DXO) was performed by enzyme-catalyzed ROP in order to avoid contamination of product polymers by toxic organometallic catalysts [92]. High molecular weight of poly(DXO) was obtained ($\overline{M_n} = 56000$; $\overline{M_w} = 112000$, 97% yield) at 60 °C for 4 h. The polymerization had the characteristics of a living polymerization, as indicated by the linearity of plots between $\overline{M_n}$ and monomer conversion, meaning that the product molecular weight could be controlled by the stoichiometry of the reactants. Similarly, Nishida et al. [91] carried out enzymatic ROP of 1,4-dioxan-2-one at 60 °C catalyzed by Novozym 435 that resulted in a polymer with $\overline{M_w} = 41000$ in 77% yield.

Enzymes have also been used to catalyze the ring-opening polymerization of cyclic carbonate monomers in order to synthesize polycarbonates [89, 95, 100, 105]. Lipases from *Candida antarctica*, porcine pancreas, *Pseudomonas cepacia* (PS-30), *Pseudomonas fluorescens*, *Candida cylindracea*, *Mucor miehei* (MAP), and *Rhizomucor miehei* (lipozyme-IM) were evaluated as catalysts for the bulk polymerization of trimethylene carbonate (TMC, 1,3-dioxan-2-one) [95]. Of these catalysts, immobilized CALB (Novozym 435) was found to be most effective. In one example, Novozym 435-catalyzed polymerization of TMC at 70 °C for 120 h gave 97% monomer conversion to poly(TMC) with $\overline{M_n} = 15000$, without decarboxylation during propagation [95]. Similarly Matsumura et al. [100] reported that poly(TMC) of extraordinarily high molecular weight ($\overline{M_w} = 156000$) was obtained by using low quantities of porcine pancreatic lipase (0.1 wt %) as the catalyst at very high reaction temperature (100 °C). In contrast to this, Kobayashi et al. [105] reported the formation of low molecular weight poly(TMC) (80% yield, $\overline{M_n} = 800$) using porcine pancreatic lipase (50 wt %) as catalyst at 75 °C for 72 h. The fact that TMC is known to thermally polymerize in the absence of a catalyst can possibly be used to explain the discrepancy in results.

The lipase-catalyzed polymerization of the disubstituted TMC analog 5-methyl-5-benzyloxycarbonyl-1,3-dioxan-2-one (MBC) was also studied [89]. The bulk polymerization, catalyzed by *Pseudomonas fluorescens* lipase for 72 h at 80 °C, gave 97% monomer conversion and product in 97% yield with $\overline{M_n} = 6100$. The benzyl ester protecting groups of poly(MBC) were removed with Pd/C in ethyl acetate to give the corresponding functional polycarbonate with pendant carboxylic acid groups.

Enantiomerically pure functional polycarbonate was synthesized from a novel seven-membered cyclic carbonate monomer derived from naturally occurring L-tartaric acid [99]. The ROP catalyzed by Novozym 435 was performed in bulk, at 80 °C, for 48 h to afford optically active polycarbonate with $\overline{M_n} = 15500$ g/mol and PDI 1.7 (Scheme 4.17). Hydroxy group functionality in the carbonate chain was achieved by deprotection of the ketal group. The polycarbonates have potential in biomedical applications.

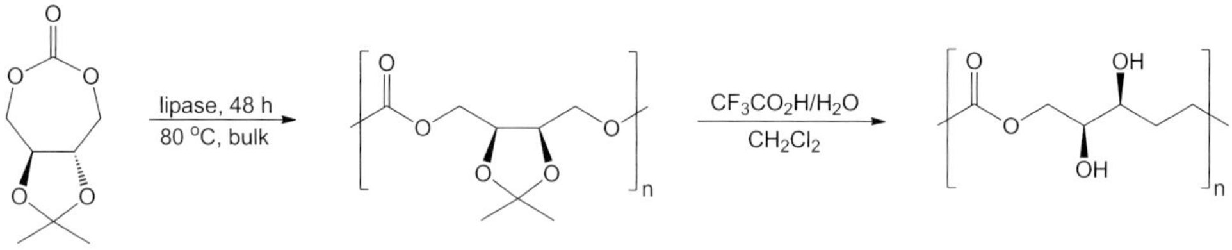

Scheme 4.17 Enzymatic polymerization of seven-membered cyclic carbonate monomer from L-tartaric acid.

The enantioselective ROP of 3-methyl-4-oxa-6-hexanolide (MOHEL) was catalyzed in bulk at 60 °C [34]. A comparison of the initial rate of poly(MOHEL) formation from the (R) and (S) antipodes showed that the (S) enantiomer had an initial rate that was seven times larger. Lipase from *Pseudomonas aeruginosa* and *Pseudomonas fluorescens* catalyzed the polymerization of (S)-MOHEL but not (R)-MOHEL (Scheme 4.18).

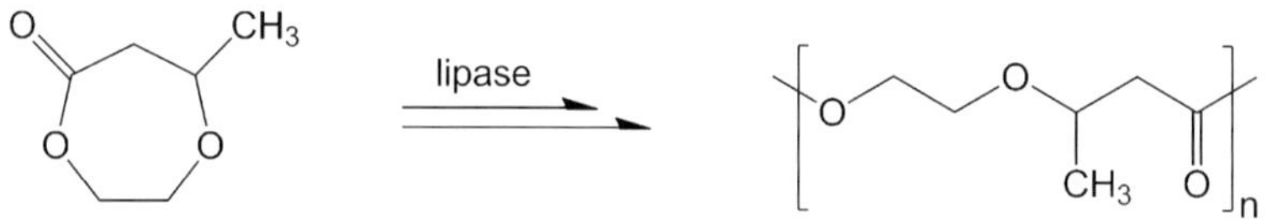

Scheme 4.18 Lipase-catalyzed ring-opening polymerization of MOHEL.

Enzymatic ROP of 2-oxo-12-crown-4-ether (OC) was studied by Meijer and coworkers [106]. OC is different from other ether containing lactone monomers previously studied as it combines high hydrophilicity with a large ring size. Using Novozym 435 as catalyst, at 60 °C for 90 min in a mixture of toluene and tri-*t*-butylbenzene, homopolymerization of OC was successfully accomplished giving poly(OC) in yields >95% with $\overline{M_n}$ and $\overline{M_w}/\overline{M_n}$ values of 3400 and 2.1, respectively.

Substituted oxo-crown-ethers were studied as the starting monomers for cross-linked polymers gels. For instance, the CALB-catalyzed ROP of 2-methylene-4-

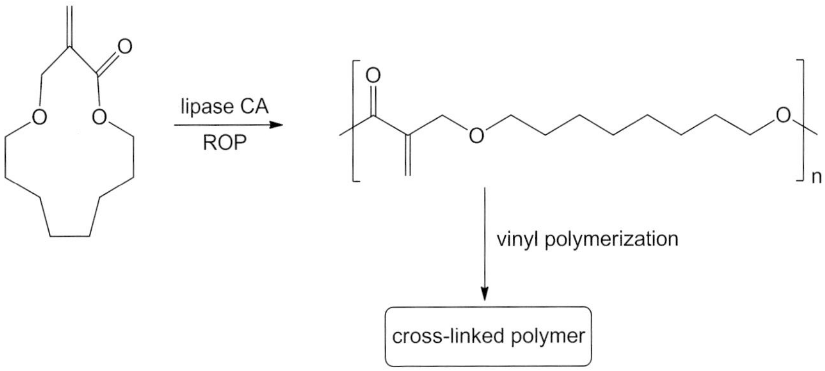

Scheme 4.19 Lipase-catalyzed ring-opening polymerization of 2-methylene-4-oxa-12-dodecanolide.

oxa-12-dodecanolide at 75 °C for 24 hours in toluene yielded a polyester with $\overline{M_n}$ and $\overline{M_w}/\overline{M_n}$ values of 8100 and 1.9, respectively, having reactive exo-methylene group in the main chain (Scheme 4.19). Obtained polyesters, containing vinylene groups, were postmodified by vinyl polymerization induced by anionic and radical initiators to give polymer gels [93, 94].

An ester-amide polymer was prepared by ROP, catalyzed by *Porcine pancreatic lipase* (5 wt %) of a six-membered cyclic depsipeptide, 3(S)-isopropylmorpholine-2,5-dione (IPDM), in bulk at 100 °C, with $\overline{M_n} = 17500$, $\overline{M_w} = 18500$, and 66% yield [103].

A compilation of cyclic ester related polymers, the enzymes used, and the corresponding citation(s) is given in Table 4.2.

4.5
Enzymatic Ring-Opening Copolymerizations

Copolymerization of lactones allows the tuning of polymer properties while introducing new challenges to enzyme-catalyzed ROP such as understanding relationships between comonomer reactivity ratios, transesterification and copolymer microstructure (Scheme 4.20).

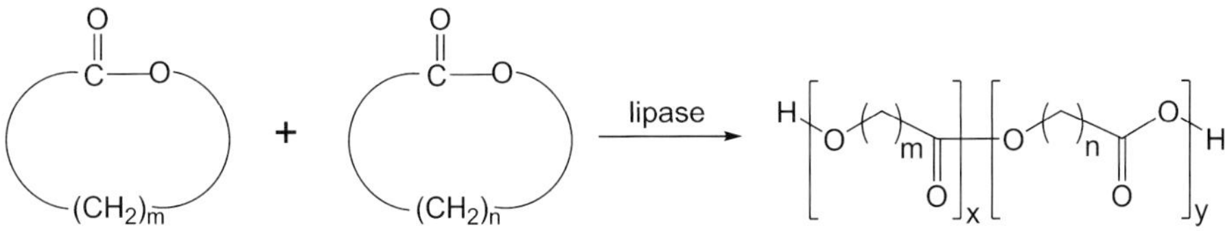

Scheme 4.20 Ring-opening copolymerization of lactone monomers.

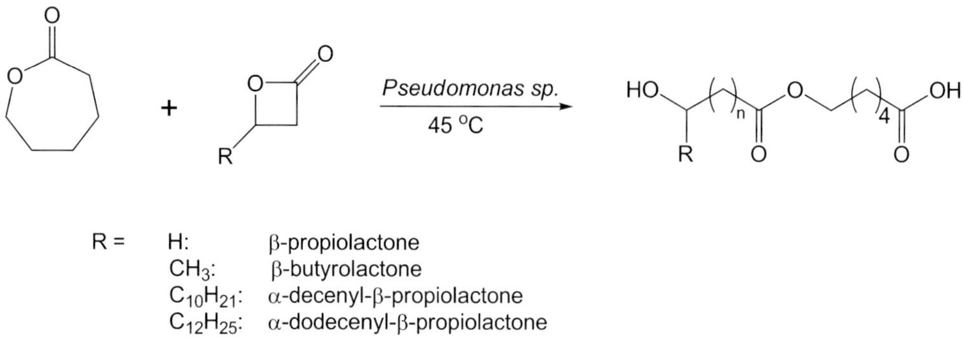

Scheme 4.21 Ring-opening copolymerization of ε-caprolactone and β-propiolactone and its derivative.

R =
- H: β-propiolactone
- CH$_3$: β-butyrolactone
- C$_{10}$H$_{21}$: α-decenyl-β-propiolactone
- C$_{12}$H$_{25}$: α-dodecenyl-β-propiolactone

Kobayashi and coworkers [107] first studied the enzyme-catalyzed copolymerization of β-propiolactone and ε-CL. Furthermore, ring-opening copolymerizations of PDL with δ-VL, ε-CL, DDL and UDL using the lipase from *Pseudomonas fluorescens*, in bulk at 60–75 °C for 240 h were performed [73]. Low molecular weight copolymers ($\overline{M_n}$ ranging 1200 to 6300) that tended to be block-like were obtained in >95% yields. By using Novozym 435 copolymerizations of ε-CL and PDL at 70 °C for 45 min were conducted by Gross and coworkers [77]. High yields (about 88%) and molecular weights ($\overline{M_n}$ about 20 000) were obtained. According to the calculations, the authors revealed that PDL polymerization is 13 times faster than that of ε-CL. Nevertheless, random copolymers were formed. This was attributed to the fact that, in addition to catalyzing chain propagation, Novozym 435 is also actively catalyzing polymer–polymer transacylation or transesterifications reactions.

Dong and coworkers [27] reported copolymerizations (bulk, 45 °C, 20 days) catalyzed by the lipase from *Pseudomonas sp.* (40 mg of lipase/0.1 mmol of monomer) of ε-CL with some cyclic and linear monomers. Among the copolymerizations performed, that of ε-CL with cyclopentadecanolide gave the highest product $\overline{M_n}$ (8400, yield 67%). The molecular weights of copolymers of ε-CL with lactones were higher than those of copolymers prepared from the corresponding linear hydroxyesters (Scheme 4.21).

Copolymerization of δ-VL with ε-CL using lipase from *Pseudomonas fluorescens*, and copolymerization of 8-OL with ε-CL and DDL using immobilized form of CALB, were reported by Kobayashi and coworkers [80]. In this later report, copolymerization was performed in isooctane at 60 °C for 48 h, and, according to ^{13}C NMR analysis, random-structured copolymers were obtained with $\overline{M_n}$ values up to 9000 and yields up to 97%.

Similarly, using Novozym 435 in toluene or diphenyl ether at 70 °C under nitrogen for 26 h, ω-PDL and *p*-dioxanone (DO) copolymerizations were carried out (Scheme 4.22), using various PDL/DO feed ratios, to give poly(PDL-*co*-DO) with

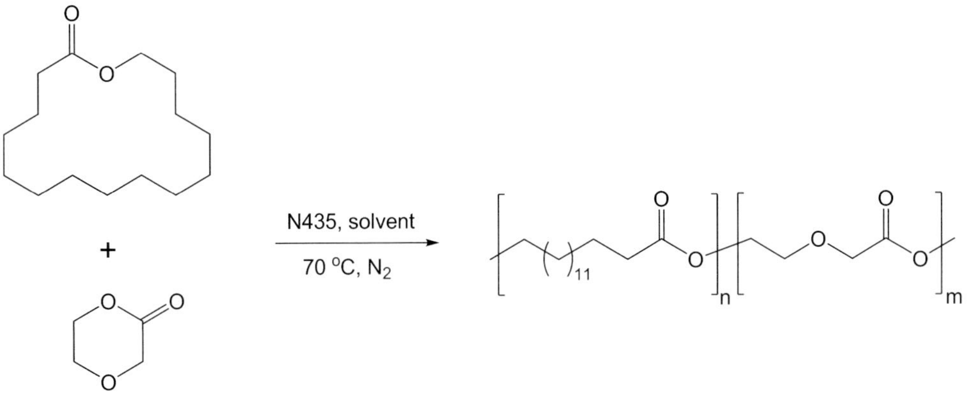

Scheme 4.22 Ring-opening copolymerization between PDL and DO catalyzed by Novozym 435.

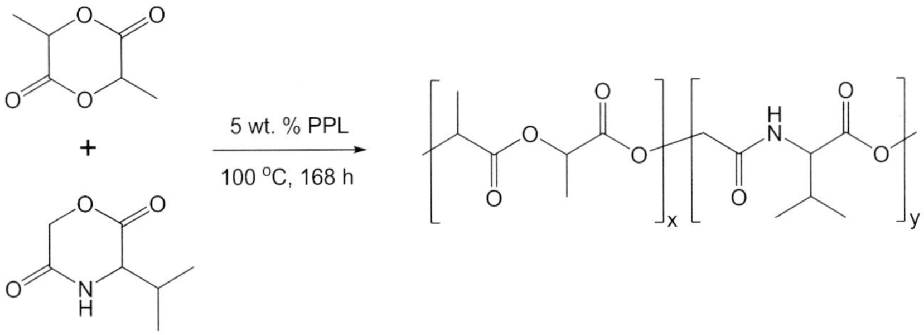

Scheme 4.23 Lipase-catalyzed ring-opening copolymerization of 3(S)-isopropylmorpholine-2,5-dione (IPDM) and D,L-lactide (DLLA).

random repeat unit structures and high molecular weights ($11300 > \overline{M_n} > 29100$; $107000 > \overline{M_w} > 18900$) in 51–87 wt % yields [108]. During the copolymerization reaction, PDL was found to be more reactive than DO, resulting in higher PDL/DO unit ratios in polymer chains than the corresponding PDL/DO monomer feed ratios. However, due to the ability of Novozym 435 to catalyze polymer–polymer transesterification reactions, ^{1}H and ^{13}C NMR analysis showed that poly(PDL-co-DO) of varying compositions had nearly random sequences of PDL and DO units with a slight tendency toward alternating arrangements.

Feng et al. [103] investigated copolymerizations of 3(S)-isopropylmorpholine-2,5-dione (IPDM) with D,L-lactide (DLLA), using porcine pancreatic lipase as catalyst at 100 °C for 168 h. By varying the feed composition, copolymers with different yields (between 13 and 57%) and molecular weights (between 8600 and 18100 g mol^{-1}) were obtained (Scheme 4.23).

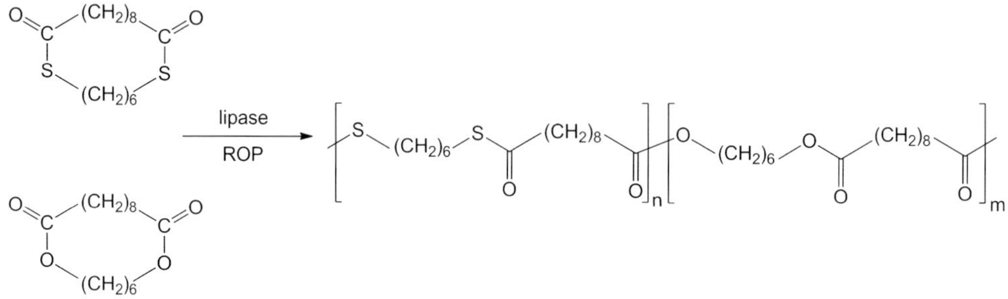

Scheme 4.24 Ring-opening copolymerization of cyclic thioester and cyclic ester using *Candida antarctica* lipase.

Furthermore, using Novozym 435 as catalyst, 2-oxo-12-crown-4-ether (OC) was copolymerized with PDL giving copolyesters with random sequence distributions [106].

Kato et al. [109] explored Novozym 435-catalyzed ring-opening copolymerization of cyclic(hexanedithiol-sebacate) and cyclic(hexanediol-sebacate) (Scheme 4.24). The reaction was performed in bulk at 120 °C. Direct polycondensation of hexane-1,6-dithiol and dimethyl sebacate yielded a poly(hexanedithiolsebacate) with an $\overline{M_w}$ of 10 000 after 48 h, that is about 1/10th the $\overline{M_w}$ of the polymer produced by the ring-opening copolymerization method, in which the $\overline{M_w}$ was 115 000 ($\overline{M_w}/\overline{M_n} = 2.3$) in 90.1% yield. The authors explained these results by the lack of methanol production in ring-opening copolymerization, meaning that the reverse reaction is suppressed, which results in the formation of a product with a higher molecular weight. Furthermore, thermal analysis by DSC showed that the T_m and T_c values for poly(hexanedithiol-sebacate) were higher (108.8 and 85.6, respectively) than those of the ester analog poly(hexanediol-sebacate) (74.8 and 40.0, respectively). Indeed, it is well known that sulfur-containing polymer analogs of corresponding oxygen-containing polymers have relatively higher melting temperatures [110]. Furthermore, the rigidity of polythioester chains was found to be greater than that of the corresponding polyester based on fusion entropy, ΔS_u, a parameter related to chain flexibility.

Multiarm heteroblock star-type copolymers of poly(lactic acid) (PLA) and poly(ε-CL), poly(LA-*co*-ε-CL) were prepared via a chemoenzymatic route [90]. Firstly, ROP of ε-CL was initiated regioselectively from 6-OH site of ethyl glucopyranoside (EGP) (see also above) using porcine pancreatic lipase as catalyst followed by termination of the EGP-PCL-OH terminus by lipase-catalyzed acetylated using vinyl acetate. Subsequently, Sn-catalyzed ROP of lactide was initiated from 2-, 3- and 4-OH groups of EGP to give a copolymer consisting of one poly(ε-CL) arm with $\overline{M_n} = 1300$ and three PLA arms so that the $\overline{M_n}$ of the final product was 11 500 (Scheme 4.25).

Biodegradable polyesters were synthesized via ring-opening copolymerizations of various oxiranes (glycidyl phenyl ether, benzyl glycidate, glycidyl methyl ether, styrene oxide) and various dicarboxylic anhydrides (succinic anhydride, phthalic

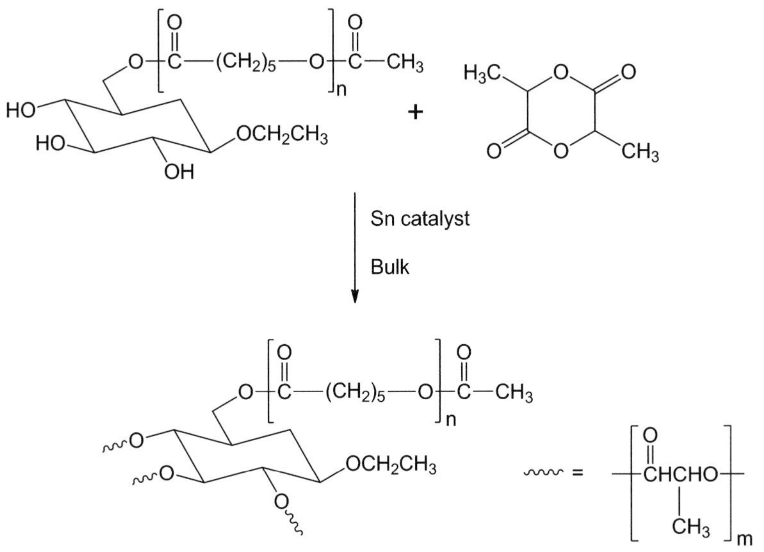

Scheme 4.25 Synthesis of multiarm heteroblock star-type copolymer via chemoenzymatic route.

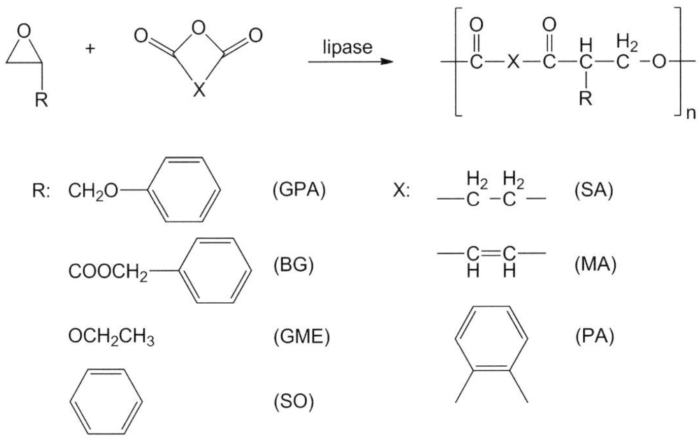

Scheme 4.26 Basic enzymatic polymerization of oxiranes (Glycidyl phenyl ether: GPE; benzyl glycidate: BG; glycidyl methyl ether: GME; styrene oxide: SO) and dicarboxylic anhydrides (succinic anhydride: SA; maleic anhydride: MA; phthalic anhydride: PA).

anhydride, maleic anhydride) catalyzed by various lipases (Porcine pancreatic lipase, *Candida rugosa* lipase, *Pseudomonas* sp. lipase, *Pseudomonas fluorescens* lipase, CALB) in stepwise copolymerizations (Scheme 4.26) [111]. The maximum molecular weight was obtained in a stepwise reaction forming either a carboxy or hydroxy end group. This procedure resulted in a polyester with $\overline{M}_w = 13500$

($\overline{M_w}/\overline{M_n} = 1.4$) Furthermore, a NMR study of the polymer structure showed that it contained only a small fraction of ether linkages.

Similar to this, Matsumura et al. [112] studied copolymerizations of succinic anhydride with oxiranes (glycidyl phenyl ether and benzyl glycidate). Copolymerization between succinic anhydride and glycidyl phenyl ether, using porcine pancreatic lipase as catalyst, in bulk reactions at 80 °C for 7 days resulted in a polyester in 80% yield ($\overline{M_w} = 4900$; $\overline{M_w}/\overline{M_n} = 2.4$).

There has been a significant effort to copolymerize TMC with lactones and other carbonate monomers. Matsumura et al. performed copolymerizations of lactide with TMC using porcine pancreatic lipase at 100 °C for 168 h [113]. They obtained random copolymers with $\overline{M_w}$ values up to 21 000. However, since trimethylene carbonate is known to thermally polymerize at 100 °C (see above), the extent of polymerization that occurs due to activation of monomers at the lipase catalytic triad versus by thermal or other chemical processes is not known [95]. Lipase AK-catalyzed copolymerizations of 1,3-dioxan-2-one (TMC) with 5-methyl-5-benzyloxycarbonyl-1,3-dioxan-2-one (MBC) were carried out in bulk at 80 °C for 72 h (Scheme 4.27). Although TMC reacted more rapidly than MBC, the product isolated at 72 h appeared to have a random repeat unit distribution [102]. Similarly, using Novozym 435 in toluene at 70 °C, TMC/PDL copolymerizations were performed and gave random copolymers.

Varying the feed ratio of the comonomers allowed regulation of the copolymer composition. The isolated yield and $\overline{M_n}$ of poly(PDL-co-43 mol % TMC) formed after 24 h (feed 2 : 1 PDL : TMC) was 90% and 30 900 g/mol, respectively. Thus far, an alternative chemical route to random poly(PDL-co-TMC) is not known. For example, PDL/TMC copolymerizations with chemical catalyst such as stannous octanoate, methylaluminoxane, and aluminum triisoproxide resulted either in homo-poly(TMC) or block copolymers of poly(TMC-co-PDL) [114]. Chemical

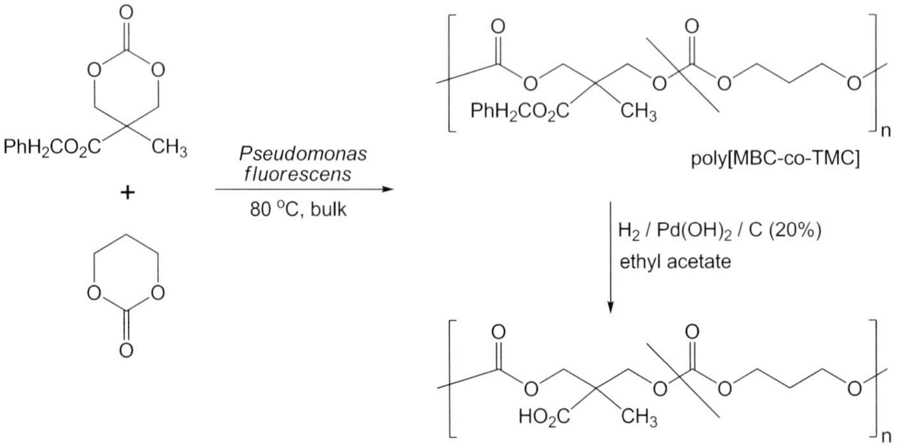

Scheme 4.27 *Pseudomonas fluorescens* lipase-catalyzed synthesis of poly (MBC-co-TMC).

catalysts have thus far favored TMC over PDL polymerization. In contrast, by lipase catalysis, PDL was more rapidly polymerized than TMC. Thus, herein lie important differences in the inherent catalytic properties of lipases as opposed to chemical catalysts that can be exploited to give unique copolymers.

In addition, the lipase-catalyzed copolymerization of PDL with a sugar carbonate (IPXTC) in toluene at 70 °C was studied [115]. Novozym 435 was found to be the most effective lipase catalyst based on its ability to form PDL/IPXTC copolymers. For example, by this method, poly(PDL-co-19 mol % IPXTC) was prepared in 38% isolated yield in 5 days with $\overline{M_n}$ 4070. The copolymer formed consisted of PDL blocks with random interruptions by IPXTC units or short segments.

Enzymatic ring-opening copolymerization of 5-benzyloxy-trimethylene carbonate (BTMC) and 1,4-dioxan-2-one (DON) was investigated using immobilized porcine pancreas lipase (IPPL) on silica particles [116]. A series of copolymers with different compositions were successfully synthesized in bulk at 150 °C. The BTMC monomer had higher reactivity in comparison with the DON monomer, which led to higher BTMC contents in the copolymers than that in the feed. The hydrophilicity of poly(BTMC-co-DON) increased along with the DON content.

Furthermore, the ring-opening co-polymerization of BTMC with 5,5-dimethyl-trimethylene carbonate (DTC) by immobilized porcine pancreatic lipase (0.1 wt%) catalyzed in bulk copolymerization at 150 °C for 24 h [117]. Under these conditions, the highest molecular weight of poly(BTMC-co-DTC) of $\overline{M_n} = 26\,400$ was obtained, with 83% monomer conversion.

A degradable triblock copolymer, poly(trimethylene carbonate)-block-poly[poly(ethylene glycol)-co-cyclic acetal]-block-poly(trimethylene carbonate) (PTMC-b-PECA-b-PTMC), was obtained via chemo-enzymatic approach [118]. The synthesized triblock copolymer consists of a degradable hydrophilic PECA (α,ω-glycol synthesized chemically) and an amorphous hydrophobic PTMC (lipase CA-catalyzed polymerization of TMC).

A compilation of the lipase-catalyzed lactone, cyclic ester related monomers and copolymers, the enzymes used, and the corresponding citation(s) are given in Table 4.4.

Table 4.4 Lipase-catalyzed ROP copolymerization.

Enzyme	Comonomer (1) /comonomer (2)	Reference
AK lipase	1,3-dioxan-2-one / 5-methyl-5-benzyloxycarbonyl-1,3-dioxan-2-one	[102]
Candida antarctica lipase A	cyclic (hexanedithiol-sebacate) / corresponding ester monomer	[110]
Candida antarctica lipase B	ε-CL / PDL	[77]
	8-OL / ε-CL	[80]
	8-OL / DDL	[80]

Table 4.4 *Continued*

Enzyme	Comonomer (1) /comonomer (2)	Reference
	PDL / *p*-dioxanone	[108]
	2-oxo-12-crown-4-ether / PDL	[106]
	PDL / TMC	[114]
	PDL / sugar carbonate (IPXTC)	[115]
	5-benzyloxy-trimethylene carbonate / 1,4-dioxan-2-one	[116]
	poly-(butylene carbonate) / poly(butylene succinate)	[57]
	cyclic (hexanedithiol-sebacate) / corresponding ester monomer	[109]
Porcine pancreatic lipase	5-benzyloxy-trimethylene carbonate / 5,5-dimethyl-trimethylene carbonate	[117]
	lactide / TMC	[113]
	poly(lactic acid) (PLA) / poly(ε-CL)	[90]
	succinic anhydride / glycidyl phenyl ether	[112]
	3(S)-isopropylmorpholine-2,5-dione / D,L-lactide	[103]
	succinic anhydride / benzyl glycidate	[112]
Pseudomonas fluorescens	β-propiolactone / ε-CL	[107]
	PDL / DDL	[73]
	PDL / UDL	[73]
	PDL / δ-VL	[73]
	PDL / ε-CL	[73]
	δ-VL / ε-CL	[80]
Pseudomonas sp.	ε-CL / ethyl lactate	[27]
	ε-CL / lactide	[27]
	ε-CL / γ-butyrolactone	[27]
	ε-CL / ethyl 4-hydroxybutyrate	[27]
	ε-CL / cyclopentadecanolide	[27]
	ε-CL / ethyl 15-hydroxypentadecanoate	[27]
	ε-CL / lactide / cyclopentadecanolide	[27]

4.6
Combination of Condensation and Ring-Opening Polymerization

It was shown that lipases can catalyze enzymatic ROP and polycondensation simultaneously. This lipase ability was employed in order to obtain various polyesters.

The copolymer of 12-hydroxydodecanoic acid/β-butyrolactone was synthesized at 45 °C in toluene, using *Porcine pancreatic* lipase [29]. After 72 h, the molecular weight of the obtained copolymer in 70% yield was $\overline{M_n} = 1800$. Electrospray

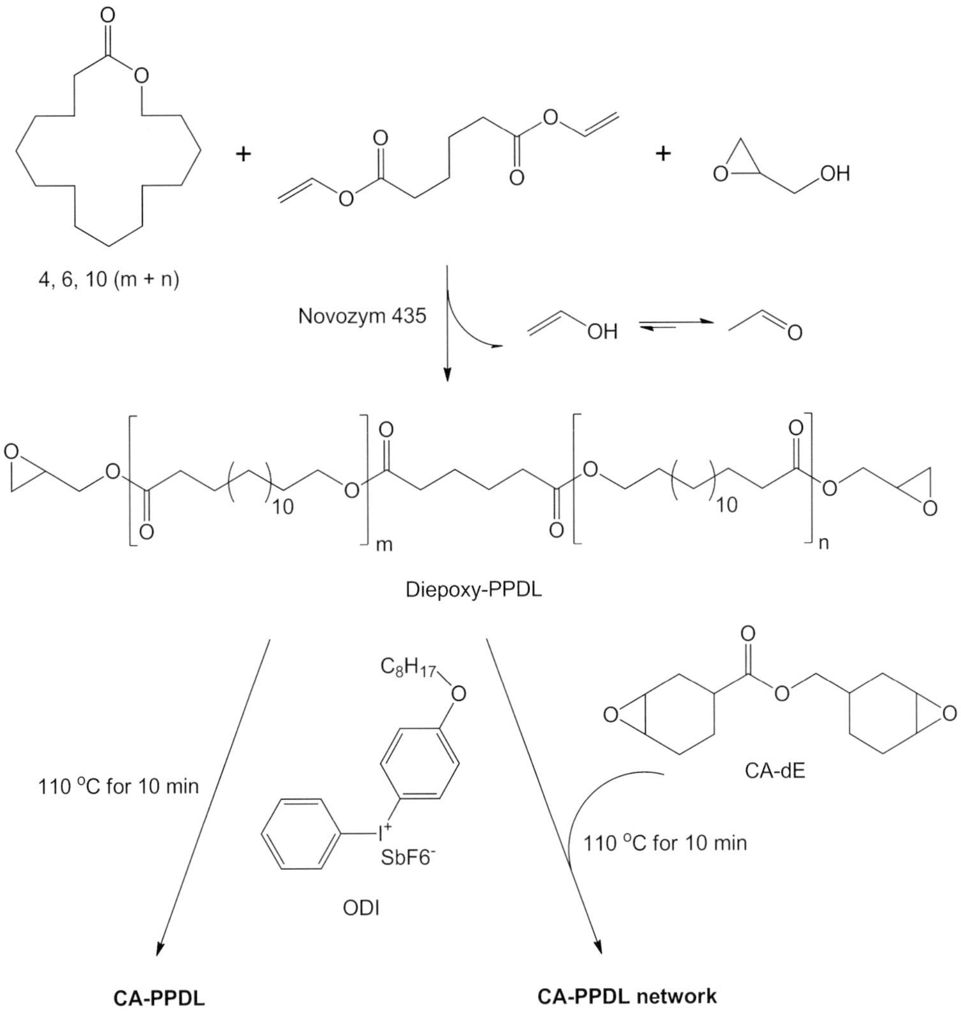

Scheme 4.28 Lipase-catalyzed one-pot synthesis of semicrystalline diepoxy functional macromonomers based on glycidol, pentadecalactone and adipic acid.

ionization mass spectrometry (ESI-MS) of the copolymer showed that the chain segments formed contained various compositions of 3-hydroxybutanoate and 12-hydroxydodecanoate units.

Eriksson and coworkers [119] performed a CALB one-pot procedure to synthesize semicrystalline diepoxy functional macromonomers based on glycidol, pentadecalactone and adipic acid. Diepoxy-PPDL was synthesized in toluene at 60 °C for 24 h, and by changing the stoichiometry of the building blocks, macromonomers in around 90% yield with controlled molecular weight from 1400 to 2700 were prepared (Scheme 4.28).

Iwata et al. [120] used Novozym 435 to catalyze copolymerizations of ε-CL with 11-mercaptoundecanoic acid (11MU) and 3-mercaptopropionic acid (3MP). The same authors also demonstrated that Novozym 435 catalyzed transesterifications between poly(-CL) and 11-mercaptoundecanoic acid or 3-mercaptopropionic acid in o-xylene.

4.7
Conclusion

This chapter provides numerous examples of significant advancements documented in the literature describing cell-free enzyme-catalyzed polymerizations, predominantly using lipases as catalysts. Polymerization reactions occurred by (i) condensations; (ii) ring-opening homo- and copolymerizations; and (iii) combination of condensation and ring-opening polymerization.

Lipase-catalyzed synthesis of polyesters came to the focus after two major breakthroughs: in 1984, novel lipase-catalyzed polycondensation to give oligoesters; in 1993, discovery of the catalysis for ring-opening polymerization of lactones.

Since then, many papers have appeared that have sought to explore the potential of this method using a wide variety of monomer-types. This work has largely been driven by the well-known benefits of enzymes over chemical catalysts. Advantages that enzyme-catalysts bring to polymer chemistry include: (i) extremely high selectivity and activity; (ii) mild reaction conditions; (iii) compatibilities with other chemical catalysts; (iv) none or low amount of by-products; (v) enzymes are metal-free, non-toxic and renewable; (vi) enzyme immobilization can provide increased activity, stability, easy recovery and reusability; (vii) protection and deprotection chemistry steps are no longer needed. Although various aspects of enzymatic polymer synthesis can be described as *environmentally friendly*, compared with conventional chemical synthesis, a number of problems still exist in transferring these polyester synthesis methods from the laboratory to industrial processes. This situation is mostly due to the high costs of enzymes and their need for improved stability so that they can be re-used over many reaction cycles. Therefore, researchers must continue to define where enzymes provide significant advantages relative to traditional chemical processes and develop improved enzyme catalysts. The yard-stick will always be the need for enzyme-catalyzed processes to provide cost-competitive products with similar or improved performance. Certainly the mild conditions of enzyme-catalyzed transformations will save in energy costs. Furthermore, enzyme selectivity can reduce by-product formation that increases cost. In addition, cost savings of enzyme-catalyzed processes can be realized in the development of safer processes.

References

1 Berzelius, J. (1847) Rapport annel de l'Institut geologique de Hongrie. 26.
2 Berthelot, M.M. (1853) Sur les combinaisons de la glycérine avec les acides. *C. R.*, **37**, 398–405.
3 Van Bemmelen, J. (1856) Über die Einwirkung der Bernsteinsäure und Citronesäure auf Glycerin. *J. Prakt. Chem.*, **69**, 84–93.
4 Smith, W. (1901) A monthly record for all interested in chemical manufactures: part 1. *J. Soc. Chem. Ind. Trans.*, **20** (11), 1075–1106.
5 Kienle, R.H., and Hovey, A.G. (1929) The polyhydric alcohol-polybasic acid reaction. I. Glycerol-phthalic anhydride. *J. Am. Chem. Soc.*, **51** (2), 509–519.
6 Carothers, W.H., and Arvin, J.A. (1929) Studies of polymerization and ring formation. II. Poly-esters. *J. Am. Chem. Soc.*, **51** (8), 2560–2570.
7 Carothers, W.H., and Hill, J.W. (1932) Studies of polymerization and ring formation. XI. The use of molecular evaporation as a means for propagating chemical reactions. *J. Am. Chem. Soc.*, **54** (4), 1557–1559.
8 Carothers, W.H., and Hill, J.W. (1932) Studies of polymerization and ring formation. XII. Linear superpolyesters. *J. Am. Chem. Soc.*, **54** (4), 1559–1566.
9 Carothers, W.H., and Hill, J.W. (1932) Studies of polymerization and ring formation. XV. Artificial fibers from synthetic linear condensation superpolymers. *J. Am. Chem. Soc.*, **54** (4), 1579–1587.
10 Okumara, S., Iwai, M., and Tominaga, Y. (1984) Synthesis of ester oligomer by Aspergillus niger lipase. *Agric. Biol. Chem.*, **48** (11), 2805–2808.
11 Gutman, A.L., Zuobi, K., and Boltansky, A. (1987) Enzymatic lactonisation of γ-hydroxyesters in organic solvents. Synthesis of optically pure γ-methylbutyrolactones and γ-phenylbutyrolactone. *Tetrahedron Lett.*, **28** (33), 3861–3864.
12 Knani, D., Gutman, A.L., and Kohn, D.H. (1993) Enzymatic polyesterification in organic media. Enzyme-catalyzed synthesis of linear polyesters. I. Condensation polymerization of linear hydroxyesters. II. Ring-opening polymerization of ε-caprolactone. *J. Polym. Sci. Part A Polym. Chem.*, **31** (5), 1221–1232.
13 Uyama, H., and Kobayashi, S. (1993) Enzymatic ring-opening polymerization of lactones catalyzed by lipase. *Chem. Lett.*, **22** (7), 1149–1150.
14 Gross, R.A., Kumar, A., and Kalra, B. (2001) Polymer synthesis in vitro enzyme catalysis. *Chem. Rev.*, **101** (7), 2097–2124.
15 Kobayashi, S., and Makino, A. (2009) Enzymatic polymer synthesis: an opportunity for green polymer chemistry. *Chem. Rev.*, **109** (11), 5288–5353.
16 Kobayashi, S. (2009) Recent development in lipase-catalyzed synthesis of polyesters. *Macromol. Rapid Commun.*, **30** (4), 237–266.
17 Albertsson, A.C. (2002) *Degradable Aliphatic Polyesters Springer*, vol. 157 of Advances in polymer science, Heidelberg, Berlin.
18 Kobayashi, S., Uyama, H., and Kimura, S. (2001) Enzymatic polymerization. *Chem. Rev.*, **101** (12), 3793–3818.
19 Kobayashi, S. (1999) Enzymatic polymerization: a new method of polymer synthesis. *J. Polym. Sci. Part A Polym. Chem.*, **37** (16), 3041–3056.

20 Kobayashi, S., Uyama, H., and Ohmae, M. (2001) Enzymatic polymerization for precision polymer synthesis. *Bull. Chem. Soc. Jpn.*, **74** (4), 613–635.

21 Matsumura, S. (2002) Enzyme-catalyzed synthesis and chemical recycling of polyesters. *Macromol. Biosci.*, **2** (3), 105–126.

22 Uyama, H., and Kobayashi, S. (2002) Enzyme-catalyzed polymerization to functional polymers. *J. Mol. Catal. B Enzym.*, **19–20**, 117–127.

23 Varma, I.K., Albertsson, A.C., Rajkhowa, R., and Srivastava, R.K. (2005) Enzyme catalyzed synthesis of polyesters. *Prog. Polym. Sci.*, **30** (10), 949–981.

24 Uyama, H., and Kobayashi, S. (2006) Enzymatic synthesis of polyesters via polycondensation. *Adv. Polym. Sci.*, **194**, 133–158.

25 Mahapatro, A., Kumar, A., and Gross, R.A. (2004) Mild, solvent-free ω-hydroxy acid polycondensations catalyzed by *Candida antarctica* lipase B. *Biomacromolecules*, **5** (1), 62–68.

26 O'Hagan, D., and Zaidi, N. (1993) Polymerisation of 10-hydroxydecanoic acid with the lipase from *Candida cylindracea*. *J. Chem. Soc. Perkin Trans. 1*, (20), 2389–2390.

27 Dong, H., Wang, H.D., Cao, S.G., and Shen, J.C. (1998) Lipase-catalyzed polymerization of lactones and linear hydroxyesters. *Biotechnol. Lett.*, **20** (10), 905–908.

28 O'Hagan, D., and Zaidi, N.A. (1994) Enzyme-catalysed condensation polymerization of 11-hydroxyundecanoic acid with lipase from *Candida cylindracea*. *Polymer*, **35** (16), 3576–3578.

29 Jedlinski, Z., Kowalczuk, M., Adamus, G., Sikorska, W., and Rydz, J. (1999) Novel synthesis of functionalized poly(3-hydroxybutanoic acid) and its copolymers. *Int. J. Biol. Macromol.*, **25** (1–3), 247–253.

30 Olsson, A., Lindstrom, M., and Iversen, T. (2007) Lipase-catalyzed synthesis of an epoxy-functionalized polyester from the suberin monomer cis-9,10-epoxy-18-hydroxyoctadecanoic acid. *Biomacromolecules*, **8** (2), 757–760.

31 Matsumura, S., and Takahashi, J. (1986) Enzymatic synthesis of functional oligomers, 1 lipase catalyzed polymerization of hydroxy acids. *Makromolekulare Chem. Rapid Commun.*, **7** (6), 369–373.

32 Ebata, H., Toshima, K., and Matsumura, S. (2007) Lipase-catalyzed synthesis and curing of high-molecular-weight polyricinoleate. *Macromol. Biosci.*, **7** (6), 798–803.

33 Pavel, K., and Ritter, H. (1996) Enzymes, in *Polymer Synthesis* (eds R.A. Gross, D.L. Kaplan, and G. Swift), American Chemical Society, Washington, DC, p. 200.

34 Kobayashi, S., Uyama, H., and Namekawa, S. (1998) In vitro biosynthesis of polyesters with isolated enzymes in aqueous systems and organic solvents. *Polym. Degrad. Stab.*, **59** (1–3), 195–201.

35 Linko, Y.Y., Wang, Z.L., and Seppala, J. (1995) Lipase-catalyzed linear aliphatic polyester synthesis in organic solvent. *Enzyme. Microb. Technol.*, **17** (6), 506–511.

36 Kato, M., Toshima, K., and Matsumura, S. (2005) Preparation of aliphatic poly(thioester) by the lipase-catalyzed direct polycondensation of 11-mercaptoundecanoic acid. *Biomacromolecules*, **6** (4), 2275–2280.

37 Yang, Y., Lu, W., Zhang, X., Xie, W., Cai, M., and Gross, R.A. (2010) Two-step biocatalytic route to biobased functional polyesters from ω-carboxy fatty acids and diols. *Biomacromolecules*, **11** (1), 259–268.

38 Rodney, R.L., Allinson, B.T., Beckman, E.J., and Russell, A.J. (1999) Enzyme-catalyzed polycondensation reactions for the synthesis of aromatic polycarbonates and polyesters. *Biotechnol. Bioeng.*, **65** (4), 485–489.

39 Feder, D., and Gross, R.A. (2010) Exploring chain length selectivity in HIC-catalyzed polycondensation reactions. *Biomacromolecules*, **11** (3), 690–697.

40 Binns, F., Harffey, P., Roberts, S.M., and Taylor, A.J. (1998) Studies of lipase-catalyzed polyesterification of an unactivated diacid/diol system.

J. Polym. Sci. Part A Polym. Chem., **36** (12), 2069–2080.

41. Kumar, A., Kulshrestha, A.S., Gao, W., and Gross, R.A. (2003) Versatile route to polyol polyesters by lipase catalysis. *Macromolecules*, **36** (22), 8219–8221.

42. Nara, S.J., Harjani, J.R., Salunkhe, M.M., Mane, A.T., and Wadgaonkarb, P.P. (2003) Lipase-catalysed polyester synthesis in 1-butyl-3-methylimidazolium hexafluorophosphate ionic liquid. *Tetrahedron Lett.*, **44** (7), 1371–1373.

43. Mesiano, A.J., Beckman, E.J., and Russell, A.J. (2000) Biocatalytic synthesis of fluorinated polyesters. *Biotechnol. Prog.*, **16** (1), 64–68.

44. Chaudhary, A.K., Lopez, J., Beckman, E.J., and Russell, A.J. (1997) Biocatalytic solvent-free polymerization to produce high molecular weight polyesters. *Biotechnol. Prog.*, **13** (3), 318–325.

45. Uyama, H., Kuwabara, M., Tsujimoto, T., and Kobayashi, S. (2003) Enzymatic synthesis and curing of biodegradable epoxide-containing polyesters from renewable resources. *Biomacromolecules*, **4** (2), 211–215.

46. Veld, M.A.J., Palmans, A.R.A., and Meijer, E.W. (2007) Selective polymerization of functional monomers with Novozym 435. *J. Polym. Sci. Part A Polym. Chem.*, **45** (24), 5968–5978.

47. Fu, H., Kulshrestha, A.S., Gao, W., and Gross, R.A. (2003) Physical characterization of sorbitol or glycerol containing aliphatic copolyesters synthesized by lipase-catalyzed polymerization. *Macromolecules*, **36** (26), 9804–9808.

48. Uyama, H., Inaka, K., and Kobayashi, S. (2000) Lipase-catalyzed synthesis of aliphatic polyesters by polycondensation of dicarboxylic acids and glycols in solvent-free system. *Polym. J.*, **32** (5), 440–443.

49. Azim, H., Dekhterman, A., Jiang, Z., and Gross, R.A. (2006) *Candida antarctica* lipase B-catalyzed synthesis of poly(butylene succinate): shorter chain building blocks also work. *Biomacromolecules*, **7** (11), 3093–3097.

50. Linko, Y.Y., Lamsa, M., Wu, X., Uosukanum, E., Seppala, J., and Linko, P. (1998) Biodegradable products by lipase biocatalysis. *J. Biotechnol.*, **66** (1), 41–50.

51. Linko, Y.Y., Wang, Z.L., and Seppala, J. (1995) Lipase-catalyzed synthesis of poly(1,4-butyl sebacate) from sebacic acid or its derivatives with 1,4-butanediol. *J. Biotechnol.*, **40** (2), 133–138.

52. Uyama, H., Shigeru, Y., and Kobayashi, S. (1999) Enzymatic synthesis of aromatic polyesters by lipase-catalyzed polymerization of dicarboxylic acid divinyl esters and glycols. *Polym. J.*, **31** (4), 380–383.

53. Wallace, J.S., and Morrow, C.J. (1989) Biocatalytic synthesis of polymers. II. Preparation of [AA-BB]x polyesters by porcine pancreatic lipase catalyzed transesterification in anhydrous, low polarity organic solvents. *J. Polym. Sci. Part A Polym. Chem.*, **27** (10), 3271–3284.

54. Uyama, H., and Kobayashi, S. (1994) Lipase-catalyzed polymerization of divinyl adipate with glycols to polyesters. *Chem. Lett.*, **23** (9), 1687–1690.

55. Jiang, Z., Liu, C., Xie, W., and Gross, R.A. (2007) Controlled lipase-catalyzed synthesis of poly(hexamethylenecarbonate). *Macromolecules*, **40** (22), 7934–7943.

56. Morrow, C.J., and Wallace, J.S. (1992) US Patent 5,147,791; pp. 993.

57. Jiang, Y., Liu, C., and Gross, R.A. (2008) Lipase-catalyzed synthesis of aliphatic poly(carbonate-*co*-esters). *Macromolecules*, **41** (13), 4671–4680.

58. Warwel, S., Demes, C., and Steinke, G. (2001) Polyesters by lipase-catalyzed polycondensation of unsaturated and epoxidized long-chain α,ω-dicarboxylic acid methyl esters with diols. *J. Polym. Sci. Part A Polym. Chem.*, **39** (10), 1601–1609.

59. Hilker, I., Rabani, G., Verzijl, G.K.M., Palmans, A.R.A., and Heise, A. (2006) Chiral polyesters by dynamic kinetic resolution polymerization. *Angew. Chem. Int. Ed.*, **45** (13), 2130–2132.

60 Patil, D.R., Rethwisch, D.G., and Dordick, J.S. (1991) Enzymatic synthesis of a sucrose-containing linear polyester in nearly anhydrous organic media. *Biotechnol. Bioeng.*, **37** (7), 639–646.

61 Kulshrestha, A.S., Gao, W., and Gross, R.A. (2005) Glycerol copolyesters: control of branching and molecular weight using a lipase catalyst. *Macromolecules*, **38** (8), 3193–3204.

62 Kulshrestha, A.S., Gao, W., Fu, H.Y., and Gross, R.A. (2007) Synthesis and characterization of branched polymers from lipase-catalyzed trimethylolpropane copolymerizations. *Biomacromolecules*, **8** (6), 1794–1801.

63 Hu, J., Gao, W., Kulshrestha, A., and Gross, R.A. (2006) "Sweet polyesters": lipase-catalyzed condensation–polymerizations of alditols. *Macromolecules*, **39** (20), 6789–6792.

64 Li, G., Yao, D., and Zong, M. (2008) Lipase-catalyzed synthesis of biodegradable copolymer containing malic acid units in solvent-free system. *Eur. Polym. J.*, **44** (4), 1123–1129.

65 Wang, Y.F., Lalonde, J.J., Momongan, M., Bergbreiter, D.E., and Wong, C.H. (1988) Lipase-catalyzed irreversible transesterifications using enol esters as acylating reagents: preparative enantio- and regioselective syntheses of alcohols, glycerol derivatives, sugars and organometallics. *J. Am. Chem. Soc.*, **110** (21), 7200–7205.

66 Chaudhary, A.K., Beckman, E.J., and Russell, A.J. (1998) Nonequal reactivity model for Biocatalytic polytransesterification. *AIChE J.*, **44** (3), 753–764.

67 Athawale, V.D., and Gaonkar, S.R. (1994) Enzymatic synthesis of polyesters by lipase catalysed polytrans-esterification. *Biotechnol. Lett.*, **16** (2), 149–154.

68 MacDonald, R.T., Pulapura, S.K., Svirkin, Y.Y., Gross, R.A., Kaplan, D.L., Akkara, J., Swift, G., and Wolk, S. (1995) Enzyme-catalyzed ε-caprolactone ring-opening polymerization. *Macromolecules*, **28** (1), 73–78.

69 Henderson, L.A., Svirkin, Y.Y., Gross, R.A., Kaplan, D.L., and Swift, G. (1996) Enzyme-catalyzed polymerizations of ε-caprolactone: effects of initiator on product structure, propagation kinetics, and mechanism. *Macromolecules*, **29** (24), 7759–7766.

70 Kumar, A., and Gross, R.A. (2000) *Candida antartica* lipase B catalyzed polycaprolactone synthesis: Effects of organic media and temperature. *Biomacromolecules*, **1** (1), 133–138.

71 Xu, J., Gross, R.A., Kaplan, D.L., and Swift, G. (1996) Chemoenzymatic synthesis and study of poly(α-methyl-β-propiolactone) stereocopolymers. *Macromolecules*, **29** (13), 4582–4590.

72 Namekawa, S., Suda, S., Uyama, H., and Kobayashi, S. (1999) Lipase-catalyzed ring-opening polymerization of lactones to polyesters and its mechanistic aspects. *Int. J. Biol. Macromol.*, **25** (1–3), 145–151.

73 Uyama, H., Kikuchi, H., Takeya, K., and Kobayashi, S. (1996) Lipase-catalyzed ring-opening polymerization and copolymerization of 15-pentadecanolide. *Acta Polymerica*, **47** (8), 357–360.

74 Geyer, U., Klemm, D., Pavel, K., and Ritter, H. (1995) Group transfer polymerization of methyl methacrylate. *Makromolekulare Chem. Rapid Ommunications*, **6** (5), 337–339.

75 Uyama, H., Kikuchi, H., Takeya, K., and Kobayashi, S. (1995) Enzymatic ring-opening polymerization of lactones to polyesters by lipase catalyst: unusually high reactivity of macrolides. *Bull. Chem. Soc. Jpn.*, **68** (1), 56–61.

76 Bisht, K.S., Henderson, L.A., Gross, R.A., Kaplan, D.L., and Swift, G. (1997) Enzyme-catalyzed ring-opening polymerization of ω-pentadecalactone. *Macromolecules*, **30** (9), 2705–2711.

77 Kumar, A., Kalra, B., Dekhterman, A., and Gross, R.A. (2000) Efficient ring-opening polymerization and copolymerization of ε-caprolactone and ω-pentadecalactone catalyzed by *Candida antartica* lipase B. *Macromolecules*, **33** (17), 6303–6309.

78 Nobes, G.A.R., Kazalauskas, R.J., and Marchessault, R.H. (1996) Lipase-catalyzed ring-opening polymerization of lactones: a novel route to poly(hydroxyalkanoate)s. *Macromolecules*, **29** (14), 4829–4833.

79 Matsumura, S., Beppu, H., Tsukuda, K., and Toshima, K. (1996) Enzyme-catalyzed ring-opening polymerization of β-propiolactone. *Biotechnol. Lett.*, **18** (9), 1041–1046.

80 Kobayashi, S., Uyama, H., Namekawa, S., and Hayakawa, H. (1998) Enzymatic ring-opening polymerization and copolymerization of 8-octanolide by lipase catalyst. *Macromolecules*, **31** (17), 5655–5659.

81 Uyama, H., Takeya, K., Hoshi, N., and Kobayashi, S. (1995) Lipase-catalyzed ring-opening polymerization of 12-dodecanolide. *Macromolecules*, **28** (21), 7046–7050.

82 Divakar, S. (2004) Porcine pancreas lipase catalyzed ring-opening polymerization of ε-caprolactone. *J. Macromol. Sci. Part A Pure. Appl. Chem.*, **41** (5), 537–546.

83 Cordova, A., Iversen, T., and Hult, K. (1999) Lipase-catalyzed formation of end-functionalized poly(ε-caprolactone) by initiation and termination reactions. *Polymer*, **40** (24), 6709–6721.

84 Dong, H., Cao, S.G., Li, Z.Q., Han, S.P., You, D.L., and Shen, J.C. (1999) Study on the enzymatic polymerization mechanism of lactone and the strategy for improving the degree of polymerization. *J. Polym. Sci. Part A Polym. Chem.*, **37** (9), 1265–1275.

85 Mei, Y., Kumar, A., and Gross, R.A. (2002) Probing water-temperature relationships for lipase-catalyzed lactone ring-opening polymerizations. *Macromolecules*, **35** (14), 5444–5448.

86 Kobayashi, S., Takeya, K., Suda, S., and Uyama, H. (1998) Lipase-catalyzed ring-opening polymerization of medium-size lactones to polyesters. *Macromol. Chem. Phys.*, **199** (8), 1729–1736.

87 Hunsen, M., Azim, A., Mang, H., Wallner, S.R., Ronkvist, A., Xie, W.C., and Gross, R.A. (2007) A cutinase with polyester synthesis activity. *Macromolecules*, **40** (2), 148–150.

88 Bisht, K.S., Deng, F., Gross, R.A., Kaplan, D.L., and Swift, G. (1998) Ethyl glucoside as a multifunctional initiator for enzyme-catalyzed regioselective lactone ring-opening polymerization. *J. Am. Chem. Soc.*, **120** (7), 1363–1367.

89 Al-Azemi, T.F., and Bisht, K.S. (1999) Novel functional polycarbonate by lipase-catalyzed ring-opening polymerization of 5-methyl-5-benzyloxycarbonyl-1,3-dioxan-2-one. *Macromolecules*, **32** (20), 6536–6540.

90 Deng, F., Bisht, K.S., Gross, R.A., and Kaplan, D.A. (1999) Chemoenzymatic synthesis of a multiarm poly(lactide-co-ε-caprolactone). *Macromolecules*, **32** (15), 5159–5161.

91 Nishida, H., Yamashita, M., Nagashima, M., Endo, T., and Tokiwa, Y. (2000) *J. Polym. Sci. Polym. Chem. Ed.*, **38** (9), 1560–1567.

92 Srivastava, R.K. (2005) High-molecular-weight poly(1,5-dioxepan-2-one) via enzyme-catalyzed ring-opening polymerization. *J. Polym. Sci. Part. A Polym. Chem.*, **43** (18), 4206–4216.

93 Uyama, H., Kobayashi, S., Morita, M., Habaue, S., and Okammoto, Y. (2001) Chemoselective ring-opening polymerization of a lactone having exo-methylene group with lipase catalysis. *Macromolecules*, **34** (19), 6554–6556.

94 Habaue, S., Asai, M., Morita, M., Okammoto, Y., Uyama, H., and Kobayashi, S. (2003) Chemospecific ring-opening polymerization of α-methylenemacrolides. *Polymer*, **44** (18), 5195–5200.

95 Bisht, K.S., Svirkin, Y.Y., Henderson, L.A., Gross, R.A., Kaplan, D.L., and Swift, G. (1997) Lipase-catalyzed ring-opening polymerization of trimethylene carbonate. *Macromolecules*, **30** (25), 7735–7742.

96 Kullmer, K., Kikuchi, H., Uyama, H., and Kobayashi, S. (1998) Lipase-catalyzed ring-opening polymerization of α-methyl-δ-valerolactone and α-methyl-ε-caprolactone. *Macromol. Rapid Commun.*, **19** (2), 127–130.

97 Peeters, J.W., van Leeuwen, O., Palmans, A.R.A., and Meijer, E.W. (2005) Lipase-catalyzed ring-opening polymerizations of 4-substituted ε-caprolactones: mechanistic considerations. *Macromolecules*, **38** (13), 5587–5592.

98 van Buijtenen, J., van As, B.A.C., Verbruggen, M., Roumen, L., Vekemans, J.A.J.M., Pieterse, K., Hilbers, P.A.J., Hulshof, L.A., Palmans, A.R.A., and Meijer, E.W. (2007) Switching from S- to R-selectivity in the *Candida antarctica* lipase B-catalyzed ring-opening of ω-methylated lactones: tuning polymerizations by ring size. *J. Am. Chem. Soc.*, **129** (23), 7393–7398.

99 Wu, R., Al-Azemi, T.F., and Bisht, K.S. (2008) Functionalized polycarbonate derived from tartaric acid: enzymatic ring-opening polymerization of a seven-membered cyclic carbonate. *Biomacromolecules*, **9** (10), 2921–2928.

100 Matsumura, S., Tsukada, K., and Toshima, K. (1997) Enzyme-catalyzed ring-opening polymerization of 1,3-dioxan-2-one to poly(trimethylene carbonate). *Macromolecules*, **30** (10), 3122–3124.

101 Svirkin, Y.Y., Xu, J., Gross, R.A., Kaplan, D.L., and Swift, G. (1996) Enzyme-catalyzed stereoelective ring-opening polymerization of α-methyl-β-propiolactone. *Macromolecules*, **29** (13), 4591–4597.

102 Al-Azemi, T.F., Harmon, J.P., and Bisht, K.S. (2000) Enzyme-catalyzed ring-opening copolymerization of 5-methyl-5-benzyloxycarbonyl-1,3-dioxan-2-one (MBC) with trimethylene carbonate (TMC): synthesis and characterization. *Biomacromolecules*, **1** (3), 493–500.

103 Feng, Y., Klee, D., and Höcker, H. (2004) Lipase catalyzed copolymerization of 3(S)-isopropylmorpholine-2,5-dione and D,L-lactide. *Macromol. Biosci.*, **4** (6), 587–590.

104 Matsumura, S., Beppu, H., Nakamura, K., Osanai, S., and Toshima, K. (1996) Preparation of poly(β-malic acid) by enzymatic ring-opening polymerization of benzyl β-malolactonate. *Chem. Lett.*, **25** (9), 795–796.

105 Kobayashi, S., Kikuchi, H., and Uyama, H. (1997) Lipase-catalyzed ring-opening polymerization of 1,3-dioxan-2-one. *Macromol. Rapid Commun.*, **18** (7), 575–579.

106 van der Mee, L., Antens, A., van de Kruijs, B., Palmans, A.R.A., and Meijer, E.W.J. (2006) Oxo-crown-ethers as comonomers for tuning polyester properties. *J. Polym. Sci. Part A Polym. Chem.*, **44** (7), 2166–2176.

107 Namekawa, S., Uyama, H., and Kobayashi, S. (1996) Lipase-catalyzed ring-opening polymerization and copolymerization of β-propiolactone. *Polym. J.*, **28** (8), 730–731.

108 Jiang, Z., Azim, H., and Gross, R.A. (2007) Lipase-catalyzed copolymerization of ω-pentadecalactone with p-dioxanone and characterization of copolymer thermal and crystalline properties. *Biomacromolecules*, **8** (7), 2262–2269.

109 Kato, M., Toshima, K., and Matsumura, M. (2007) Enzymatic synthesis of polythioester by the ring-opening polymerization of cyclic thioester. *Biomacromolecules*, **8** (11), 3590–3596.

110 Lottia, N., Siracusab, V., Finellia, L., Marchesea, P., and Munaria, A. (2006) Sulphur-containing polymers: Synthesis and thermal properties of novel polyesters based on dithiotriethylene glycol. *Eur. Polym. J.*, **42** (12), 3374–3382.

111 Soeda, Y., Okamoto, T., Toshima, K., and Matsumura, S. (2002) Enzymatic ring-opening polymerization of oxiranes and dicarboxylic anhydrides. *Macromol. Biosci.*, **2** (9), 429–436.

112 Matsumura, S., Okamoto, T., Tsukada, K., and Toshima, K. (1998) Novel lipase-catalyzed ring-opening copolymerization of oxiranes and succinic anhydride forming polyesters bearing functional groups. *Macromol. Rapid Commun.*, **19** (6), 295–298.

113 Matsumura, S., Tsukada, K., and Toshima, K. (1999) The effect of the chain length of polynucleotides

on their binding with platinum complexes. *Int. J. Biol. Macromol.*, **25** (2–3), 161–166.
114 Kumar, A., Garg, K., and Gross, R.A. (2001) Copolymerizations of ω-pentadecalactone and trimethylene carbonate by chemical and lipase catalysis. *Macromolecules*, **34** (11), 3527–3533.
115 Mahapatro, A., Kumar, A., and Gross, R.A. (2000) Control of polyester chain scission by lipase-catalysis. *Polym. Preprint*, **41** (2), 1826–1827.
116 He, F., Jia, H.L., Liu, G., Wang, Y.P., Feng, J., and Zhuo, R.Z. (2006) Enzymatic synthesis and characterization of novel biodegradable copolymers of 5-benzyloxy-trimethylene carbonate with 1,4-dioxan-2-one. *Biomacromolecules*, **7** (8), 2269–2273.
117 He, F., Wang, Y., Feng, J., Zhuo, R., and Wang, X. (2003) Synthesis of poly[(5-benzyloxy-trimethylene carbonate)-co-(5,5-dimethyl-trimethylene carbonate)] catalyzed by immobilized lipase onsilica particles with different size. *Polymer*, **44** (11), 3215–3219.
118 Kaihara, S., Fisher, J.P., and Matsumura, S. (2009) Chemo-enzymatic synthesis of degradable PTMC-*b*-PECA-*b*-PTMC triblock copolymers and their micelle formation for pH-dependent controlled release. *Macromol. Biosci.*, **9** (6), 613–621.
119 Eriksson, M., Fogelström, L., Hult, K., Malmström, E., Johansson, M., Trey, S., and Martinelle, M. (2009) Enzymatic one-pot route to telechelic polypentadecalactone epoxide: synthesis, UV curing, and characterization. *Biomacromolecules*, **10** (11), 3108–3113.
120 Iwata, S., Toshima, K., and Matsumura, S. (2003) Enzyme-catalyzed preparation of aliphatic polyesters containing thioester linkages. *Macromol. Rapid Commun.*, **24** (7), 467–471.
121 Magolin, A.L., Creene, J.Y., and Klibanov, A.M. (1987) Stereoselective oligomerizations catalyzed by lipases in organic solvents. *Tetrahedron Lett.*, **28** (15), 1607–1609.

5
Enzyme-Catalyzed Synthesis of Polyamides and Polypeptides

H. N. Cheng

5.1
Introduction

The synthesis of polypeptides and polyamides by chemical means is well known and well established [1–4]. Nevertheless, there have been substantial efforts to use enzymes as catalysts for these reactions [5]. The motivation is to find improved processes that are more environmentally friendly and that involve lower reaction temperatures, thereby decreasing energy usage. In general, enzymatic catalysis entails milder reaction conditions but can be more difficult for the synthesis of high-molecular-weight polymers. However, several promising reports have recently appeared. This review will be confined to the synthesis of polyamides and polypeptides using enzymes that are commercially available or isolated from appropriate organisms. Thus, production of proteins *in vivo* and in cell-free extracts is not covered, nor are polyesters and poly(ester-amides).

Prior to 2000 enzymatic methods for polyamides and polypeptides have mostly been used for the synthesis of oligomers (with degree of polymerization, DP, about 2–8) and condensation of large protein fragments. After 2000, several articles have appeared that reported the synthesis of higher-molecular-weight polymers. This field is evolving, with new approaches and new methodologies still being developed.

Three general approaches involving isolated enzymes have been used for polyamide or polypeptide synthesis. The first approach uses protease and other proteolytic enzymes. The second approach uses lipases and esterases. These two approaches account for most of the papers in the literature. The proteases and the lipases used tend to have relatively broad substrate specificity and can be applied to the synthesis of different types of polyamides and polypeptides. The third approach includes enzymes other than proteases and lipases. For example, in protein synthesis there has been much research into the use of cell-free extracts to produce proteins *in vivo*. In a few cases, the enzyme(s) responsible for the synthesis have been isolated and used to produce the same or similar proteins *in vitro*. Selected publications have been included in this review.

5.2
Catalysis via Protease

Proteases have been known to catalyze the synthesis of peptides for quite some time, and there are several excellent recent reviews on this topic [6–8]. In general, there are six classes of proteases: (1) serine proteases (e.g., trypsin, chymotrypsin, and subtilisin); (2) threonine proteases; (3) cysteine protease (such as papain and bromelain); (4) aspartate proteases (e.g., pepsin); (5) glutamate proteases; and (6) metal proteases (e.g., thermolysin). Protease-catalyzed peptide synthesis can proceed through either thermodynamically or kinetically controlled mechanism [6, 7]. For thermodynamically controlled mechanism (applicable to all classes of proteases), synthesis is in equilibrium with hydrolysis. The main drawbacks are that the reaction rates and product yields are often low, a high amount of enzyme is needed, and reaction conditions must be used that drive the equilibrium towards synthesis. Kinetically controlled synthesis is limited only to serine, cysteine and threonine proteases. As shown in Figure 5.1, water and amine compete for reaction with the acyl–enzyme complex, the former leading to hydrolysis and the latter to peptide formation. The success of peptide synthesis depends on the relative kinetic rates of these reactions (see also Chapter 14). In this case, enzyme specificity, substrate concentration, mode of carboxylic group activation, temperature, pH, and reaction medium can all impact on product yield and molecular weight.

It has been noted [6] that despite their favorable catalytic properties, proteases are not necessarily ideal catalysts for peptide synthesis. Hydrolysis is always a concern which prevents the formation of high-molecular-weight polymers. A number of strategies have been designed to optimize reactions, such as improved proteases, substrate modification, modification of the physical state of reactants,

Thermodynamically controlled synthesis

$$R\text{-COOH} + NH_2\text{-R}' \rightleftharpoons R\text{-CONH-R}' + H_2O$$

Kinetically controlled synthesis

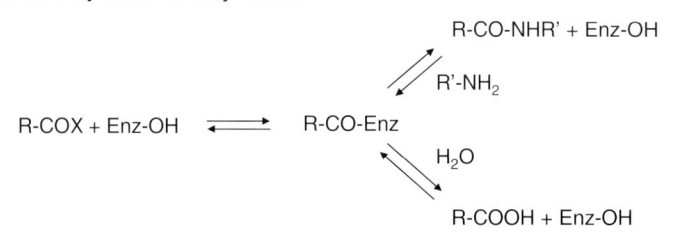

$$R\text{-COX} + Enz\text{-OH} \rightleftharpoons R\text{-CO-Enz} \begin{array}{c} \xrightarrow{R'\text{-}NH_2} R\text{-CO-NHR}' + Enz\text{-OH} \\ \xrightarrow{H_2O} R\text{-COOH} + Enz\text{-OH} \end{array}$$

Figure 5.1 Protease-catalyzed synthesis of polypeptides (after ref [7]).

immobilized or crosslinked enzymes (see also Chapters 2 and 3), and use of organic, ionic liquids (see also Chapter 13), and frozen systems [7, 8].

In view of the above considerations, it is not surprising that most of the protease papers in the past 12 years deal with the synthesis of oligomers. Some recent examples include the synthesis of dipeptides by chymotrypsin and subtilisin Carlsberg [9], by neutrase codeposited with sorbitol onto polyamide support [10], by immobilized chymotrypsin [11] with the help of a kinetic control [12], by Boilysin and thermolysin activated by salts [13], by chymotrypsin in frozen aqueous solutions [14], papain and immobilized cell extracts [15, 16]. The synthesis of tripeptide was reported with trypsin and chymotrypsin [17]. A number of dipeptides, tripeptides, and tetrapeptides were synthesized through the use of subtilisin immobilized on poly(vinyl alcohol) cryogel [18]. Papain was used to polymerize tyrosine ethyl ester hydrochloride with DP 4–9 [19]. A large number of papers were published on oligopeptides from cholecystokinin (CCK), which is a polypeptide hormone with gastrointestinal, biliary, and brain activities, including tripeptides [20–22], tetrapeptide, pentapeptide [23–25], and octapeptide [26–28].

A large area of research includes the homopolymers of glutamic and aspartic acids. These are biodegradable polyelectrolytes that have potential industrial applications. In 1990 Uemura *et al.* used papain modified with poly(ethylene glycol) to produce oligopeptides from dialkyl L-aspartate and L-glutamate in benzene [29]. Subsequent reports dealt with oligopeptide synthesis from dialkyl glutamate hydrochloride with papain and α-chymotrypsin [30] and dialkyl aspartate with alkalophilic proteinase from *Streptomyces* sp. [31]. Later: Soeda *et al.* [32] reported the preparation of poly(ethyl aspartate) with a protease from *Bacillus subtilis* in organic media, where M_w ranged from 1000 to 3800 Da. Uyama *et al.* [33] reported the synthesis of polymers and copolymers of ethyl L-glutamate with papain and bromelain. However, papain-catalyzed polymerization of γ-methyl L-glutamate under similar reaction conditions was not found, supporting the regioselective production of the polymer having an α-peptide linkage. Copolymerization of diethyl glutamate and various amino acid esters with papain catalyst yielded peptide copolymers. Recently Li *et al.* [34, 35] also used papain to produce oligopeptides (DP 8–9) from diethyl glutamate. Later, Li *et al.* [36] reported protease-catalyzed co-oligomerization of γ-ethyl-L-glutamate with L-leucine ethyl ester. The average DP observed varied from 6.5 to 8.5.

An interesting area of research deals with peptides of non-natural amino acids. Poly-L-pentenyl glycine, poly-L-propargyl glycine, and poly-L-allyl glycine were synthesized chemically and enzymatically using subtilisin Carlsberg. Whereas higher molecular weights (DP = 40–160) were achieved with chemical means, the protease-catalyzed polymers have DP = 8–12 [37]. In a separate work involving the non-natural amino acid allyglycine (Ag), subtilisin Carlsberg was used to synthesize the tetrapeptide ester L-Ag-L-Phe-L-Phe-LAg-OEt in several miscible aqueous/organic solvent systems. Mass spectrometry indicated that the octapeptide ester (Ag-Phe-Phe-Ag)$_2$OEt was formed in all cases and that the dodecapeptide ester (Ag-Phe-Phe-Ag)$_3$OEt was formed in one case [38].

In a non-polypeptide example, Gu, et al. [39] attempted to use four proteases (chymotrypsin, trypsin, subtilisin, and papain) to polymerize water-soluble polyamides from dimethyl adipate and diethylene triamine. However, only oligoamides were obtained.

5.3
Catalysis via Lipase

Lipases are also known to catalyze amide formation. Thus in 1987 Margolin and Klibanov [40] showed the use of porcine pancreatic lipase (PPL) for dipeptide synthesis in organic solvents. West and Wong [41] investigated several lipases and esterases for dipeptide and tripeptide synthesis, with variable yields observed. Later, Kawashiro et al. [42] also studied the dimerization of phenylalanine with PPL.

In 1998 So, et al. [43] screened 15 different commercial lipases for the synthesis of dipeptides from D-amino acids, and found only PPL to be effective. They also found certain amino acids to be better acyl acceptors, and L-isomers of the same D-amino acids showed similar reactivity for peptide synthesis, but the D-isomers resulted in better yields.

In a patent application filed in 2000 (granted in 2004) Cheng et al. [44] used selected commercial lipases to synthesize polyamides from diesters and diamines with the following general structures:

$$R_1OC(O)\text{-}R\text{-}C(O)OR_2 \quad H_2N\text{-}[R_3\text{-}X\text{-}R_4\text{-}NH]_n\text{-}H$$

where R_1 and R_2 are methyl, ethyl, or other convenient leaving groups, R, R_3 and R_4 are alkyl groups that may contain olefinic double bonds or heteroatoms, X may be O, CH_2, NH or S, and n is the number of the repeating unit. For $n = 1$, the reaction can be depicted in Scheme 5.1.

The polyamides reported had molecular weights that varied from 3000 to over 10 000 Daltons. This patent is the first report of the formation of a high-molecular-weight polyamide from a lipase-catalyzed process.

Several examples of polyamides made with this process are given in Tables 5.1 and 5.2 [44, 45]. The enzymatic process is milder in temperature and produces narrower molecular weight distributions relative to the chemical process. An additional advantage is that some polyamides that are not easily synthesized via the chemical process, can be made via the enzymatic process, for example, the polyamides derived from dialkyl malonate, dialkyl fumarate, and dialkyl maleate.

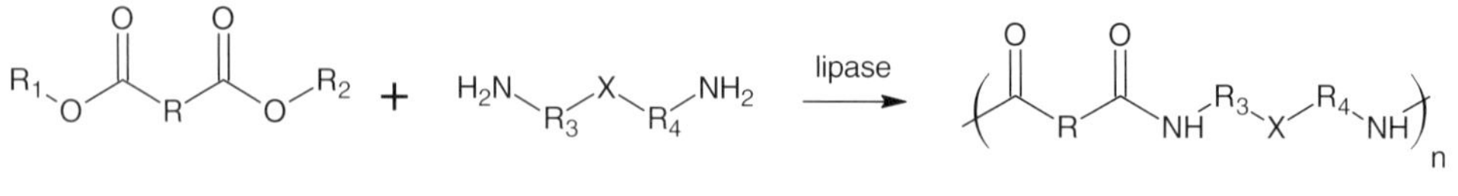

Scheme 5.1

Table 5.1 Lipase-catalyzed polyamides made with diethylene triamine and different dialkyl esters.

Dialkyl ester	Rxn temperature	M_w	M_w/M_n	Enzyme used
Adipate	90 °C	8400	2.73	Novozym 435
Malonate	80 °C	8000	2.10	Novozym 435
Malonate	70 °C	15790	1.83	Palatase®
Phenylmalonate	90–100 °C	3600	1.85	Novozym 435
Fumarate	50 °C	3060	1.85	Novozym 435

Table 5.2 Lipase-catalyzed polyamides made with dialkyl adipate and different diamines.

Diamine	Rxn temperature	M_w	M_w/M_n	Enzyme used
NH_2-terminated triethylene glycol	70 °C	4540	2.71	Novozym 435
Triethylene tetraamine	80 °C	8000	2.10	Novozym 435
Diethylene triamine	90 °C	8400	2.73	Novozym 435

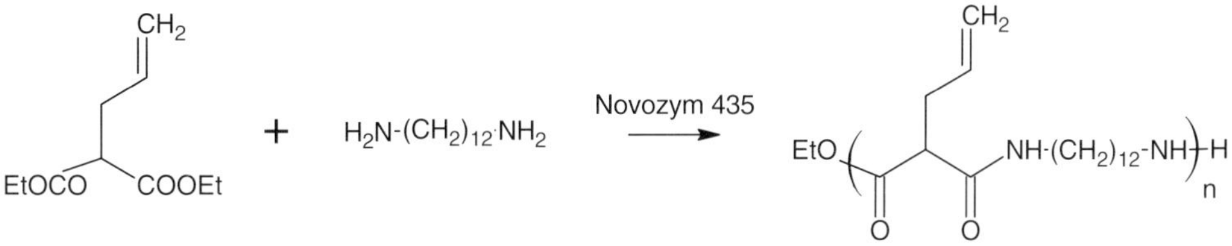

Scheme 5.2

Separately, Gu et al. [46] reported the use of lipases to facilitate the synthesis of a family of polyaminoamides. The polyaminoamides were made by a two-step reaction, where the first step was Michael addition of 1 mole of a polyamine and 2 moles of an acrylic compound to form an amine-containing diester, and the second step was the polymerization of additional polyamine with the resulting diester prepolymer at 70–140 °C or in the presence of an enzyme at 60–80 °C.

In 2005, Azim, Sahoo, and Gross [47] reported the use of immobilized lipase B from *Candida antarctica* (Novozym® 435) as a catalyst for direct formation of amide bonds between dialkyl esters and diamines under mild reaction conditions. Oligoamides were obtained with DP up to 9. The example of the reaction of diethyl allylmalonate and 1,12-dodecanediamine is shown in Scheme 5.2, and the oligoamide structure elucidated by ^{1}H NMR.

In 2005, Panova et al. [48] filed a patent application (granted in 2009) where they carried out a detailed study using lipases to produce cyclic amide oligomers

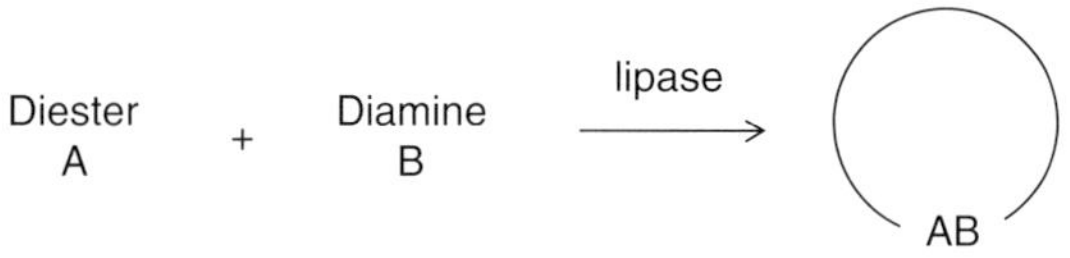

Scheme 5.3

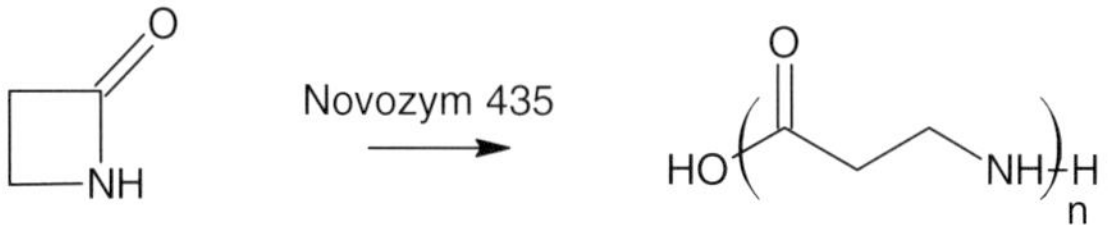

Diester A + Diamine B →(lipase) cyclic AB

Scheme 5.4

2-azetidinone →(Novozym 435) HO-[C(=O)-CH₂-CH₂-NH]ₙ-H

from diesters and diamines. The cyclic amide oligomers were useful for the subsequent production of higher-molecular-weight polyamides (Scheme 5.3).

Also in 2005 Kong et al. [49] filed a patent application on a process for preparing an aqueous polyamide dispersion by lipase-catalyzed polycondensation of a diamine and a dicarboxylic acid in aqueous medium. In a separate patent application [50], they reported the process for preparing an aqueous polyamide dispersion by lipase-catalyzed reaction of an aminocarboxylic acid compound in aqueous medium.

In an elegant recent work, Loos et al. [51, 52] reported the synthesis of poly(β-alanine) via lipase-catalyzed ring-opening of 2-azetidinone (Scheme 5.4) (see Chapter 14 for details on the polymerization mechanism). After removal of cyclic side products and low-molecular-weight species, pure linear poly(β-alanine) was obtained. The average DP of the polymer obtained was limited to 8 because of the solubility of the polymer in the reaction medium. Control experiments with β-alanine as substrate confirmed that the ring structure of the 2-azetidinone was necessary to obtain the polymer.

5.4
Catalysis via Other Enzymes

The use of cell-free extracts to produce proteins has been known for a long time. Early work in this area has been reviewed [53–55]. However, it is much less common to isolate the enzymes, sometimes to purify them, and use them for *in vitro* production of proteins. A successful example is a publication by Fruton et al. [56], who demonstrated the action of dipeptidyl transferase as a polymerase. Using this enzyme, they made a number of oligopeptides.

One relatively recent example is the cyanobacterial nitrogen reserve cyanophycin (multi-L-arginyl-poly-L-aspartic acid). This polymer is of interest because the oligo-arginyl polyaspartic acid may function as a biodegradable polyelectrolyte in selected applications. This polymer is synthesized *in vivo* by the enzyme cyanophycin synthetase, and studies of production, biosynthesis and mechanism have been published [57–59] (see also Chapter 10 for more detailed information). Furthermore, the genes for this enzyme have been identified, cloned and sequenced, and the enzyme isolated and purified [60]. The *in vitro* activity of the enzyme depends on ATP, primers, and both substrates, L-arginine and L-aspartic acid. In addition to native cyanophycin, the purified enzyme has been found to accept a modified cyanophycin containing less arginine, arginyl aspartic acid dipeptide, and poly-α,β-DL-aspartic acid as primers and also incorporated β-hydroxyaspartic acid instead of L-aspartic acid or L-canavanine instead of L-arginine at a significant rate. The lack of specificity of this thermostable enzyme with respect to primers and substrates, the thermal stability of the enzyme, and the finding that the enzyme is suitable for *in vitro* production of cyanophycin make it an interesting candidate for biotechnological processes.

In contrast to cyanophycin, the biosynthesis of poly-γ-glutamate appears to involve a coupled system composed of various enzymes [61]. *In vitro* synthesis of poly-γ-glutamate via isolated enzymes has not been reported. An article has been published on the isolation of folylpoly-α-glutamate synthetase and its *in vitro* reaction to add glutamate residues to polyglutamate substrates via α-COOH linkages [62].

A different type of polypeptide has been made by polymerizing tyrosine and its derivatives with peroxidase, laccase, or bilirubin oxidase [19, 63]. The resulting polymers do not have the α-peptide structure, but consist of polyphenols with mixtures of phenylene and oxyphenylene units. Tyranosinase is also known to oxidize tyrosine, which can lead to polymerization reactions to form melanin [64].

Recently an enzyme was found in planktonic marine cyanobacterium *Prochlorococcus* MIT9313 that could transform up to 29 different linear peptides into a library of polycyclic, conformationally constrained products with highly diverse ring topologies [65].

5.5
Comments

In the literature a majority of the publications involving enzyme-catalyzed synthesis of polypeptides and polyamides use proteases and related proteolytic enzymes. The data indicate that proteases are good catalysts for the synthesis of oligomers. For high-molecular-weight polymers, the use of protease is more difficult and less reported.

In contrast, lipases have been used for the synthesis of both oligoamides and polyamides. Under suitable reaction conditions, lipases appear to be good catalysts for higher-molecular-weight polymers for both condensation and ring-opening polymerizations. Optimization is still needed, and further developments will hopefully extend the scope of the reaction.

As for enzymes other than proteases and lipases, only a few examples of enzyme-catalyzed synthesis of polyamides can be gleaned from the literature.

A potentially active area is the *in vitro* synthesis of proteins via dipeptidyl transferase and cyanophycin synthetase. In specific cases these enzymes may perhaps be useful. Potential handicaps of this approach in the commercial context are the lack of general availability and higher costs of these two enzymes.

(Names of products are necessary to report factually on available data; however, the USDA neither guarantees nor warrants the standards of the products, and the use of the name USDA implies no approval of the products to the exclusion of others that may also be suitable.)

References

1 Mitchell, A.R. (2008) Bruce Merrifield and solid-phase peptide synthesis: a historical assessment. *Biopolymers*, **90** (3), 175–184.
2 Pennington, M.W., and Dunn, B.M. (eds) (1994) *Peptide Synthesis Protocol*, Humana Press.
3 For example: (a) Deming, T. J. (2007) Synthetic polypeptides for biomedical applications. *Prog. Polymer Sci.*, **32**, 858–875.; (b) Aliferis, T., Iatrou, H., Hadjichristidis, N. (2004) Living polypeptides. *Biomacromolecules*, **5**(5),1653–1656.
4 Examples of books on this topic include: (a)Deopura, B.L., Alagirusamy, R., Joshi, M., and Gupta, B. (eds) (2008) *Polyesters and Polyamides*, Taylor and Francis.; (b)Jones, J. (1991) *The Chemical Synthesis of Peptides*, International Series of Monographs on Chemistry, vol. **23**, Oxford Science Publications, Clarendon Press, Oxford.
5 General reviews of polymer biocatalysis can be found in: (a) Cheng, H.N., and Gross, R.A. (eds) (2008) *Polymer Biocatalysis and Biomaterials II*, ACS Symposium Series 999, American Chemical Society.; (b) Cheng, H.N., and Gross, R.A. (eds) (2005) *Polymer Biocatalysis and Biomaterials*, ACS Symposium Series 900, American Chemical Society.; (c) Kobayashi, S., and Makino, A. (2009) Enzymatic polymer synthesis: an opportunity for green polymer chemistry. *Chem. Rev.*, **109**, 5288–5353.; (d) Kobayashi, S., Uyama, H., and Kimura, S. (2001) Enzymatic polymerization. *Chem. Rev.*, **101**, 3793–3818.; (e) Gross, R.A., Kumar, A., and Kalra, B. (2001) Polymer synthesis by *in vitro* enzyme catalysis. *Chem. Rev.*, **101**, 2097–2124.
6 Guzman, F., Barberis, S., and Illanes, A. (2007) Peptide synthesis: chemical or enzymatic. *Electron. J. Biotechnol*, **10** (2), 279–314.
7 Lombard, C., Saulnier, J., and Wallach, J.M. (2005) Recent trends in protease-catalyzed peptide synthesis. *Protein Pept. Lett.*, **12**, 621–629.
8 Kumar, D., and Bhalla, T.C. (2005) Microbial proteases in peptide synthesis: approaches and applications. *Appl. Microbiol. Biotechnol.*, **68**, 726–736.
9 Sergeeva, M.V., Paradkar, V.M., and Dordick, J.S. (1997) Peptide synthesis using proteases dissolved in organic solvents. *Enzyme Microb. Technol.*, **20** (8), 623–628.
10 Clapes, P., Pera, E., and Torres, J.L. (1997) Peptide bond formation by the industrial protease, neutrase, in organic media. *Biotechnol. Lett.*, **19** (10), 1023–1026.
11 Barros, R.J., Wehtje, E., and Adlercreutz, P. (1999) Enhancement of immobilized protease catalyzed dipeptide synthesis by the presence of insoluble protonated nucleophile. *Enzyme Microb. Technol.*, **24** (8–9), 480–488.
12 Barros, R.J., Wehtje, E., and Adlercreutz, P. (2001) Modeling the performance of immobilized alpha-chymotrypsin catalyzed peptide synthesis in acetonitrile medium. *J. Mol. Catal. B Enzym.*, **11** (4–6), 841–850.

13 Kühn, D., Dürrschmidt, P., Mansfeld, J., and Ulbrich-Hofman, R. (2002) Boilysin and thermolysin in dipeptide synthesis: a comparative study. *Biotechnol. Appl. Biochem.*, **36**, 71–76.

14 Salam, S.M.A., Kagawa, K., and Kawashiro, K. (2006) alpha-Chymotrypsin-catalyzed peptide synthesis in frozen aqueous solution using N-protected amino acid carbamoylmethyl esters as acyl donors. *Tetrahedron Asymmetry*, **17** (1), 22–29.

15 Morcelle, S.R., Barberis, S., Priolo, N., Caffini, N.O., and Clapes, P. (2006) Comparative behaviour of proteinases from the latex of Carica papaya and Funastrum clausum as catalysts for the synthesis of Z-Ala-Phe-OMe. *J. Mol. Catal. B Enzym.*, **41** (3–4), 117–124.

16 Quiroga, E., Priolo, N., Obregon, D., Marchese, J., and Barberis, S. (2008) Peptide synthesis in aqueous-organic media catalyzed by proteases from latex of Araujia hortorum (Asclepiadaceae) fruits. *Biochem. Eng. J.*, **39** (1), 115–120.

17 So, J.E., Shin, J.S., and Kim, B.G. (2000) Protease-catalyzed tripeptide (RGD) synthesis. *Enzyme Microb. Technol.*, **26** (2–4), 108–114.

18 Belyaeva, A.V., Bacheva, A.V., Oksenoit, E.S., Lysogorskaya, E.N., Lozinskii, V.I., and Filippova, I.Y. (2005) Peptide synthesis in organic media with the use of subtilisin 72 immobilized on a poly(vinyl alcohol) cryogel. *Russ. J. Bioorganic Chem.*, **31** (6), 529–534.

19 Fukuoka, T., Tachibana, Y., Tonami, H., Uyama, H., and Kobayashi, S. (2002) Enzymatic polymerization of tyrosine derivatives. Peroxidase- and protease-catalyzed synthesis of Poly(Tyrosine)s with different structures. *Biomacromolecules*, **3** (4), 768–774.

20 Capellas, M., Benaiges, M.D., Caminal, G., Gonzalez, G., LopezSantin, J., and Clapes, P. (1996) Enzymatic synthesis of a CCK-8 tripeptide fragment in organic media. *Biotechnol. Bioeng*, **50** (6), 700–708.

21 Xiang, H., and Eckstein, H. (2004) Enzymatic synthesis of a CCK-8 tripeptide fragment. *Chin. J. Chem.*, **22** (10), 1138–1141.

22 Guo, L., Zhang, L.Z., Lu, Z.M., and Xu, Z. (2003) Synthesis of a CCK-8C-terminal tripeptide derivative catalyzed by immobilized enzyme. *Acta Chim. Sin.*, **61** (3), 406–410.

23 Meng, L.P., Joshi, R., and Eckstein, H. (2006) Application of enzymes for the synthesis of the Cholecystokinin Pentapepticle (CCK-5). A contribution to green chemistry. *Chim. Oggi-Chem. Today*, **24** (3), 50–53.

24 Xiang, H., Xiang, G.Y., Lu, Z.M., Guo, L., and Eckstein, H. (2004) Total enzymatic synthesis of cholecystokinin CCK-5. *Amino Acids*, **27** (1), 101–105.

25 Capellas, M., Caminal, G., Gonzalez, G., LopezSantin, J., and Clapes, P. (1997) Enzymatic condensation of cholecystokinin CCK-8 (4-6) and CCK-8 (7-8) peptide fragments in organic media. *Biotechnol. Bioeng.*, **56** (4), 456–463.

26 Fite, M., Clapes, P., Lopez-Santin, J., Benaiges, M.D., and Caminal, G. (2002) Integrated process for the enzymatic synthesis of the octapeptide PhAcCCK-8. *Biotechnol. Prog.*, **18** (6), 1214–1220.

27 Ruiz, S., Feliu, J.A., Caminal, G., Alvaro, G., and LopezSantin, J. (1997) Reaction engineering for consecutive enzymatic reactions in peptide synthesis: application to the synthesis of a pentapeptide. *Biotechnol. Prog.*, **13** (6), 783–787.

28 Joshi, R., Meng, L.P., and Eckstein, H. (2008) Total enzymatic synthesis of the cholecystokinin octapeptide (CCK-8). *Helv. Chim. Acta*, **91** (6), 983–992.

29 Uemura, T., Fujimori, M., Lee, H.-H., Ikeda, S., and Aso, K. (1990) Polyethylene glycol-modified papain catalyzed oligopeptide synthesis from the esters of L-aspartic and L-glutamic acids in benzene. *Agric. Biol. Chem.*, **54** (9), 2277–2281.

30 Aso, K., Uemura, T., and Shiokawa, Y. (1988) Protease-catalyzed synthesis of oligo-L-glutamic acid from L-glutamic acid diethyl ester. *Agric. Biol. Chem.*, **52** (10), 2443–2449.

31 Matsumura, S., Tsushima, Y., Otozawa, N., Murakami, S., Toshima, K., and Swift, G. (1999) Enzyme-catalyzed polymerization of L-aspartate. *Macromol. Rapid Commun.*, **20** (1), 7–11.

32 Soeda, Y., Toshima, K., and Matsumura, S. (2003) Sustainable enzymatic preparation of polyaspartate using a bacterial protease. *Biomacromolecules*, **4**, 196–203.

33 Uyama, H., Fukuoka, T., Komatsu, I., Watanabe, T., and Kobayashi, S. (2002) Protease-catalyzed regioselective polymerization and copolymerization of glutamic acid diethyl ester. *Biomacromolecules*, **3** (2), 318–323.

34 Li, G., Vaidya, A., Viswanathan, K., Cui, J., Xie, W., Gao, W., Gross, R.A. (2006) Rapid regioselective oligomerization of L-glutamic acid diethyl ester catalyzed by papain. *Macromolecules*, **39**, 7915–7921.

35 Li, G., Vaidya, A., Xie, W., Gao, W., and Gross, R.A. (2008) Enzyme-catalyzed oligopeptide synthesis: Rapid regioselective oligomerization of L-glutamic acid diethyl ester catalyzed by papain. *ACS Symp. Ser.*, **999**, 294–308.

36 Li, G., Kodandaraman, V., Xie, E.C., and Gross, R.A. (2009) Protease-catalyzed synthesis of co-oligopeptides consisting of glutamate and leucine residues. *ACS Polym. Preprints*, **50** (2), 60–61.

37 Guinn, R.M., Margot, A.O., Taylor, J.R., Schumacher, M., Clark, D.S., and Blanch, H.W. (1995) Synthesis and characterization of polyamides containing unnatural amino acids. *Biopolymers*, **35** (5), 503–512.

38 Falender, C.A., Blanch, H.W., and Clark, D.S. (1995) Enzymatic oligomerization of the tetrapeptide ester allylglycine-phenylalanine-phenylalanine-allylglycine ethyl ester. *Biocatal. Biotransformation*, **13** (2), 131–139.

39 Gu, Q.-M., Maslanka, W.W., and Cheng, H.N. (2006) Enzymatic synthesis of polyamides. *ACS Polym. Preprints*, **47** (2), 234–235.

40 Margolin, A.L., and Klibanov, A.M. (1987) Peptide synthesis catalyzed by lipases in anhydrous organic solvents. *J. Am. Chem. Soc.*, **109**, 3802–3804.

41 West, J.B., and Wong, C.-H. (1987) Use of nonprotease in peptide synthesis. *Tetrahedron Lett.*, **28**, 1629–1632.

42 Kawashiro, K., Kaiso, K., Minato, D., Sugiyama, S., and Hayashi, H. (1993) Lipase-catalyzed peptide synthesis using Z-amino acid esters as acyl donors in aqueous water-miscible organic solvents. *Tetrahedron*, **49**, 4541–4548.

43 So, J.E., Kang, S.H., and Kim, B.G. (1998) Lipase-catalyzed synthesis of peptides containing D-amino acid. *Enzyme Microb. Technol.*, **23** (3–4), 211–215.

44 Cheng, H.N., Gu, Q.-M., and Maslanka, W.W. (2004) Enzyme-catalyzed polyamides and compositions and porcesses of preparing and using the same. US Patent 6,677,427.

45 Gu, Q.-M., Maslanka, W.W., and Cheng, H.N. (2008) Enzyme-catalyzed polyamides and their derivatives. *ACS Symp. Ser.*, **999**, 309–319.

46 Gu, Q.-M., Michel, A., Cheng, H.N., Maslanka, W.W., and Staib, R.R. (2003) Methyl acrylate-diamine based polyamide resins and processes for producing the same. US Patent 6,667,384.

47 Azim, A., Azim, H., Sahoo, B., and Gross, R.A. (2005) Lipase-catalyzed oligoamide synthesis. *ACS PMSE Preprints*, **93**, 743–744.

48 Panova, A., Dicosimo, R., Brugel, E.G., and Tam, W. (2009) Enzymatic production of macrocyclic amide oligomers. US Patent 7,507,560.

49 Kong, X.-M., Yamamoto, M., and Haring, D. (2008) Method for producing an aqueous polyamide dispersion. US Patent Application 2008/0275182 A1.

50 Kong, X.-M., Yamamoto, M., and Haring, D. (2008) Method for producing an aqueous polyamide dispersion. US Patent Application 2008/0167418 A1.

51 Schwab, L.W., Kroon, R., Schouten, A.J., and Loos, K. (2008) Enzyme-catalyzed ring-opening polymerization of unsubstituted β-lactam. *Macromol. Rapid Commun.*, **29**, 794–797.

52 Schwab, L.W., Kroon, R., Schouten, A.J., and Loos, K. (2009) Enzymatic catalyzed ring-opening polymerization of unsubstituted β-lactam. *ACS Polym. Preprints*, **50** (2), 19–20.

53 Zubay, G. (1973) *In vitro* synthesis of protein in microbial systems. *Annu. Rev. Genet*, **7**, 267–287.

54 Spirin, A.S., Baranov, V.I., Ryabova, L.A., Ovodov, S.Y., and Alakhov, Y.B. (1988) A continuous cell-free translation system capable of producing polypeptides in high yield. *Science*, **242**, 1162–1164.

55 Swartz, J., and Kim, D.M. (2002) Vitro protein synthesis using glycolytic intermediates as an energy source. US Patent 6,337,191.

56 Heinrich, C.P., and Fruton, J.S. (1968) The action of dipeptidyl transferase as a polymerase. *Biochemistry*, **7** (10), 3556–3565.

57 Berg, H., Ziegler, K., Piotukh, K., Baier, K., Lockau, W., and Volkmer-Engert, R. (2000) Biosynthesis of the cyanobacterial reserve polymer multi-L-arginyl-poly-L-aspartic acid (cyanophycin): Mechanism of the cyanophycin synthetase reaction studied with synthetic primers. *Eur. J. Biochem.*, **267**, 5561–5570.

58 Joentgen, W., Groth, T., Steinbuechel, A., Hai, T., and Oppermann, F.B. (2001) Polyasparaginic acid homopolymers and copolymers, biotechnical production and use therof. US Patent 6,180,752.

59 Ziegler, K., Deutzmann, R., and Lockau, W. (2002) Cyanophycin synthetase-like enzymes of non-cyanobacterial eubacteria: characterization of the polymer produced by a recombinant synthetase of *Desulfitobacterium hafniense*. *Z. Naturforsch.*, **57c**, 522–529.

60 Hai, T., Oppermann-Sanio, F.B., and Steinbuechel, A. (2002) Molecular characterization of a thermostable cyanophycin synthetase from the thermophilic cyanobacterium *Synechococcus sp.* strain MA19 and *in vitro* synthesis of cyanophycin and related polyamides. *Appl. Environ. Microbiol.*, **68** (1), 93–101.

61 Ashiuchi, M., and Misono, H. (2002) Biochemistry and molecular genetics of poly-γ-glutamate synthesis. *Appl. Microbiol. Biotechnol.*, **59**, 9–14.

62 Ferone, R., Singer, S.C., and Hunt, D.F. (1986) In vitro synthesis of α-carboxyl-linked folylpolyglutamates by an enzyme preparation from *Escherichia coli*. *J. Biol. Chem.*, **261** (35), 16363–16371.

63 Uyama, H., and Kobayashi, S. (2003) Enzymatic synthesis of polyphenols. *Curr. Org. Chem*, **7** (13), 1387–1397.

64 For example, (a) Tatsuma, T., and Sato, T. (2004) Self-wiring from tyrosinase to an electrode with redox polymers. *J. Electroanalytical Chem.*, **572** (1), 15–19.; (b) Kobayashi, T., Urabe, K., Winder, A., Jimenez-Cervantes, C., Imokawa, G., Brewington, T., Solano, F., Garcia-Borron, J.C., and Hearing, V.J. (1994) Tyrosinase related protein 1 (TRP1) functions as a DHICA oxidase in melanin biosynthesis. *EMBO J.*, **13** (24), 5818–5825.

65 Li, B., Sher, D., Kelly, L., Shi, Y., Huang, K., Knerr, P.J., Joewono, I., Rusch, D., Chisholm, S.W., van der Donk, W.A. (2010) Catalytic promiscuity in the biosynthesis of cyclic peptide secondary metabolites in planktonic marine cyanobacteria. *Proc. Natl. Acad. Sci. USA*, **107** (23),10430–10435.

6
Enzymatic Polymerization of Vinyl Polymers

Frank Hollmann

6.1
Introduction

Probably the first example reporting enzymatic polymerization of vinyl monomers was reported in 1951 by Parravano [1] on the oxidase-initiated polymerization of methyl methacrylate. However, it was only in the 1990s that this reaction type was further investigated. Ever since then, enzymatic polymerizations of vinyl monomers have been experiencing a steadily growing popularity [2–5]. Inspired by the very mild reaction conditions combined with the usually high selectivity, significant benefits of biocatalysis have also been assumed for polymer synthesis.

It has been demonstrated that enzyme-catalyzed vinyl polymerizations enable significant control over polymer properties such as molecular weight, polydispersity and yield. Even though a convincing demonstration of enzyme-derived stereoselectivity is missing so far, this approach exhibits an enormous potential for environmentally benign and economically feasible production of tailored polymers.

This chapter introduces the most predominant enzyme classes used for vinyl polymerization so far. An overview the current mechanistic understanding as well as selected practical examples are given. Complementary to the content of this chapter, detailed enumerations of polymer characteristics can be obtained from some excellent reviews [2–5].

6.2
General Mechanism and Enzyme Kinetics

It is now generally accepted that the so-called 'enzymatic polymerization' or 'enzyme-catalyzed polymerization' of vinyl monomers is best described as enzyme-initiated polymerization (Figure 6.1, initiation). Thus, the principal difference between so-called chemical polymerizations (radical formation by photochemical or thermal homolysis of the initiator precursors) and enzymatic

Biocatalysis in Polymer Chemistry. Edited by Katja Loos
Copyright © 2011 WILEY-VCH Verlag GmbH & Co. KGaA, Weinheim
ISBN: 978-3-527-32618-1

(1) Initiation

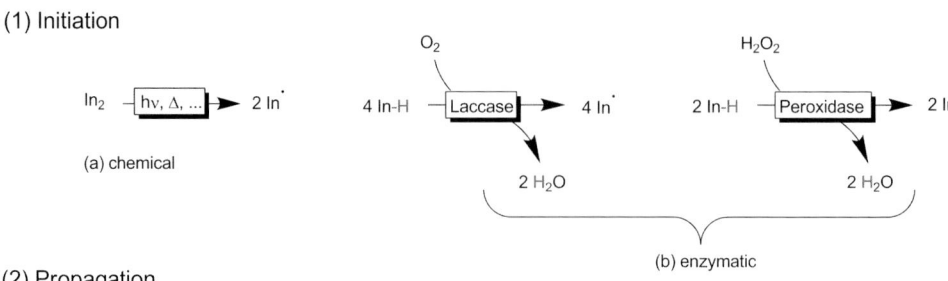

$In_2 \xrightarrow{h\nu, \Delta, ...} 2\ In^\bullet$

(a) chemical

$4\ In\text{-}H \xrightarrow{O_2,\ Laccase} 4\ In^\bullet + 2\ H_2O$

$2\ In\text{-}H \xrightarrow{H_2O_2,\ Peroxidase} 2\ In^\bullet + 2\ H_2O$

(b) enzymatic

(2) Propagation

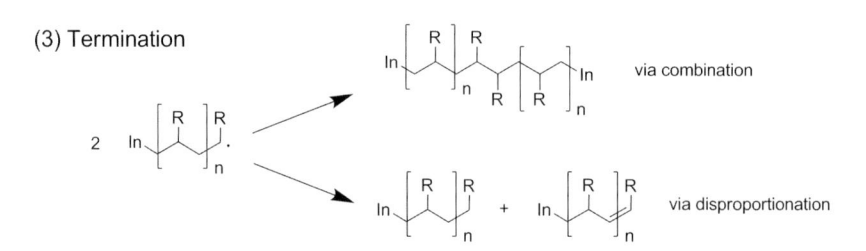

$In^\bullet + CH_2=CHR \longrightarrow In\text{-}CH_2\text{-}CHR^\bullet \longrightarrow In\text{-}(CH_2\text{-}CHR)_n\text{-}...$

(3) Termination

$2\ In\text{-}(CH_2\text{-}CHR)_n^\bullet \longrightarrow$
- $In\text{-}(CH_2\text{-}CHR)_n\text{-}(CHR\text{-}CH_2)_n\text{-}In$ via combination
- $In\text{-}(CH_2\text{-}CHR)_n\text{-}H + In\text{-}(CH_2\text{-}CHR)_n$ via disproportionation

Figure 6.1 Radical chain polymerization. The predominant difference between so-called chemical and enzymatic polymerizations lies in the initiation reaction (1).

polymerization (oxidative radical generation) lies predominantly in the initiation step. The follow-up reactions such as chain propagation and termination (Figure 6.1) are essentially identical.

Among the various radical-forming enzyme systems (Table 6.1), peroxidases and (to a lesser extend) laccases have been used as bio-initiators and will be discussed in detail here. Both systems catalyze an overall hydrogen abstraction (either as such or in a sequence of single electron transfer and deprotonation) from their respective substrates producing the initiator radical. The reducing equivalents liberated from the initiator precursor are transferred to suitable electron acceptors. In the case of peroxidases, hydrogen peroxide serves as electron acceptor, whereas laccases use molecular oxygen. In both cases water is the final product.

The biocatalyst represents the biggest cost factor in the reaction scheme. Therefore it is highly desirable to optimally use the enzyme, that is, providing reaction conditions under which it is most active and stable. General parameters such as reaction temperature, pH, and organic solvents and their influence on enzyme activity and stability have to be thoroughly studied. Apart from these, substrate concentration is an efficient but also double-edged handle to influence the rate

Table 6.1 Enzymatic initiation reactions discussed in this chapter.

Enzyme system (E.C. number) [6]	Initiation reaction	Reference in this chapter
Alcohol oxidase (E.C.1.1.3.13)	$O_2 + CH_3OH \longrightarrow H_2CO + H_2O_2 \xrightarrow{Fe^{II}\text{-EDTA}} \boxed{HO^{\bullet}/HO_2^{\bullet}}$	[5]
Laccase (E.C. 1.10.3.2)	$O_2 + R\text{-CO-CH}_2\text{-CO-}R \longrightarrow H_2O + \boxed{R\text{-CO-}\overset{\bullet}{C}H\text{-CO-}R}$	[4]
	$O_2 + CH_2\text{=}CR\text{-CHO} \longrightarrow H_2O + \boxed{{}^{\bullet}CH\text{=}CR\text{-CHO}}$	[4]
Lipoxygenase (E.C. 1.13.11.12)	unknown	[5]
Peroxidase (E.C. 1.11.1)	$H_2O_2 + R\text{-CO-CH}_2\text{-CO-}R \longrightarrow H_2O + \boxed{R\text{-CO-}\overset{\bullet}{C}H\text{-CO-}R}$	[3]
	$H_2O_2 + CH_2\text{=}CR\text{-CHO} \longrightarrow H_2O + \boxed{{}^{\bullet}CH\text{=}CR\text{-CHO}}$	[3]
Xanthine oxidase (E.C.1.17.1.3)	$O_2 + \text{hypoxanthine} \longrightarrow \text{xanthine} + \boxed{O_2^{\bullet-}}$	[5]

of an enzymatic reaction. Often, enzyme activity shows a saturation-dependency on the substrate concentration applied. According to the kinetic model by Michaelis and Menten (Figure 6.2) [7] this can be rationalized by assuming a reversible binding of the substrate to the enzyme's active site prior (irreversible) conversion. Two parameters, K_M and v_{max} suitably describe this type of kinetics. The so-called Michaelis constant (K_M) is a measure of the affinity of the substrate to the enzyme active site; it corresponds to the substrate concentration at which half-maximal enzyme rate is observed. v_{max} corresponds to the maximal enzymatic conversion rate under given conditions. It is worth mentioning that both laccases and peroxidases act on two substrates (laccases on the initiator precursor and O_2; peroxidases on the initiator precursor and H_2O_2) for each of which Michaelis–Menten kinetics can be assumed.

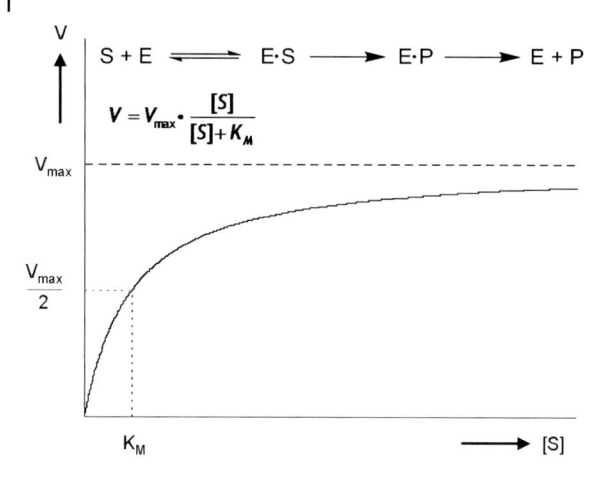

Figure 6.2 Rate-dependence of an enzymatic reaction on the substrate concentration [7]. V: Enzyme rate; E: Enzyme; S: Substrate; ES: Enzyme–substrate complex (formed reversibly); EP: Enzyme–product complex.

$$S + E \rightleftharpoons E \cdot S \rightarrow E \cdot P \rightarrow E + P$$

$$V = V_{max} \cdot \frac{[S]}{[S] + K_M}$$

As becomes obvious from Figure 6.2, enzymatic activity increases with substrate concentration in saturation-type dependence. Thus, on the one hand mere increase of the substrate concentration above a certain level (usually 5–10-fold K_M) does not significantly accelerate the enzymatic initiation reaction. On the other hand, however, high substrate concentrations can dramatically impair enzyme stability (see below). For example, H_2O_2 is not only the oxidant in the peroxidase-initiated polymerization (and therewith a substrate) but also a known inactivator of enzymes. O_2, being the terminal electron acceptor for laccases and therewith inevitable from the initiation reaction, inhibits the chemical polymerization reaction. Also, the solvent properties of many initiators frequently used can impair biocatalysts activity and stability. Hence, for efficient use of the biocatalyst, precisely fine-tuned reaction conditions based on a thorough knowledge of enzyme kinetics and stability is inevitable. Unfortunately, most investigations so far have touched this issue rather superficially.

As initiators predominantly ß-diketones (especially 2,4-pentanedione, acetylacetone, Acac) have been reported (see Figure 6.7 below). Enzymatic H-atom abstraction results in a mesomerically stabilized radical which initiates the polymerization mechanism [8]. A detailed discussion on the influence of the type of mediator and concentration can be found under Section 6.3.2.3.

6.3
Peroxidase-Initiated Polymerizations

Peroxidase-catalyzed radical formation is by far the most popular and best characterized initiation reaction for enzymatic polymerizations. Particularly horseradish peroxidase (HRP) has been used for a variety of polymerization reactions.

6.3.1
Mechanism of Peroxidase-Initiated Polymerization

The utility of peroxidases originates from their natural function as radical initiators for lignin degradation [9] and defense against pathogens via reactive radical species [10]. Ferriprotoporphyrin constitutes the catalytically active prosthetic group of the predominant heme-dependent peroxidases (Figure 6.3).

Hydrogen peroxide (H_2O_2) as well as organic hydroperoxides (R-OOH) serve as stoichiometric oxidants to form two radical equivalents (Figure 6.4). The first step

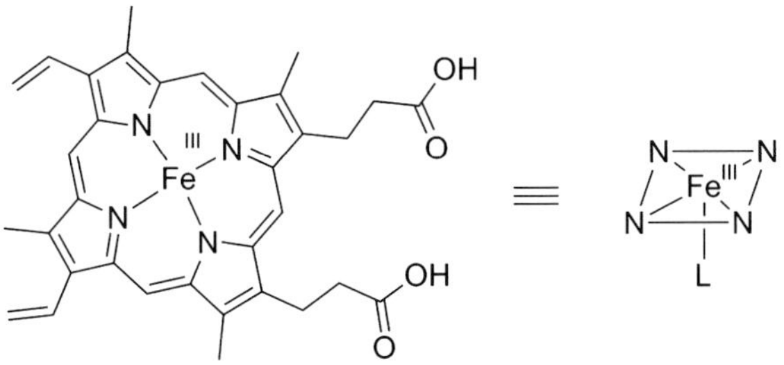

Figure 6.3 Ferriprotoporphyrin IX is the catalytically active prosthetic group of heme-dependent peroxidases. L = histidine (except for chloroperoxidase(CPO) from *Caldariomyces fumago* where L = cysteine).

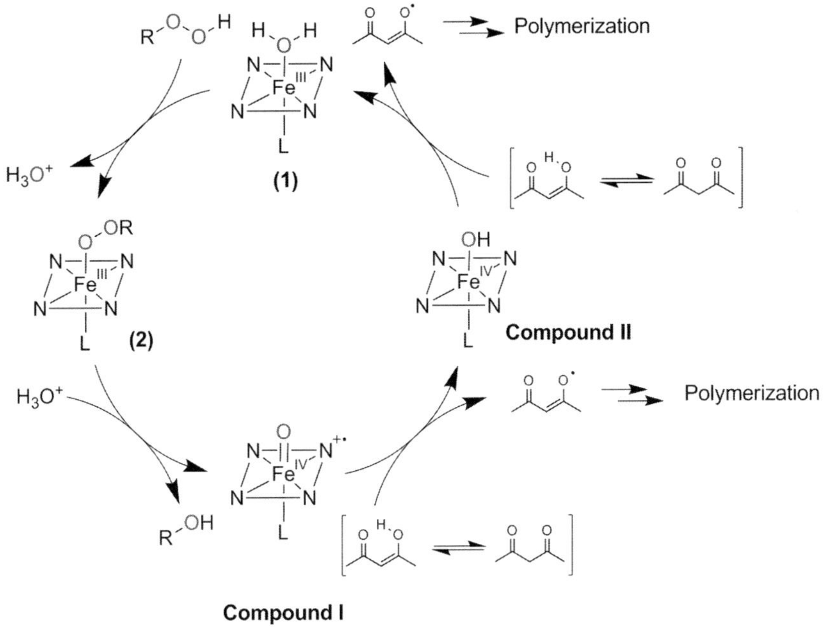

Figure 6.4 The catalytic mechanism of peroxidase-catalyzed radical formation.

of peroxidase-catalysis involves two-electron oxidation of the ferriheme resting state (**1**) via replacement of the water ligand (yielding intermediate (**2**)) Water (or alcohol) elimination produces the catalytically active Compound I. Formally being Fe^V, Compound I is best described as oxyferryl embedded in a porphyrin π–radical cation [11]. In the presence of an electron donor (e.g., Acac) Compound I looses one oxidizing equivalent producing Compound II and one free radical (Acac•). Compound II abstracts one further electron thereby returning into the resting state (**1**) and liberating a second radical (Acac•). Overall, peroxidases function as electron relays transforming a two-electron transfer step (reduction of peroxide to water) into two subsequent single electron transfer steps producing two radical species.

The role of Acac• in the polymerization reaction has been clarified by ESR spectroscopy [8]. It was demonstrated that the initial Acac• radical formed is consumed in the course of the polymerization reaction and substituted by another, C-centered radical attributed to the growing polyacrylamide chain. Using Acac or dibenzoylmethane as initiators, NMR studies revealed the presence of the initiator in the polymer product [12].

6.3.2
Influence of the Single Reaction Parameters

The influence of the single reaction parameters on the efficiency of the peroxidase-catalyzed initiation reaction and polymer properties has been subject of various studies discussed below. Particularly, [H_2O_2], [Acac], and [O_2] have been investigated in detail.

6.3.2.1 Enzyme Concentration

The radical formation rate and thereby the radical chain initiation rate is directly related to the biocatalyst concentration. Thus, the peroxidase concentration is a very efficient handle to control the polymer weight: the higher the biocatalyst concentration, the lower the average polymer weight (Figure 6.5) [13]. At very high enzyme concentration, the initiation reaction is so predominant that only oligomers are formed. Thus, the overall polymer yield is low even though conversion might be quantitative. Under denaturing conditions, low enzyme concentrations manifest in low conversion due to poor enzyme stability (see below) [14].

6.3.2.2 Hydrogen Peroxide Concentration

H_2O_2 plays an ambivalent role in the peroxidase-initiated polymerization reaction. On the one hand, it serves as oxidant and therefore cannot be omitted from the reaction scheme. Above a minimal value, [H_2O_2] hardly influences polymer yield and weight [12, 13, 15]. On the other hand, if [H_2O_2] is too high, polymerization is inhibited twofold: (i) H_2O_2 is a known inactivator of heme [16, 17]. It was proposed that Compound II reacts with another equivalent of H_2O_2 instead of returning to the resting state as depicted in Figure 6.4. The resulting compound III decomposed along various pathways leading to irreversible enzyme inactivation

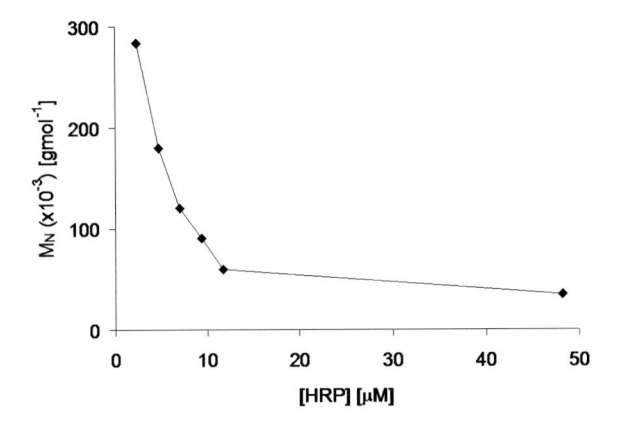

Figure 6.5 Representative example of the influence of [HRP] on the average polymer weight (M_N) [13].

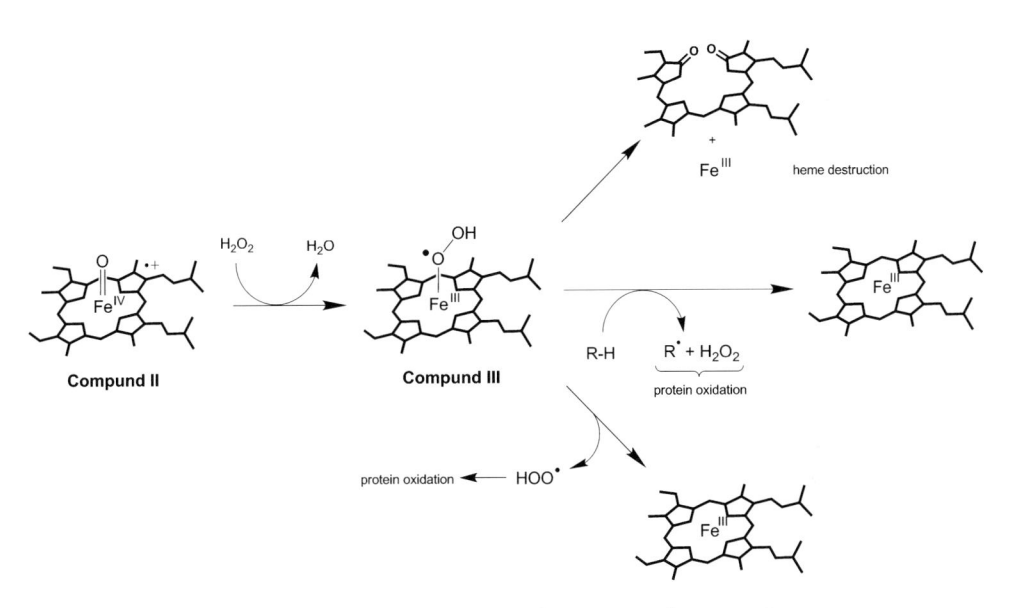

Figure 6.6 Alternative oxidative degradation pathways in the presence of excess H_2O_2. (Adapted from Valderrama et al. [17]).

(Figure 6.6). (ii) Catalase-like activity of peroxidases [18] at high [H_2O_2] results in the generation of O_2 thereby inhibiting the chemical polymerization reaction. This effect is frequently observed in form of a lag-phase in the course of the polymerization reaction [14, 15, 19–22]. During this lag phase, remaining O_2 is consumed, probably by reaction with Acac˙. It should, however, be emphasized that the exact mechanism remains to be elucidated. Polymerization only proceeds if O_2 formed by the catalase-activity has been consumed [23]. Currently, the inhibitory effect of O_2 on radical polymerizations is unclear. Possibly, fast disproportionation of suboxide (O_2^-) into O_2 and H_2O_2 [24] might account for the observed radical quenching.

6.3.2.3 Mediator and Mediator Concentration

Like enzyme concentration, choice of mediator and its concentration can have a major impact on the polymer properties. Kaplan and coworkers demonstrated the influence that the choice of ß-diketone can exceed on yield and polymer characteristics (Table 6.2) [12]. So far, detailed studies clarifying the influence of ß-diketone structure on enzyme activity are missing. Steric and electronic effects influence both, the rate of the enzymatic initiation reaction as well as the rate of the chain growth reaction [12, 25].

It is generally assumed that the enol form represents the actual substrate for the enzymatic radical formation reaction [26]. Thus, the state of the keto–enol equilibrium determines the actual substrate concentration. Principally, an unfavorable equilibrium could be compensated by increasing either the ß-diketone concentration or the pH value. Such measures, however, might be counteracted by decreasing enzyme activity at higher pH values and/or decreased enzyme stability in the presence of high concentrations of the comparably polar ß-diketones [27]. Principally, electron withdrawing substituents should increase the enzyme rate by shifting the keto-enol equilibrium and lowering the redox potential on the one hand. On the other hand, the resulting radical might be too stable due to very low lying SOMO energy (Figure 6.7).

Furthermore, steric properties can significantly influence binding of the substrate to the enzyme active site (influencing K_M) and/or orientation of the substrate in the transition state (influencing v_{max}).

Obviously, the ratio of monomer to initiator has a strong influence on the average size of the resulting polymer. The number of growing polymer chains, competing for the remaining monomer reservoir, increases with initial initiator concentration (Figure 6.8) [13].

There is an ongoing discussion about the necessity of ß-diketones to initiate the polymerization reaction. Earlier reports on ß-diketone-free polymerization of acrylamide [1, 28] could not be reproduced by others [22, 25]. One potential solution of this apparent discrepancy may be found in the very high H_2O_2 concentration in the mediator-free polymerizations resulting in 'unusual' heme-iron species (see Figure 6.6). Thus, highly reactive peroxidase-species might initiate the polymerization reaction by direct H-atom abstraction from the monomer. Alternatively, oxidative degradation of heme resulting of Fe-release into the reaction medium might account for catalytic, Fenton-like generation of reactive oxygen species (ROS) initiating the polymerization reaction. Astonishingly, this issue has not been investigated in detail so far, as it would enable initiator-free polymerization reactions. Such reactions, however, would be highly desirable from an environmental point-of-view.

Table 6.2 Influence of ß-diketone on the HRP-initiated polymerization of styrene.

Initiator	Yield [%]	M_w [×10^{-3} gmol^{-1}]	PD
Styrene polymerization[12]			
acetylacetone	16.7	26.9	2.07
1,3-cyclohexanedione	59.4	67.6	1.98
dihydrofuran-2,4-dione	41.1	50.9	2.22
1-phenyl-1,3-butanedione	14.1	80.1	1.96
1,3-diphenyl-1,3-propanedione	14.4	96.5	2.16
chroman-2,4-dione	14.5	57.2	1.64
Acrylamide polymerization[25]			
acetylacetone	93	124	2.5
2-acetylcyclohexanone	84	56.3	2.9
2-acetylcyclopentanone	76	5.1	4.4
1,3-cyclohexanedione	72	27	3.3
2-methyl-1,3-cyclohexanedione	86	9.8	3.9
1,3-cyclopentanedione	78	84.5	2.7
2-methyl-1,3-cyclopentanedione	38	10.5	3.9

Figure 6.7 ß-Diketones are the most popular radical initiators for chemoenzymatic radical polymerization reactions.

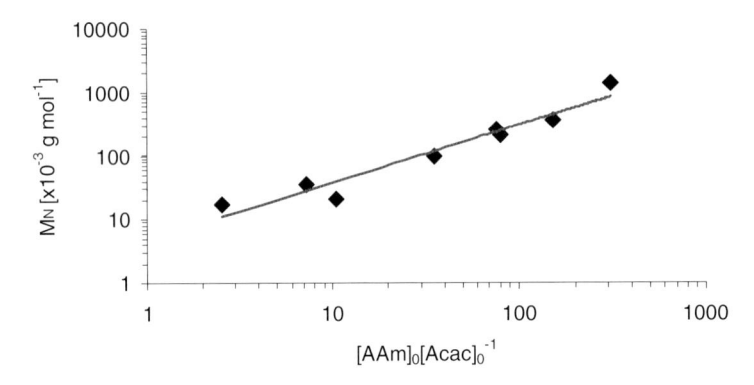

Figure 6.8 Correlation between $[AAm]_0/[Acac]_0$ on the average polymer weight (AAm = acrylamide) [13].

6.3.2.4 Miscellaneous

Especially in case of hydrophobic monomers, solubility becomes a limiting factor for the efficiency of enzyme-initiated polymerizations. Biocatalyst solubilization in organic media may be a viable solution as proposed recently by Li and coworkers [29]. Polar organic solvents can be used to some extend to increase the monomer solubility in the reaction medium [12, 30]. However, the delicate balance between enhanced monomer solubility on the one hand and decreased enzyme stability in the presence of water soluble organic solvents on the other hand [27] has to be considered. Emulsion polymerizations have been reported for highly unpolar substrates such as styrenes [14, 21, 31] or use of unconventional reaction media such as supercritical CO_2 [19].

Gross and coworkers examined the influence of surfactants on the peroxidase-initiated polymerization of acrylamide [21]. Irrespective of the nature of surfactant (cationic, anionic, neutral), they found a significant rate acceleration, which was mostly attributed to a decreased duration of the lag-phase. Interestingly, in chemical polymerization, an activating effect was observed only for anionic surfactants while cationic polar head groups inhibited the polymerization [32, 33].

Enzyme immobilization is an attractive method to stabilize the enzyme against denaturating reaction conditions and to make the biocatalyst recyclable [34–37]. In combination with peroxidase-initiated polymerizations, this has been reported recently by Zhao et al. [38]. Incorporation of HRP into a hydrogel increased both storage and thermal stability of the enzyme; also resistance against H_2O_2 was increased to some extent. Furthermore, the immobilized enzyme preparation could be reused at least four times, albeit with significant activity losses after each cycle.

6.3.3
Selected Examples for Peroxidase-Initiated Polymerizations

To date, approximately 40 publications reporting on peroxidase-catalyzed polymerizations are available. The biocatalyst of choice for these works clearly is HRP. Other catalysts such as soybean peroxidase [30], manganese peroxidase [8], or hematin [15] have been used scarcely. Furthermore, most of these deal with the polymerization of acrylamide and styrene as a model compounds (Table 6.3).

Polymerization of vinyl monomers proceeds much slower than the 'natural' polymerization of for example, phenolic compounds. This can be exploited for the chemoselective polymerization of vinyl-substituted phenols and aminophenols (see Chapter 7) [41, 43]. Kobayashi and coworkers reported the selective polymerization of methacrylate-esters [41]. Thus, phenol-esters were selectively polymerized via the phenolic moieties leaving the methacrylate functionality unmodified (Figure 6.9). The latter was only attacked in the absence of polymerizable phenols.

Another interesting example was reported by Singh and Kaplan [40, 42]. In a bi-enzymatic cascade reaction ascorbate-modified poly(methacrylates) and poly(styrenes) were produced (Figure 6.10). In the first step activated esters of *p*-vinyl benzoic acid were transesterified into corresponding ascorbyl esters using the commercially available *Candida antarctica* lipase B (CALB). Polymerization is initiated by the system HRP/H_2O_2/Acac yielding a novel class of hydrophilic polymers with antioxidant properties which might have interesting applications in food and cosmetic products. However, the use of trifluorethyl esters as starting material appears to be a current limitation of this approach from an environmental point of view.

A very creative application of the system HRP/H_2O_2/Acac for graft polymerization of acryl amide on starch was reported recently by Biswas and coworkers [39]. Thus, a biocatalytic alternative route to the established cerium(IV)-based routes to poly(acrylate)-grafted starch with applications as superabsorbers [44] or performance additives in paper making [45] or textile sizing [46] was reported. Under non-optimized conditions grafting efficiencies of up to 65% are reported. The

Table 6.3 Overview over literature known examples of HRP-initiated polymerizations.

Monomer	Remarks / Reference
acrylamide (CH₂=CH-CO-NH₂)	General [13, 16, 20, 22]; Mediator-free [28]; Various β-diketone mediators [25]; Immobilized enzyme [38]; Water in $^{sc}CO_2$ polymerizations [19]; Surfactants [21]; Poly acrylamide starch graft polymers [39].
methyl methacrylate	[30]
2-hydroxyethyl methacrylate	Mediator-free [28]
ascorbyl methacrylate	[40]
sodium acrylate	Significantly slower than acrylamide polymerization [21]
2-phenylethyl methacrylate	Mediator-free, corresponding phenols were selectively polymerized via the phenolic moiety [41]
styrene	Emulsion polymerization [14, 31]
ascorbyl 4-vinylbenzoate	[42]
substituted styrene (R)	[12, 15]

authors suggest covalent attachment of the poly(acrylamide) chain to the starch backbone via H-abstraction at the glycosidic C-atom (Figure 6.11).

Obviously, the biocatalytic grafting mechanism differs significantly from the chemical mechanism [47] and further studies will be necessary to clarify the mechanism.

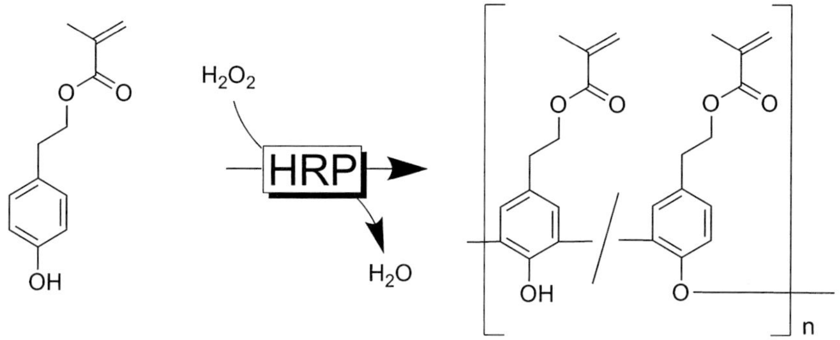

Figure 6.9 Chemoselective polymerization of bifunctional (phenol/vinyl) monomer units using HRP [41].

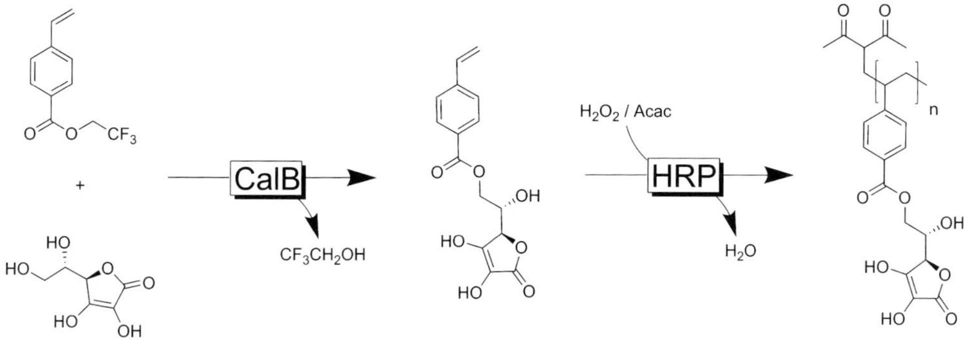

Figure 6.10 Bi-enzymatic cascade reaction to ascorbyl-modified poly(styrenes) [40, 42].

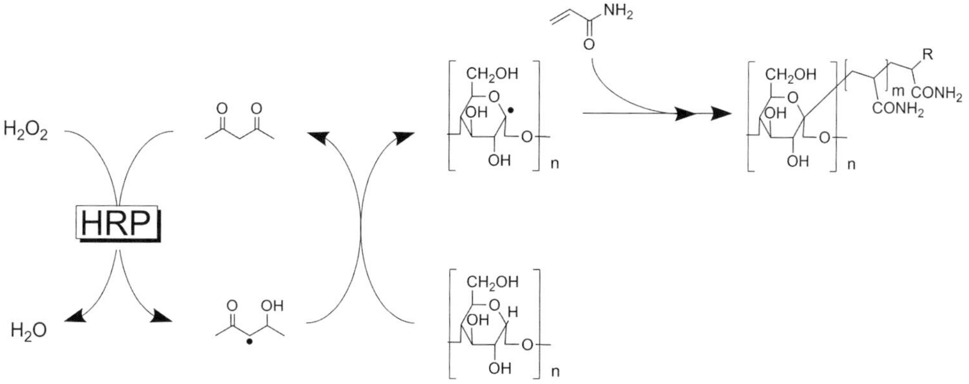

Figure 6.11 Proposed mechanism for the HRP-initiated poly(acrylamide)-grafting of starch [39].

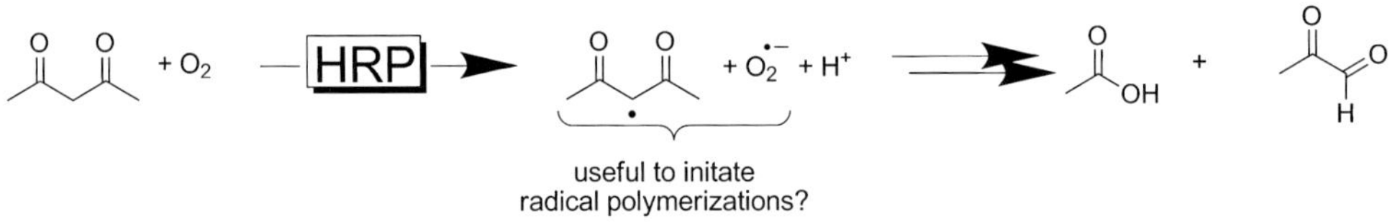

Figure 6.12 A putative useful alternative, H_2O_2-independent access to Acac-radicals to initiate vinyl polymerizations [48].

Recently, Ximenes and coworkers reported on the H_2O_2-independent Acac oxidation by HRP in the presence of molecular oxygen [48]. Intermediate formation of Acac· and suboxide was postulated (Figure 6.12) which might potentially be usefully exploited for the initiation of radical polymerizations. If feasible, such a pathway might open up new possibilities while circumventing the limitations of H_2O_2-promoted chain initiation reactions mentioned above. It remains to be shown whether such a mechanism might be fruitfully exploited for peroxidase-initiated polymerization reactions.

6.4
Laccase-Initiated Polymerization

Laccases (E.C. 1.10.3.2) belong to the so-called blue-copper oxidases predominantly found in fungi but also in plants and insects [49, 50]. Their physiological roles vary depending on the host organism from sclerotization in insects to the formation of UV-resistant spores in some bacteria. White-rot fungi excrete laccases to facilitate the delignification processes. The various applications of laccases to polymerize phenolic monomers and conducting polymers are discussed in detail in Chapters 7 and 8.

Laccases catalyze hydrogen abstraction reactions from phenolic and related substrates resulting in corresponding phenoxy radicals. Laccases contain four copper ions classified in one T1 copper ion and a T2/T3 cluster. It has been shown that the T1 site is the primary redox center accepting electrons from the electron donors. Thus, the fully oxidized laccase is transformed via four successive, fast single electron transfer (SET) steps into the fully reduced laccase. Molecular oxygen interacts with the fully reduced (T2/T3) cluster via a fast 2-electron-transfer process. The resulting peroxide is tightly bound so that H_2O_2 release prior the second 2-electron-transfer is efficiently prevented. As a result, the fully oxidized form of laccases comprising a μ_3-oxo-bridged trinuclear (T2/T3) site is formed. This structure is thermodynamically relatively stable and provides the driving force for the overall process and also provides an efficient electron transfer bridge to regenerate the fully reduced state (Figure 6.13).

Next to their 'natural' phenolic substrates, laccases also accept a range of non-phenolic substrates such as ABTS, syringaldazine, and TEMPO and derivates

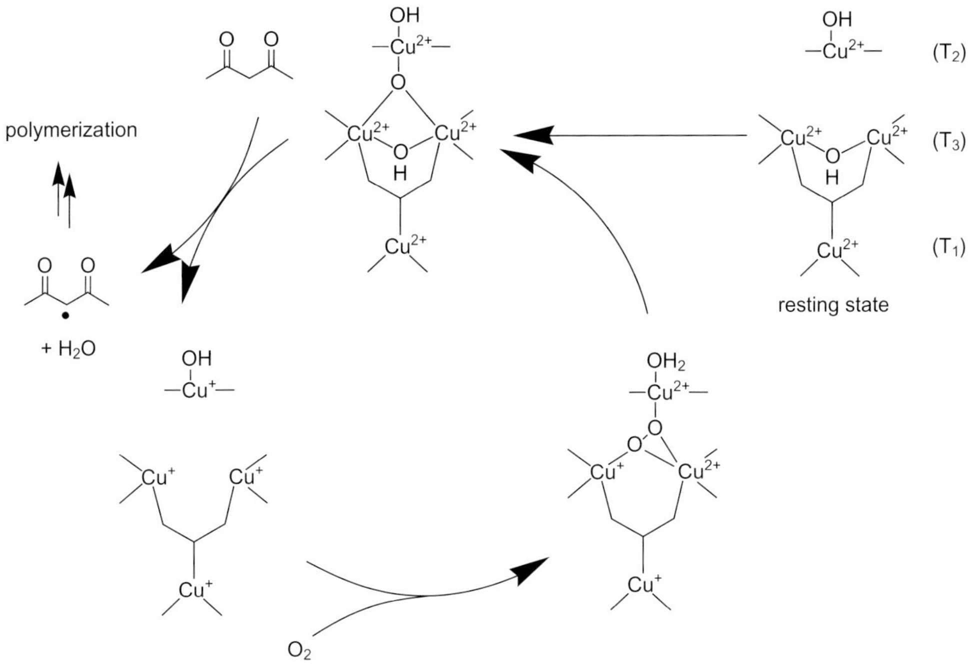

Figure 6.13 Schematic mechanism for the aerobic H-atom abstraction from Acac: [51–53].

[50]. Especially the latter found some interest in the so-called laccase-mediator-system (LMS) for selective, chemo-enzymatic oxidation of alcohols [50, 53–66]. ß-Diketones were established as laccase substrates about a decade ago [67], paving the way to laccase-initiated polymerization of vinyl monomers [68, 69]. Interestingly enough, the number of publications dealing with laccase-initiated polymerization falls back significantly behind peroxidase-initiated polymerization studies (Section 6.3). This is astonishing insofar as the laccase-based systems appear to be significantly easier. One apparent advantage is that laccases use molecular oxygen instead of hydrogen peroxide as oxidant. Thereby, the hazardous side-effects of the latter (Section 6.3.2.2) can be circumvented leading to simpler and more robust polymerization protocols.

The influence of [O_2], [laccase] on the performance of the laccase-initiated polymerization of vinyl monomers is essentially the same as observed with peroxidases (6.3.2) [69]. Thus, the polymer characteristics are inversely correlated with [laccase] and [monomer]. [Acac] exceeds only in minor effect on the polymer weight with the exception of very low [Acac] [69]. Here, no polymerization was observed unless the [O_2] was reduced significantly. This indicates an O_2-dependent irreversible inactivation mechanism of the Acac radical. The mechanism of this inactivation however remains to be elucidated. On the other hand, [O_2] may be an efficient handle to control the average molecular weight of the resulting

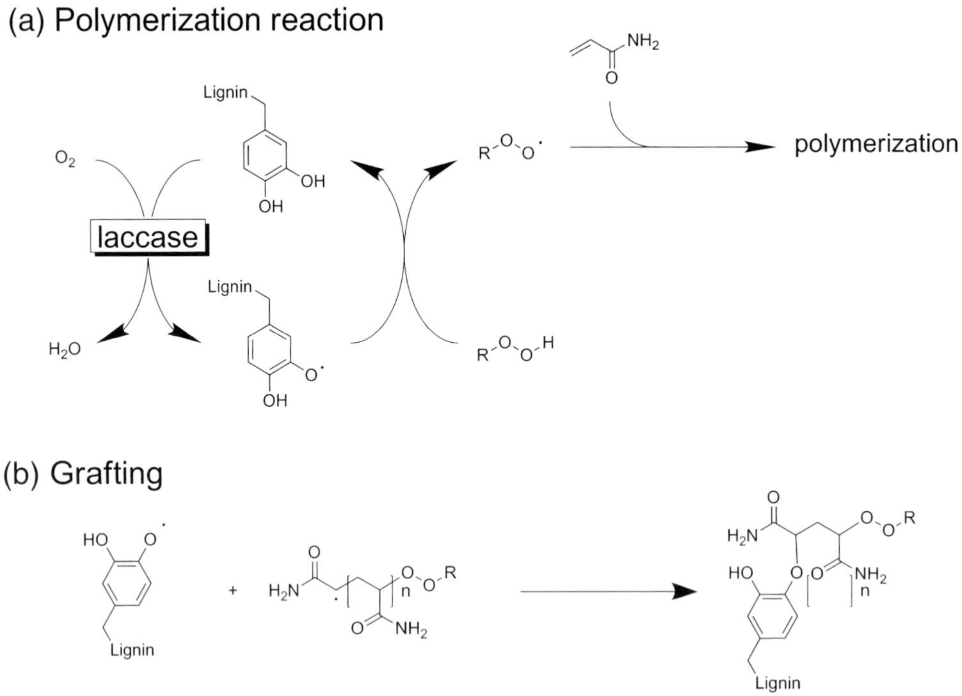

(a) Polymerization reaction

(b) Grafting

Figure 6.14 Proposed mechanism of laccase/hydroperoxide polymerization grafting of acrylamide onto lignin [72].

polymer. In contrast to peroxidases, efficient polymerization using laccases as mediators was only efficient at elevated temperatures (>50 °C). It remains to be clarified whether this is due to poor activity of laccases with ß-diketones, decreased O_2-solubility, or a combination of both [68, 69].

Kobayashi and coworkers reported on the mediator-free direct polymerization of acryl amide using the laccase from *Pycnoporus coccineus* suggesting a simplified polymerization scheme [67]. Using the laccase from *Myceliophthora thermophilia*, this was not achieved [68, 69] and a follow-up study validating Kobayashi's initial study is still awaited.

Poly(acrylamide) grafting onto lignin has been reported in the presence of organic hydroperoxides [70, 71]. The proposed mechanism (Figure 6.14) comprises laccase-catalyzed H-atom abstraction from phenolic lignin residues. The resulting phenoxy radicals interact with the peroxides generation alkoxy- and hydroperoxy-radicals which are supposed to initiate the acrylamide polymerization process. It is suggested that grafting of poly(acrylamide) onto lignin occurs by combination of the respective radicals.

Overall, it can be asserted that laccases as initiation catalysts represent an interesting and promising, yet largely unexplored, alternative to the well-established peroxidase-based systems.

6.5
Miscellaneous Enzyme Systems

In addition to the peroxidase and laccases discussed above, a range of other, radical-forming enzymes have been reported in the context of vinyl polymerizations (Table 6.1).

As early as 1951, Parravano demonstrated the polymerization of methyl methacrylate using xanthine oxidase (XO)-catalyzed oxidation of formaldehyde to formic acid thereby producing suboxide (O_2^-) to initiate the polymerization reaction [1]. He also suggested direct partially reduced XO to directly initiate the polymerization process with H-abstraction from methyl methacrylate. The latter assumption however was not confirmed by later studies of Derango et al. [28] as suboxide dismutase (catalyzing the disproportionation of suboxide into O_2 and H_2O_2) efficiently inhibited XO-initiated polymerization of various acrylates (Figure 6.15).

In the same contribution, the authors also introduce a very interesting chemoenzymatic initiation system based on alcohol oxidase in combination with iron salts. Here, H_2O_2 originating for example, from the oxidation of methanol to formaldehyde undergoes Fenton-formation [73] of hydroxyl radicals to initiate polymerization (Figure 6.16).

Kobayashi and coworkers later evaluated various other oxidases observing efficient polymerization with sarcosine oxidase whereas only little polymerization

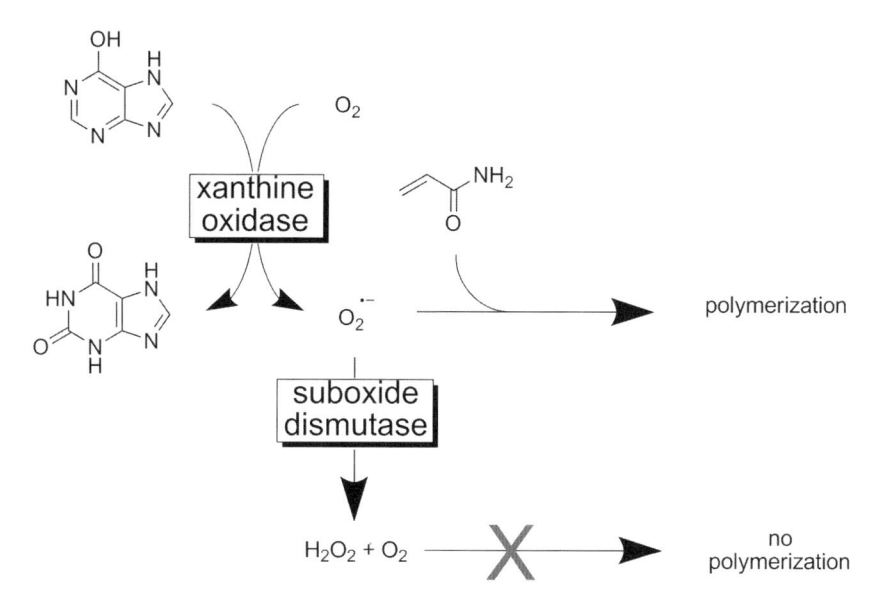

Figure 6.15 Xanthine oxidase (XO)-catalyzed formation of suboxide radicals to initiate acrylamide-polymerization. In the presence of suboxide dismutase no polymerization was observed.

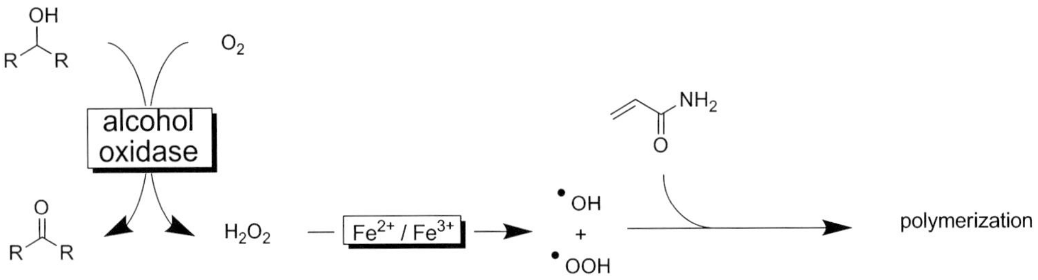

Figure 6.16 Putative mechanism for alcohol oxidase / iron salt-initiated polymerization of acrylamide.

was observed with diamine oxidase. Furthermore, XO and other oxidases such as glucose-, bilirubin-, and choline oxidase were found to be inactive in an initial enzyme screening [68]. Very interestingly, in this work also indications for a potential usefulness of lipoxygenase (from soybean) were found. Further investigations confirming the potential of lipoxygenase are missing so far.

6.6
The Current State-of-the-Art and Future Developments

Enzymatic reactions are enjoying increasing attention as alternatives to classical chemical routes. 'The hallmark of enzyme catalysis is its superior catalytic power and high selectivity under mild reaction conditions.' [5] Furthermore, enzyme catalysis generally exhibits high chemo-, regio-, and enantioselectivity. In the context of vinyl polymerizations these advantages come into play only to a limited extend. All approaches described in this chapter comprise the enzyme-initiated polymerizations wherein the role of the enzyme is confined to the generation of a starter radical. Thus, enzyme-initiated polymerizations comprise all characteristics of a classical chemical polymerization. Putative influences of biocatalysts on the stereochemical outcome of the polymerization reaction are most likely to be attributed to the lower polymerization temperatures.

Admittedly, enzymatic routes offer various handles to control polymer weight with enzyme concentration being the most efficient one. Furthermore, enzymes are derived from renewable feedstock and therefore can be considered being more benign than many classical initiators. But it should be kept in mind that so far enzymes are still comparatively expensive catalysts which, in combination with the bulk-nature of the accessible polymers, makes industrial applications quite unlikely. At the present stage, the major limitation of the current state of the art in enzyme-initiated polymerizations is the poor activity of laccases and peroxidases on ß-diketones requiring too high catalyst loadings to be of commercial interest. Future research should focus on this aspect by (i) evaluating better-suited initiators and/or (ii) improving the catalytic activity of the biocatalyst towards β-diketones. For the latter, enzyme engineering may come up with some significantly improved enzymes overcoming this limitation. Furthermore, mechanistic aspects have not been addressed sufficiently yet. Further investigations in this direction will put the basis for an in-depth understanding of the limiting factors and enable design of more efficient reactions. Also the evaluation of alternative routes, for example, alternative enzyme systems or exploiting unconventional reactivities of the established systems, will open up new possibilities.

Overall, enzyme-initiated polymerization of vinyl monomers represents a promising alternative approach to the established chemical routes. Implementation on the industrial scale will not occur on a short-term but may eventually result in significantly greener production routes (though a full life cycle analysis will be necessary, as enzymatic reactions are not per se greener than their chemical counterparts). Furthermore, it represents another example that enzyme catalysis is not confined to its traditional playgrounds such as chiral molecules.

References

1 Parravano, G. (1951) *J. Am. Chem. Soc.*, **73**, 183–184.
2 Singh, A., and Kaplan, D.L. (2002) *J. Polym. Environ.*, **10**, 85–91.
3 Singh, A., and Kaplan, D. (2006) *Enzyme-Catalyzed Synthesis of Polymers*, vol. 194 (eds S. Kobayashi, H. Ritter, and D. Kaplan), Springer, Berlin / Heidelberg, pp. 211–224.
4 Kobayashi, S., Uyama, H., and Kimura, S. (2001) *Chem. Rev.*, **101**, 3793–3818.
5 Gross, R.A., Kumar, A., and Kalra, B. (2001) *Chem. Rev.*, **101**, 2097–2124.
6 Brenda (2009) http://www.brenda-enzymes.info (accessed 7 July 2009).
7 Michaelis, L., and Menten, M.L. (1913) *Biochem. Z.*, **49**, 333–369.
8 Iwahara, K., Hirata, M., Honda, Y., Watanabe, T., and Kuwahara, M. (2000) *Biotechnol. Lett.*, **22**, 1355–1361.
9 Everse, J., Everse, K.E., and Grisham, M.B. (1990) *Peroxidases in Chemistry and Biology*, CRC Press, Boca Raton.
10 Kawano, T. (2003) *Plant Cell Rep.*, **21**, 829–837.
11 van Rantwijk, F., and Sheldon, R.A. (2000) *Curr. Opin. Biotechnol.*, **11**, 554–564.
12 Singh, A., Ma, D., and Kaplan, D.L. (2000) *Biomacromolecules*, **1**, 592–596.
13 Durand, A., Lalot, T., Brigodiot, M., and Maréchal, E. (2001) *Polymer*, **42**, 5515–5521.
14 Qi, G.G., Jones, C.W., and Schork, F.J. (2006) *Biomacromolecules*, **7**, 2927–2930.
15 Singh, A., Roy, S., Samuelson, L., Bruno, F., Nagarajan, R., Kumar, J., John, V., and Kaplan, D.L. (2001) *J. Macromol. Sci. Part A*, **38**, 1219–1230.
16 Durand, A., Lalot, T., Brigodiot, M., and Maréchal, E. (2000) *Polymer*, **41**, 8183–8192.
17 Valderrama, B., Ayala, M., and Vazquez-Duhalt, R. (2002) *Chem. Biol.*, **9**, 555–565.
18 Hiner, A.N.P., Hernández-Ruiz, J., Williams, G.A., Arnao, M.B., García-Cánovas, F., and Acosta, M. (2001) *Arch. Biochem. Biophys.*, **392**, 295–302.
19 Villarroya, S., Thurecht, K.J., and Howdle, S.M. (2008) *Green Chem.*, **10**, 863–867.
20 Lalot, T., Brigodiot, M., and Maréchal, E. (1999) *Polym. Int.*, **48**, 288–292.
21 Kalra, B., and Gross, R.A. (2002) *Green Chem.*, **4**, 174–178.
22 Emery, O., Lalot, T., Brigodiot, M., and Maréchal, E. (1997) *J. Polym. Sci. A Polym. Chem.*, **35**, 3331–3333.
23 Collinson, E., Dainton, F.S., and McNaughton, G.S. (1957) *Faraday Trans.*, **53**, 476–488.

24 Massey, V. (1994) *J. Biol. Chem.*, **269**, 22459–22462.
25 Teixeira, D., Lalot, T., Brigodiot, M., and Marechal, E. (1999) *Macromolecules*, **32**, 70–72.
26 Baader, W.J., Bohne, C., Cilento, G., and Dunford, H.B. (1985) *J. Biol. Chem.*, **260**, 10217–10225.
27 Laane, C., Boeren, S., Vos, K., and Veeger, C. (1987) *Biotechnol. Bioeng.*, **30**, 81–87.
28 Derango, R., Chiang, L.-C., Dowbenko, R., and Lasch, J. (1992) *Biotechnol. Tech.*, **6**, 523–526.
29 Angerer, P.S., Studer, A., Witholt, B., and Li, Z. (2005) *Macromolecules*, **38**, 6248–6250.
30 Kalra, B., and Gross, R.A. (2000) *Biomacromolecules*, **1**, 501–505.
31 Shan, J., Kitamura, Y., and Yoshizawa, H. (2005) *Coll. Polym. Sci.*, **284**, 108–111.
32 Behari, K., Raja, G.D., and Agarwal, A. (1989) *Polymer*, **30**, 726–731.
33 Gupta, K.C., Verma, M., and Behari, K. (1986) *Macromolecules*, **19**, 548–551.
34 Hanefeld, U., Gardossi, L., and Magner, E. (2009) *Chem. Soc. Rev.*, **38**, 453–468.
35 Iyer, P.V., and Ananthanarayan, L. (2008) *Proc. Biochem.*, **43**, 1019–1032.
36 Sheldon, R.A. (2007) *Adv. Synth. Catal.*, **349**, 1289–1307.
37 Mateo, C., Palomo, J.M., Fernandez-Lorente, G., Guisan, J.M., and Fernandez-Lafuente, R. (2007) *Enzyme Microb. Technol.*, **40**, 1451–1463.
38 Zhao, Q., Sun, J., Ren, H., Zhou, Q., and Lin, Q. (2008) *J. Polym Sci. A Polym. Chem.*, **46**, 2222–2232.
39 Shogren, R.L., Willett, J.L., and Biswas, A. (2009) *Carboh. Polym.*, **75**, 189–191.
40 Singh, A., and Kaplan, D.L. (2004) *J. Polym. Sci. A Polym. Chem.*, **41**, 1377–1386.
41 Uyama, H., Lohavisavapanich, C., Ikeda, R., and Kobayashi, S. (1998) *Macromolecules*, **31**, 554–556.
42 Singh, A., and Kaplan, D.L. (2003) *Adv. Mat.*, **15**, 1291–1294.
43 Goretzki, C., and Ritter, H. (1998) *Macromol. Chem. Phys.*, **199**, 1019–1024.
44 Kiatkamjornwong, S., Mongkolsawat, K., and Sonsuk, M. (2002) *Polymer*, **43**, 3915–3924.
45 Yeng, W.S., Tahir, P.M., Chiang, L.K., Yunus, W.M.Z.W., and Zakaria, S. (2004) *J. Appl. Polym. Sci.*, **94**, 154–158.
46 Hebeish, A., El-Rafie, M.H., Higazy, A., and Ramadan, M. (1996) *Starch*, **48**, 175–179.
47 Fakhru'L-Razi, A., Qudsieh, I.Y.M., Yunus, W.M.Z.W., Ahmad, M.B., and Ab Rahman, M.Z. (2001) *J. Appl. Polym. Sci.*, **82**, 1375–1381.
48 Rodrigues, A.P., Fonseca, L.M., de Faria Oliveira, O.M., Brunetti, I.L., and Ximenes, V.F. (2006) *Biochim. Biophys.*, **1760**, 1755–1761.
49 Kunamneni, A., Camarero, S., Garcia-Burgos, C., Plou, F., Ballesteros, A., and Alcalde, M. (2008) *Microb. Cell Fact.*, **7**, 32.
50 Riva, S. (2006) *Trends Biotechnol.*, **24**, 219–226.
51 Morozova, O., Shumakovich, G., Gorbacheva, M., Shleev, S., and Yaropolov, A. (2007) *Biochemistry (Mosc.)*, **72**, 1136–1150.
52 Christenson, A., Dimcheva, N., Ferapontova, E.E., Gorton, L., Ruzgas, T., Stoica, L., Shleev, S., Yaropolov, A.I., Haltrich, D., Thorneley, R.N.F., and Aust, S.D. (2004) *Electroanal*, **16**, 1074–1092.
53 Witayakran, S., and Ragauskas, A.J. (2009) *Adv. Synth. Catal.*, **351** (9), 1187–1209.
54 Baminger, U., Ludwig, R., Galhaup, C., Leitner, C., Kulbe, K.D., and Haltrich, D. (2001) *J. Mol. Catal. B Enzym.*, **11**, 541–550.
55 Ludwig, R., Ozga, M., Zamocky, M., Peterbauer, C., Kulbe, K.D., and Haltrich, D. (2004) *Biocatal. Biotransf.*, **22**, 97–104.
56 Baiocco, P., Barreca, A.M., Fabbrini, M., Galli, C., and Gentili, P. (2003) *Org. Biomol. Chem.*, **1**, 191–197.
57 Aksu, S., Arends, I.W.C.E., and Hollmann, F. (2009) *Adv. Synth. Catal.*, **351**, 1211–1216.
58 Wells, A., Teria, M., and Eve, T. (2006) *Biochem. Soc. Trans.*, **34**, 304–308.
59 Widsten, P., and Kandelbauer, A. (2008) *Enzyme Microb. Technol.*, **42**, 293–307.
60 Morozova, O.V., Shumakovich, G.P., Shleev, S.V., and Yaropolov, Y.I. (2007) *Appl. Biochem. Microbiol.*, **43**, 523–535.

61 Couto, S.R., and Herrera, J.L.T. (2006) *Biotechnol. Adv.*, **24**, 500–513.
62 Husain, Q. (2006) *Crit. Rev. Biotechnol.*, **26**, 201–221.
63 Fabbrini, M., Galli, C., Gentili, P., and Macchitella, D. (2001) *Tetraherdon Lett.*, **42**, 7551–7553.
64 Arends, I.W.C.E., Li, Y.-X., Ausan, R., and Sheldon, R.A. (2006) *Tetrahedron*, **62**, 6659–6665.
65 Sheldon, R.A., Arends, I.W.C.E., ten Brink, G.J., and Dijksman, A. (2002) *Acc. Chem. Res.*, **35**, 774–781.
66 Sheldon, R.A., and Arends, I.W.C.E. (2004) *Adv. Synth. Catal.*, **346**, 1051–1071.
67 Ikeda, R., Tanaka, H., Uyama, H., and Kobayashi, S. (1998) *Macromol. Rapid Commun.*, **19**, 423–425.
68 Tsujimoto, T., Uyama, H., and Kobayashi, S. (2001) *Macromol. Biosci.*, **1**, 228–232.
69 Hollmann, F., Gumulya, Y., Toelle, C., Liese, A., and Thum, O. (2008) *Macromolecules*, **41**, 8520–8524.
70 Mai, C., Milstein, O., and Hüttermann, A. (1999) *Appl. Microbiol. Biotechnol.*, **51**, 527–531.
71 Mai, C., Milstein, O., and Hüttermann, A. (2000) *J. Biotechnol.*, **79**, 173–183.
72 Witayakran, S., and Ragauskas, A.J. (2009) *Enzyme Microb. Technol.*, **44**, 176–181.
73 Goldstein, S., Meyerstein, D., and Czapski, G. (1993) *Free Radical Biol. Med.*, **15**, 435–445.

7
Enzymatic Polymerization of Phenolic Monomers
Hiroshi Uyama

7.1
Introduction

Phenol-formaldehyde resins using prepolymers such as novolaks and resols are widely used in industrial fields. These resins show excellent toughness and temperature-resistant properties, but the general concern over the toxicity of formaldehyde has resulted in limitations on their industrial preparation and use. Therefore, an alternative process for the synthesis of phenolic polymers avoiding the use of formaldehyde is strongly desired.

In living cells, various oxidoreductases play an important role in maintaining the metabolism of living systems. Most of oxidoreductases contain low valent metals as their catalytic center. *In vitro* enzymatic oxidoreductions have afforded functional organic materials. Some oxidoreductases such as peroxidase, laccase, and polyphenol oxidase have received much attention as catalysts for oxidative polymerizations of phenol derivatives to produce novel polyaromatics [1–10]. This chapter deals with enzymatic oxidative polymerization of phenolic compounds.

7.2
Peroxidase-Catalyzed Polymerization of Phenolics

Peroxidase is the most often used catalyst for enzymatic oxidative polymerizations. Peroxidase is an enzyme whose catalysis is an oxidation of a donor to an oxidized donor by the action of hydrogen peroxide, liberating two water molecules (see also Chapters 6 and 8). Horseradish peroxidase (HRP) is a single-chain β–type hemoprotein that catalyzes the decomposition of hydrogen peroxide at the expense of aromatic proton donors. HRP has an Fe-containing porphyrin-type structure and is well known to catalyze coupling of a number of phenol and aniline derivatives using hydrogen peroxide as oxidant. The catalytic cycle of HRP for a phenol substrate is shown in Scheme 7.1.

The peroxidase-catalyzed oxidative polymerization of phenols proceeds fast in aqueous solutions, giving rise to the formation of oligomeric compounds. However,

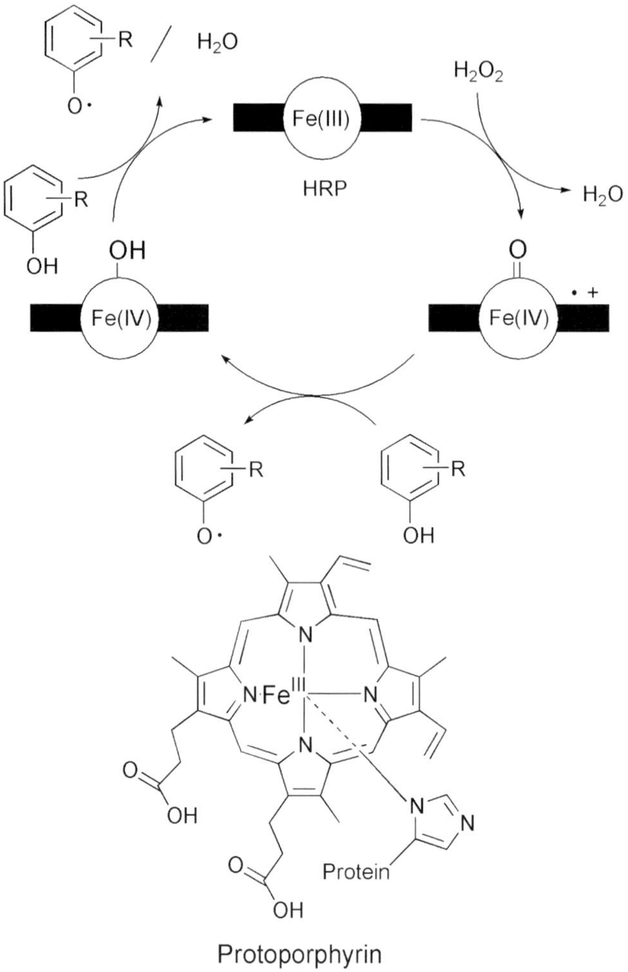

Scheme 7.1

the resulting oligomers have not well been characterized, since most of them show low solubility towards common organic solvents and water. In 1987, the enzymatic efficient synthesis of phenolic polymers was first achieved by using a mixture of water and water-miscible solvents such as 1,4-dioxane, acetone, DMF, and methyl formate as solvent [11]. The HRP-catalyzed polymerization of *p*-phenylphenol proceeded at room temperature and during this process, powdery polymers were precipitated, which were readily collected after the polymerization. The reaction medium composition greatly affected the molecular weight, and the polymer with the highest molecular weight (2.6×10^4) was obtained in 85% 1,4-dioxane.

In the case of phenol, the simplest and most important phenolic compound in industrial fields, conventional polymerization catalysts afford an insoluble product

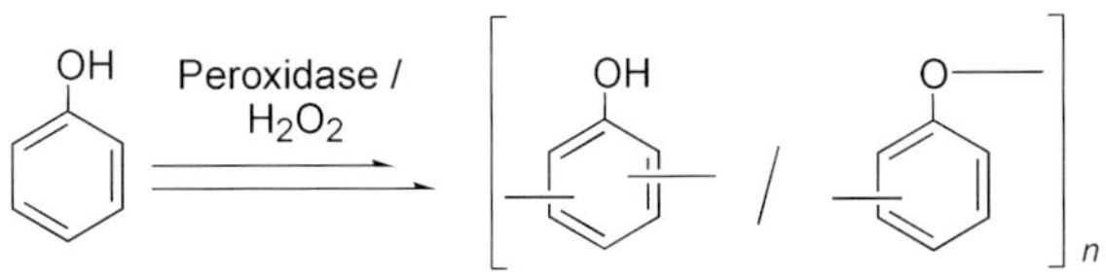

Scheme 7.2

with non-controlled structure since phenol is a multifunctional monomer for oxidative polymerization [12]. Phenol was subjected to the oxidative polymerization using HRP or soybean peroxidase (SBP) as catalyst in a mixture of 1,4-dioxane and buffer, yielding a polymer consisting of phenylene and oxyphenylene units (Scheme 7.2). The polymer showed low solubility; it was partly soluble in DMF and DMSO, and insoluble in other common organic solvents [13, 14]. On the other hand, aqueous methanol afforded a DMF-soluble polymer with molecular weight of 2100–6000 in good yields [15, 16]. The solubility of the resulting polymer strongly depended on the buffer pH and content of the mixed solvent. The resulting phenolic polymer showed relatively high thermal stability and no clear glass transition temperature (T_g) was observed below 300 °C. The molecular weight of the phenolic polymers could be controlled by copolymerization with 2,4-dimethylphenol [17].

Control of the polymer structure was achieved by solvent engineering. The ratio of phenylene and oxyphenylene units was strongly dependent on the solvent composition. In the HRP-catalyzed polymerization of phenol in a mixture of methanol and buffer, the oxyphenylene unit increased by increasing the methanol content, while the buffer pH scarcely influenced the polymer structure [15, 16].

The polymerization behaviors and properties of the phenolic polymers depended on the monomer structure, solvent composition, and enzyme origin. In the HRP-catalyzed polymerization of p-n-alkylphenols in aqueous 1,4-dioxane, the polymer yield increased as the chain length of the alkyl group increased from 1 to 5 [18, 19]. HRP catalyzed the oxidative polymerization of all cresol isomers [20], whereas among o-, m-, and p-isopropylphenol isomers, only p-isopropylphenol polymerized by HRP catalysis. Poly(p-n-alkylphenol)s prepared in aqueous 1,4-dioxane showed low solubility toward organic solvents. On the other hand, a soluble oligomer with molecular weight lower than 1000 was formed from p-ethylphenol using aqueous DMF as a solvent [21]. Poly(p-t-butylphenol) enzymatically synthesized in aqueous 1,4-dioxane showed T_g and melting point (T_m) at 182 °C and 244 °C, respectively.

In the case of the HRP-catalyzed polymerization of p-substituted phenols in an equivolume mixture of a polar organic solvent and phosphate buffer (pH 7), the regioselectivity was influenced by the monomer substituents and the solvent nature [22, 23]. The hydrophobic nature of the monomer substituent and the organic solvent (evaluated as π and log P, respectively) strongly affected the polymer structure. A significant first-order correlation between these parameters and the polymeric structure was observed and the phenolic polymers in a wide range of the unit ratio between the phenylene and oxyphenylene units (from 94/6 to 4/96) were obtained, indicating that the regioselectivity can be controlled by varying the solvent and substituent nature, yielding poly(phenylene) or poly(oxyphenylene).

Efficient phenolic polymer production was achieved by the peroxidase-catalyzed polymerization of m-alkyl substituted phenols in aqueous methanol [24]. The ratio of methanol to buffer greatly affected the yields and the molecular weight of the polymer. The enzyme source greatly affected the polymerization pattern of m-substituted monomers. Using SBP catalyst, the polymer yield increased as a function of the bulkiness of the substituent, whereas the opposite tendency was observed when HRP was the catalyst.

The self-association of m-cresol in aqueous organic solvents was examined to elucidate the solvent effect on the enzymatic polymerization of phenols [25]. Clustering of m-cresol in these solvents was observed by UV absorption spectroscopy and mass spectrometry for clusters. The pattern of the clustering formation in the solvents of different composition was significantly related to the results of the enzymatic polymerization in these mixed solvents, suggesting that the clustering of the phenol monomer in the water-organic solvent mixtures affords the phenolic polymer more efficiently than that in the absence of the organic solvent.

Ionic liquids are effective as co-solvents for the enzymatic oxidative polymerization of phenols [26]. In a mixture of a phosphate buffer and 1-butyl-3-methylimidazolium tetrafluoroborate or 1-butyl-3-methylpyridinium tetrafluoroborate, p-cresol, 1-naphthol, 2-naphthol, and p-phenylphenol were polymerized by SBP to give polymers in good yields.

Peroxidase catalyzed the oxidative polymerization of fluorinated phenols, 3- and 4-fluorophenols, and 2,6-difluorophenol in an aqueous organic solvent, yielding fluorine-containing polymers [27]. Elimination of fluorine atom partly took place during the polymerization giving polymers with complicated structures.

Various bisphenol derivatives were subjected to the peroxidase-catalyzed polymerization. The HRP-catalyzed polymerization of 4,4′-biphenol produced a thermally stable polymer [28]. A phenolic polymer soluble in acetone, DMF, DMSO, and methanol was formed from bisphenol-A via peroxidase catalysis [29]. The polymer was produced in higher yields using SBP as a catalyst and showed T_g at 154 °C. Peroxidase also induced the polymerization of an industrial product, bisphenol-F, consisting of 2,2′-, 2,4′-, and 4,4′-dihydroxydiphenylmethanes [30]. Under the selected reaction conditions, the quantitative formation of a soluble phenolic polymer was achieved. Among the isomers, 2,4′- and 4,4′-dihydroxydiphenylmethanes were polymerized to give the corresponding polymers in high yields, whereas no polymerization of the 2,2′-isomer occurred. In the case of 4,4′-dihydroxyphenyl monomers, the bridge structure enormously affected the polymerization behaviors and the thermal properties of the resulting polymers.

A bienzymatic system was developed as catalyst for the oxidative polymerization of phenol [31]. The HRP-catalyzed polymerization of phenol in the presence of glucose oxidase and glucose gave polymer in a moderate yield, in which hydrogen peroxide was formed *in situ* by the oxidative reaction of glucose catalyzed by glucose oxidase. In this system, no successive addition of hydrogen peroxide was involved.

The enzymatic polymerization of phenols in aqueous solutions often resulted in low yield of the insoluble polymer. The peroxidase-catalyzed polymerization of phenol took place in presence of 2,6-di-O-methyl-α-cyclodextrin (DM-α-CD) in buffer [32]. Only a catalytic amount of DM-α-CD was necessary to induce the polymerization efficiently. Even for water-insoluble m-substituted phenols, the addition of 2,6-di-O-methyl-β-cyclodextrin (DM-β-CD) enabled the enzymatic polymerization in a buffer [33]. The water-soluble complex of the monomer and DM-β-CD was formed, which was polymerized by HRP to give a soluble polymer. Coniferyl alcohol was oxidatively polymerized in the presence of α-CD in an aqueous solution [34].

Polyethylene glycol (PEG) was found to act as a template for an oxidative polymerization of phenol in water [35, 36]. The presence of the PEG template in an aqueous medium greatly improved the regioselectivity of the polymerization, yielding a phenol polymer with a phenylene unit content higher than 90%. During the reaction, polymer was produced in high yields as precipitates in complexing with PEG. The molecular weight of PEG strongly affected the polymer yield. The unit molar ratio of the phenolic polymer and PEG was ca. 1:1. The FT-IR and DSC analyses exhibited the formation of the miscible complex of the phenolic polymer and PEG by hydrogen-bonding interaction. PEG monododecyl ether, a commercially available nonionic surfactant, was also a good template for the polymerization of phenol in water [37]. By using PEG-poly(propylene glycol) (PPG)-PEG triblock copolymer (Pluronic) with a high PEG content as template, a phenolic polymer with ultrahigh molecular weight ($M_w > 10^6$) was formed [38]. The regioselectivity was also high (phenylene unit content of 86%). From other phenols, high molecular weight polymers were obtained as well.

Numerical and Monte Carlo simulations of the peroxidase-catalyzed polymerization of phenols were demonstrated [39]. The monomer reactivity, molecular weight and its index were simulated for precise control of the polymerization of bisphenol-A. In aqueous 1,4-dioxane, aggregates from p-phenylphenol were detected by difference UV absorption spectroscopy [40]. Such aggregate formation might elucidate the specific solvent effects in the enzymatic polymerization of phenols.

The proposed polymerization mechanism is shown in Scheme 7.3. A phenoxy free radical is first formed, then two molecules of the radical species dimerize via coupling. Since peroxidase often does not recognize larger molecules, a radical transfer reaction between a monomeric phenoxy radical and a phenolic polymer takes place to give the polymeric radical species. In the propagation step, such propagating radicals are subjected to oxidative coupling, producing polymers of higher molecular weight.

The mechanistic study on the regioselectivity of the HRP-catalyzed oxidative polymerization was performed by using *in situ* NMR spectroscopy [41, 42]. In the polymerization of 8-hydroxyquinoline-5-sulfonate, the 2, 4, and 7-positions were involved in the oxidative coupling with the order of preference being $7 \geq 2 > 4$. The polymerizability of phenols via HRP catalysis was evaluated by the initial reaction rate [43]. Phenols with electron-donating groups were consumed much

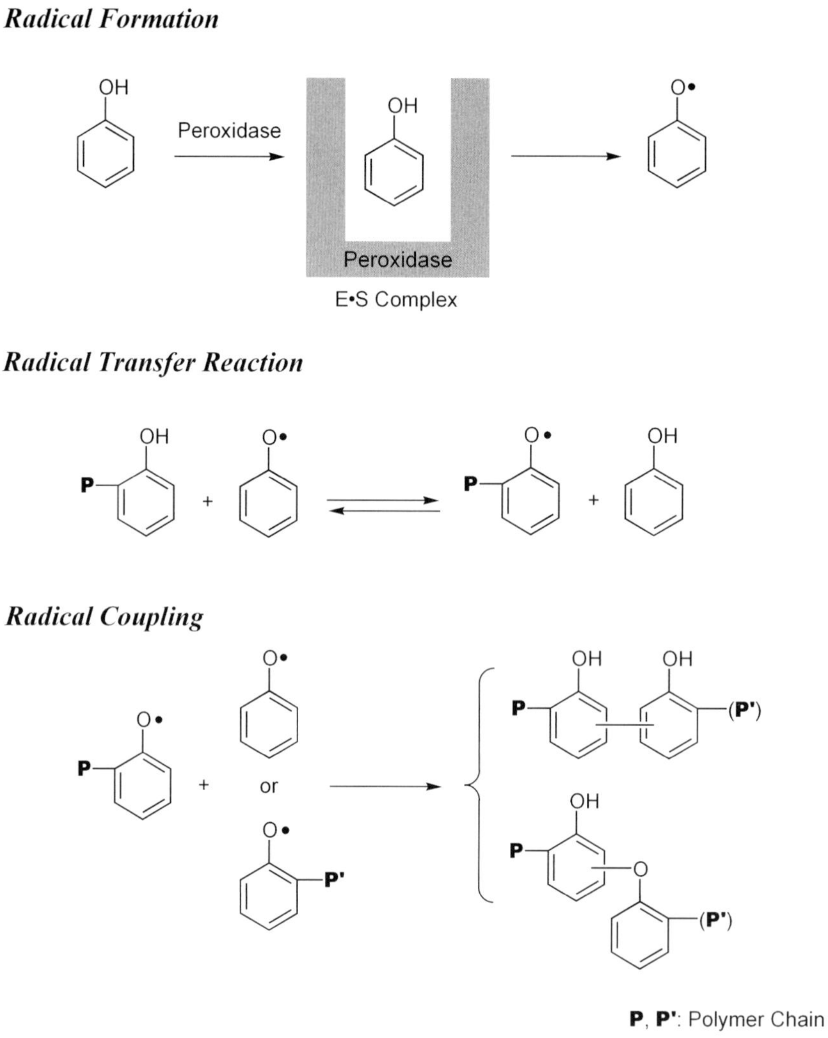

Radical Formation

Radical Transfer Reaction

Radical Coupling

P, P': Polymer Chain

Scheme 7.3

faster than those with electron-withdrawing groups. The reaction rate of para- or meta-substituted phenols was larger than that of ortho-substituted ones.

7.3
Peroxidase-Catalyzed Synthesis of Functional Phenolic Polymers

Peroxidase-catalyzed polymerization of phenols provided a new methodology for functional polymeric materials. Poly(oxy-2,6-dimethyl-1,4-phenylene) (poly(phenylene oxide), PPO) is widely used as a high-performance engineering

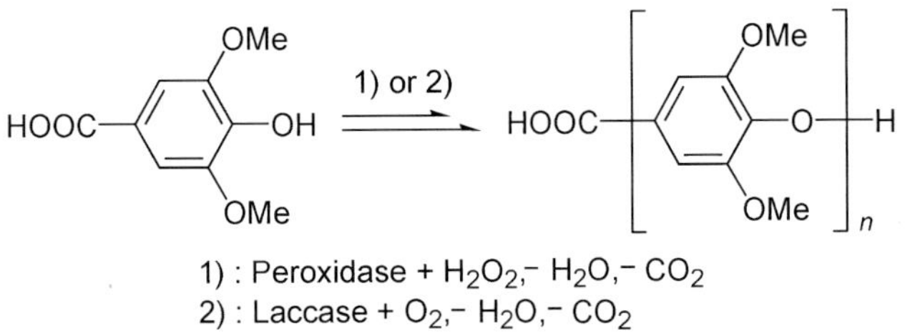

7.3 Peroxidase-Catalyzed Synthesis of Functional Phenolic Polymers

Scheme 7.4

1) : Peroxidase + H_2O_2, – H_2O, – CO_2
2) : Laccase + O_2, – H_2O, – CO_2

plastic, since the polymer has excellent chemical and physical properties, for example, high T_g (ca. 210 °C) and mechanical toughness [12]. PPO was first prepared from 2,6-dimethylphenol monomer using a copper/amine catalyst system [44]. The HRP-catalyzed polymerization of 2,6-dimethylphenol resulted in polymer consisting of exclusively oxy-1,4-phenylene units, while small amount of Mannich-base and 3,5,3'5'-tetramethyl-4,4'-diphenoquinone units are always contained in the chemically prepared PPO. Another PPO derivative was enzymatically obtained from syringic acid (Scheme 7.4) [45]. Both HRP and SBP were active for the polymerization involving elimination of carbon dioxide and hydrogen from the monomer to give polymer with molecular weight up to 1.3×10^4. 4-Hydroxy-3,5-dimethylbenzoic acid was also polymerized to give PPO, on the other hand, the polymerization of non-substituted 4-hydroxybenzoic acid did not occur under similar reaction conditions.

Formation of α-hydroxy-ω-hydroxyoligo(oxy-1,4-phenylene)s was observed in the HRP-catalyzed oxidative polymerization of 4,4'-oxybisphenol in aqueous methanol [46]. During the reaction, the redistribution and/or rearrangement of the quinone-ketal intermediate take place, involving the elimination of hydroquinone to give oligo(oxy-1,4-phenylene)s (Scheme 7.5).

HRP catalysis induced a chemoselective polymerization of a phenol derivative having a methacryloyl group [47]. Only the phenol moiety was polymerized to give a polymer having the methacryloyl group in the side chain. The resulting polymer was readily cured thermally and photochemically (Scheme 7.6). A phenol with an acetylenic substituent in the meta position was also chemoselectively polymerized by HRP to give a polymer bearing acetylenic groups (Scheme 7.7) [48]. For comparison, the reaction of the monomer using a copper/amine catalyst, a conventional catalyst for an oxidative coupling, was performed, producing a diacetylene derivative exclusively. The resulting polymer was converted to a carbon polymer in much higher yields than enzymatically synthesized poly(m-cresol), suggesting a large potential as precursor of functional carbon materials.

Hydroquinone mono-PEG ether was polymerized by HRP in aqueous 1,4-dioxane [49]. High ionic conductivities (4×10^{-5} S cm^{-1}) were found in the film consisting of the lithiated phenolic polymer and PEG. The surface resistivity of poly(p-phenylphenol) doped with nitrosylhexafluorophosphate was around 10^5 Ω [11]. The iodine-doped thin film of poly(phenol-co-tetradecyloxyphenol) showed a

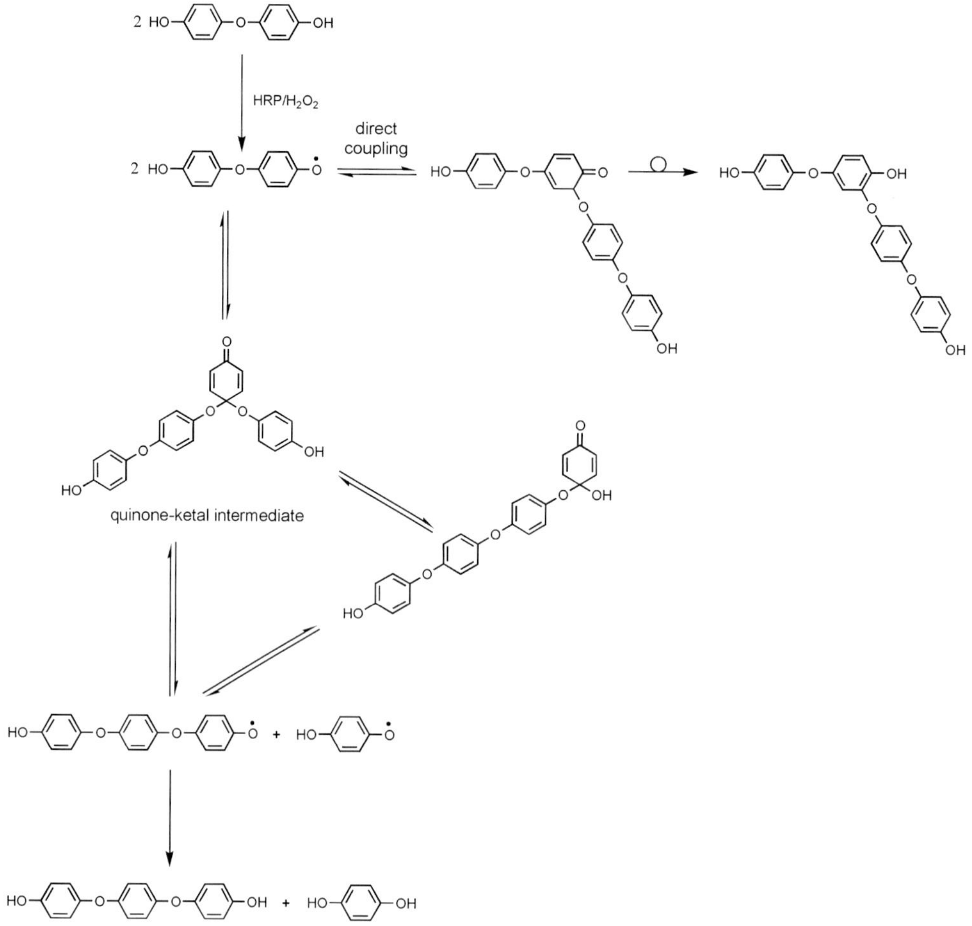

Scheme 7.5

conductivity of $10^{-2}\,\text{S}\,\text{cm}^{-1}$, which was much larger than that of the powdery polymer obtained in aqueous 1,4-dioxane [50]. The third-order optical nonlinearity (χ^3) of this film was 10^{-9} esu. An order of magnitude increase in the third-order nonlinear optical properties was observed in comparison with that prepared in the aqueous organic solution.

Phenolic copolymers containing fluorophores (fluoroscein and calcein) were synthesized by SBP catalysis and used as array-based metal-ion sensor [51]. Selectivity and sensitivity for metal ions could be controlled by changing the polymer components. A combinatorial approach was made for efficient screening of specific sensing of the metals.

A natural phenolglucoside, 4-hydroxyphenyl β-D-glucopyranoside (arbutin), was subjected to regioselective oxidative polymerization using a peroxidase catalyst in a buffer solution, yielding the water-soluble polymer consisting of

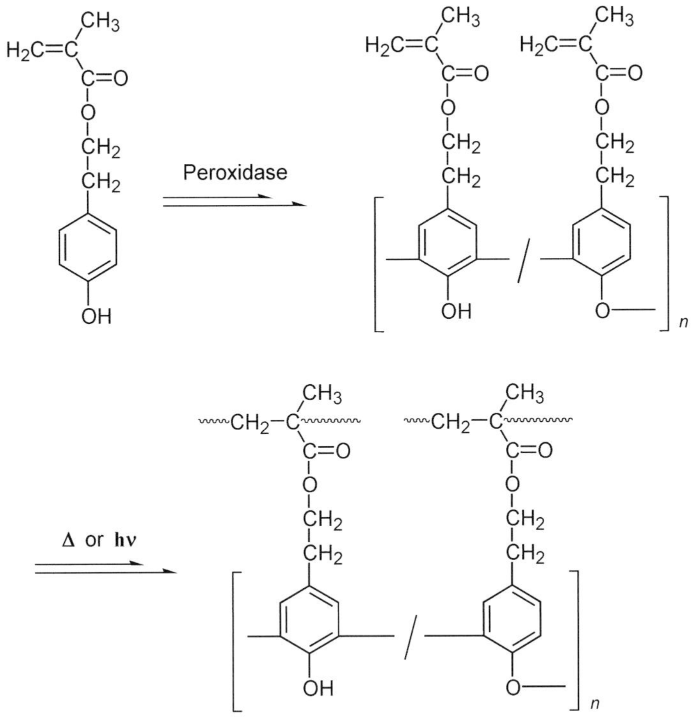

Scheme 7.6

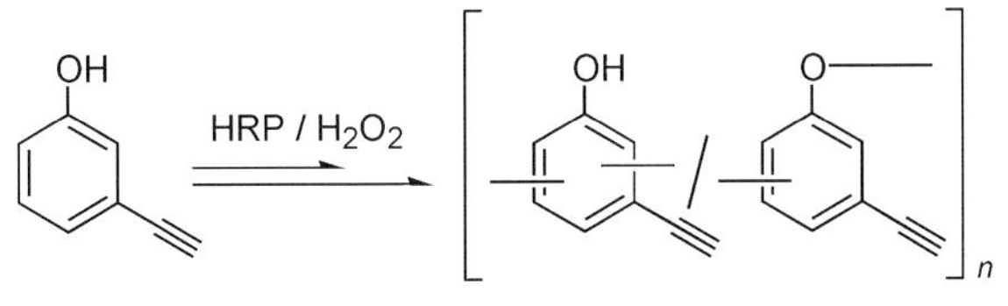

Scheme 7.7

2,6-phenylene units, in turn converted to poly(hydroquinone) by acidic deglycosylation (Scheme 7.8) [52]. The resulting polymer was used for a glucose sensor exploiting its good redox properties [53]. Another route for the chemoenzymatic synthesis of poly(hydroquinone) was the SBP-catalyzed polymerization of 4-hydroxyphenyl benzoate, followed by alkaline hydrolysis [54].

New positive-type photoresist systems based on enzymatically synthesized phenolic polymers were developed [55]. The polymers from the bisphenol monomers exhibited high photosensitivity, comparable with a conventional cresol novolak. Furthermore, this photoresist showed excellent etching resistance. The oxidative polymerization of bisphenol-A proceeded by fungal peroxidase from *Coprinus cinereus* (CiP) in aqueous isopropanol [56]. CiP also catalyzed the oxidative

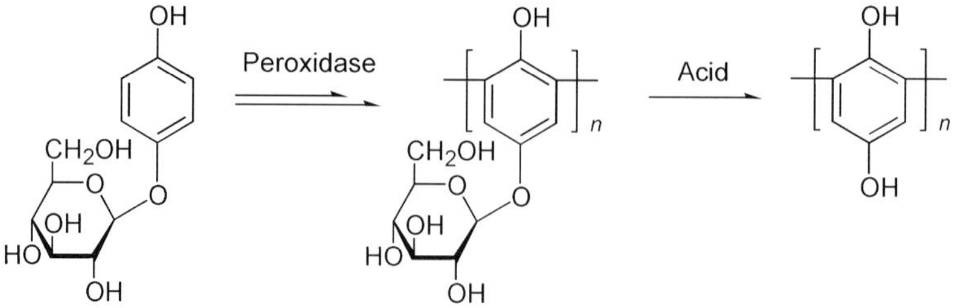

Scheme 7.8

copolymerization of methylene tri *p*-cresol (MTPC) and *m*-cresol in aqueous acetone [57]. Poly(bisphenol-A) and poly(MTPC-*co*-*m*-cresol) with a high hydroxyl value showed high dissolution characteristics which is useful for the application of positive-type photoresists. A novel photoactive azopolymer, poly(4-phenylazophenol), was synthesized using HRP catalyst [58]. The reversible *trans* to *cis* photoisomerization of the azobenzene group with long relaxation time was observed.

A polynucleoside with an unnatural polymeric backbone was synthesized by SBP-catalyzed oxidative polymerization of thymidine 5′-*p*-hydroxyphenylacetate [59]. Chemoenzymatic synthesis of a new class of poly(amino acid), poly(tyrosine) containing no peptide bonds, was achieved by the peroxidase-catalyzed oxidative polymerization of tyrosine ethyl esters, followed by alkaline hydrolysis [60].

Phenolic polymers synthesized via HRP catalysis were used for metal ion adsorption [61]. Base metal ions adsorbed to poly(alkylcatechol)s, but not to poly(*m*-cresol). Gold ions were selectively adsorbed to poly(3-methylcatechol) and poly(4-methylcatechol), leading to the production of gold particles.

In vitro synthesis of lignin, a typical phenolic biopolymer, was attempted by the HRP-catalyzed terpolymerization of *p*-coumaryl alcohol, coniferyl alcohol, and sinapyl alcohol (14:80:6 mol%) in extremely dilute aqueous solutions at pH 5.5 [62]. Dialysis membrane method was applied to the polymerization of coniferyl and sinapyl alcohols, yielding insoluble polymeric materials [63]. Coniferyl alcohol was polymerized by HRP in aqueous acetone to give insoluble polymer [64]. In the presence of a small amount of lignin component, the molecular weight distribution became much broader than that in the absence of lignin [65]. HRP catalyzed the polymerization of soluble oligomeric lignin fragments to yield insoluble polymeric precipitates [66].

Morphology of the enzymatically synthesized phenolic polymers was controlled under the selected reaction conditions. Monodisperse polymer particles in the sub-micron range were produced by the HRP-catalyzed dispersion polymerization of phenol in a mixture of 1,4-dioxane and phosphate buffer (3:2 v/v) using poly(vinyl methyl ether) as stabilizer [67–69]. The particle size could be controlled by the stabilizer concentration and solvent composition. Thermal treatment of these particles afforded uniform carbon particles. The particles could be obtained from various phenol monomers such as *m*-cresol and *p*-phenylphenol.

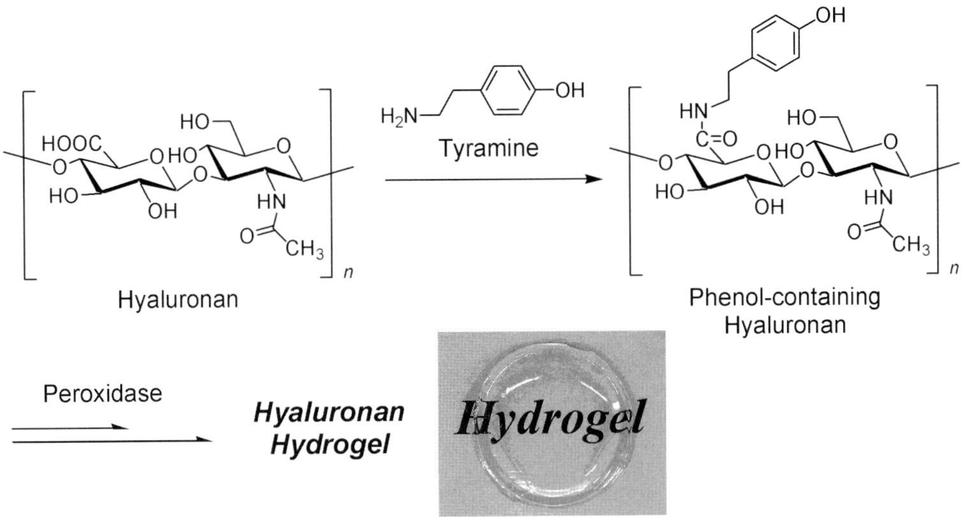

7.3 Peroxidase-Catalyzed Synthesis of Functional Phenolic Polymers | 175

Scheme 7.9

Particles of the enzymatically synthesized phenolic polymers were also formed by reverse micellar polymerization. A thiol-containing polymer was synthesized by peroxidase-catalyzed copolymerization of p-hydroxythiophenol and p-ethylphenol in reverse micelles [70]. CdS nanoparticles were attached to the copolymer to give polymer-CdS nanocomposites. The reverse micellar system was also effective for the enzymatic synthesis of poly(2-naphthol) consisting of quinonoid structure [71], which showed a fluorescence characteristic of the naphthol chromophore. Amphiphilic higher alkyl ester derivatives were enzymatically polymerized in a micellar solution to give surface-active polymers at the air-water interface [72, 73].

Nano-scale polymer patterning was reported to be fabricated by the enzymatic oxidative polymerization of caffeic acid on 4-aminothiolphenol-functionalized gold surface with dip-pen nanolithography technique [74].

HRP was used as catalyst for the crosslinking of peptides (soy proteins and wheat gliadin) [75, 76]. Tyrosine residues of the proteins were subjected to enzymatic oxidative coupling, yielding a network of peptide chains. The treatment increased the tensile strength of the materials. A phenol-containing hyaluronan derivative, which was obtained by the introduction of tyramine into the side chain of hyaluronan, was intermolecularly coupled by HRP to yield a crosslinked hydrogel (Scheme 7.9) [77]. The mechanical strength and gelation rate were independently controlled by selecting the curing conditions. The mechanical strength was precisely tuned by the concentration of hydrogen peroxide to control the degradation rate. Subcutaneous injection showed that the rapid gelation could prevent diffusion of the injected polymer solution and ensure localized gelation at the injection site [78].

This injectable hydrogel system was applied to protein delivery [79]. Negatively charged α-amylase was used as model protein The different release rate was accomplished by controlling the crosslinking density of the hydrogel and the activity of the released α-amylase remained. The injectable bone cement composed of nanocrystalline apatite and crosslinked hyaluronan conjugates was developed by the peroxidase-catalyzed crosslinking of the phenol-containing hyaluronan derivative in the presence of carbonate-substituted apatite and hydroxyapatite pastes [80]. Even in the presence of these inorganics, the enzymatic crosslinking proceeded quickly and the mechanical strength of the cement was controlled by varying the feed ratio and crosslinking conditions.

The tyrosine residue of proteins was directly attached to the peptide chain; thus, the reactivity for the enzymatic crosslinking is not high. Thus, a polypeptide bearing the highly reactive group was designed. 3-(4-Hydroxyphenyl)propionic acid was introduced into the side chain of gelatin by carbodiimide / active ester-mediated coupling reaction [81]. This enzymatically crosslinkable gelatin was applied as substance of hydrogel fibers for homogeneous immobilization of cells.

Some of naturally occurring polysaccharides contain a ferulic ester group in the side chain and they are subjected to enzymatic crosslinking [82, 83]. Pectin extracts containing feruloylated rhamnogalacturonan and feruloylated arabinan groups were crosslinked by HRP to form hydrogels.

Blends of enzymatically synthesized poly(bisphenol-A) and poly(p-t-butylphenol) with poly(ε-caprolactone) (poly(ε-CL)) were examined [84]. FT-IR analysis showed the expected strong intermolecular hydrogen bonding interaction between the phenolic polymer with poly(ε-CL). A single T_g was observed for the blend, and the value increased as a function of the polymer content, indicating their good miscibility in the amorphous state. In the blend of enzymatically synthesized poly(4,4′-oxybisphenol) with poly(ε-CL), both polymers were also miscible in the amorphous phase [85]. The crystallinity of poly(ε-CL) decreased by poly(4,4′-oxybisphenol).

7.4
Laccase-Catalyzed Polymerization of Phenolics

Laccase is a protein containing copper as its active site and uses oxygen as an oxidizing agent (see also Chapters 6 and 8). An oxidative polymerization of phenol and its derivatives was performed using laccase as catalyst without hydrogen peroxide in aqueous organic solvents at room temperature under air [86]. Laccase derived from *Pycnoporus coccineus* (PCL) efficiently induced the polymerization to produce phenolic polymers consisting of a mixture of phenylene and oxyphenylene units. The unit ratio of the polymer could be precisely controlled by selection of the solvent nature and the monomer substituent.

PCL and laccase derived from *Myceliophthore* (ML) were active for the polymerization of syringic acid to give PPO with molecular weight up to 1.8×10^4 [87]. Enzymatic synthesis of PPO was also achieved from 2,6-dimethylphenol using

PCL catalyst [44]. The polymerization of 1-naphthol using laccase from *Trametes versicolor* (TVL) proceeded in an aqueous acetone to give the polymer with molecular weight of several thousands [88].

Coniferyl alcohol was polymerized by laccase catalyst. The polymerization behavior depended on the origin of the enzyme. PCL and laccase from *Coriolus versicolar* showed high catalytic activity to give the dehydrogenative insoluble polymer, whereas very low catalytic activity was observed in laccase from *Rhus vernicifera* Stokes [64]. The increase of the molecular weight was observed in the treatment of soluble lignin using TVL catalyst [89, 90].

A combination of ML and wood vinegar was used to treat the shell surface of boiled eggs [91]. The laccase catalysis enhanced the sterilization effect of phenolic compounds (e.g., guaiacol) contained in the wood vinegar and resulted in prevention of microbial infection of the boiled eggs after storage for 21 days at 40 °C and 75% relative humidity.

Lignophenols having a linear structure were obtained by the surface reaction of a native lignin and phenols in sulfuric acid. Laccase catalyzed the oxidative polymerization of lignocatechol in a mixture of ethanol and phosphate buffer to give the crosslinked polymer [92]. The product showed high affinity for bovine serum albumin and glucoamylase.

7.5
Enzymatic Preparation of Coatings

Urushi is a typical Japanese traditional coating showing excellent toughness and brilliance for a long period. In the early days of this century, pioneering works by Majima revealed that the main important components of urushi are 'urushiols', whose structure is a catechol derivative directly linked to unsaturated hydrocarbon chains consisting of a mixture of monoenes, dienes, and trienes at 3- or 4-position of catechol [93, 94]. Film formation of urushiols proceeds under air at room temperature without organic solvents; hence, urushi seems very desirable for coating materials from the environmental standpoint.

Besides urushiol, similar natural phenolic lipids, laccol from Taiwan and Vietnam and thitsiol from Thailand and Myanmar, are produced in Southeast Asian countries. They are inexpensive and often used for primer coating. The laccase-catalyzed crosslinking of urushiol, laccol, and thitsiol was examined in the presence of a protein hydrolysate [95]. Laccase (PCL and ML) efficiently catalyzed the crosslinking of urushiol and laccol to produce the film with high gloss surface and hardness.

In vitro enzymatic hardening reaction of catechol derivatives bearing an unsaturated alkenyl group at 4-position of the catechol ring proceeded using laccase (PCL) as catalyst to give the crosslinked film showing excellent dynamic viscoelasticity [96].

Fast drying hybrid urushi was developed [97]. Kurome urushi was reacted with silane-coupling agents possessing an amino, epoxy or isocyanate group, resulting

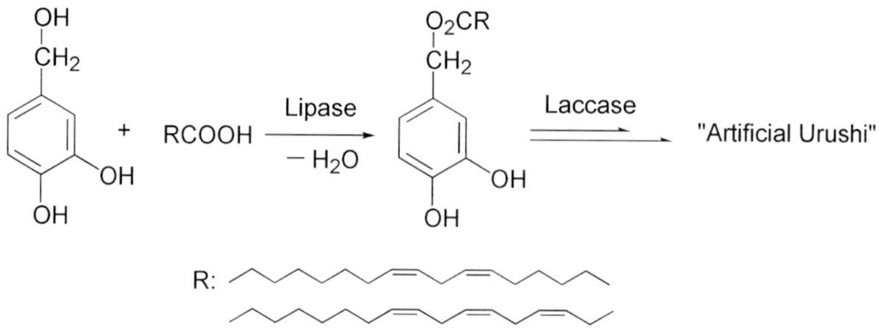

Scheme 7.10

in the shorter curing time of urushi. A water-in-oil type emulsion was prepared by three-roll mill from Vietnam urushi, a protein hydrolysate, and ML [98]. The three-roll mill afforded the emulsion to produce a coating film having high gloss and good physical properties.

So far, few modeling studies of urushi have been attempted, mainly due to the difficult chemical synthesis of the urushiol. New urushiol analogs have been developed by the convenient synthetic process for the preparation of 'artificial urushi' (Scheme 7.10) [99–102]. The urushiol analogs were synthesized by using lipase as catalyst in a single step. These compounds were cured using PCL as catalyst in the presence of acetone powder (AP, an acetone-insoluble part of the urushi sap containing mainly polysaccharides and glycoproteins) under mild reaction conditions without use of organic solvents, yielding the brilliant film ('artificial urushi') with a high gloss surface. Starch-urea phosphate (SP), a synthetic material, was also available as a substitute of AP for *in vitro* enzymatic curing of the urushiol analogs, although the film hardness was smaller than that obtained in the presence of AP. The use of SP as the third component provided the artificial urushi from exclusively synthetic compounds. The viscoelastic properties of the artificial urushi were similar to those of natural urushi.

Cardanol, a main component obtained by thermal treatment of cashew nut shell liquid (CNSL), is a phenol derivative having mainly the meta substituent of a C15 unsaturated hydrocarbon chain with one to three double bonds as the major constituent. Since CNSL is nearly one-third of the total nut weight, much CNSL is obtained as by-product from mechanical processes for the edible use of the cashew kernel. Only a small part of cardanol obtained in the production of cashew kernel is used in industrial fields, though it has various potential industrial uses such as resins, friction-lining materials, and surface coatings. Therefore, development of new applications for cardanol is very attractive.

A new crosslinkable polymer was synthesized by the SBP-catalyzed polymerization of cardanol [103]. When HRP was used as catalyst for cardanol polymerization, the reaction took place in the presence of a redox mediator (phenothiazine derivative) to give the polymer [104].

Peroxidase (CiP) oxidatively polymerized cardol, a major component of CNSL [105]. The efficient polymer production was achieved in an equivolume mixture of t-butylphenol and phosphate buffer (pH 7). The resulting polymer was rapidly cured at room temperature to give the hardened, dry and dark brown color coating. Anacardic acid, separated from CNSL, was also oxidatively polymerized by SBP in the presence of phenothiazine-10-propionic acid to give a polymer with molecular weight of several thousands [106]. The coating of the resulting polymer was effective for antibiofouling against both Gram-positive and Gram-negative bacteria.

Laccase was used to coat flax fibers. The enzymatic treatment of lignocellulosic surface of the fibers was examined in the presence of various phenols [107]. Bacterial growth of *Bacillus subtilis* and *Staphylococcus aureus* was reduced significantly using ferulic acid.

7.6
Enzymatic Oxidative Polymerization of Flavonoids

Bioactive polyphenols are present in a variety of plants and used as important components of human and animal diets [108, 109]. Flavonoids are a broad class of low molecular weight secondary plant polyphenolics, which are benzo-γ-pyrone derivatives consisting of phenolic and pyrane rings. Their biological and pharmacological effects including antioxidant, anti-mutagenic, anti-carcinogenic, antiviral and anti-inflammatory properties have been demonstrated in numerous human, animal and *in vitro* studies.

Major components of polyphenols in green tea are flavanols, commonly known as catechins; the major catechins in green tea are (+)-catechin, (−)-epicatechin (EC), (−)-epigallocatechin (EGC), (−)-epicatechin gallate (ECG), and (−)-epigallocatechin gallate (EGCG) (Scheme 7.11). Numerous biological activities have been reported for green tea and its contents, among them, the preventive effects against cancer are most notable. The HRP-catalyzed polymerization of catechin was carried out in an equivolume mixture of 1,4-dioxane and buffer (pH 7) to give the polymer with molecular weight of 3.0×10^3 in 30% yield [110]. Using methanol as cosolvent improved the polymer yield and molecular weight [111].

In the polymerization of catechin by using laccase (ML) as catalyst, the reaction conditions were examined in detail [112]. A mixture of acetone and acetate buffer (pH 5) was suitable for the efficient synthesis of soluble poly(catechin) with high molecular weight. The mixed ratio of acetone greatly affected the yield, molecular weight, and solubility of the polymer. The polymer synthesized in 20% acetone showed low solubility toward DMF, whereas the polymer obtained in the acetone content less than 5% was completely soluble in DMF. In the UV-Vis spectrum of poly(catechin) in methanol, a broad peak centered at 370 nm was observed. In alkaline solution, this peak was red-shifted and the peak intensity became larger than that in methanol. In the ESR spectrum of the enzymatically synthesized poly(catechin), a singlet peak at $g = 1.982$ was detected, whereas the catechin monomer possessed no ESR peak.

Superoxide anion scavenging activity of the enzymatically synthesized poly(catechin) was evaluated. Poly(catechin), synthesized by HRP catalyst, greatly

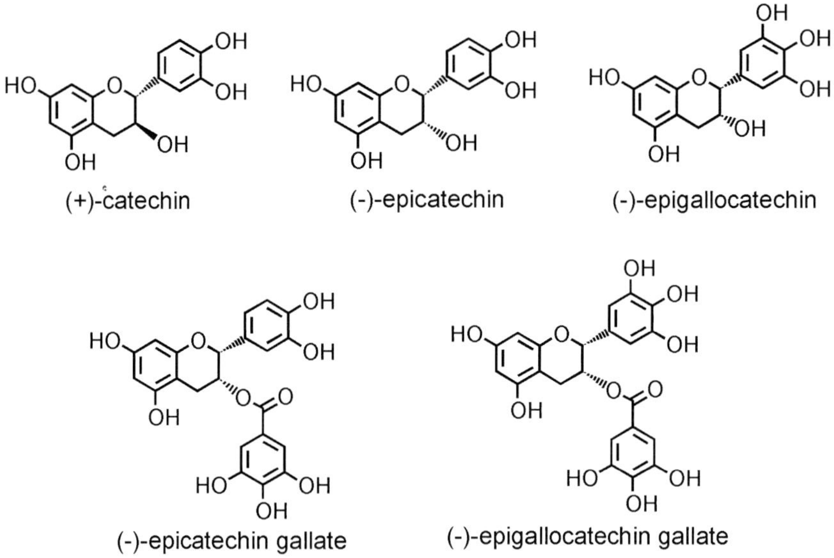

(+)-catechin

(-)-epicatechin

(-)-epigallocatechin

(-)-epicatechin gallate

(-)-epigallocatechin gallate

Scheme 7.11

scavenged superoxide anion in a concentration dependent manner, and almost completely scavenged at 200 μM of a catechin unit concentration [111]. The laccase-catalyzed synthesized poly(catechin) also showed excellent antioxidant properties [112]. It was known that catechin has pro-oxidant property in concentrations lower than 300 μM. These results demonstrated that the enzymatically synthesized poly(catechin) possessed much higher potential for superoxide anion scavenging, compared with intact catechin. Furthermore, the enzymatically synthesized poly(catechin) also showed much greater inhibition activity against human low-density lipoprotein (LDL) oxidation in a concentration dependent manner, comparing to the catechin monomer.

EGCG is a major ingredient of green tea possessing powerful antioxidant activity and cancer-chemopreventive activity due to actions of radical scavenging, enzyme inhibition and metal chelation. The polymer obtained by the laccase-catalyzed oxidative polymerization of EGCG showed much higher superoxide anion scavenging activity than the EGCG monomer and enzymatically synthesized poly(catechin) [113].

Xanthine oxidase (XO) is not only an important biological source of reactive oxygen species but also the enzyme responsible for the formation of uric acid associated with gout leading to painful inflammation in the joints. The XO inhibition effect by the enzymatically synthesized poly(catechin) increased as an increasing concentration of catechin units, while the monomeric catechin showed almost negligible inhibition effect in the same concentration range [111, 112]. This markedly amplified XO inhibition activity of poly(catechin) was considered to be due to effective multivalent interaction between XO and the condensed catechin units in the poly(catechin).

Poly(EGCG) also showed further excellent inhibition effect for XO. The XO inhibition effect of EGCG monomer was quite low with inhibition less than about 5% over a range of tested concentrations. In contrast, poly(EGCG) showed greatly amplified XO inhibition effect in a concentration dependent manner [113]. Moreover, the inhibition of poly(EGCG) was higher than that of allopurinol, frequently used commercial inhibitor for gout treatment. Thus, poly(EGCG) is expected as one of leading candidates of therapeutic molecules against various diseases induced by free radicals and/or enzymes including gout. Kinetic analysis showed that poly(EGCG) is an uncompetitive inhibitor of XO.

Mutans streptococci are the major pathogenic organisms of dental caries in humans. The pathogenicity is closely related to production of extracellular, water-insoluble glucans from sucrose by glucosyltransferase and acid release from various fermentable sugars. Poly(catechin) obtained by HRP catalyst in a phosphate buffer (pH 6) markedly inhibited glucosyltransferase from *Streptococcus sorbrinus* 6715 [114], whereas the inhibitory effect of catechin for this enzyme was very low.

Rutin is one of the most commonly found flavonol glycosides identified as vitamin P with quercetin and hesperidin. An oxidative polymerization of rutin using ML as catalyst was examined in a mixture of methanol and buffer to produce a flavonoid polymer [115]. Under selected conditions, the polymer with molecular weight of several thousands was obtained in good yields. The resulting polymer was readily soluble in water, DMF, and DMSO, although rutin monomer showed very low water solubility. UV measurement showed that the polymer had broad transition peaks around 255 and 350 nm in water, which were red-shifted in an alkaline solution. ESR measurement showed the presence of a radical in the polymer.

The antioxidant activity of rutin was greatly amplified by the laccase-catalyzed oxidative coupling; poly(rutin) also showed much higher scavenging capacity toward superoxide anion than rutin. The protection effects of rutin and poly(rutin) against endothelial cell damage caused by radical generator was reported [115]. Poly(rutin) enhanced cell viability with higher protection effects against the oxidative damage than that of the rutin monomer at the low concentration. In particular, in the high concentration, the polymer exhibited further raised protection relating to a concentration increase. In contrast, the monomer induced fatal cytotoxicity by itself at the same concentration. These results imply that poly(rutin) is a more potent chain-breaking antioxidant when scavenging free radicals in an aqueous system than the monomer.

Peroxidase-catalyzed polymerization of various flavonoids was investigated in an equivolume mixture of 1,4-dioxane and pH 8 buffer [110]. Flavonols (quercetin) as well as isoflavones (diadzein and 5,6,4′-trihydroxyisoflavone) were subjected to the enzymatic oxidative polymerization to produce the flavonoid polymers. From diadzein, the polymer soluble in DMF and DMSO was formed in a high yield.

7.7
Concluding Remarks

In this chapter, enzymatic oxidative polymerizations are reviewed. Advantages for the enzymatic synthesis of useful phenolic polymers are summarized as follows [116]: (i) the polymerization of phenols proceeds under mild reaction conditions without use of toxic reagents (environmentally benign process); (ii) phenol monomers having various substituents are polymerized to give a new class of functional polyaromatics; (iii) the structure and solubility of the polymer can be controlled by changing the reaction conditions; (iv) the procedures of the polymerization as well as the polymer isolation are very facile. As described above, various functional polymers have been developed by using specific catalysis of enzymes, which may be satisfied for the increasing demands in the production of high-performance polymers in material science. Further development of the enzymatic synthesis of phenolic polymers are expected to provide a future essential technology in chemical industry.

References

1 Kobayashi, S., Shoda, S., and Uyama, H. (1995) *Adv. Polym. Sci.*, **121**, 1.
2 Kobayashi, S., Uyama, H., and Ohmae, M. (2001) *Bull. Chem. Soc. Jpn.*, **74**, 613.
3 Gross, R.A., Kumar, A., and Kalra, B. (2001) *Chem. Rev.*, **101**, 2097.
4 Kobayashi, S., Uyama, H., and Kimura, S. (2001) *Chem. Rev.*, **101**, 3793.
5 Uyama, H., and Kobayashi, S. (2002) *J. Mol. Catal. B Enzym.*, **19–20**, 117.
6 Gross, R.A., and Kalra, B. (2002) *Science*, **202**, 803.
7 Kobayashi, S., and Uyama, H. (2002) *Curr. Org. Chem.*, **6**, 209.
8 Reihmann, M., and Ritter, H. (2006) *Adv. Polym. Sci.*, **194**, 1.
9 Uyama, H., and Kobayashi, S. (2006) *Adv. Polym. Sci.*, **194**, 51.
10 Uyama, H. (2006) *Macromol. Biosci.*, **7**, 410.
11 Dordick, J.S., Marletta, M.A., and Klibanov, A.M. (1987) *Biotechnol. Bioeng.*, **30**, 31.
12 Hay, A.S. (1998) *J. Polym. Sci., Polym. Chem. Ed.*, **36**, 505.
13 Uyama, H., Kurioka, H., Kaneko, I., and Kobayashi, S. (1994) *Chem. Lett.*, 423.
14 Uyama, H., Kurioka, H., Sugihara, J., and Kobayashi, S. (1996) *Bull. Chem. Soc. Jpn.*, **69**, 189.
15 Oguchi, T., Tawaki, S., Uyama, H., and Kobayashi, S. (1999) *Macromol. Rapid Commun.*, **20**, 401.
16 Oguchi, T., Tawaki, S., Uyama, H., and Kobayashi, S. (2000) *Bull. Chem. Soc. Jpn.*, **73**, 1389.
17 Mita, N., Tawaki, S., Uyama, H., and Kobayashi, S. (2001) *Polym. J.*, **33**, 374.
18 Kurioka, H., Komatsu, I., Uyama, H., and Kobayashi, S. (1994) *Macromol. Rapid Commun.*, **15**, 507.
19 Uyama, H., Kurioka, H., Sugihara, J., Komatsu, I., and Kobayashi, S. (1997) *J. Polym. Sci. Polym. Chem. Ed.*, **35**, 1453.
20 Uyama, H., Kurioka, H., Sugihara, J., Komatsu, I., and Kobayashi, S. (1995) *Bull. Chem. Soc. Jpn.*, **68**, 3209.
21 Ayyagari, M.S., Marx, K.A., Tripathy, S.K., Akkara, J.A., and Kaplan, D.L. (1995) *Macromolecules*, **28**, 5192.
22 Mita, N., Tawaki, S., Uyama, H., and Kobayashi, S. (2002) *Chem. Lett.*, 402.
23 Mita, N., Tawaki, S., Uyama, H., and Kobayashi, S. (2004) *Bull. Chem. Soc. Jpn.*, **77**, 1523.

24 Tonami, H., Uyama, H., Kobayashi, S., and Kubota, M. (1999) *Macromol. Chem. Phys.*, **200**, 2365.
25 Oguchi, T., Wakisaka, A., Tawaki, S., Tonami, H., Uyama, H., and Kobayashi, S. (2002) *J. Phys. Chem. B*, **106**, 1421.
26 Eker, B., Zagorevski, D., Zhu, G., Linhardt, R.J., and Dordick, J.S. (2009) *J. Mol. Catal. B Enzym.*, **59**, 177.
27 Ikeda, R., Maruichi, N., Tonami, H., Tanaka, H., Uyama, H., and Kobayashi, S. (2000) *J. Macromol. Sci. Pure Appl. Chem.*, **A37**, 983.
28 Kobayashi, S., Kurioka, H., and Uyama, H. (1996) *Macromol. Rapid Commun.*, **17**, 503.
29 Kobayashi, S., Uyama, H., Ushiwata, T., Uchiyama, T., Sugihara, J., and Kurioka, H. (1998) *Macromol. Chem. Phys.*, **199**, 777.
30 Uyama, H., Maruichi, N., Tonami, H., and Kobayashi, S. (2002) *Biomacromolecules*, **3**, 187.
31 Uyama, H., Kurioka, H., and Kobayashi, S. (1997) *Polym. J.*, **27**, 190.
32 Mita, N., Tawaki, S., Uyama, H., and Kobayashi, S. (2002) *Macromol. Biosci.*, **2**, 215.
33 Tonami, H., Uyama, H., Kobayashi, S., Reihmann, M., and Ritter, H. (2002) *e-Polymers*, (003), 1.
34 Nakamura, R., Matsuhsita, Y., Umemoto, K., Usuhi, A., and Fukushima, K. (2006) *Biomacromolecules*, **7**, 1929.
35 Kim, Y.-J., Uyama, H., and Kobayashi, S. (2003) *Macromolecules*, **36**, 5058.
36 Kim, Y.-J., Uyama, H., and Kobayashi, S. (2004) *Polym. J.*, **36**, 992.
37 Kim, Y.-J., Uyama, H., and Kobayashi, S. (2004) *Macromol. Biosci.*, **4**, 497.
38 Kim, Y.-J., Shibata, K., Uyama, H., and Kobayashi, S. (2008) *Polymer*, **49**, 4791.
39 Ryu, K., MaEldoon, J.P., Pokora, A.R., Cyrus, W., and Dordick, J.S. (1993) *Biotechnol. Bioeng.*, **42**, 807.
40 Liu, W., Ma, L., Wang, J.D., Jiang, S.M., Cheng, Y.H., and Li, T.J. (1995) *J. Polym. Sci. Polym. Chem. Ed.*, **33**, 2339.
41 Alva, K.S., Marx, K.A., Kumar, J., and Tripathy, S.K. (1997) *Macromol. Rapid Commun.*, **18**, 133.
42 Alva, K.S., Samuelson, L., Kumar, J., Tripathy, S.K., and Cholli, A.L. (1998) *J. Appl. Polym. Sci.*, **70**, 1257.
43 Xu, Y.-P., Huang, G.-L., and Yu, Y.-T. (1995) *Biotechnol. Bioeng.*, **47**, 117.
44 Ikeda, R., Sugihara, J., Uyama, H., and Kobayashi, S. (1996) *Macromolecules*, **29**, 8702.
45 Ikeda, R., Sugihara, J., Uyama, H., and Kobayashi, S. (1998) *Polym. Int.*, **47**, 295.
46 Fukuoka, T., Tonami, H., Maruichi, N., Uyama, H., Kobayashi, S., and Higashimura, H. (2000) *Macromolecules*, **33**, 9152.
47 Uyama, H., Lohavisavapanich, C., Ikeda, R., and Kobayashi, S. (1998) *Macromolecules*, **31**, 554.
48 Tonami, H., Uyama, H., Kobayashi, S., Fujita, T., Taguchi, Y., and Osada, K. (2000) *Biomacromolecules*, **1**, 149.
49 Mandal, B.K., Walsh, C.J., Sooksimuang, T., and Behroozi, S.J. (2000) *Chem. Mater.*, **12**, 6.
50 Bruno, F.F., Akkara, J.A., Kaplan, D.L., Sekher, P., Marx, K.A., and Tripathy, S.K. (1995) *Ind. Eng. Chem. Res.*, **34**, 4009.
51 Wu, X., Kim, J., and Dordick, J.S. (2000) *Biotechnol. Prog.*, **16**, 513.
52 Wang, P., Martin, D., Parida, S., Rethwisch, D.G., and Dordick, J.S. (1995) *J. Am. Chem. Soc.*, **117**, 12885.
53 Wang, P., Amarasinghe, S., Leddy, J., Arnold, M., and Dordick, J.S. (1998) *Polymer*, **39**, 123.
54 Tonami, H., Uyama, H., Kobayashi, S., Rettig, K., and Ritter, H. (1999) *Macromol. Chem. Phys.*, **200**, 1998.
55 Kadota, J., Fukuoka, T., Uyama, H., Hasegawa, K., and Kobayashi, S. (2004) *Macromol. Rapid Commun.*, **25**, 441.
56 Kim, Y.H., An, E.S., Park, S.Y., Lee, J.-O., Kim, J.H., and Song, B.K. (2007) *J. Mol. Catal. B Enzym.*, **44**, 149.
57 Kim, Y.H., An, E.S., and Song, B.K. (2009) *J. Mol. Catal. B Enzym.*, **56**, 227.
58 Liu, W., Bian, S., Li, L., Samuelson, L., Kumar, J., and Tripathy, S. (2000) *Chem. Mater.*, **12**, 1577.
59 Wang, P., and Dordick, J.S. (1998) *Macromolecules*, **31**, 941.

60 Fukuoka, T., Tachibana, Y., Tonami, H., Uyama, H., and Kobayashi, S. (2002) *Biomacromolecules*, **3**, 768.
61 Kawakita, H., Hamamoto, K., Ohto, K., and Inoue, K. (2009) *Ind. Eng. Chem. Res.*, **48**, 4440.
62 Freudenberg, K. (1965) *Science*, **148**, 595.
63 Tanahashi, M., and Higuchi, T. (1981) *Wood. Res.*, **67**, 29.
64 Okusa, K., Miyakoshi, T., and Chen, C.-L. (1996) *Horzforschung*, **50**, 15.
65 Guan, S.-Y., Mlynár, J., and Sarkanen, S. (1997) *Phytochemistry*, **45**, 911.
66 Guerra, A., Ferraz, A., Cotrim, A.R., and Silva, F.T. (2000) *Enzyme Microb. Technol.*, **26**, 315.
67 Uyama, H., Kurioka, H., and Kobayashi, S. (1995) *Chem. Lett.*, 795.
68 Kurioka, H., Uyama, H., and Kobayashi, S. (1998) *Polym. J.*, **30**, 526.
69 Uyama, H., Kurioka, H., and Kobayashi, S. (1999) *Colloids Surf A Physicochem. Eng. Aspects*, **153**, 189.
70 Premachandran, R., Banerjee, S., John, V.T., McPherson, G.L., Akkara, J.A., and Kaplan, D.L. (1997) *Chem. Mater.*, **9**, 1342.
71 Premachandran, R.S., Banerjee, S., Wu, X.-K., John, V.T., McPherson, G.L., Akkara, J., Ayyagari, M., and Kaplan, D. (1996) *Macromolecules*, **29**, 6452.
72 Sarma, R., Alva, K.S., Marx, K.A., Tripathy, S.K., Akkara, J.A., and Kaplan, D.L. (1996) *Mater. Sci. Engi.*, **C4**, 189.
73 Marx, K.A., Zhou, T., and Sarma, R. (1999) *Biotechnol. Prog.*, **15**, 522.
74 Xu, P., Uyama, H., Whitten, J.E., Kobayashi, S., and Kaplan, D.L. (2005) *J. Am. Chem. Soc.*, **127**, 11745.
75 Stuchell, Y.M., and Krochta, J.M. (1994) *J. Food Sci.*, **59**, 1332.
76 Michon, T., Wang, W., Ferrasson, E., and Guéguen, J. (1999) *Biotechnol. Bioeng.*, **63**, 449.
77 Kurisawa, M., Chung, J.E., Yang, Y.Y., Gao, S.J., and Uyama, H. (2005) *Chem. Commun.*, 4312.
78 Lee, F., Chung, J.E., and Kurisawa, M. (2008) *Soft Matter*, **4**, 880.
79 Lee, F., Chung, J.E., and Kurisawa, M. (2009) *J. Control. Release*, **134**, 186.
80 Pek, Y.S., Kurisawa, M., Gao, S., Chung, J.E., and Ying, J.Y. (2009) *Biomaterials*, **30**, 822.
81 Hu, M., Kurisawa, M., Deng, R., Teo, C.-M., Schumacher, A., Thong, Y.-X., Wang, L., Schumacher, K.M., and Ying, J.Y. (2009) *Biomaterials*, **30**, 3523.
82 Robertson, J.A., Faulds, C.B., Smith, A.C., and Waldron, K.W. (2008) *J. Agric. Food Chem.*, **56**, 1720.
83 Oosterveld, A., Beldman, G., and Voragen, A.G.J. (2000) *Carbohydr. Res.*, **328**, 199.
84 He, Y., Li, J., Uyama, H., Kobayashi, S., and Inoue, Y. (2001) *J. Polym. Sci. Polym. Phys. Ed.*, **39**, 2898.
85 Li, J., Fukuoka, T., He, Y., Uyama, H., Kobayashi, S., and Inoue, Y. (2006) *J. Appl. Polym. Sci.*, **101**, 149.
86 Mita, N., Tawaki, S., Uyama, H., and Kobayashi, S. (2003) *Macromol. Biosci.*, **3**, 253.
87 Ikeda, R., Uyama, H., and Kobayashi, S. (1996) *Macromolecules*, **29**, 3053.
88 Aktas, N., Kibarer, G., and Tanyolac, A. (2000) *J. Chem. Technol. Biotechnol.*, **75**, 840.
89 Milstein, O., Hüttermann, A., Majcherczyk, A., and Schulze, K. (1993) *J. Biotechnol.*, **30**, 37.
90 Milstein, O., Hüttermann, A., Fründ, R., and Lüdemann, H.-D. (1994) *Appl. Microbiol. Biotechnol.*, **40**, 760.
91 Sakaguchi, H., Uyama, N., and Uyama, H. (2007) *Anim. Sci. J.*, **78**, 668.
92 Yoshida, T., Lu, R., Han, S., Hattori, K., Katsuta, T., Takeda, K., Sugimoto, K., and Funaoka, M. (2009) *J. Polym. Sci. Part A Polym. Chem.*, **47**, 824.
93 Majima, R. (1909) *Ber. Dtsch. Chem. Ges.*, **42B**, 1418.
94 Majima, R. (1922) *Ber. Dtsch. Chem. Ges.*, **55B**, 191.
95 Tsujimoto, T., Ando, N., Oyabu, H., Uyama, H., and Kobayashi, S. (2007) *J. Macromol. Sci. Part A Pure Appl. Chem.*, **44**, 1055.
96 Terada, M., Oyabu, H., and Aso, Y. (1994) *J. Jpn. Soc. Colour Mater.*, **66**, 681.

97 Nagase, K., Lu, R., and Miyakoshi, T. (2004) *Chem. Lett.*, **33**, 90.
98 Ando, N., Oyabu, H., Tsujimoto, T., and Uyama, H. (2006) *J. Jpn. Colour Mater.*, **79**, 438.
99 Kobayashi, S., Ikeda, R., Oyabu, H., Tanaka, H., and Uyama, H. (2000) *Chem. Lett.*, 1214.
100 Ikeda, R., Tsujimoto, T., Tanaka, H., Oyabu, H., Uyama, H., and Kobayashi, S. (2000) *Proc. Acad. Jpn.*, **76B**, 155.
101 Ikeda, R., Tanaka, H., Oyabu, H., Uyama, H., and Kobayashi, S. (2001) *Bull. Chem. Soc. Jpn.*, **74**, 1067.
102 Kobayashi, S., Uyama, H., and Ikeda, R. (2001) *Chem. Eur. J.*, **7**, 4754.
103 Ikeda, R., Uyama, H., and Kobayashi, S. (2001) *Polym. J.*, **33**, 540.
104 Won, K., Kim, Y.H., An, E.S., Lee, Y.S., and Song, B.K. (2004) *Biomacromolecules*, **5**, 1.
105 Park, S.Y., Kim, Y.H., Won, K., and Song, B.K. (2009) *J. Mol. Catal. B Enzym.*, **57**, 312.
106 Chelikani, R., Kim, Y.H., Yoon, D.-Y., and Kim, D.-S. (2009) *Appl. Biochem. Biotechnol.*, **157**, 263.
107 Schroeder, M., Aichernig, N., Guebitz, G.M., and Kokol, V. (2007) *Biotechnol. J.*, **2**, 334.
108 Bravo, L. (1998) *Nutr. Rev.*, **56**, 317.
109 Selman, J.S., Swiercz, H.R., and Skrzypczak-Jankun, E. (1997) *Nature*, **387**, 561.
110 Mejias, L., Reihmann, M.H., Sepulveda-Boza, S., and Ritter, H. (2002) *Macromol. Biosci.*, **2**, 24.
111 Kurisawa, M., Chung, J.E., Kim, Y.-J., Uyama, H., and Kobayashi, S. (2003) *Biomacromolecules*, **4**, 469.
112 Kurisawa, M., Chung, J.E., Uyama, H., and Kobayashi, S. (2003) *Macromol. Biosci.*, **3**, 758.
113 Kurisawa, M., Chung, J.E., Uyama, H., and Kobayashi, S. (2004) *Chem. Commun.*, 294.
114 Hamada, S., Kontani, M., Hosono, H., Ono, H., Tanaka, T., Ooshima, T., Mitsunaga, T., and Abe, I. (1996) *FEMS Microbiol. Lett.*, **143**, 35.
115 Kurisawa, M., Chung, J.E., Uyama, H., and Kobayashi, S. (2003) *Biomacromolecules*, **4**, 1394.
116 Uyama, H., and Kobayashi, S. (1999) *Chemtech*, **29** (10), 22.

8
Enzymatic Synthesis of Polyaniline and Other Electrically Conductive Polymers

Rodolfo Cruz-Silva, Paulina Roman, and Jorge Romero

8.1
Introduction

Since their discovery [1], intrinsically conducting polymers (ICPs) have changed our way of thinking about polymers. With properties such as light weight, organic functionality, and distinctive electronic behavior, new applications were soon devised for these materials. Polyaniline (PANI) is one of the most interesting ICPs because of its chemical stability, tunable conductivity, and interesting electrochromic behavior [2]. This polymer is usually synthesized by chemical oxidation of the monomer in strong acidic media. Besides PANI, polythiophenes and polypyrrole are other important ICPs that have also been synthesized by oxidative polymerization. These polymers have been extensively used in several applications in the fields of organic electronics [3], sensors [4], actuators [5], separation technologies [6], and energy storage [7], among others, and new applications for them are found continuously.

Enzymatic polymerizations have become an interesting alternative route for polymer synthesis [8]. They are generally carried out under milder conditions than similar synthetic routes. Enzymes are natural catalysts which are non-toxic, biodegradable, energy efficient and obtained from renewable resources. In this chapter, we review the different strategies that have been used for the enzymatic synthesis of PANI. Xu *et al.* [9] reviewed published work regarding enzymatic polymerization of aniline derivatives from its beginning until 2004. However, more recently, several groups have developed either new approaches for the enzymatic synthesis of PANI or different methods inspired by it. Indeed, the influence of the enzymatic synthesis of PANI can be seen in the development of several biocatalytic and biomimetic synthetic routes not only for the synthesis of PANI, but also for other important conductive polymers, such as polypyrrole (PPy) and poly(3,4-ethylendioxythiophene) (PEDOT).

8.2
PANI Synthesis Using Templates

8.2.1
Polyanion-Assisted Enzymatic Polymerization

Peroxidase activity (see Chapters 6 and 7 for general peroxidase mechanism) is optimal at near neutral pH, therefore the first attempts to synthesize PANI encountered the problem of branching and crosslinking during polymerization because of preferential ortho-coupling and ring multisubstitution [10]. On the other hand, several studies of chemical polymerization had shown that para-coupling of aniline radicals were dominant below pH 4.0; unfortunately, horseradish peroxidase (HRP) was also quickly inactivated at this pH. Another prevailing idea was that partially organic media were needed in order to obtain high M_w polymers derived from aromatic amines [11–13], even though a polymerization study carried out by chemical oxidation of aniline in the presence of organic solvents showed the opposite [14]. Both the processability and the control on the polymer structure were dramatically improved when Samuelson et al. [15] introduced the template-assisted enzymatic polymerization of aniline in 1998 (Figure 8.1). They showed that sulfonated polystyrene (SPS), an anionic water soluble polymer, could be used as a template to induce the para-coupling of aniline radicals, and to provide solubility to the resulting PANI/polyanion complex.

In the following year a couple of studies detailing the role of the template in the enzymatic polymerization of aniline [16], and a complete study of the enzymatic polymerization of aniline [17] were published. These studies showed that a 'local' pH environment was provided by the template, and that the template was crucial in order to obtain the electrically conductive linear form of the polymer. Rapidly, several other negative polyelectrolytes, such as poly(vinylsulfonic) acid [18], polyvinylphosphonic acid [19] and deoxyribonucleic acid (DNA) [20], were assayed as templates in enzymatic polymerization. Datta et al. [21] studied in detail the use of DNA as a template for the enzymatic synthesis of PANI oligomers and found that PANI induced distortion on the DNA molecule and that the critical length for the oligomers, in order to show PANI features, was four aniline units. One interesting feature of the template-assisted enzymatic polymerization was that apparently the enzyme induced specific helical conformation to the enzymatically synthesized PANI, according to Thiyagarajan et al. [22]

Huh et al. [23] synthesized a copolymer of aniline and 3-aminobenzeneboronic acid by using HRP as catalyst and SPS as template. The resulting poly(aniline-co- aminobenzeneboronic acid) has a green color, characteristic of the electrically conductive emeraldine salt form of PANI, and a 1:2 boronic acid to aniline ratio. This boronic-acid-containing PANI was able to complex with glucose, as indicated by the UV-Vis and potentiometric studies. Boronic-acid-containing PANIs, prepared by chemical oxidation, were well known due to their sugar-sensing properties [24], however, the enzymatically synthesized boronic-acid-containing PANI showed superior sensitivity. Nabid and Entezami have recently studied several

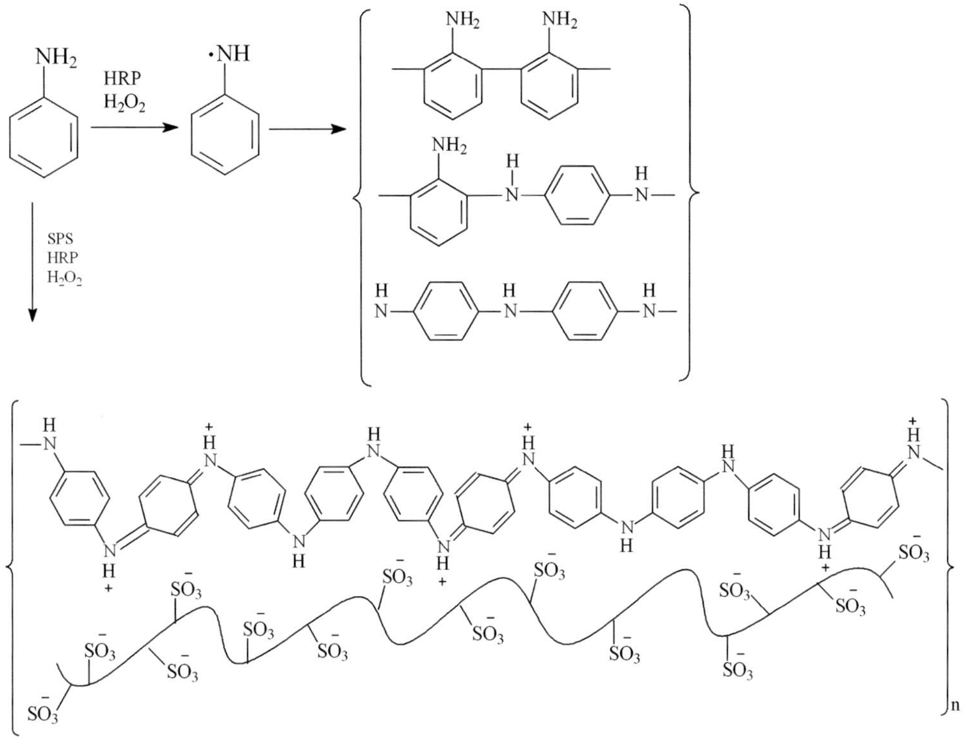

Figure 8.1 Two different pathways for aniline enzymatic polymerization. On top, traditional ortho-coupling. Below, template-assisted enzymatic polymerization of aniline. (Reproduced with permission from [15]. Copyright © (1998) American Chemical Society).

monomers used during PANI enzymatic synthesis using SPS as templates. In a couple of papers, they studied N-substituted anilines and alkoxyanilines as monomers in SPS-templated PANI synthesis. In the former study, the complete series of N-substituted aniline, starting from the methyl and ending with the butyl group was studied [25]. The enzymatic polymerizations of the homopolymers were carried out in the presence of SPS as anionic template and using HRP as catalyst. The spectroscopic features of the PANIs obtained through enzymatic polymerization were similar to previously reported N-substituted PANIs prepared by chemical oxidation [26]. Titration of the polymers in aqueous media showed the formation of a broad polaronic band, consistent with an insulator to conductive state transition. Apparently, the bulky alkyl N-substitution did not affect the water dispersability of the homopolymer/SPS complexes. In the second study, four different alkoxyanilines, the ortho- and meta-substituted anisidine and phenetidine, were enzymatically polymerized using SPS as template in order to obtain the corresponding homopolymers [27]. The four monomers were able to homopolymerize enzymatically, however, spectroscopic properties of their SPS complexes were strongly dependent on the group position, and just slightly dependent on the kind of substituent. The electronic absorption spectra showed that the ortho-substituted monomers resulted in polymers with reversible transition upon doping, nevertheless, the spectroscopic features of these polymers were different from the same homopolymers obtained through chemical oxidation.

8.2.2
Polycation-Assisted Templated Polymerization of Aniline

Self-doped PANI are very interesting due to their unique electrochemical behavior; unlike PANI, the self-doped polymer remains in its doped state in near neutral or alkaline media [28]. Fully self-doped PANIs are not easy to synthesize due to the lower reactivity of acid-functionalized anilines. Kim et al. [29, 30] introduced an alternative approach in the template-assisted enzymatic polymerization of aniline. Previously, only polyanionic templates had been used for PANI synthesis. However, acid-functionalized anilines bear a net anionic charge in aqueous solution, and attempts to use SPS as template with carboxyl-functionalized aniline resulted in red-brown colored polymers with no polaron transitions, regardless of the synthetic conditions. The use of polycationic templates, such as those shown in Figure 8.2 allowed the synthesis of linear and electrically conductive PANIs with self-doping ability due to the doping effect of the carboxyl groups present in the polymer backbone.

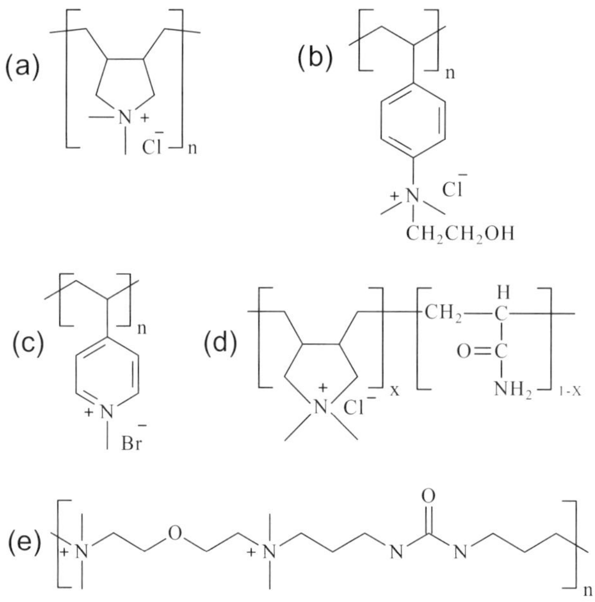

Figure 8.2 Polycationic templates used during enzymatic synthesis of self-doped PANI. (Reproduced with permission from Kim et al. [29]. Copyright © (2007) Taylor & Francis).

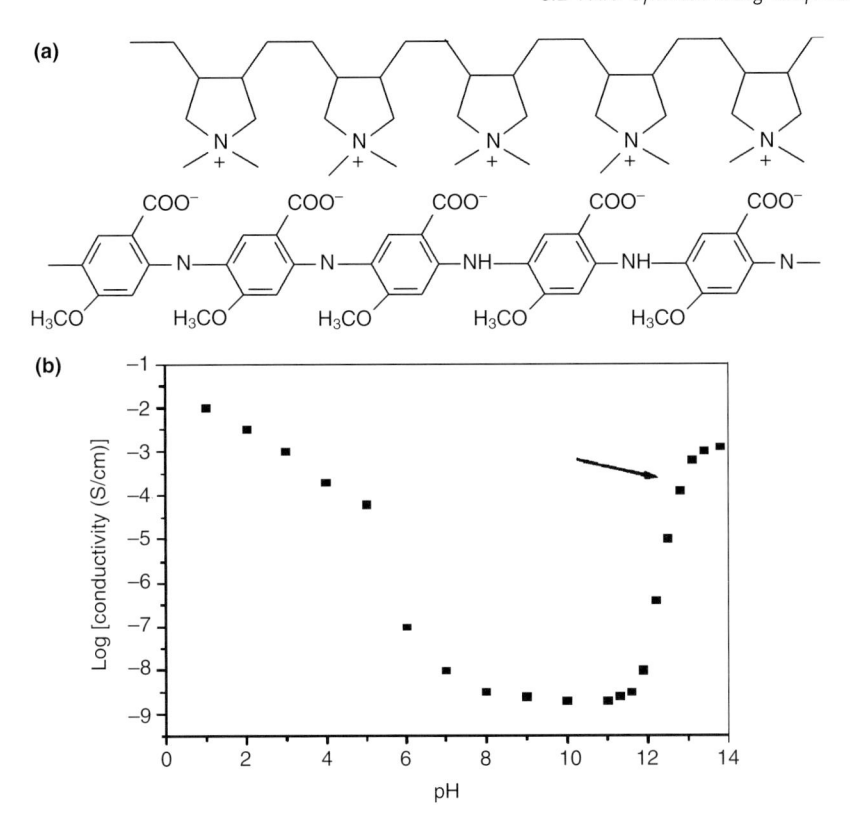

Figure 8.3 (a) PANI complex of carboxylated self-doped PANI/PDADMAC complex of (b) electrical conductivity behavior of carboxylated self-doped PANI/PDADMAC complex at different pH. (Reproduced with permission from Kim et al. [30]. Copyright © (2006) American Chemical Society).

The reactions were carried out in even higher pH conditions (pH 4–6) than those employed in SPS-templated PANI synthesis. Electrical conductivity as high as $0.3\,S\,cm^{-1}$ was achieved for a self-doped carboxylated PANI templated with poly(diallyldimethyl ammonium chloride) (PDADMAC) after additional doping with hydrochloric acid. Another unusual feature of the self-doped carboxylated PANI/PDADMAC complex (Figure 8.3a), was its electrical conductivity behavior at high pH (Figure 8.3b). Electrical conductivity of PANI generally decreases dramatically above pH 3 due to de-doping of the polymer. However, the carboxylated self-doped PANI shows a gradual decrease of conductivity from pH 1 to pH 5. As the pH increases above 5, a dramatic decrease is observed due to de-doping, but between pH 11 and pH 13 the PANI experiences a dramatic increase in electrical conductivity due to an electronic transition that is evident by their UV-Vis spectra. This behavior was ascribed to a *n*-doping process most likely related to the strong positive/negative interaction of the cationic template and the anionic enzymatically synthesized polymer.

8.3
Synthesis of PANI in Template-Free, Dispersed and Micellar Media

8.3.1
Template-Free Synthesis of PANI

Most of the papers dealing with enzymatic synthesis of PANI took the template-assisted approach and only a couple of papers were devoted to the development of template-free methods. In our group we took advantage of the higher stability of soybean peroxidase (SBP) at low pH to synthesize PANI in an aqueous environment at pH 3.0 using toluene sulfonic acid (TSA) as doping agent [31]. Spectroscopic studies indicated that the PANI prepared under these reaction conditions was very similar to that obtained by traditional chemical oxidation. Kim et al. [32] polymerized three different aniline monomers using HRP at pH 3.0 in partially organic media using ethanol as cosolvent. Branching and crosslinking was avoided during polymerization by introducing methyl and methoxy groups in the ortho positions of the aniline monomer. It was found that the optimum concentration of ethanol was 15% wt. Peroxyacetic acid was used as a mild oxidant instead of hydrogen peroxide. FTIR spectra showed that the de-doped form of the PANI was the leucoemeraldine form of the PANI, which is the reduced form of the polymer.

8.3.2
Synthesis in Dispersed Media

Water-dispersible PANI colloids have been studied as an interesting material due to their potential applications in coatings, blends and separation technologies. Usually, PANI dispersions are prepared by chemical polymerization in aqueous media using a water soluble polymer as steric stabilizer [33]. PANI enzymatic polymerization is very attractive for the synthesis of PANI colloids due to its smooth and rate-controlled oxidation stage [31]. Unlike SPS-templated polymerization, in dispersion polymerization the stabilizer is attached mainly to the surface of the colloids, providing water dispersability without doping the PANI. In this way, undoped PANI dispersions can be obtained. Takamuku et al. reported the use of nonionic polymers, such as polyvinylpyrrolidone, poly(ethylene oxide) and poly(vinyl alcohol) (PVA) as stabilizers during aniline enzymatic polymerization in aqueous media [34]. Our group studied the enzymatic synthesis of PANI colloids using fully hydrolyzed PVA, poly(N-isopropylacrylamide) and chitosan as steric stabilizers, and different acids, such as camphor sulfonic acid (CSA) or TSA, as doping agents (See Figure 8.4) [35]. We found a strong dependence of the morphology of the colloids on the doping acid used during synthesis, due to changes in the hydrophobicity of the PANI due to the different acids. The combination of XPS, elemental analysis and FTIR showed that most of the stabilizer remains on the surface of the colloids. We later expanded the study of the enzymatic synthesis of colloids to other steric stabilizers such as

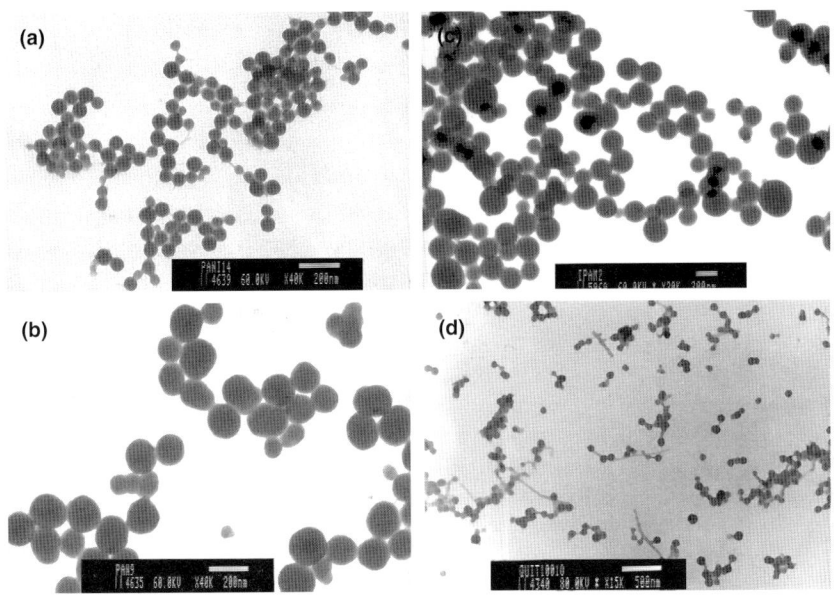

8.3 Synthesis of PANI in Template-Free, Dispersed and Micellar Media | 193

Figure 8.4 PANI colloids obtained by enzymatic polymerization of aniline in presence of different steric stabilizers. (a) Partially hydrolyzed PVA, (b) fully hydrolyzed PVA, (c) PNIPAM, and (d) chitosan. The scale bar in (a), (b), and (c) is 200 nm, and in (d) is 500 nm. (Reproduced with permission from [36]. Copyright © (2007) American Chemical Society).

poly(N-isopropylacrylamide) and chitosan [36, 37]. The resulting nanoparticles acquired the smart behavior from the stabilizer, thereby its flocculation could be induced by changing the pH or temperature of the aqueous dispersions. The smart behavior of these colloids was very different from that of PANI colloids synthesized using PVA as the steric stabilizer.

8.3.3
Enzymatic Synthesis of PANI Using Anionic Micelles as Templates

Besides polyanions such as SPS, anionic surfactants were able to provide the negative charges needed to template the formation of electrically conductive PANI. Indeed, micelles of sodium dodecylbenzensulfonate (SDBS) were used successfully as templates to induce the formation of linear and electrically conductive PANI [38]. On the other hand, cationic and non-ionic surfactants did not provide the negative charges and low pH environment needed for the synthesis of conducting PANI, as expected. Using a similar approach, dodecyl diphenyloxide disulfonate (DODD) was used as a bifunctional template for the enzymatic synthesis of PANI [39]. UV-Vis and electrical conductivity studies showed that the electrically-conducting form of PANI was only obtained above the critical micellar concentration (CMC) of DODD. The bifunctional nature of the DODD provided good properties to the synthesized PANI, such as good water dispersability and higher electrical conductivity than that obtained with dodecylbenzensulfonic acid (DBSA) under similar conditions. Very recently, vesicles made of SDBS and decanoic acid were used as soft templates for the enzymatic synthesis of PANI [40]. The reaction was carried out using HRP-H_2O_2 as oxidizer under pH 4.3 buffer. Fluorescence measurements indicate that the enzyme binds to the vesicles' surface. Control reactions carried out in the absence of the vesicles confirm that the anionic surface of the template is essential to guide the enzymatic polymerization in order to obtain the electrically conductive form of PANI. Self-doped PANI was enzymatically synthesized by using the surface of cationic micelles prepared using cetyltrimethylammonium bromide (CTAB) as templates [41]. The use of cationic micelles, which had not been useful to template the synthesis of conducting PANI, was possibly due to the negative charge of the monomer, 3-amino-4-methoxybenzensulfonic acid. UV-Vis and electrical conductivity studies of the resulting polyaniline confirmed the templating mechanism of the positively charged micelle on the negatively charged monomer.

8.4
Biomimetic Synthesis of PANI

Hydrogen peroxide is a very attractive oxidizer for PANI synthesis because it is cheap, safe to handle and generates water as by-product. However, direct oxidation of aniline by this reagent does not proceed at a rate high enough to result in high quality PANI. Peroxidases increase the oxidation rate by several orders of magnitude making this reaction more attractive, but the relatively high cost of enzymes is a major drawback of this method. For this reason, inspired by the enzymatic polymerization of aniline, several research groups have developed 'biomimetic catalysts'. These are molecules that resemble the active site of oxidoreductases and catalyze the polymerization of aniline by using hydrogen peroxide.

8.4.1
Hematin and Iron-Containing Porphyrins

Iron-protoporphyrin IX, also known as the heme group, is the active site of several peroxidases. Its structure is a macrocyclic ligand with four pyrroles coordinated to an iron(II) molecule located in the center and two dependent carboxylic groups that provide solubility in alkaline media. The heme group is also common to other enzymes, such as hemoglobin and myoglobin. It has been known for a long time that this group has peroxidase activity [42], but just recently this molecule was used for polymerization of phenolic compounds. However, its use for PANI synthesis is prevented by the insolubility of this compound in acidic media, which is required in order to synthesize the electrically-conducting form of the polymer. Roy et al. [43] functionalized the hematin with a soluble poly(ethyleneglycol) (PEG) chain in order to solubilize this molecule in an acidic aqueous environment. Later this PEG-modified hematin was used to catalyze the polymerization of aniline in the presence of ligninsulfonic acid, a naturally occurring template. The PEG-modified hematin showed catalytic activity in the pH range studied (1.0 to 4.0). The resulting polymer was characterized by UV-Vis and FTIR and TGA studies, indicating that the conductive form of PANI was obtained. Sulfonated porphyrins are water soluble over a wide range of pH, thus they can be used directly for PANI synthesis. Electrically-conducting PANI-SPS was synthesized by Nabid et al. [44] using a sodium tetra(p-sulfonatophenyl)porphyrin as biomimetic catalyst. The polymerization was carried out in acidic environment and optimum pH conditions were found at pH between 3.0 and 4.0. Special attention was given to the synthetic conditions to avoid polymerization of aniline by the Fenton reagent produced at low pH, due to Fe released from the biomimetic compound. Not only anionic porphyrins, but also cationic porphyrins, including manganese- and cobalt-containing porphyrins, have been successfully used [45]. In fact, Mn and Co tetrapyridylporphyrins showed higher catalytic activity compared with their iron analog. Cationic porphyrins were catalytically active in a pH range from 1 to 5, giving the best results in terms of yield, at pH 2.0. FTIR and UV-Vis, as well as CV, showed that PANI synthesized using cationic porphyrins were similar to those prepared by chemical oxidation.

8.4.2
Heme-Containing Proteins

Hemoglobin is another heme-containing oxygen-transport protein which is known to display peroxidase-like activity. Hu et al. [46] have shown that this protein can be used as a peroxidase mimic in the synthesis of PANI templated by SPS. The reaction was carried out in an acidic buffer at pH between 1.0 and 4.0, using typical reaction conditions, such as those described for SPS-templated PANI synthesis [17]. The resulting PANI was spectroscopically similar to that prepared using peroxidase under similar reaction conditions, although the conductivity was slightly lower, in the $10^{-3}\,\mathrm{S\,cm^{-1}}$ range. The catalytic activity of hemoglobin was successfully applied in micellar conditions [47], by using sodium dodecyl sulfate (SDS), an anionic surfactant. By working above and below the CMC it was shown that in the same way that in enzymatic reactions [38], the SDS provides an anionic template in order to induce the para-coupling of aniline radicals. This effect of the anionic surfactant on the PANI structure was evident by using cationic and anionic surfactants, which resulted in non-electrically conducting materials.

8.5
Synthesis of PANI Using Enzymes Different From HRP

Horseradish peroxidase has been extensively studied in enzymatic polymerization reactions, in part because it is widely available, has a high catalytic activity towards aniline, and its mechanism of oxidation is relatively well understood, although some details are still under debate. HRP is usually sold as a mixture of isoenzymes, and is usually employed without further purification. However, in recent years, several other enzymes have been employed in the enzymatic polymerization of aniline.

8.5.1
Other Peroxidases

HRP is not the optimum peroxidase for carrying out aniline polymerization because this reaction should be carried at low pH (4.5), and HRP quickly becomes inactivated at this pH. For this reason, other peroxidases with higher stability under acidic conditions have been studied. Sakharov et al. [48, 49] studied the polymerization of aniline by using royal palm tree peroxidase, which had been reported to have an outstanding stability over a wide range of pH. The enzymatic polymerization was carried out using SPS as at pH 3.5. Later, our group reported the higher stability of SBP compared to HRP in the template-free synthesis of PANI [31]. The reaction was carried out in aqueous media, low temperature, and pH 3.0. However, during the polymerization there is a fast decrease in the pH and most of the reaction proceeds at a pH between 2.2 and 3. At the end of the reaction no activity was shown by the remaining SBP showing that the enzyme was inactivated during the reaction, but precisely due to the low pH the PANI obtained was of great quality, with a chemical structure very similar to that chemically synthesized. Caramyshev et al. [50] screened five different enzymes for aniline polymerization at pH 2.8 in presence of poly(2-acrylamido-3-methyl-1-propanesulfonic acid) (PAMPS) as template. They found that palm tree peroxidase (PTP), was the most efficient at low pH, followed by SBP. The measured activity of the other three enzymes, HRP-C, and cationic and anionic peanut peroxidase, after 140 min of incubation in 20 mM buffer at pH 2.8 at 25 °C, was below 10% of the maximum activity. PTP and SBP on the other hand, kept 80% and 40%, respectively, of the its maximum activity. It is not only peroxidases from vegetable origin that have been employed successfully in PANI synthesis. Kim et al. [51] used recombinant *Coprinus cinereus* peroxidase, a fungal enzyme, in PANI synthesis. In order to synthesize PANI as nanofibers an interfacial enzymatic method was adopted, with carbon tetrachloride as the organic phase, and CSA as doping agent in the aqueous phase, similar to that developed by Huang and Kaner [52]. The product of the enzymatic interfacial polymerization (see Figure 8.5) consisted of PANI nanofibers with a morphology very similar to that shown by PANI nanofibers prepared by chemical method.

UV-Vis spectroscopic studies showed that PANI enzymatically synthesized was able to be de-doped and re-doped by treatment with aqueous ammonia solution and CSA solution, respectively, which allowed using these nanofibers in ammonia sensing devices.

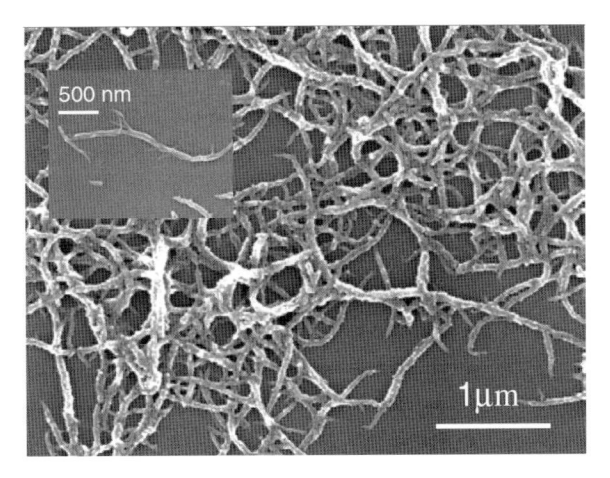

Figure 8.5 PANI nanofibers prepared by Kim et al. (Reproduced with permission from [51] [55b]. Copyright © (2005) IOP Science).

8.5.2
Synthesis of PANI Using Laccase Enzymes

Laccase are copper-containing enzymes found in many plants, fungi and microorganisms (see also Chapters 6 and 7). It has been long known that laccases react with phenol and phenol derivatives performing their oxidation while simultaneously reducing oxygen dissolved in the reaction media. Laccases are heavily used in industrial processes, are relatively economic and highly stable. For these reasons its application in PANI synthesis is highly attractive. Karamyshev showed that fungal laccase from *Coriolus hirsutus* could be used instead of HRP in the SPS-templated synthesis of PANI [53]. The involvement of oxygen in the oxidation reaction of aniline was evident by analyzing the effect of stirring on the kinetics of polymerization. Conductivity as high as $2 \times 10^{-4}\,\mathrm{S\,cm^{-1}}$ was obtained, a value lower than chemically synthesized PANI but typical of SPS-templated PANI samples. UV-Vis and CV characterization showed that PANI synthesized was the electrically-conductive and redox-active form. Another fungal laccase, obtained from *Trametes hirsuta* [54] was used in the synthesis of optically active PANI colloids. Unlike HRP peroxidase, laccase from trametes hirsuta kept almost 50% of its initial activity after being incubated for 24 h in buffer at pH 2.5. In further research from the same group this type of laccase was immobilized on cellulose and used to prepare water dispersible PANI templated by PAMPS [55]. Immobilized laccase improved slightly its stability as compared to native laccase and a value as high as 45% of the initial activity was preserved after incubation in buffer pH 3.5 at 4°C, conditions ideal for the synthesis of PANI. The possibility of reusing the immobilized catalyst gives great technological potential to this synthetic route. UV-Vis characterization of PANI prepared by laccase of *Trametes hirsuta* showed also the typical features of electrically conductive linear PANI, which is ascribed to the relatively low pH employed in these reactions. Indeed, among enzymatic reactions, the lower pH conditions and the longest reaction times can only be achieved by these enzymes extremely resistant to low pH conditions. Curveto *et al.* [56] used several protein extracts from different types of white-rot fungi, that presumably contained at least lignin peroxidase, manganese peroxidase and phenoloxidase laccase, among other enzymes. Among the white-rot fungi studied were *Lentinula edodes, Ganoderma lucidum, Pleurotus florida, P. sajor-caju, and Trametes versicolor*. Aniline polymerization by the different extracts was carried out in buffered aqueous media at pH 4.0. The best results in terms of electrochemical activity and UV-Vis spectroscopic features were obtained when *P. sajor-caju* extracts were used at low (5 °C) or room temperature. The use of crude extracts, instead of the highly purified enzyme, is a more economic alternative and therefore attractive from the industrial point of view.

8.5.3
Synthesis of PANI Using Other Enzymes

Glucose oxidase is an oxidoreductase that catalyzes the oxidation of β-D-glucose into D-glucono-1,5-lactone, which then is hydrolyzed in aqueous media to gluconic acid, while reducing molecular oxygen to hydrogen peroxide. It is a very important enzyme, which is usually employed in glucose biosensors. Ramanavicius *et al.* [57] took advantage of the enzymatically produced hydrogen peroxide to oxidize aniline in a broad pH conditions (Figure 8.6), ranging from 2.0 to 9.0. Optimum conditions for hydrogen peroxide production were close to physiological conditions, although is highly likely that PANI synthesized under this conditions consists primarily of oligomers and ortho-linked/crosslinked structures.

Longoria *et al.* [58] recently reported the oxidation of highly chlorinated aniline by Chloroperoxidase from *Calderomyces fumago*. The specific activity shown by this enzyme was two orders of magnitude higher than that of HRP. Polymeric

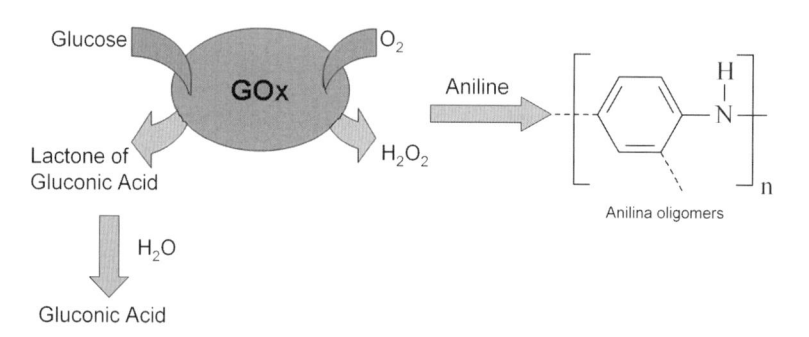

Figure 8.6 Schematic representation of the biocatalytic polymerization of aniline in presence of glucose oxidase, (Reproduced with permission from [57]. Copyright © (2007) Elsevier).

insoluble products were reported as the main material obtained from the reaction, reaching typically between 87 and 95% of the transformed compounds. Although poly(halogenated anilines) might not reach high electrical conductivities, its polymerization represents an alternative pathway to remove aromatic chlorinated compounds from wastewater.

8.6
PANI Films and Nanowires Prepared with Enzymatically Synthesized PANI

One of the major drawbacks of PANI is its processability, which is not as good as that of thermoplastics. Unfortunately, many of the PANI applications require its use either as thin film, free-standing film, fiber or nanowires. Several methods to prepare films, which were first studied with chemically synthesized PANI, such as *in situ* chemical deposition, casting using solvents, and layer by layer assemblies, have been adapted in order to process PANI using the enzymatically synthesized polymer.

8.6.1
In Situ Enzymatic Polymerization of Aniline

During the aqueous oxidative chemical polymerization of aniline, a thin film is usually deposited in the walls of the reaction vessel. Due to the amphiphilic nature of the monomer, this film is produced on either hydrophilic or hydrophobic substrates, thus researchers quickly applied this technique to coat a wide variety of materials [59]. The *in situ* chemical deposition of PANI has been widely studied, and most researchers agree with the mechanism proposed by Stejskal *et al.* [60]. However, aniline enzymatic oxidation is not autocatalytic and therefore, the film growth process is different. Our group studied the deposition of PANI in a wide variety of substrates using a template-free approach [61] and the resulting PANI was characterized mainly through XPS and UV-Vis spectroscopy. It was found that the influence of the surface groups was more drastic because the enzymatic reaction is carried out at higher pH than chemical aniline polymerization. For instance, silicon based substrates, such as fuming silica, silica gel, and glass, which are known to have an acidic surface induced the formation of emeraldine salt, whereas templates such as wollastonite, and mordenite, that contain alkaline ions, produced crosslinked/branched PANI similar to that obtained at near neutral pH. Surface area, enzyme type (SBP or HRP) and pore size apparently had no effect. Recently, our group studied the enzymatic deposition of PANI films by combining quartz crystal microbalance (QCM) and open circuit potential (OCP) measurements and the resulting films were characterized [62]. We found that thinner films with redox electrochemical activity and granular morphology can grow on silicon, indium tin oxide (ITO) and glass, although its adhesion to the substrate is lower than that of chemically synthesized PANI films. However, by comparing both techniques the importance of the autocatalytic oxidation of aniline for film growth was evidenced. Nabid et al. studied the enzymatic polymerization of aniline in the presence of titanium dioxide nanoparticles [63] and γ-alumina nanosheets [64]. In the case of the polymer deposited on TiO_2, HRP was used as catalyst and SPS was used as an aiding template. The SPS doped PANI was deposited on the surface of the TiO_2 particles resulting in a core-shell morphology. The polymer was characterized by UV-Vis and CV, suggesting that the conductive and electroactive form of the polymer was obtained. A comparison of the effect of the template during PANI enzymatic deposition on the Al_2O_3 indicated that the polymer deposited using the polymeric template gave better results in terms of electrical conductivity and coating efficiency.

8.6.2
Immobilization of HRP on Surfaces

8.6.2.1 Surface Confinement of the Enzymatic Polymerization
Another approach to obtain PANI films by enzymatic polymerization is to confine the enzymatic reaction to the surface of a substrate. One of the earliest works [65] showed that adsorbed bilirubin oxidase (BOD) was able to polymerize aniline resulting in a film. In order to do this, BOD was first adsorbed on several substrates that were later immersed in a buffer solution containing aniline. The film was characterized by FTIR, CV, and elemental analysis confirming the presence of redox active PANI. In a similar approach, Alvarez et al. [66] modified polyethylene by plasma, and later successfully attached covalently HRP to the surface of the film. These films showed peroxidase activity and PANI was obtained by the surface initiated polymerization and deposited as a film. XPS characterization shows the formation of a very thin poyaniline film. In both cases, the proximity of the surface and the nascent aniline radicals are the driving force that resulted in the formation of a PANI film. More recently, Tracktenberg et al. [67] synthesized PANI films by using a SPS template bearing photocrosslinkable functional groups. Two different methods were employed, the first one was the traditional template-assisted polymerization using this photocrosslinkable polyanion. This reaction resulted in water soluble PANI that was cast and later cured by UV irradiation. The second approach involved the patterning of the polyanion by radiation induced crosslinking on a polyester film surface. After being immersed in the reaction media containing aniline, the monomer was adsorbed on the template and later, in presence of peroxidase and hydrogen peroxide solution, films of PANI doped with crosslinked SPS were obtained. This work showed the synthesis and simultaneous patterning of enzymatically synthesized PANI films. In a similar approach, Sfez et al. [68] deposited ultra-thin PANI films on silicon surfaces by using a sequential electrostatic self-assembly. First, the surface of the oxidized silicon substrate was modified with ammonium persulfate, then a thin layer of SPS was assembled on the surface by electrostatic interaction of the polyanion with the positively charged surface. Finally, a thin layer of positively charged aniline monomers were assembled. Even though the layer was not completely flat, AFM measurements indicate that the layer was compact and

homogeneus, following the surface relief produced by the previous polyanion layer, suggesting that this layer serves as an immobilized template for the aniline monomers. XPS provided evidence of polymer formation, and interestingly, after polymerization a significant change in the morphology was observed, that is ascribed to chemical reaction of the surface. This work proves that HRP can be catalytically active even for monomers partially immobilized on the surface, although the mechanism was not completely elucidated.

8.6.2.2 Nanowires and Thin Films by Surface-Confined Enzymatic Polymerization

PANI nanowires are 1D nanostructures which are very interesting from the nanotechnology point of view, due to their potential use in sensors and nanoscale electronic devices. DNA can be considered as an anionic polyelectrolyte that can behave as a template for PANI enzymatic polymerization in solution [20]. Double stranded DNA can be attached to aminated surfaces by electrostatic self assembly, providing a linear template for further aniline electrostatic assembly. By using this method followed by enzymatic polymerization of the assembled aniline, Ma et al. [69] and Nickels et al. [70] showed that stretched DNA molecules were able to template the enzymatic synthesis of PANI nanowires, according to the mechanism showed in Figure 8.7.

Unlike previous results that showed that enzymatically synthesized PANI templated by DNA in solution results in agglomeration [20], the stretched DNA on silicon surface lead to nanowires of remarkably homogeneity. The I-V characterization showed that this nanowires have a slight rectifying behavior characteristic of semiconductors. On the other hand, Ma et al. [69] showed that this conductive nanowires can be used as sensors with good signal to noise ratio and a very short response time, as expected due to their low dimensionality. Since, PANI/DNA complexes are electroactive, Gao et al. [71] took advantage of the DNA-templated PANI deposition to prepare electrochemical sensors by measuring the current generated during the PANI/DNA complex oxidation/reduction (see Figure 8.8). First, a bioaffinity surface on a gold electrode was constructed using single stranded DNA. The complementary single-stranded DNA was further assembled

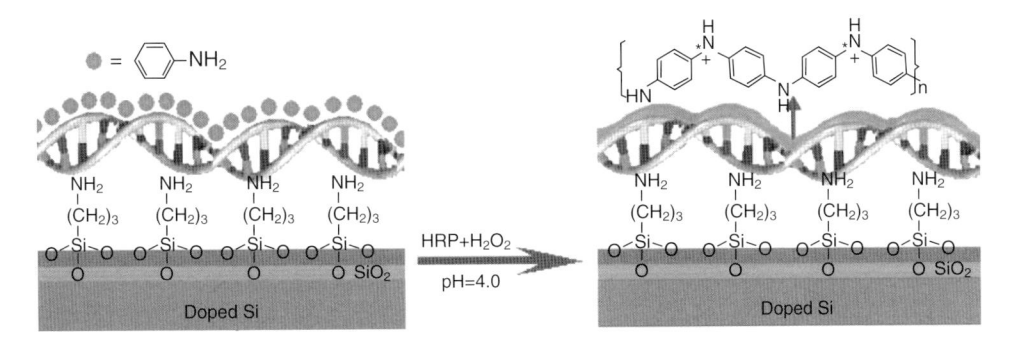

Figure 8.7 Enzymatic synthesis of PANI nanowires attached to silicon surface (Reproduced with permission from[69]. Copyright © (2004) American Chemical Society).

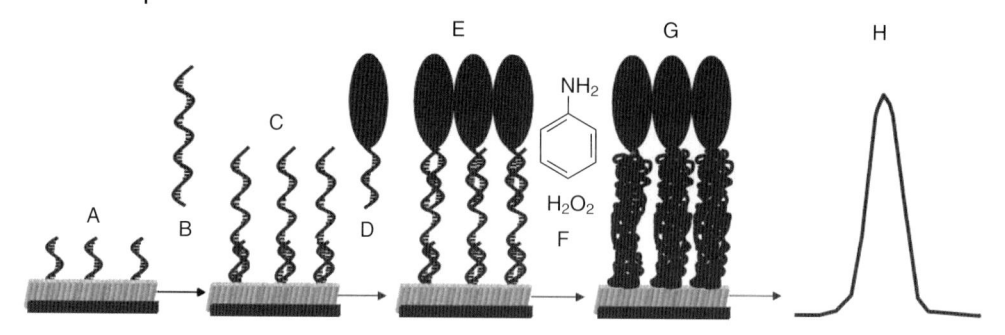

Figure 8.8 Schematic representation of a DNA biosensor based on the DNA-templated enzymatic polymerization of aniline (Reproduced with permission from[71]. Copyright © (2007) Wiley-VCH Verlag).

on the surface, generating a highly negative surface due to the phosphonic acid groups, and provided negative groups for the aniline cation assembly. After incubation with an oligonucleotide detection probe labeled with HRP, the enzyme assembled on the surface of the electrode and mediated the enzymatic polymerization of aniline in a fourth step. This method, although complex, is reproducible and in principle, the detection limit can be lowered by increasing the incubation time before sampling. Indeed, the HRP/ PANI system works as a signal amplifier, much like in colorimetric detection in ELISA methods.

8.6.3
PANI Fibers Made with Enzymatically-Synthesized PANI

The ability of a polymer to be spun into a fiber has been considered a proof of the concept of polymer processability for many years. Wanga et al. [72] have prepared enzymatically synthesized PANI fibers. First, the water-soluble and conductive form of PANI was synthesized using SPS as template and HRP as catalyst. Later, the resulting aqueous dispersion was used to prepare fibers by using a dry-spinning method. The microstructure, mechanical properties, and electrical conductivity of these fibers were improved by traditional fiber techniques such as hot drawing.

8.6.4
Layer-by-Layer and Cast Films of Enzymatically-Synthesized PANI

Enzymatically synthesized PANI has been deposited in thin films of controlled thickness by the layer by layer assembly technique. Aniline was first polymerized using a template-free approach, de-doped and dissolved in dimethylacetamide. After filtration, this solution was further diluted with water at pH 2.5 (adjusted with hydrochloric acid) and filtered again, resulting in the formation of a stable ultrafine aqueous colloidal dispersion. This dispersion was used in the layer-by-layer assembly of thin films with a bacterial poly(γ-glutamic acid) [73] or a poly(phenyl-ethynylene) bearing soft glycol ester segments [74] by alternate absorption. In both cases, a linear dependence of the absorbance with the number of layers indicates a constant increase of thickness using this self assembly technique. The analysis by AFM indicates that films with granular morphology, which arise most likely from the ultrafine colloids, were obtained and potential applications as hole injector layer and nanocapacitors were proposed. On the other hand, unlike PANI prepared by template-assisted polymerization, PANI prepared by template-free polymerization, can be de-doped and obtained as emeraldine base. This allows the enzymatically synthesized PANI to be dissolved in N-methyl-2-pyrrolidinone (NMP) in concentrations high enough to form films by casting, apparently very similar to those prepared using chemically synthesized PANI [75], however, since the enzymatic synthesis is carried at higher pH, the polymer contains a relatively high number of chain defects which together with some sulfonic acid traces resulting from incomplete de-doping, promoted the crosslinking and thermal degradation of the polymer during the casting. Nevertheless, it was shown that enzymatically synthesized PANI can be processed in a similar way to the parent chemically synthesized PANI in order to prepare free standing films that can find application in gas separation technologies.

8.7
Enzymatic and Biocatalytic Synthesis of Other Conductive Polymers

Besides PANI, PPy and polythiophene and their derivatives are very important families of conductive polymers due to their good properties, great chemical stability and large number of potential applications. However, although their enzymatic syntheses are not as straightforward as that of PANI some reactions pathways have been developed.

8.7.1
Enzymatic and Biocatalytic Synthesis of Polypyrrole

Despite the great success of aniline enzymatic polymerization, advances in polypyrrole (PPy) enzymatic synthesis have been slower due to the low catalytic activity shown by most oxidoreductases towards pyrrole. Nevertheless, enzymatic polymerization of pyrrole remains attractive because its polymerization by using hydrogen peroxide or oxygen, would result in a clean and environmentally friendly process. Ramanavicius et al. [76] carried out the pyrrole polymerization in mild pH (6.0) by using the hydrogen peroxide generated in situ by a glucose oxidase enzyme obtained from Penicillium vitale. The progress of the reaction was followed spectrophotometrically. The resulting polymer has a colloidal morphology and due to the high pH employed during the synthesis, is highly likely that several chain defects are present in the chemical structure and consequently the electrical conductivity should be relatively low. A similar approach was employed by

Cui et al. [77] to prepare an enzyme-PPy-carbon nanotube nanocomposite at pH 7.0 by using the hydrogen peroxide generated *in situ* by a lactate oxidase enzyme. The morphologies of the composites was studied by AFM and SEM microscopy. The pH used in this work is very high for conductive PPy synthesis, nevertheless, the presence of carbon nanotubes should improve considerably the conductivity of the composite. However, these polymerizations initiated by glucose and lactate, took place by traditional chemical oxidation, although the generation of the oxidizer, that is, the hydrogen peroxide, was generated *in situ* via an enzymatic pathway. Indeed, the polymerization rate was very low in both cases and the reaction times were up to a few days in both cases, in agreement with previous works that show that pyrrole polymerization by reaction with hydrogen peroxide is a very slow reaction in absence of a catalyst, such as iron salts. Fortunately, pyrrole monomer in aqueous acidic medium has a lower oxidation potential than aniline (600 mV), therefore its oxidation is favored, so instead of relying in a direct oxidation of the monomer by the enzyme, one alternative is to use the oxidative potential of a better enzyme substrate. These reactions where there is no direct reaction between the monomer and the enzyme should be refereed as biocatalytic rather than enzymatic polymerizations. Indeed, the use of redox mediators in enzymatic polymerization has been employed in order to polymerize some monomers that were not able to react directly with the enzyme due to bulky substituents [78]. Laccase from *Trametes versicolor* is a well-known copper-containing oxidoreductase that catalyzes the oxidation of a broad group of organic compounds. Song and Palmore showed that the activity of this enzyme towards pyrrole is low, but enough to form PPy [79]. This fact was discovered by observation of black PPy precipitates in mixtures of laccase and pyrrole monomer. The amount of product diminished significantly when this mixtures where purged with argon to remove the oxygen, a clear indicative that oxygen was involved in the mechanism of PPy formation. However, the activity of laccase towards pyrrole, ca. $0.11\,\mu M\,min^{-1}$, was very low and even comparable to the direct oxidation of pyrrole with dissolved oxygen. The polymerization reaction was considerably improved after the addition of 2,2′-azino-bis(3-ethylbenzothiazoline-6-sulfonate) diammonium salt (ABTS), a much better substrate of laccase, with a measured activity at least 500 times higher than that towards pyrrole. After being oxidized by the laccase, the ABTS forms a relatively stable radical with an oxidizing potential of about 0.9 V. This potential is high enough to react with pyrrole, oxidizing this compound and reducing at the same time, accordingly to the mechanism shown in Figure 8.9.

The function of ABTS is not only a redox mediator, but also reacts with the resulting PPy as dopant, and provides additional redox activity to this conductive polymer. ABTS is a very good substrate of laccases and peroxidases, however, higher specific oxidation rates can be achieved by using the HRP-H_2O_2 instead of the laccase-O_2 systems. In addition, hydrogen peroxide can be dissolved more quickly in water than gaseous oxygen. Kupriyanovich et al. [80] took advantage of this to obtain electrically conductive PPy in aqueous media at pH 4.5 by using the HRP-H_2O_2-ABTS catalytic system. Control reactions carried out in absence

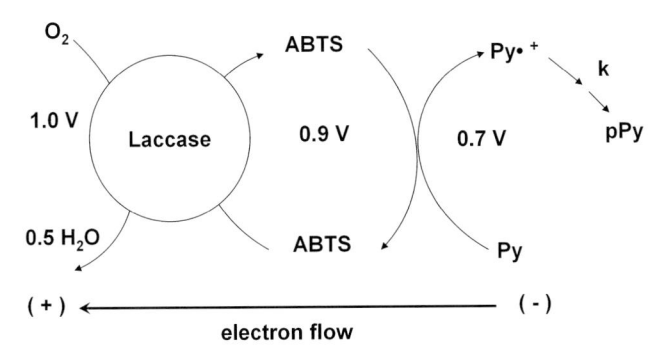

Figure 8.9 Oxidation mechanism of pyrrole by using the O_2/laccase/ABTS biocatalytic pathway (Reproduced with permission from[79]. Copyright © (2005) American Chemical Society).

of the enzyme showed that the pyrrole polymerization was enzyme mediated. Conductivities as high as $5.1 \times 10^{-3}\,S\,cm^{-1}$ were obtained. Nabid and Entezami previously had published the enzymatic synthesis of water-soluble PPy at pH 2.0 with a long reaction time, however HRP is quickly inactivated at this pH and no data is provided regarding the yield in the absence of enzyme, therefore a non-enzymatic reaction between the pyrrole and the H_2O_2 cannot be ruled out. In our group, we also studied the synthesis of PPy powders, water dispersible colloids, and films by the biocatalytic polymerization of pyrrole [81]. The resulting polymer was spectroscopically very similar to that prepared by chemical oxidation using ferric chloride as oxidizing agent. The electrical conductivity of the samples was as high as $1.6 \times 10^{-3}\,S\,cm^{-1}$ as measured by the four point probe technique. This value is slightly lower than the conductivity of the chemically synthesized PPy, probably because of the chain defects produced because the sample is prepared at higher pH as compared to the one chemically synthesized. The enzyme mediated polymerization of pyrrole was particularly useful to obtain well defined colloids with great stability, probably because of the controlled oxidation rate. Recently Mazur et al. [82] encapsulated laccase within PPy shells by taking advantage of the laccase induced polymerization of pyrrole. The polymer was characterized by FTIR and SEM whereas the morphology of the microcontainers was characterized by SEM. Activity evaluation of the laccase showed that most of the activity of the enzyme was preserved.

8.7.2
Enzymatic and Biocatalytic Synthesis of Polythiophenes

Polythiophenes are another interesting family of conductive polymers that have broad potential applications and technological relevance. However, unlike pyrrole and aniline, the oxidation potential of thiophene is relatively high and is almost insoluble in aqueous media. For this reason, there are only reports dealing with the enzymatic polymerization of (3,4-ethylendioxythiophene) (EDOT), one of the most important polythiophenes. This monomer is slightly soluble in aqueous medium and has a relatively low oxidation potential as compared to other thiophenes. The polymer, namely poly(3,4-ethylendioxythiophene) (PEDOT), is usually prepared by chemical oxidation in presence of SPS in order to be obtained in a water dispersible form. Bruno et al. [83] used a chemically modified hematin as biomimetic catalyst for the synthesis of homopolymers and copolymers of pyrrole and EDOT. The reaction was carried out in presence of SPS as template and in order to provide water solubility to the polymers. UV-Vis, XPS and FTIR spectroscopic characterization and electrical measurements indicate the presence of the electrically conducting form of the polymers. Rumbau et al. [84] reported the enzymatic synthesis of water soluble PEDOT using HRP in aqueous system, at low temperature (4 °C) and low pH (2.0) conditions, using SPS as dispersant/doping agent. Control experiments showed that the enzyme was involved in the reaction, although there is not conclusive evidence that the reaction is carried out in a truly enzymatic pathway. We must point out that the pH is low enough to quickly inactivate HRP, although it is proposed that the enzyme can withstand these conditions by complexation with the oil droplets of the monomer, which are produced due to the low solubility of EDOT in aqueous media. In fact, the preferential solubility of the HRP in the monomer phase, and the solubility of the product (PEDOT-PSS complex) in the aqueous media, was used in a further work to recycle the HRP by performing the polymerization in a biphasic media [85]. Nevertheless, even if the enzyme was inactivated due to the low pH, another possibility is a non-enzymatic oxidation due to the hydrogen peroxide in the reaction media and the prosthetic Fe-porphyrin group. Indeed, as we mentioned before, Bruno et al. have shown that hematin derivatives can catalyze EDOT and pyrrole polymerizations [83]. On the other hand, Nagarajan et al. [86], working with SBP, which is a more resistant enzyme to low pH conditions, shown that water soluble PEDOT can be obtained by using tertiophene as a redox mediator that behaves also as comonomer and is a better substrate of the peroxidase than EDOT (Figure 8.10). Another advantage is that, due to the chemical similarity between the tertiophene and the PEDOT, the polymer backbone of the resulting polymer consists exclusively of thiophene rings, which ensure high electrical conductivity of the sample, because the conjugation of the polymer is not reduced.

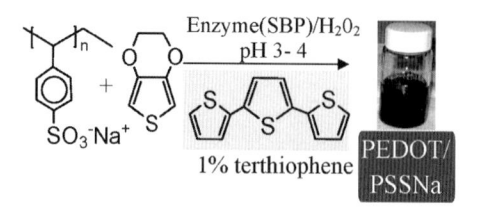

Figure 8.10 PEDOT enzymatic polymerization by using terthiophene as substrate/redox mediator (Reproduced with permission from [86]. Copyright © (2008) American Chemical Society).

8.8
Conclusions

In recent years, the synthesis of PANI and other ICPs has been successfully carried out by enzymatic oxidative polymerization under milder conditions than those traditionally used in chemical oxidation processes. In addition, these catalysts are non-toxic, biodegradable, and obtained from renewable resources. Nevertheless, there are still many issues to solve in order to make enzymatic synthesis of ICPs attractive for industrial applications. One disadvantage of the enzymatic polymerizations is still its economic side, due to the high price of the enzymes as compared with the chemical oxidizers. However, it is expected that advances in biotechnology engineering will impact the production of oxidoreductases in order to make these catalysts affordable and easily available in large quantities, so they would be employed not only in research but also in industrial-scale production. Another impact of the study of enzymatic polymerizations is on the development of synthetic and semi-synthetic catalysts for polymerization inspired on the active site of oxidoreductases. Iron, copper, and other transition metal coordination compounds are being tested as enzyme mimics in oxidative polymerization reactions. Hematin is perhaps the best example of this kind of compound, since it has been already used as catalyst in the synthesis of PANI and PEDOT with good results. Due to the interest in green chemistry and environmentally friendly polymerization routes, in conjunction with advances in biotechnology engineering, we can expect an increase in the use of enzymes for the synthesis of ICPs.

References

1. Shirakawa, H., Louis, E.J., MacDiarmid, A.G., Chiang, C.K., and Heeger, A.J. (1977) *J. Chem. Soc. Chem. Commun.*, 578.
2. Lux, F. (1994) *Polymer*, **35**, 2915.
3. Yoo, J.E., Krekelberg, W.P., Sun, Y., Tarver, J.D., Truskett, T.M., and Loo, Y.-L. (2009) *Chem. Mater.*, **21**, 1948.
4. Virji, S., Kaner, R.B., and Weiller, B.H. (2005) *Chem. Mater.*, **17**, 1256.
5. Wang, H.L., Gao, J., Sansiñena, J.M., and McCarthy, P. (2002) *Chem. Mater.*, **14**, 2546.
6. Pellegrino, J., Radebaugh, R., and Mattes, B.R. (1996) *Macromolecules*, **29**, 4985.
7. Ryu, K.S., Jeong, S.K., Joo, J., and Kim, K.M. (2007) *J. Phys. Chem. B*, **111**, 731.
8. Kobayashi, S., Uyama, H., and Kimura, S. (2001) *Chem. Rev.*, **101**, 3793.
9. Xu, P., Singh, A., and Kaplan, D.L. (2006) *Adv. Polym. Sci.*, **194**, 69.
10. Lim, C.H., and Yoo, Y.J. (2000) *Process Biochem.*, **36**, 233.
11. Shan, J., and Cao, S. (2000) *Polym. Adv. Tech.*, **11**, 288.
12. Shan, J., Han, L., Bai, F., and Cao, S. (2003) *Polym. Adv. Tech.*, **14**, 330.
13. Diaz, A., Fenzel-Alexander, D., Wollmann, D., and Eisenberg, A. (1991) *J. Polym. Sci. Part B Polym. Phys.*, **29**, 1559.
14. Adams, P.N., Laughlin, P.J., Monkman, A.P., and Kenwright, A.M. (1996) *Polymer*, **37**, 3411.
15. Samuelson, L.A., Anagnostopoulos, A., Shridhara Alva, K., Kumar, J., and Tripathy, S.K. (1998) *Macromolecules*, **31**, 4376.

16 Liu, W., Cholli, A.L., Nagarajan, R., Kumar, J., Tripathy, S., Bruno, F.F., and Samuelson, L. (1999) *J. Am. Chem. Soc.*, **121**, 11345.

17 Liu, W., Kumar, J., Tripathy, S., Senecal, K.J., and Samuelson, L. (1999) *J. Am. Chem. Soc.*, **121**, 71.

18 Shen, Y., Sun, J., Wu, J., and Zhou, Q. (2005) *J. Appl. Polym. Sci.*, **96**, 814.

19 Nagarajan, R., Tripathy, S., Kumar, J., Bruno, F.F., and Samuelson, L.A. (2000) *Macromolecules*, **33**, 9542.

20 Nagarajan, R., Liu, W., Kumar, J., Tripathy, S.K., Bruno, F.F., and Samuelson, L.A. (2001) *Macromolecules*, **34**, 3921.

21 Datta, B., Schuster, G.B., McCook, A., Harvey, S.C., and Zakrzewska, K. (2006) *J. Am. Chem. Soc.*, **128**, 14428.

22 Thiyagarajan, M., Samuelson, L.A., Kumar, J., and Cholli, A.L. (2003) *J. Am. Chem. Soc.*, **125**, 11502.

23 PilHo, H., Seong-Cheol, K., Younghoon, K., Yanping, W., Singh, J., Kumar, J., Samuelson, L.A., Bong-Soo, K., Nam-Ju, J., and Jang-Oo, L. (2007) *Biomacromolecules*, **8**, 3602.

24 Deore, B.A., Yu, I., and Freund, M.S. (2004) *J. Am. Chem. Soc.*, **126**, 52.

25 Nabid, M.R., and Entezami, A.A. (2005) *Polym. Adv. Technol.*, **16**, 305.

26 Cataldo, F., and Maltese, P. (2002) *Eur. Polym. J.*, **38**, 1791.

27 Nabid, M.R., Sedghi, R., and Entezami, A.A. (2007) *J. Appl. Polym. Sci.*, **103**, 3724.

28 Yue, J., and Epstein, A.J. (1990) *J. Am. Chem. Soc.*, **112**, 2800.

29 Kim, S.C., Kumar, J., Bruno, F.F., and Samuelson, L.A. (2006) *J. Macromol. Sci. Part A Pure Appl. Chem.*, **43**, 2007.

30 Kim, S.-C., Sandman, D., Kumar, J., Bruno, F.F., and Samuelson, L.A. (2006) *Chem. Mater.*, **18** (9), 2201.

31 Cruz-Silva, R., Romero-García, J., Angulo-Sánchez, J.L., Ledezma-Pérez, A., Arias-Marín, E., Moggio, I., and Flores-Loyola, E. (2005) *Eur. Polym. J.*, **41**, 1129.

32 Kim, S.-C., Huh, P., Kumar, J., Kim, B., Lee, J.-O., Bruno, F.F., and Samuelson, L.A. (2007) *Green Chem.*, **9**, 44.

33 Stejskal, J., Spirkova, M., Riede, A., Helmstedt, M., Mokreva, P., and Prokes, J. (1999) *Polymer*, **40**, 2487.

34 Takamuku, S., Takeoka, Y., and Rikukawa, M. (2003) *Synthetic Met.*, **135–136**, 331.

35 Cruz-Silva, R., Ruiz-Flores, C., Arizmendi, L., Romero-García, J., Arias-Marin, E., Moggio, I., Castillon, F., and Farias, M.H. (2006) *Polymer*, **47**, 1563.

36 Cruz-Silva, R., Arizmendi, L., Del-Angel, M., and Romero-Garcia, J. (2007) *Langmuir*, **23**, 8.

37 Cruz-Silva, R., Escamilla, A., Nicho, M.E., Padron, G., Ledezma-Perez, A., Arias-Marin, E., Moggio, I., and Romero-Garcia, J. (2007) *Eur. Polym. J.*, **43**, 3471.

38 Liu, W., Kumar, J., Tripathy, S., and Samuelson, L.A. (2002) *Langmuir*, **18** (25), 9696–9704.

39 Rumbau, V., Pomposo, J.A., Alduncin, J.A., Grande, H., Mecerreyes, D., and Ochoteco, E. (2007) *Enzyme Microb. Technol.*, **40**, 1412.

40 Guo, Z., Regger, H., Kissner, R., Ishikawa, T., Willeke, M., and Walde, P. (2009) *Langmuir*, **25**, 11390.

41 Kim, S.-C., Kim, D., Lee, J., Wang, Y., Yang, K., Kumar, J., Bruno, F.F., and Samuelson, L.A. (2007) *Macromol. Rapid Commun.*, **28**, 1356.

42 Woggon, W.-D. (2005) *Acc. Chem. Res.*, **38**, 127.

43 Roy, S., Fortier, J.M., Nagarajan, R., Tripathy, S., Kumar, J., Samuelson, L.A., and Bruno, F.F. (2002) *Biomacromolecules*, **3**, 937.

44 Nabid, M.R., Sedghi, R., Jamaat, P.R., Safari, N., and Entezami, A.A. (2006) *J. Appl. Polym. Sci.*, **102**, 2929.

45 Nabid, M.R., Zamiraei, Z., Sedghi, R., and Safari, N. (2009) *Reactive Funct. Polym.*, **69**, 319.

46 Hu, X., Zhang, Y.-Y., Tang, K., and Zou, G.-L. (2005) *Synthetic Met.*, **150**, 1.

47 Hu, X., Shu, X.-S., Li, X.-W., Liu, S.-G., Zhang, Y.-Y., and Zou, G.-L. (2006) *Enzyme Microb. Technol.*, **38** (5), 675.

48 Sakharov, I.Y., Vorobiev, A.C., and Castillo-Leon, J.J. (2003) *Enzyme Microb. Technol.*, **33**, 661.

49 Sakharov, I.Y., Ouporov, I.V., Vorobiev, A.K., Roig, M.G., Pletjushkina, O.Y. (2004) *Synthetic Met.*, **142**, 127.

50 Caramyshev, A.V., Evtushenko, E.G., Ivanov, V.F., Barceló, A.R., Roig, M.G.,

Shnyrov, V.L., Huystee, R.B., Kurochkin, I.N., Vorobiev, A.K., and Sakharov, I.Y. (2005) *Biomacromolecules*, **6** (3), 1360–1366.

51 Kim, B.-K., Kim, Y.H., Won, K., Chang, H., Choi, Y., Kong, K.-J., Rhyu, B.W., Kim, J.-J., and Lee, J.-O. (2005) *Nanotechnology*, **16**, 1177.

52 Huang, J., and Kaner, R.B. (2004) *J. Am. Chem. Soc.*, **126** (3), 851.

53 Karamyshev, A.V., Shleev, S.V., Koroleva, O.V., Yaropolov, A.I., and Sakharov, I.Y. (2003) *Enzyme Microb. Technol.*, **33** (5), 556.

54 Vasil'eva, I.S., Morozova, O.V., Shumakovich, G.P., Shleev, S.V., Sakharov, I.Y., and Yaropolov, A.I. (2007) *Synthetic Met.*, **157**, 684.

55 Vasil'eva, I.S., Morozova, O.V., Shumakovich, G.P., and Yaropolov, A.I. (2009) *Appl. Biochem. Microbiol.*, **45** (1), 27–30.

56 Curvetto, N.R., Figlas, D., Brandolin, A., Saidman, S.B., Rueda, E.H., and Ferreira, M.L. (2006) *Biochem. Eng. J.*, **29**, 191.

57 Kausaite, A., Ramanaviciene, A., and Ramanavicius, A. (2009) *Polymer*, **50**, 1846.

58 Longoria, A., Tinoco, R., and Vázquez-Duhalt, R. (2008) *Chemosphere*, **72**, 485.

59 Malinauskas, A. (2001) *Polymer*, **42**, 3957.

60 Sapurina, I., Riede, A., and Stejskal, J. (2001) *Synthetic Met.*, **123**, 503.

61 Flores-Loyola, E., Cruz-Silva, R., Romero-García, J., Angulo-Sánchez, J.L., Castillon, F.F., and Farías, M.H. (2007) *Mater. Chem. Phys.*, **105**, 136.

62 Carrillo, N., Leon, U., Avalos, T., Nicho, M.E., Serna, S., Castillon, F.F., Farías, M.H. and Cruz-Silva, R. (2010) Submitted to Electrochim. Acta.

63 Reza, N.M., Maryam, G., Bayandori, M.A., Rassoul, D., and Roya, S. (2008) *Int. J. Electrochem. Sci.*, **3**, 1117.

64 Nabid, M.R., Golbabaee, M., Moghaddam, A.B., Mahdavian, A.R., and Amini, M.M. (2009) *Polym. Composites*, **30**, 841.

65 Aizawa, M., Wang, L., Shinohara, H., and Ikariyama, Y. (1990) *J. Biotechnol.*, **14**, 301.

66 Alvarez, S., Manolache, S., and Denes, F. (2003) *J. Appl. Polym. Sci.*, **88**, 369.

67 Trakhtenberg, S., Hangun-Balkir, Y., Warner, J.C., Bruno, F.F., Kumar, J., Nagarajan, R., and Samuelson, L.A. (2005) *J. Am. Chem. Soc.*, **127**, 9100.

68 Sfez, R., Peor, N., Cohen, S.R., Cohen, H., and Yitzchaik, S. (2006) *J. Mater. Chem.*, **16**, 4044.

69 Ma, Y., Zhang, J., Zhang, G., and He, H. (2004) *J. Am. Chem. Soc.*, **126**, 7097.

70 Nickels, P., Dittmer, W.U., Beyer, S., Kotthaus, J.P., and Simmel, F.C. (2004) *Nanotechnology*, **15**, 1524.

71 Gao, Z., Raffea, S., and Lim, L.H. (2007) *Adv. Mater.*, **19**, 602.

72 Wanga, X., Schreuder-Gibsonb, H., Downey, M., Tripathy, S., and Samuelson, L. (1999) *Synthetic Met.*, **107**, 117.

73 Espinosa-Gonzalez, C., Moggio, I., Arias-Marin, E., Romero-García, J., Cruz-Silva, R., Moigne, J.L., and Ortiz-Cisneros, J. (2003) *Synthetic Met.*, **139**, 155.

74 Barrientos, H., Moggio, I., Arias-Marin, E., Ledezma, A., and Romero, J. (2007) *Eur. Polym. J.*, **43**, 1672.

75 Cruz-Silva, R., Romero-García, J., Angulo-Sánchez, J.L., Flores-Loyola, E., Farías, M.H., Castillón, F.F., and Díaz, J.A. (2004) *Polymer*, **45**, 4711.

76 Ramanavicius, A., Kausaite, A., Ramanaviciene, A., Acaite, J., and Malinauskas, A. (2006) *Synthetic Met.*, **156**, 409.

77 Cui, X., Li, C.M., Zang, J., Zhou, Q., Gan, Y., Bao, H., Guo, J., Lee, V.S., and Moochhala, S.M. (2007) *J. Phys. Chem. C*, **111**, 2025.

78 Won, K., Kim, Y.H., An, E.S., Lee, Y.S., and Song, B.K. (2004) *Biomacromolecules*, **5**, 1.

79 Song, H.-K., and Palmore, G.T.R. (2005) *J. Phys. Chem. B*, **109**, 19278.

80 Kupriyanovich, Y.N., Sukhov, B.G., Medvedeva, S.A., Mikhaleva, A.I., Vakul'skaya, T.I., Myachina, G.F., and Trofimov, B.A. (2008) *Mendeleev Communications*, **18**, 56.

81 Cruz-Silva, R., Amaro, E., Escamilla, A., Nicho, M.E., Sepulveda-Guzman, S., Arizmendi, L., Romero-Garcia, J., Castillon-Barraza, F.F., and Farias, M.H. (2008) *J. Colloid Interface Sci.*, **328**, 263.

82 Mazur, M., Krywko-Cendrowska, A., Krysinski, P., and Rogalski, J. (2009) *Synthetic Met.*, **159**, 1731.
83 Bruno, F.F., Fossey, S.A., Nagarajan, S., Nagarajan, R., Kumar, J., and Samuelson, L.A. (2006) *Biomacromolecules*, **7**, 586.
84 Rumbau, V., Pomposo, J.A., Eleta, A., Rodriguez, J., Grande, H., Mecerreyes, D., and Ochoteco, E. (2007) *Biomacromolecules*, **8**, 315.
85 Sikora, T., Marcilla, R., Mecerreyes, D., Rodriguez, J., Pomposo, J.A., and Ochoteco, E. (2009) *J. Polym. Sci. [A1]*, **47**, 306.
86 Nagarajan, S., Kumar, J., Bruno, F.F., Samuelson, L.A., and Nagarajan, R. (2008) *Macromolecules*, **41**, 3049.

9
Enzymatic Polymerizations of Polysaccharides

Jeroen van der Vlist and Katja Loos[1]

9.1
Introduction

Oligo- and polysaccharides are important macromolecules in living systems, showing their multifunctional characteristics in the construction of cell walls, energy storage, cell recognition and their immune response.

Polysaccharides are an abundant source of raw materials that are interesting due to the biodegradable, biocompatible and renewable character. Saccharides are expected to play an increasingly bigger role as raw material in the future and to replace petrol-based materials. Already, polysaccharides find their way into many disparate fields of industry; a short overview is given in Table 9.1.

The great structural variety of saccharides results from the diverse stereochemistry of the monosaccharide building blocks and from the enormous number of intersugar linkages that they can form. This feature makes sugar molecules suitable for mediating many biological processes, but it greatly complicates their synthesis.

Conventional chemical synthetic approaches are, in many cases, inadequate to provide substantial quantities of saccharides. The difficulties arise from realizing complete regio- and stereo-control of the glycosylating process. At present, no such methods are available because, in chemical synthesis, most of the difficulties arise from the laborious regio- and stereochemical control. Most synthetic approaches are therefore based on the modification or degradation of naturally occurring polysaccharides resulting in less than perfect products.

Saccharide synthesis *in vivo* is a very complex process that is not regulated by universally conserved codes, and has not been automated *in vitro* yet. Ultimately, in order to provide useful quantities of materials a combination of chemical and enzymatic techniques is necessary.

---

1) In memoriam Beate Pfannemüller: her work inspired much of the research reviewed in this chapter.

Biocatalysis in Polymer Chemistry. Edited by Katja Loos
Copyright © 2011 WILEY-VCH Verlag GmbH & Co. KGaA, Weinheim
ISBN: 978-3-527-32618-1

Table 9.1 Polysaccharide processing industries.

Industry	Polysaccharide	Main function
Paper	Cellulose	Structural material
Food	Starch	Thickener, rheological control, texture
Biomedical, pharmaceutical	Dextran	Biocompatiblizer, arificial blood stabilizer, drug carrier
Package	Starch/cellulose derivates	Reduction of synthetic polymers, increase biodegradablility
Coating	Starch	Rheological control
Adhesive	Starch	Adhesive
Textile	Cotton	Structural material

Biocatalytic synthetic pathways are very attractive as they have many advantages such as mild reaction conditions, high enantio-, regio-, chemoselectivity and are nontoxic natural catalysts. In enzymatic methods for glycoside and saccharide synthesis no selective protection/deprotection steps are necessary and control of configuration at newly formed anomeric centers is absolute.

Two approaches have dominated enzyme-catalyzed saccharide synthesis: glycosyl transferase and glycosidase-catalyzed glycosidic bond formation. The first uses the normal biosynthetic machinery of living organisms. In the second, enzymes that normally catalyze transfer of an enzyme-bound glycosyl residue to water are induced to transfer it instead to a different acceptor.

Glycosyl transferases belong to the class of transferases (Enzyme Commission (EC), Class No. 2). These enzymes catalyze reactions in which a group is transferred from one compound to another. Groups that are transferred are Cl, aldehydic or ketonic residues, acyl, glycosyl, alkyl, nitrogenous, phosphorus and sulfur-containing groups [1].

By making use of the biosynthetic glycosyl transferases, it is possible to glycosylate carbohydrate substrates. A sugar nucleotide donor and acceptor are incubated with the appropriate glycosyl transferase that catalyzes efficient and selective transfer of the glycosyl residue to the acceptor. This method has the advantage of high efficiency and selectivity. Its major drawbacks are the requirement for a complex glycosyl donor and the relative inaccessibility of the glycosyl transferases.

From the three classes of enzymes used in polymer science so far (see Preface of this book) transferases are the least applied class of biocatalyst. Despite their potential for synthesizing interesting polymeric materials many transferases are very sensitive biocatalysts, preventing their isolation on a larger scale and/or their use for synthesizing polymer on a reasonable scale.

Glycosidases belong to the class of hydrolases (EC, Class No. 3). These enzymes catalyze the cleavage of C–O, C–N, C–C and other bonds by reactions involving the addition or removal of water [1].

Glycosidases (also called glycoside hydrolases; EC 3.2.1) catalyze the hydrolysis of the glycosidic linkage to generate smaller sugars. Together with glycosyltransferases (GTs)s, glycosidases form the major catalytic machinery for the synthesis and breakage of glycosidic bonds.

The native action of glycosidases is to hydrolyze glycosidic linkages of glucans in the presence of water. These types of enzyme-catalyzed reactions are reversible and hence the glycosidic bond formation is possible if the glycosyl substrate (the monomer) has a good leaving group and if the reaction conditions, with respect to substrate concentration, temperature and solvent quality, are well chosen. With this it became possible to synthesize *in vitro* natural polysaccharides such as cellulose, xylan, chitin, hyaluronan and chondroitin, and also of non-natural polysaccharides such as a cellulose–chitin hybrid, a hyaluronan–chondroitin hybrid, and others.

In contrast to the glycosyl transferases, the glycosidase approach uses simpler glycosyl donors which at the limit, can be the free monosaccharide. This method has the advantage of using relatively simple glycosyl donors and readily available robust enzymes. Its main disadvantage is that regioselectivity may not be observed in all cases.

In the following, approaches to synthesize well-defined polysaccharides via GT and glycosidase catalysis are discussed.

9.2
Glycosyltransferases

Glycosyltransferases are important biological catalysts in cellular systems generating complex cell surface glycans involved in adhesion and signaling processes. Recent advances in glycoscience have increased demands to access significant amount of glycans representing the glycome.

Glycosyltransferases catalyze the transfer of a sugar moiety from an activated donor sugar onto saccharide and nonsaccharide acceptors. They can be divided into the Leloir and non-Leloir types according to the type of glycosyl donors they use [2]. Non-Leloir GTs typically use glycosyl phosphates as donors, while Leloir GTs use sugar nucleotides as donors and transfer the monosaccharide with either retention (retaining enzymes) or inversion (inverting enzymes) of the configuration of the anomeric center. Most of the GTs responsible for the biosynthesis of mammalian glycoproteins and glycolipids are Leloir GTs.

Glycosyltransferases now play a key role for *in vitro* synthesis of oligosaccharides, and their bacterial genome is increasingly used for cloning and overexpression of active transferases in glycosylation reactions [3–8]. In a recent excellent review by Homann and Seibel possible ways to tailor biocatalysts by enzyme engineering and substrate engineering are summarized [9].

In the following section, enzymes in the EC 2.4 class are presented that catalyze the polymerization of polysaccharides. The Enzyme Commission classification scheme organizes enzymes according to their biochemical function in living systems. Enzymes can, however, also catalyze the reverse reaction which is very often used in biocatalytic synthesis. Therefore newer classification systems have been developed, based on the three-dimensional structure and function of the enzyme, the property of the enzyme, the biotransformation that the enzyme catalyzes etc. [10–15]. The Carbohydrate-Active enZYmes Database [13], which is currently the best database/classification system for carbohydrate active enzymes, uses an amino acid sequence-based classification and would classify some of the enzymes presented in the following as hydrolases rather than transferases (e.g., branching enzyme, sucrases and amylomaltase). Nevertheless we present these enzymes here because they are transferases according to the EC classification.

9.2.1
Phosphorylase

The most extensively used transferase in the field of polymer science is phosphorylase (systematic name: (1→4)-α-D-glucan:phosphate α-D-glucosyltransferase; EC 2.4.1.1). While this enzyme is responsible for the depolymerization of linear α-(1→4) glycosidic chains *in vivo* it can also be used to synthesize linear α-(1→4) glycosidic chains (amylose) *in vitro*.

In vivo linear α-1,4-glucans are synthesized from adenosine diphosphate (ADP)-glucose by the enzyme glycogen synthase [16–19]. The enzyme as well as the monomer are quite sensitive and therefore most researchers (at least in the field of polymer science) prefer to use phosphorylase for the synthesis of amylose.

Amylose is one component of starch which is the most abundant storage reserve carbohydrate in plants. Carbohydrates, such as starch, function as a reservoir of energy for later metabolic use. It is found in many different plant organs, including seeds, fruits, tubers and roots, where it is used as a source of energy during periods of dormancy and regrowth.

Starch granules are composed of two types of α-glucan, amylose and amylopectin, which represent approximately 98–99% of the dry weight. The ratio of the two polysaccharides varies according to the botanical origin of the starch.

Amylose is a linear molecule in which the glucose units are joined via α-(1→4) glucosyl linkages. Amylopectin is a branched molecule in which about 5% of the glucose units are joined by α-(1→6) glucosyl linkages (see Figure 9.1).

In animals, a constant supply of glucose is essential for tissues such as the brain and red blood cells, which depend almost entirely on glucose as an energy source.

The mobilization of glucose from carbohydrate storage provides a constant supply of glucose to all tissues. For this glucose units are mobilized by their sequential removal from the nonreducing ends of glycogen. For this process three enzymes are required *in vivo*:

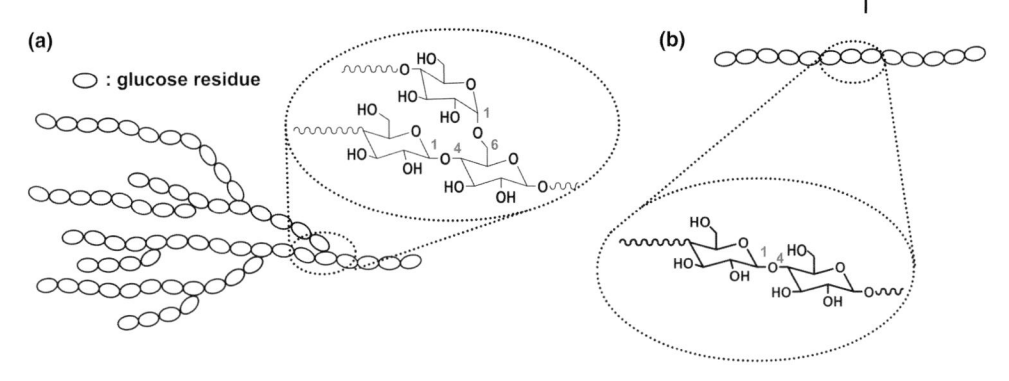

Figure 9.1 Structure of (a) amylopectin and (b) amylose.

1) Glycogen phosphorylase catalyzes glycogen phosphorolysis (bond cleavage of the α-(1→4) bonds by the substitution of a phosphate group) to yield glucose-1-phosphate.

$$\text{glycogen}(\text{Gluc})_n + P_i \rightleftarrows \text{glycogen}(\text{Gluc})_{n-1} + \text{glucose-1-phosphate}$$

Phosphorylase is just able to release glucose if the unit is at least five units away from a branching point.

2) Glycogen debranching enzyme removes α-(1→6) glycogen branches, thereby making additional glucose residues accessible to glycogen phosphorylase.

3) Phosphoglucomutase converts glucose-1-phosphate into glucose-6-phosphate which has several metabolic fates. It is for example a precursor in the pentose phosphate pathway (PPP), it can be converted to α-D-glucose and to pyruvate via the glycolysis pathway (see Figure 9.2).

However, under the appropriate conditions the catalytic action of phosphorylase can be reversed and linear synthetic amylose can be synthesized with the release of inorganic phosphate (Pi) (see Figure 9.3).

9.2.1.1 Enzymatic Polymerization of Amylose with Glycogen Phosphorylase

The existence of a phosphorylating enzyme in a higher plant was first reported by Iwanoff who observed that an enzyme he found in the germinating vetches *Vicia sativa*, liberates inorganic phosphate from organic phosphorous compounds [21]. Shortly after, the same enzyme was found in other vetches and wheat [22, 23], rice and coleseed [24], barley and malt etc. Bodńar was the first to report a progressive disappearance of inorganic phosphate (thus the reverse reaction) while incubating suspended flour from ground peas in a phosphate buffer [25]. Cori and Cori demonstrated that animal tissues contain an enzyme which acts upon glycogen as well [26–29]. Cori, Colowick and Cori suggested that the product of this reaction is α-glucopyranose-1-phosphoric acid (also called Cori-ester), which was confirmed later by Kiessling [30] and Wolfrom and Pletcher [31].

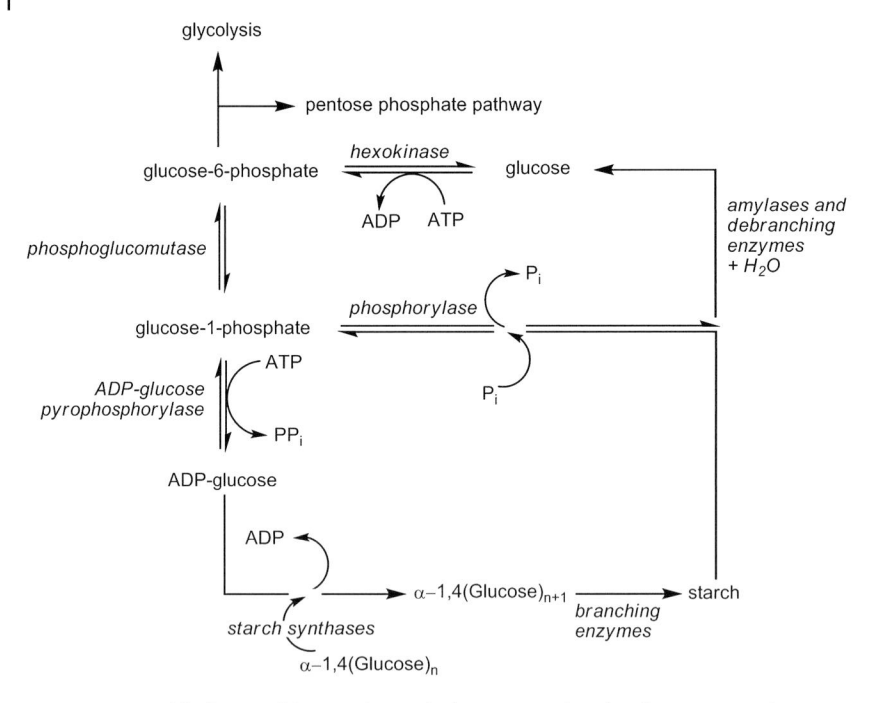

Figure 9.2 Simplified view of the starch metabolism. (Reproduced with permission from [20]. Copyright © (1998) Elsevier).

Figure 9.3 Enzymatic polymerization of amylose.

Glycogen phosphorylases belong to the group of vitamin B_6 enzymes bearing a catalytic mechanism that involves the participation of the phosphate group of pyridoxal-5′-phosphate (PLP). The proposed mechanism is a concerted one with front-side attack as can be seen in Figure 9.4 [32]. In the forward direction, for example, phosphorolysis of α-1,4-glycosidic bonds in oligo- or polysaccharides, the reaction is started by protonation of the glycosidic oxygen by orthophosphate, followed by stabilization of the incipient oxocarbonium ion by the phosphate anion and subsequent covalent binding of the phosphate to form glucose-1-phosphate. The product, glucose-1-phosphate, dissociates and is replaced by a new incoming phosphate.

In the reverse direction, protonation of the phosphate of glucose-1-phosphate destabilizes the glycosidic bond and promotes formation of a glucosyl oxocarbonium ion–phosphate anion pair. In the subsequent step the phosphate anion becomes essential for promotion of the nucleophilic attack of a terminal glucosyl

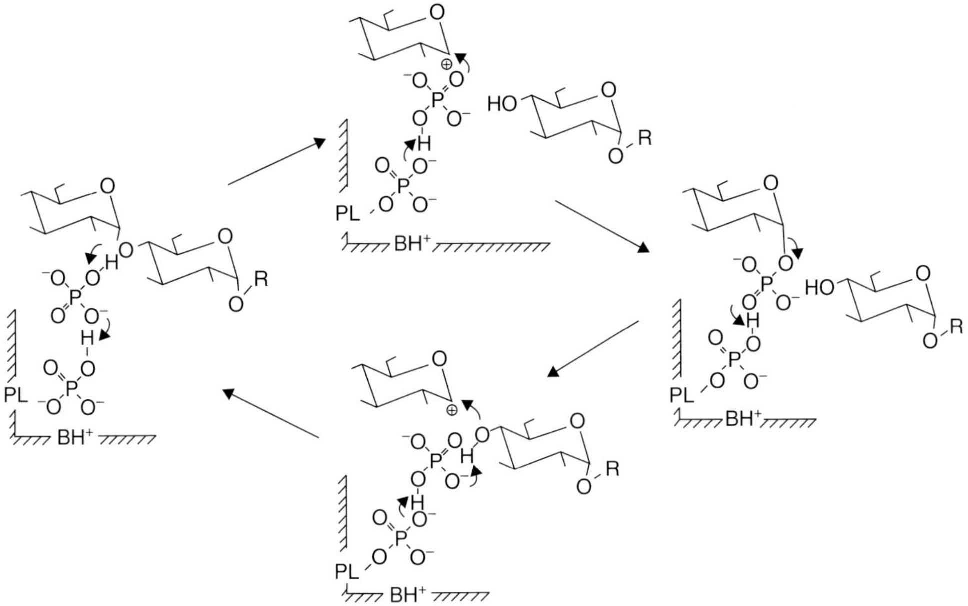

Figure 9.4 Catalytic mechanism of glycogen phosphorylases. The reaction scheme accounts for the reversibility of phosphorolysis of oligosaccharides (R) in the presence of orthophosphate (upper half) and primer-dependent synthesis in the presence of glucose-1-phosphate (lower half). PL = enzyme-bound pyridoxal; BH$^+$ = a general base contributed by the enzyme protein. (Reproduced with permission from [32]. Copyright © (1990) American Chemical Society).

residue on the carbonium ion. This sequence of reactions brings about α-1,4-glycosidic bond formation and primer elongation.

This mechanism accounts for retention of configuration in both directions without requiring sequential double inversion of configuration. It also provides for a plausible explanation of the essential role of pyridoxal-5′-phosphate in glycogen phosphorylase catalysis, as the phosphate of the cofactor, pyridoxal-5′-phosphate, and the substrate phosphates approach each other within a hydrogen-bond distance allowing proton transfer and making the phosphate of pyridoxal-5′-phosphate into a proton shuttle which recharges the substrate phosphate anion.

The fact that glycogen phosphorylase can be used to polymerize amylose was first demonstrated by Schäffner and Specht [33] in 1938 using yeast phosphorylase. Shortly after, the same behavior was also observed for other phosphorylases from yeast by Kiessling [34, 35], muscles by Cori, Schmidt and Cori [36], pea seeds [37], potatoes by Hanes [38] and preparations from liver by Ostern and Holmes [39], Cori, Cori and Schmidt [40] and Ostern, Herbert and Holmes [41]. These results opened up the field of enzymatic polymerizations of amylose using glucose-1-phosphate as monomer and can be considered the first experiments ever to synthesize biological macromolecules *in vitro*.

One of the remarkable properties of phosphorylase is that it is unable to synthesize amylose unless a primer is added (poly- or oligomaltosaccharide).

$$n(\text{glucose-1-phospate}) + \text{primer} \rightleftarrows \text{amylose} + n(\text{orthophosphate})$$

The kinetic behavior of the polymerization of amylose with potato phosphorylase with various saccharides as primers was first studied by Hanes [38]. Green and Stumpf [42] failed to detect priming action with maltose but were able to confirm all other results by Hanes. Weibull and Tiselius [43] found that the maltooligosaccharide of lowest molecular weight (MW) to exhibit priming activity was maltotriose which was confirmed by Whelan and Bailey [44], who also showed that maltotriose is the lowest member of the series of oligosaccharides to exhibit priming activity (see Figure 9.5).

Whelan and Bailey were also able to clarify the polymerization mechanism of the enzymatic polymerization with phosphorylase [44]. Their results showed that the polymerization follows a 'multi chain' scheme in contrast to a 'single chain' scheme that was also proposed by some authors. In the 'multi chain' polymerization scheme the enzyme–substrate complex dissociates after every addition step, whereas in the 'single chain' scheme each enzyme continuously increases the length of a single primer chain without dissociation.

By studying the polydispersities of amyloses obtained by enzymatic polymerization with potato phosphorylase from maltooligosaccharides of various lengths, Pfannemüller and Burchard were able to show that the reaction mechanism of the polymerization with maltotriose as primer varies from its higher homologs [45]. While the amyloses built by polymerization from maltotetraose or higher showed a Poisson distribution [46] that can be expected from a polymerization

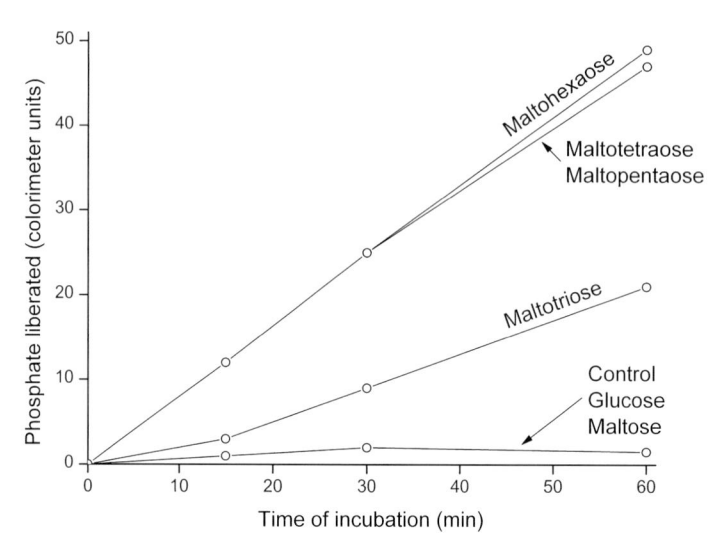

Figure 9.5 Priming activity of glucose and maltooligosaccharides in the enzymatic polymerization with potato phosphorylase and glucose-1-phosphate as monomer. (Reproduced with permission from [44]. Copyright © (1954) Portland Press Ltd.).

following a 'multi chain' scheme (random synthesis occurs and all the primer chains grow at approximately equal rates) a bimodal broad distribution was observed when maltotriose was used as primer. The authors found that in the case of maltotriose as a primer the reaction can be divided into a start reaction and the following propagation, the rate of the first reaction being 400 times slower than the rate of the propagation. Due to this start reaction not all chains start to grow at the same time which results in a broader distribution. The propagation follows again a 'multi chain' reaction scheme. Suganuma et al. [47] were able to determine the exact kinetic parameters of the synthetic as well as the phosphorolytic reaction using maltotriose and higher maltooligosaccharides as primer and were able to confirm the results of Whelan and Bailey [44], and Pfannemüller and Burchard [45].

Recently Kuriki et al. succeeded in producing glucose-1-phosphate *in situ* during the enzymatic polymerization of amylose. By using sucrose phosphorylase or cellobiose phosphorylase the monomer was produced during the polymerization from inorganic phosphate and sucrose or cellobiose respectively [48–50].

Kadokawa et al. recently reported on a very interesting two-phase system for the enzymatic polymerization of amylose via phosphorylase catalysis [51]. They were able to show that the polymerization readily proceeds in a calcium alginate hydrogel beads/DMSO system. When the calcium alginate hydrogel beads including glucose-1-phosphate, maltoheptaose, and phosphorylase were suspended in DMSO and the system was slowly stirred at 40 °C for 12 h, the reaction proceeded to produce amylose, which eluted to the DMSO solution, see Figure 9.6 for a schematic representation of the process. The obtained amylose was purified by the treatment with ion-exchange resins, and its structure was confirmed by ^{1}H-NMR spectroscopy.

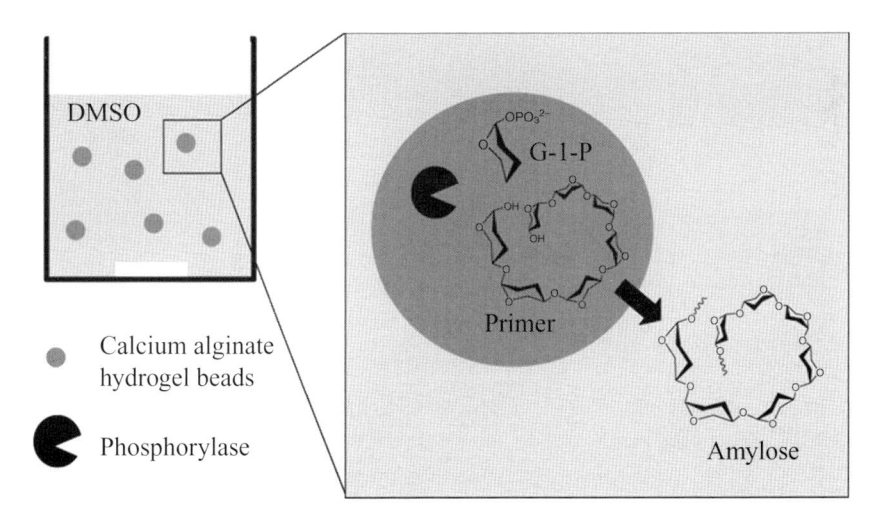

Figure 9.6 Image of the calcium alginate hydrogel beads/DMSO system. (Reproduced with permission from [51]. Copyright © (2009) Taylor & Francis).

9.2.1.2 Hybrid Structures with Amylose Blocks

The strict primer dependence of the glycogen phosphorylases makes them ideal candidates for the synthesis of hybrid structures of amylose with non-natural materials (e.g., inorganic particles and surfaces, synthetic polymers). For this, a primer functionality (maltooligosaccharide) can be coupled to a synthetic structure and subsequently elongated by enzymatic polymerization resulting in amylose blocks.

Various examples of these types of hybrid materials are reported which are outlined below.

Amylose Hybrids with Short Alkyl Chains Pfannemüller et al. showed that it is possible to obtain carbohydrate-containing amphiphiles with various alkyl chains via amide bond formation. For this, maltooligosaccharides were oxidized to the corresponding aldonic acid lactones which could subsequently be coupled to alkylamines [52–60]. Such sugar-based surfactants are important industrial products finding their applications in cosmetics, medical applications etc. [61–63]. The authors were also able to extend the attached maltooligosaccharides with enzymatic polymerization with potato phosphorylase which resulted in products with very interesting solution properties [64, 65].

Amylose Brushes on Inorganic Surfaces Amylose brushes (a layer consisting of polymer chains dangling in a solvent with one end attached to a surface is frequently referred to as a polymer brush) on spherical and planar surfaces can have several advantages, such as detoxification of surfaces etc. The modification of surfaces with thin polymer films is widely used to tailor surface properties such as wettability, biocompatibility, corrosion resistance and friction [66–68]. The advantage of polymer brushes over other surface modification methods, like self-assembled monolayers, is their mechanical and chemical robustness, coupled with a high degree of synthetic flexibility towards the introduction of a variety of functional groups.

Commonly, brushes are prepared by grafting polymers to surfaces by, for instance, chemical bonding of reactive groups on the surface and reactive end groups of the attached polymers. This 'grafting to' approach has several disadvantages as it is very difficult to achieve high grafting densities and/or thicker films due to steric crowding of reactive surface sites by already adsorbed polymers.

The so-called 'grafting from' approach (polymers are grown from initiators bound to surfaces) is a superior alternative as the functionality, density and thickness of the polymer brushes can be controlled with almost molecular precision.

The first surface-initiated enzymatic polymerization reported was the synthesis of amylose brushes on planar and spherical surfaces [69]. For this silica or silicone surfaces were modified with self-assembled monolayers of (3-aminopropyl) trimethoxysilane or chlorodimethylsilane respectively. To these functionalities oligosaccharides were added via (a) reductive amidation of the oligosaccharides to surface bound amines; (b) conversion of the oligosaccharide to the according aldonic acid lactone and reaction with surface bound amines; and (c) incorporation of a double bond to the oligosaccharide and subsequent hydrosilylation to surface bound Si–H functions. The surface-bound oligosaccharides could be enzymatically elongated with potato phosphorylase and glucose-1-phosphate as monomer to amylose chains of any desired length. The degree of polymerization could be determined by spectrometric measurement of the liberated amount of inorganic phosphate [70] which was confirmed by cleavage of the amylose brushes (either enzymatically or by prior incorporation of light sensitive spacers) and subsequent characterization of the free amylose chains. The modified surfaces of the amylose obtained showed good chiral discrimination when employed as column materials in chiral affinity chromatography. Modification of the OH-groups of the amylose brushes even enhanced the separation strength of the developed column materials (Loos, unpublished results). The results were recently confirmed by Breitinger who attached maltooligosaccharides to surfaces via acid labile hydrazide linkers and enzymatically extended the chains with potato phosphorylase [71].

Copolymers with Amylose The combination of oligo- or polysaccharides with non-natural polymeric structures opens up a novel class of materials. By varying the chain topology of the individual blocks as well as of the whole copolymer, the type of blocks, the composition etc. a complete set with tailor-made properties can be designed.

Amylose is a rod-like helical polymer consisting of α-(1→4) glycosidic units. A measurement of the stiffness of a polymer is afforded by the so-called persistence length, which gives an estimate of the length scale over which the tangent vectors along the contour of the chains backbone are correlated. Typical values for persistence lengths in synthetic and biological systems can be several orders of magnitude larger than for flexible, coil-like polymers. Rod-like polymers have been found to exhibit lyotropic liquid crystalline ordered phases such as nematic and/or layered smectic structures with the molecules arranged with their long axes nearly parallel to each other. Supramolecular assemblies of rod-like molecules are also capable of forming liquid crystalline phases. The main factor governing the geometry of supramolecular structures in the liquid crystalline phase is the anisotropic aggregation of the molecules.

Copolymeric systems with amylose are therefore systems in which at least one component is based on a conformationally rigid segment, which are generally referred to as rod–coil systems [72–75]. By combining rod-like and coil-like polymers a novel class of self-assembling materials can be produced since the molecules share certain general characteristics typical of diblock molecules and thermotropic calamitic molecules. The difference in chain rigidity of rod-like and coil-like blocks is expected to greatly affect the details of molecular packing in the condensed phases and thus the nature of thermodynamically stable morphologies in these materials. The thermodynamic stable morphology probably originates as the result of the interdependence of microsegregation and liquid cristallinity. From this point of view it is fascinating to compare the microstructures originating in solution and in the bulk for such materials.

Practical applications in which copolymers are characterized by some degree of structural asymmetry have been suggested. For instance a flexible block may be chosen as it donates a flexural compliance, whereas the more rigid portion offers tensile strength. In addition to the mechanical properties, the orientational order and the electrical conductance of certain rigid blocks could be exploited in optical and electrical devices.

Comb type and linear block copolymer systems with enzymatically synthesized amylose are reported, which are outlined below.

Comb-Type Copolymers with Amylose The first comb-like structures synthesized by enzymatic grafting *from* polymerization from a polymeric backbone were reported by Husemann et al. [76, 77]. Acetobromo oligosaccharides were covalently bound to 6-trityl-2,3-dicarbanilyl-amylose chains and subsequently elongated by enzymatic polymerization with potato phosphorylase, the result being amylopectin-like structures with various degrees of branching. Pfannemüller et al. extended this work by grafting amylose chains onto starch molecules. The modified starches were studied by the uptake of iodine and by light scattering measurements of carbanilate derivates [78] and appeared to be star-like in electron microscopy studies [79].

A full series of star-, network- and comb-like hybrid structures with oligosaccharides were synthesized by Pfannemüller et al. (see Figure 9.7) and it was

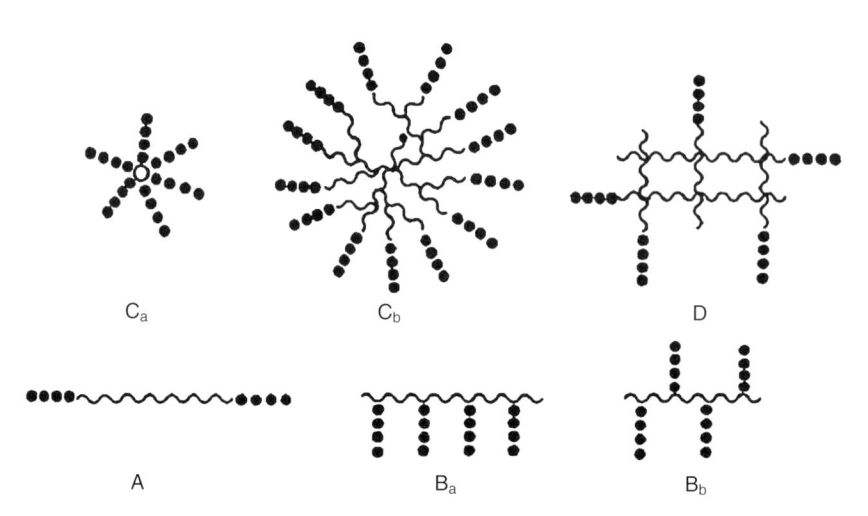

Figure 9.7 Maltotetraose hybrids with various carriers resulting in different chain architecture. A: poly(ethylene oxide); B_a and B_b: poly(acrylic acid), amylose, cellulose and other polysaccharides; C_a: cyclodextrin and multifunctional acids; C_b: amylopectin; D: crosslinked poly(acryl amide). (Reproduced with permission from [80]. Copyright © (1978) John Wiley & Sons, Inc.).

shown that the attached oligosaccharides can be extended via the enzymatic polymerization with potato phosphorylase [52–54, 80, 81].

Another type of comb-like amylose hybrids synthesized via enzymatic grafting with phosphorylase is based on polysiloxane backbones. To achieve these structures double bonds were incorporated to the reducing end of oligosaccharides which were then attached to poly(dimethylsiloxane-co-methylsiloxane) copolymers via hydrosilylation [82, 83] or to silane monomers which were subsequently polymerized to polysiloxanes [84]. Various mono-, di-, tri- and oligosaccharides were attached to siloxane backbones and their solution properties were studied with viscosimetry and static and dynamic light scattering [85]. The pendant oligosaccharide moieties could be extended with enzymatic grafting from polymerization [86, 87].

Kobayashi *et al.* succeeded in attaching maltopentaose to the *para* position of styrene and performed free radical polymerizations towards the homopolymers [88, 89] as well as copolymers with acrylamide [88]. Kobayashi *et al.* also reported on the successful attachment of maltopentaose to poly(L-glutamic acid) [90]. Kakuchi *et al.* showed that the saccharide-modified styrene monomers could also be polymerized with TEMPO-mediated controlled radical polymerization [91]. Amylose-grafted polyacetylenes were recently reported by Kadokawa *et al.* Maltooligosaccharide-grafted polyacetylene was synthesized by Rh-catalyzed polymerizaton of *N*-propargylamide monomers having a maltooligosaccharide chain [92] and by attaching maltoheptaonolactone to amine-functionalized polyacetylene [93]. Kadokawa *et al.* also succeeded in covalently attaching maltooligosaccharides to natural biopolymers such as cellulose, chitin and chitosan [94–96].

In all cases the authors could successfully elongate the attached oligosaccharide structures with enzymatic polymerization, the product being comb-type block copolymers with amylose.

Linear Block Copolymers with Amylose Various linear block copolymers of the AB, ABA and ABC type with enzymatically polymerized amylose blocks have been reported. Ziegast and Pfannemüller converted the hydroxyl end groups of poly(ethylene oxide) into amino groups via tosylation and further reaction with 2-aminoalkylthiolate [97]. To the resulting mono- and di-amino-functionalized poly(ethylene oxide) maltooligosaccharide lactones were attached and subsequently elongated to amylose via enzymatic polymerization [98]. Pfannemüller *et al.* performed a very detailed study on the solution properties of the synthesized A-B-A triblock copolymers, as they can be considered model substances for 'once-broken rod' chains [99]. With static and dynamic light scattering they found that the flexible joint between the two rigid amylose blocks has no detectable effect on the common static and dynamic properties of the chain. With dielectric measurements it however became obvious that the directional properties of the electric dipoles of the broken rigid chains showed a different behavior to the non-broken rods (pure amylose). Akyoshi *et al.* also synthesized amylose-block-poly(ethylene glycol) block copolymers via enzymatic grafting from oligosaccharide terminated

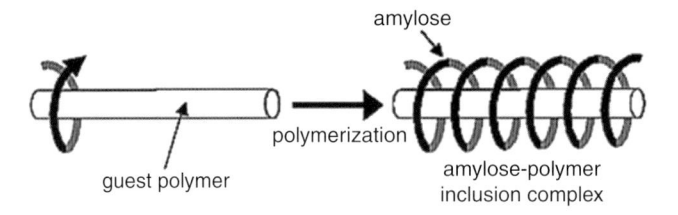

Figure 9.8 Schematic representation of the concept of vine-twining polymerization. (Reproduced with permission from [112]. Copyright © (2005) Wiley Periodicals, Inc.).

poly(ethylene oxide) and studied the solution properties of these amphiphilic block copolymers by static and dynamic light scattering [100, 101].

It was also shown that the enzymatic polymerization of amylose could be started from oligosaccharide-modified polymers that are not soluble in the medium of polymerization (aqueous buffers). Amylose-block-polystyrene block copolymers could be synthesized by attaching maltooligosaccharides to anionically-synthesized amino-terminated polystyrene and subsequent enzymatic elongation to amylose [102, 103]. Block copolymers with a wide range of MWs and copolymer composition were synthesized via this synthetic route. The solution properties of star-type as well as crew-cut micelles of these block copolymers were studied in water and THF and the according scaling laws were established [104]. In THF up to four different micellar species were detectable, some of them in the size range of vesicular structures, whereas the crew-cut micelles in water were much more defined. Bosker *et al.* studied the interfacial behavior of amylose-block-polystyrene block copolymers at the air–water interface with the Langmuir–Blodgett technique [105].

Recently two groups reported on controlled radical polymerizations started from maltooligosaccharides (ATRP- [106] and TEMPO-mediated radical polymerization [107]) which will certainly lead to new synthetic routes towards amylose-containing block copolymers.

Even though the products are not block copolymer structures the work of Kadokawa *et al.* should be mentioned here. In a process the authors named 'vine-twining polymerization' (after the way vine plant grow helically around a support rod) the enzymatic polymerization of amylose is performed in the presence of synthetic polymers in solution and the authors showed that the amylose chain grown, incorporates the polymers into its helical cavity while polymerizing, see Figure 9.8 [108–115].

9.2.2
Branching Enzyme

The formation of the α-(1→6) glucosyl branches of amylopectin and glycogen is synthesized by branching enzymes (systematic name : (1→)-α-D-glucan:(1→4)-α-D-glucan 6-α-D-[(1→4)-α-D-glucano]-transferase; EC 2.4.1.18).

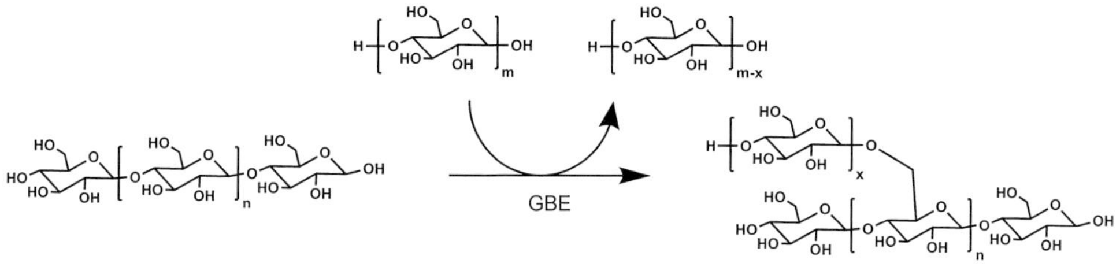

Figure 9.9 Catalytic action of the branching enzyme.

The branching enzyme itself is not able to induce polymerization. Instead it catalyzes the formation of α-(1→6) branch points by the hydrolysis of an α-(1→4) glycosidic linkage and subsequent inter- or intra-chain transfer of the nonreducing end-terminal fragment to the C-6 hydroxyl position of an α-glucan (see Figure 9.9) [116]. Depending on their source, branching enzymes have a preference for transferring different lengths of glucan chains [117–120].

The branching enzyme, also known as the Q-enzyme, has been isolated from various sources, including potato tubers [121–124], maize [125–127], teosinte [128], sorghum [129], rice [130], *Pisum sativum* [131], spinach [132, 133], mammalian muscle [134], *Escherichia coli* [116, 135] and rabbit liver [136]. The first evidence of the action of a branching enzyme was found in potato juice [137]. It is the branching enzyme that is responsible for the branched structure of glycogen as well as amylopectin. As well as the intra- and inter-chain transfer as prooven by Whelan [138] in potato branching enzyme, Takata [118, 139] proposed a third possibilty: backbiting, resulting in intramolecular cyclization of the chain.

In potatoes [140] and maize [141] at least two distinct isozymes have been identified, while only one form is found in different bacteria. Depending on the source, the branching enzyme has a preference for different donors and acceptors. Furthermore the branching enzymes may have different specificities in the length of the transferred chain. For example, starch branching enzyme (SBE) I from maize preferentially transfers long chains and is more active on amylose while the isozyme SBEII is more active on amylopectin and transfers shorter chains [142]. Many branching enzymes have been over-expressed in *E. coli* bacteria [119, 120, 139, 141–145] to study the exact mechanism in more detail.

In nature a cascade of enzyme-catalyzed reactions is involved for the biosynthesis of starch. When selecting the appropriate enzymes and reaction circumstances reactions with multiple enzymes can be performed *in vitro*. Synthetic glycogen was first synthesized *in vitro* by Cori [146] in 1943 via the cooperative action of muscle phosphorylase and branching enzymes isolated from rat liver and rabbit heart.

In this reaction phosphorylase catalyzes the polymerization of glucose-1-phosphate in order to obtain linear polysaccharide chains with α-(1→4) glycosidic linkages; the glycogen branching enzyme is able to transfer short, α-(1→4) linked, oligosaccharides from the nonreducing end of starch to an α-(1→6)

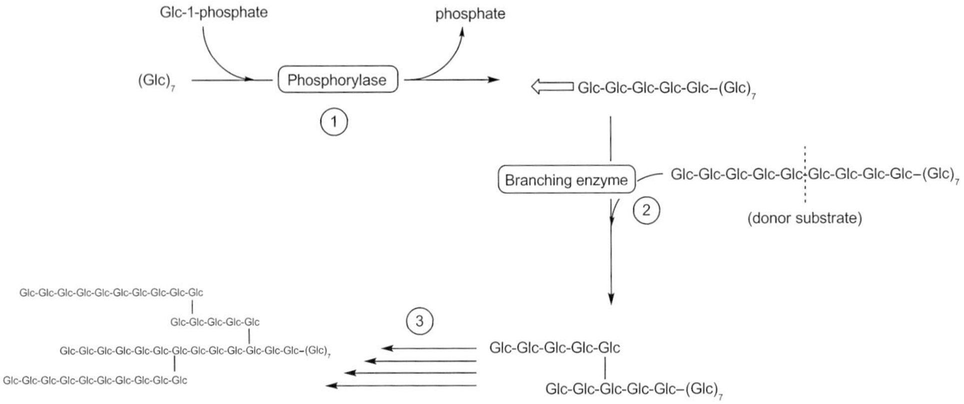

Figure 9.10 Schematic representation of the reactions catalyzed by glycogen phosphorylase and glycogen branching enzyme of *Deinococcus geothermalis* (GBE$_{Dg}$):
1. Phosphorylase transfers a glucose (Glc) residue from glucose-1-phosphate to the nonreducing end of the α-(1.4) linked primer ((Glc)$_7$) releasing an inorganic phosphate. This process continues until the length of the growing chain approaches a degree of polymerization of approximately 20 glucose residues. – 2. GBE$_{DG}$ catalyzes the formation of an α-(1.6) branch point by cleaving an α-(1.4) glycosidic linkage in the donor substrate and transferring the nonreducing end-terminal fragment of the chain to the C-6 hydroxyl position of an internal glucose residue that acts as the acceptor substrate. – 3. Reactions 1 and 2 repeat until ~80% glucose-1-phosphate is consumed, the resulting (hyper)branched 'amylose' has a degree of branching of around 11%.

position (see Figure 9.10). By combining the branching enzyme with phosphorylase it becomes possible to synthesize branched structures via an one-pot synthesis, as phosphorylase will polymerize linear amylose and the glycogen branching enzyme will introduce the branching points which are again extended by phosphorylase.

This method was repeated by others with phosphorylases and branching enzymes from various sources [118, 134, 147–153]. Waldmann [154] used sucrose phosphorylase and glucan phosphorylase in a 1-pot synthesis to produce synthetic amylose from sucrose. Sucrose phosphorylase promotes the synthesis of glucose-1-phosphate from inorganic phosphate and sucrose while glucan phosphorylase catalyzes the amylose chain formation from glucose-1-phosphate. Kuriki [48] proposed to extend this reaction by adding a branching enzyme which would result in branched α-glucans from sucrose and inorganic phosphate. Since sucrose and inorganic phosphate are cheaper than the rather expensive glucose-1-phosphate, sucrose phosphorylase may be a key enzyme to make these reactions industrially attractive. The combined action of branching enzyme with starch synthase [155, 156] and glycogen synthase [135, 157, 158] has been reported as well and resulted in synthetic glycogen.

Palomo Reixach et. al. [159] overexpressed the glycogen branching enzyme of *Deinococcus geothermalis* (GBE$_{Dg}$) in *E. coli*. The GBE$_{Dg}$ is an interesting branching

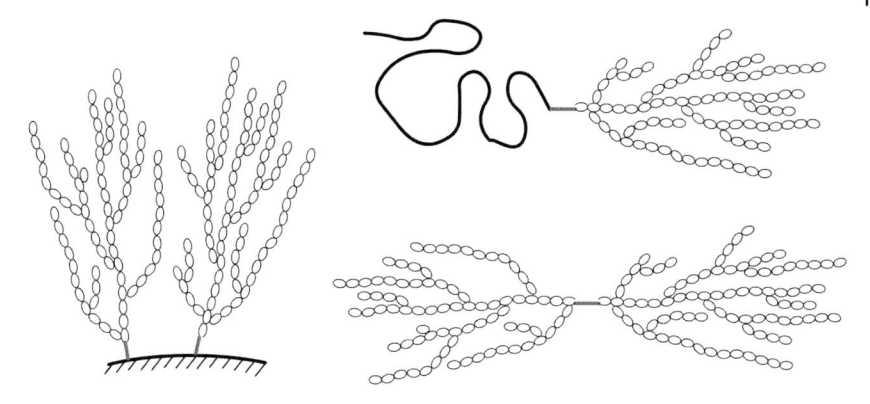

Figure 9.11 Schematic representation of hybrid structures with amylopectin.

enzyme as it transfers rather short fragments, resulting in a highly branched structure. In addition the optimum activity is in the same pH and temperature range as that of potato phosphorylase, making it possible to execute an enzymatic tandem reaction together with potato phosphorylase. The degree of polymerization can be regulated by changing the ratio glucose-1-P/primer while the source of the GBE regulates the degree of branching obtained [152]. GBE_{Dg} catalyzes the branch formation until an average degree of branching of around 11% is reached as determined with ^{1}H-NMR and MALDI–TOF MS.

The monomer-up approach opens up the possibility of enzymatically grafting branched polysaccharides from functionalized substrates in order to make hybrid materials bearing a highly branched amylose part. Following the same route we are currently synthesizing hybrid materials bearing (hyper)branched polysaccharide structures as shown in Figure 9.11 with the described tandem reaction of two enzymes.

The branched structure, high amount of functional groups and the biocompatibility of these structures make these architectures suitable for applications in the biomedical field and in the food industry.

9.2.3
Sucrase

Amylosucrase (systematic name : sucrose:(1→4)-α-D-glucan 4-α-D-glucosyltransferase; EC 2.4.1.4) was used to synthesize amylose *in vitro* from sucrose monomers by Buleon *et al.* [160, 161]. In contrast to the phosphorylase-catalyzed polymerization in Section 9.2.1.1 no primer is necessary to start this polymerization. The morphology and structure of the resulting amylose depended on the initial concentration of sucrose. When glycogen particles were used as primer, amylose-based dendritic nanoparticles were obtained from the enzymatic polymerization.

Another very valuable sucrase for the construction of well-defined polysaccharides is glucansucrase (dextransucrase; systematic name : sucrose:(1→6)-α-D-glucan 6-α-D-glucosyltransferase; EC 2.4.1.5).

Recently Dijkhuizen et al. reported on a family of glucansucrases found in *Lactobacillus reuteri*, that convert sucrose into large, heavily branched α-glucans [162]. One of these glucansucrases (GTF180) [163] produces an α-glucan with α-(1→3) and α-(1→6) glycosidic linkages. Kamerling et al. could show that the (1→3, 1→6)-α-D-glucan of *L. reuteri* strain 180 has a heterogeneous structure with no repeating units present [164]. It contains only α-D-Glc-(1→6)- units in terminal position. All -α-D-Glc-(1→3)- units were shown to be 6-substituted, and the polysaccharide is built-up from different lengths of isomaltooligosaccharides, interconnected by single α-(1→3) bridges. The unique polysaccharide structure produced was suggested to be pre-biotic [165].

The GTF180 enzyme shows large similarity with other glucansucrase enzymes, but has a relatively large N-terminal variable region. Truncation of the enzyme, by deletion of the variable region, had no effect on the linkage distribution of the α-glucan produced [162].

Seibel et al. succeeded in constructing various new complex glycoconjugates containing thioglycosidic linkages with different glycopyranosides (galactose, glucose, neuraminic acid) (branched thiooligosaccharides) by changing the chemoselectivity of the various glucansucrases from α-1,6- to α-1,2-, α-1,3- or α-1,4-linked glucose [166].

Seibel et al. could also show that mutagenesis is an effective tool for altering the regioselectivity and acceptor substrate specificity of glucansucrase glycosyltransferase R (GTFR) of *Streptococcus oralis,* a dextran-producing enzyme. By random mutagenesis they were able to switch the regioselectivity and acceptor specificity of GTFR of *S. oralis* towards synthesis of (a) various short chain oligosaccharides or (b) novel (mutan) polymers with completely altered linkages, without compromising its high transglycosylation activity [167].

In addition Dijkhuizen et al. identified and characterized a *Lactobacillus* levansucrase (systematic name : sucrose:[6)-β-D-fructofuranosyl-(2→]n α-D-glucopyranoside 6-β-D-fructosyltransferase; EC 2.4.1.10) from *Lactobacillus reuteri* strain 121 that could produce a high MW levan polysaccharide from fructose [168].

9.2.4
Amylomaltase

Glycosyltransferases are also used extensively to modify natural polysaccharides (see Chapter 16). Thermoreversible gels that retrograded reversibly – comparable to gelatin gels – from enzymatically-modified starch using amylomaltase have been reported recently [169–174].

Amylomaltase (4-α-glucanotransferase; systematic name : (1→4)-α-D-glucan: (1→4)-α-D-glucan 4-α-D-glycosyltransferase; EC 2.4.1.25) catalyzes the glucan-chain transfer from one α-1,4-glucan to another α-1,4-glucan (or to the 4-hydroxyl group of glucose) or within a single linear glucan molecule to produce a cyclic-α-1,4-glucan [169, 175, 176]. It was first found in *E. coli*, but seems to be distributed in various bacterial species with different physiological functions. In *E. coli*, amylomaltase is expressed with glucan phosphorylase from the same operon. Amylomaltase is a member of the maltooligosaccharide transport and utilization system and plays a role in converting short maltooligosaccharides into longer chains upon which glucan phosphorylase can act [177]. The name amylomaltase is used for the microbial 4-α-glucanotransferases whereas the plant counterparts are usually called disproportionating enzymes (abbreviated to D-enzymes) [169, 175, 176].

Typically, enzymatic modification of starch employs hydrolyzing enzymes such as α-amylase, pullulanase and glucoamylase which hydrolyze the α-1,4- or α-1,6-glycosidic bonds in amylose and amylopectin by first breaking the glycosidic linkage and subsequently using a water molecule as acceptor substrate. Amylomaltases also initially break the glycosidic linkage but instead of water they use another oligosaccharide as an acceptor substrate and form a new glycosidic linkage. Amylomaltases can use high MW starch as both donor and acceptor molecule and can catalyze the transfer of long α-1,4-glucan chains [178], or even highly-branched cluster units of amylopectin.

9.2.5
Hyaluronan Synthase

Hyaluronan (HA) is a nonsulfated nonepimerized linear glycosaminoglycan (GAG) existing *in vivo* as a polyanion of hyaluronic acid and composed of repeating disaccharide units of D-glucuronic acid and *N*-acetyl-D-glucosamine [GlcAβ(1→3)GlcNAcβ(1→4)] (see also Chapter 16) [179–181]. It is a major constituent of the extracellular matrix (ECM) of the skin, joints, eyes and many other tissues and organs. Despite the simple structure of this macromolecule, the complexity of its physicochemical properties and biological functions is tremendous. HA has extraordinary hydrophilic, rheological, and signaling properties and is viscoelastic. This naturally occurring biopolymer is dynamically involved in many biological processes, such as embryogenesis, inflammation, metastasis, tissue turnover, and wound healing.

The isolation, purification, and identification of nearly pure HA has been the center of scientific interest for many years. The bacterial production of HA by *Streptococcus equi* [182] and *Streptococcus zooepidemicus* [183] enabled it to be produced in larger quantities than could be achieved with extraction methods alone. HA produced by *S. equi* has a lower MW than does HA produced by *S. zooepidemicus*, which has a MW of about 1.8 to 2×10^6 Da with a yield of around 4 grams of HA per liter of the cultivated solution. At present, HA from various sources, with different degrees of purity and MWs, is available for medical applications.

Hyaluronan is synthesized by hyaluronan synthase (systematic name: alternating UDP-α-*N*-acetyl-D-glucosamine:β-D-glucuronosyl-(1→3)-[nascent hyaluronan] 4-*N*-acetyl-β-D-glucosaminyltransferase and UDP-α-D-glucuronate:*N*-acetyl-β-Dglucosaminyl-(1→4)-[nascent hyaluronan] 3-β-D-glucuronosyltransferase; EC 2.4.1.212) [184–187]. Hyaluronan synthase (HAS) is a single protein GT that is able to transfer two different monosaccharides while most GTs catalyze one glycosidic transfer reaction exclusively.

Markovitz et al. successfully characterized the HAS activity from *Streptococcus pyogenes* and discovered the enzyme's membrane localization and its requirements for sugar nucleotide precursors and Mg^{2+} [188]. DeAngelis et al. were the first to succeed in the molecular cloning and characterization of the Group A streptococcal gene encoding the protein HasA, known to be in an operon required for bacterial HA synthesis [189, 190]. Following this, sequences of the genes encoding other HAS proteins were identified using molecular biological techniques [179, 184, 186, 191–204]. However little is yet known about the structure and mechanism of HAS.

The *in vitro* synthesis of hyaluronan oligomers and polymers using HAS and UDP-sugars was reported by the group of DeAngelis. The monosaccharide units from UDP–GlcNAc and UDP–GlcA are transferred sequentially in an alternating fashion to produce the disaccharide repeats of the heteropolysaccharide. Recombinant derivatives of one HA synthase, PmHAS from the gram-negative bacterium *Pasteurella multocida* type A [201], have proven to be very useful for chemoenzymatic syntheses of both oligosaccharides [205] and polysaccharides [206, 207].

In 2004, the PmHAS was employed in synchronized, stoichiometrically-controlled polymerization reactions *in vitro* to produce monodisperse HA polysaccharide preparations [206]. Reaction synchronization is achieved by providing the HA synthase with an oligosaccharide acceptor to bypass the slow polymer initiation step *in vitro*. All HA chains are elongated in parallel and thus reach the same length, yielding a narrow size distribution population. The synthase will add all available UDP–sugar precursors to the nonreducing termini of acceptors in a nonprocessive fashion (see Figure 9.12)

Therefore, size control is possible. For example, if there are many termini (i.e., z is large), then a limited amount of UDP–sugars will be distributed among many molecules and thus result in many short polymer chain extensions. Conversely, if there are few termini (i.e., z is small), then the limited amount of UDP–sugars will be distributed among few molecules and thus result in long polymer chain extensions. With this it became possible to synthesize high MW well-defined HA polymers [208].

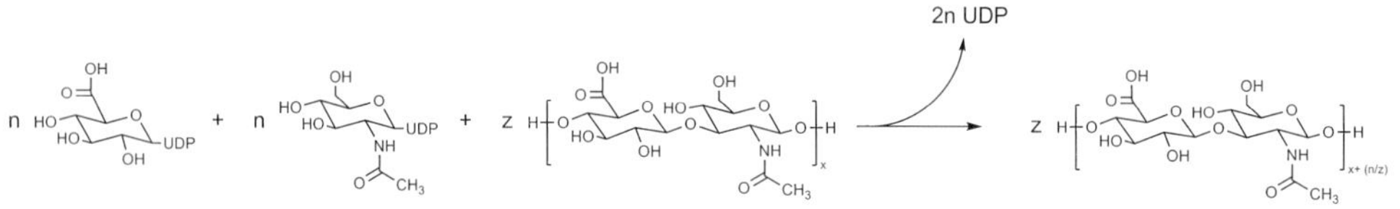

Figure 9.12 *In vitro* synthesis of monodisperse HA.

9.3 Glycosidases

The potential of glycosidases as tools for the regio- and stereospecific synthesis of glycosides was realized nearly 100 years ago [209–211].

Under conditions that favor reversal of their normal hydrolytic reaction, retaining glycosidases have been extensively used as catalysts in oligosaccharide synthesis [212–216].

The two main approaches to glycosidase-catalyzed synthesis of glycosidic linkages involve direct reversal of hydrolysis (equilibrium-controlled synthesis) and trapping of a glycosyl-enzyme intermediate (kinetically-controlled process) [217].

The equilibrium approach struggles with an unfavorable thermodynamic balance of the synthesis–hydrolysis process. High concentration of sugars, addition of organic co-solvents and elevated reaction temperatures are a requirement to achieve significant transformation with low yield.

The kinetically-controlled process depends on the more rapid trapping of an activated glycosyl–enzyme intermediate by the glycosyl acceptor than by water. Although under the right conditions glycoside formation is favored kinetically, hydrolysis either of the intermediate or of the resulting product is favored thermodynamically, and practical yields generally range from 20 to 40%.

The type of activated donor substrate used in a glycosidase-catalyzed reaction can greatly influence several important factors such as donor reactivity, solubility, reaction yields and propensity for donor self-condensation reactions. The accessibility of the donor substrates also needs to be considered.

Many glycosyl donors have been used for transglycosylation reactions, the most successful and common being aryl glycosides, oligosaccharides such as lactose, and glycosyl fluorides. One major consideration in a transglycosylation reaction is that the activated donor substrate must function as a significantly better glycosyl donor than does the product, to allow the transglycosylation product to accumulate.

Obviously, techniques that reduce the rate of hydrolysis should increase the yield of the desired transglycosylation reaction. Thus, reduction of the effective concentration of water, running the transglycosylation reactions at high concentrations and the addition of large excesses of the acceptor molecule have all been shown to improve yields.

Glycosidases have already been used extensively to catalyze the formation of polysaccharides. This field is greatly inspired by the seminal work of Professor Shiro Kobayashi and his co-workers who succeeded in synthesizing:

- cellulose by means of polymerization of β-cellobiosyl fluoride catalyzed by cellulase [218–221],
- amylose by amylase with α-D-maltosyl fluoride [222],
- xylan by xylanase with β-xylobiosyl fluoride [223],
- chitin by chitinase with N,N-diacetylchitobiose oxazoline [224–226],
- keratan by keratanase II with Galβ(1-4)GlcNAc(6S) and Gal(6S)β(1-4)GlcNAc(6S) oxazoline [227].

9 Enzymatic Polymerizations of Polysaccharides

Furthermore Kobayashi et al. were able to show that hyaluronidase is able to catalyze the *in vitro* synthesis of hyaluronan [228], chondroitin [229], chondroitin sulfate [230], their derivatives [229, 231] and of non-natural GAGs [232].

Some excellent reviews have appeared extensively covering the enzymatic polymerization of polysaccharides using glycoside hydrolases [233–245]. Two examples of this field are highlighted below; for more details please refer to these reviews.

9.3.1
Cellulase

The *in vitro* synthesis of cellulose via a nonbiosynthetic path has been achieved for the first time by condensation of β-D-cellobiosyl fluoride as substrate for cellulase (systematic name : 4-β-D-glucan 4-glucanohydrolase; EC 3.2.1.4) in a mixed solvent of acetonitrile/acetate buffer by Kobayashi et al. (see Figure 9.13) [218, 246–249].

Glycosyl fluorides have been known for a long time to be good glycosyl donors and are commonly used in transglycosylation reactions catalyzed by glycoside hydrolases or transglycosylases. Being α-fluoro ethers, glycosyl fluorides are relatively reactive compounds, particularly when activated by binding to an enzyme active site.

Glycosyl fluorides have an advantage over aryl glycosides and oligosaccharide donors related to their use as substrates, namely that they appear to be universally processed by all glycosidases. Moreover, in the case of a glycosyl fluoride donor, the major by-product of a transglycosylation reaction is fluoride ion, which is easy to separate from the transglycosylation product.

Kobayashi and coworkers proposed a mechanism for the enzymatic polymerization of β-D-cellobiosyl fluoride with cellulase following the general accepted mechanism for glycosidase-catalyzed synthesis of glycosides.

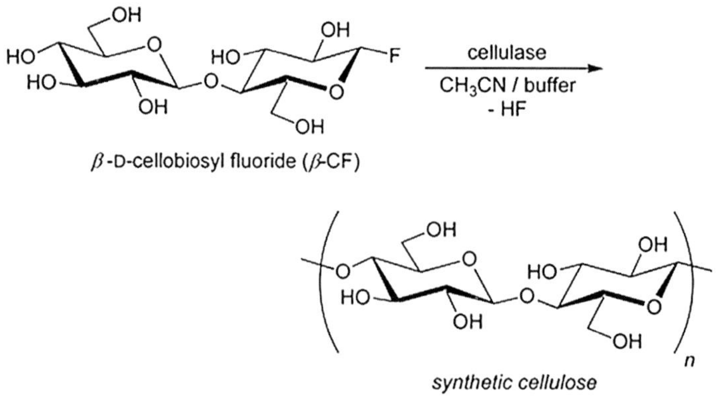

Figure 9.13 *In vitro* enzymatic polymerization of cellulose using cellulase. (Reproduced with permission from [250]. Copyright © (2005) John Wiley & Sons, Inc.).

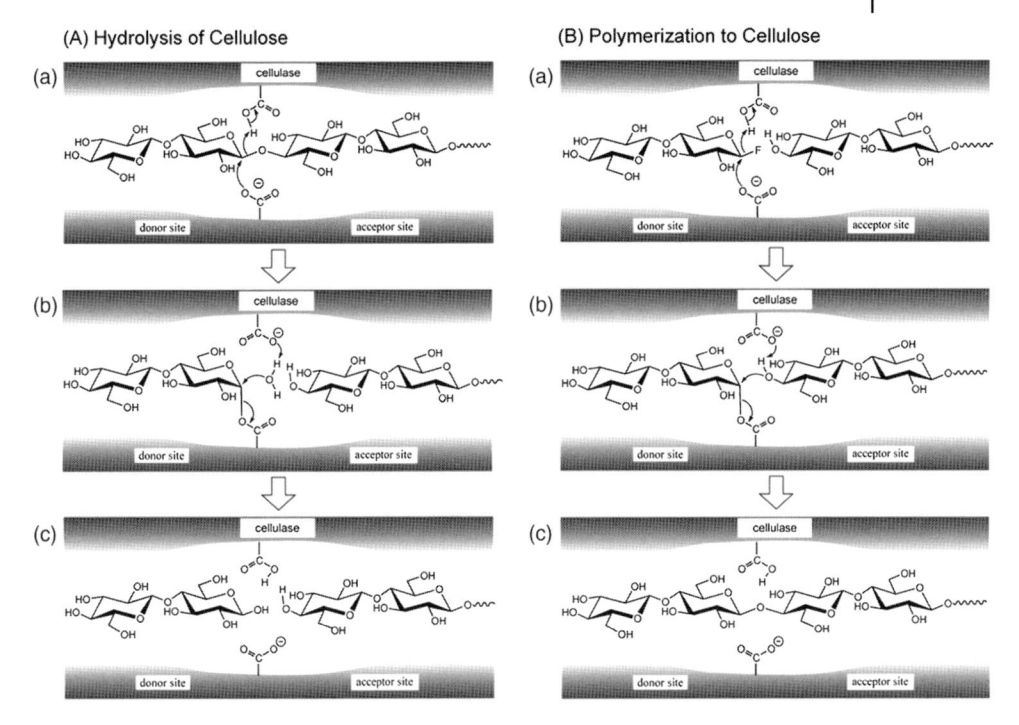

Figure 9.14 Proposed reaction mechanisms of cellulase catalysis: (A) hydrolysis of cellulose and (B) polymerization of β-CF into synthetic cellulose. (Reproduced with permission from [250]. Copyright © (2005) John Wiley & Sons, Inc.).

The native action (hydrolysis) of cellulase follows a double-displacement reaction mechanism via the formation and hydrolysis of a glycosyl–enzyme intermediate (see Figure 9.14 (A)). The canonical mechanism involves two steps with general acid–base catalysis: in the first step (glycosylation) the amino acid residue acting as a general acid protonates the glycosidic oxygen, while the deprotonated carboxylate functioning as a nucleophile attacks the anomeric center with concomitant C–O breaking of the scissile glycosidic bond leading to a covalent glycosyl–enzyme intermediate. The second deglycosylation step involves the attack by a molecule of water assisted by the conjugate base of the general acid residue, which renders the free sugar with overall retention of configuration, and the enzyme returns to its initial protonation state.

In the kinetically-controlled polymerization of β-D-cellobiosyl fluoride with cellulase the monomer is recognized at the donor site of cellulase catalyst and the fluoride anion is readily eliminated as HF via a general acid–base mechanism (Figure 9.13 (B) step a). The monomer forms a glycosyl–carboxyl intermediate with α-configuration with a similar structure to that of step b of the hydrolysis (stage b). The C-4 hydroxy group of the disaccharide monomer or the growing chain end located at the acceptor site can now attack the anomeric carbon from the β-side, and a new β(1-4) glycosidic linkage is formed (step c).

Kitaoka et al. were able to show that it is possible to synthesize cellulose from cellobiose with a cellulase/surfactant (dioleyl-N-D-glucona-L-glutamate) complex in nonaqueous LiCl/DMAc media [251]. With this they obtained cellulose with a higher degree of polymerization (>100) than previously reported from nonfluorinated monomers. In this approach the product of enzymatic polymerization (cellulose) has a higher solubility in the used solvent (lithium chloride in dimethylacetamide) while the surfactant stabilizes cellulose, retaining its activity during the course of polymerization.

β-Xylobiosyl fluoride was shown to be polymerizable by cellulase as well, resulting in xylan oligomers [223].

9.3.2
Hyaluronidase

Hyaluronidase (HAase; systematic name : hyaluronate 4-glycanohydrolase; EC 3.2.1.35) is a glycoside hydrolase that cleaves (1-4)-β-N-acetylhexosaminide linkages in hyaluronan, chondroitin, chondroitin sulfate.

It is known that the hydrolysis reaction proceeds via an oxazolinium ion intermediate. Based on this Kobayashi and coworkers succeeded in designing a disaccharide oxazoline transition state analog self-condensing substrate to revert the native action of HAase towards synthesis of hyaluronan and derivatives thereof (see Figure 9.15).

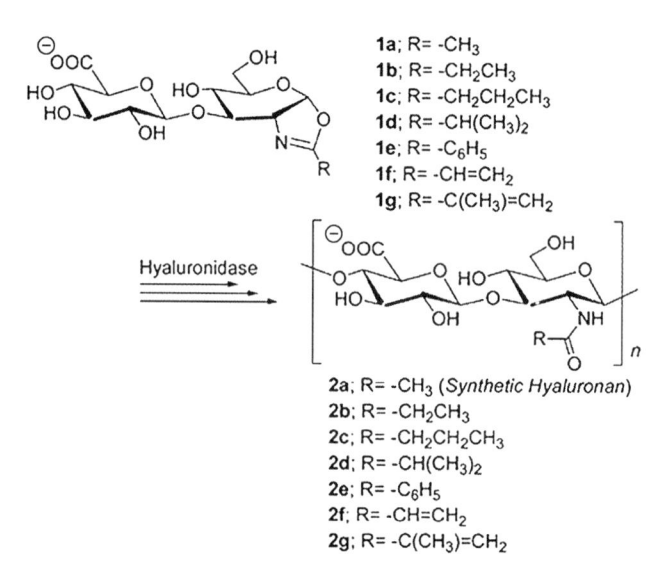

1a; R= -CH$_3$
1b; R= -CH$_2$CH$_3$
1c; R= -CH$_2$CH$_2$CH$_3$
1d; R= -CH(CH$_3$)$_2$
1e; R= -C$_6$H$_5$
1f; R= -CH=CH$_2$
1g; R= -C(CH$_3$)=CH$_2$

2a; R= -CH$_3$ (Synthetic Hyaluronan)
2b; R= -CH$_2$CH$_3$
2c; R= -CH$_2$CH$_2$CH$_3$
2d; R= -CH(CH$_3$)$_2$
2e; R= -C$_6$H$_5$
2f; R= -CH=CH$_2$
2g; R= -C(CH$_3$)=CH$_2$

Figure 9.15 Enzymatic polymerization of various monomers (series 1) to synthetic hyaluronan and its derivatives (series 2). (Reproduced with permission from [231]. Copyright © (2005) American Chemical Society).

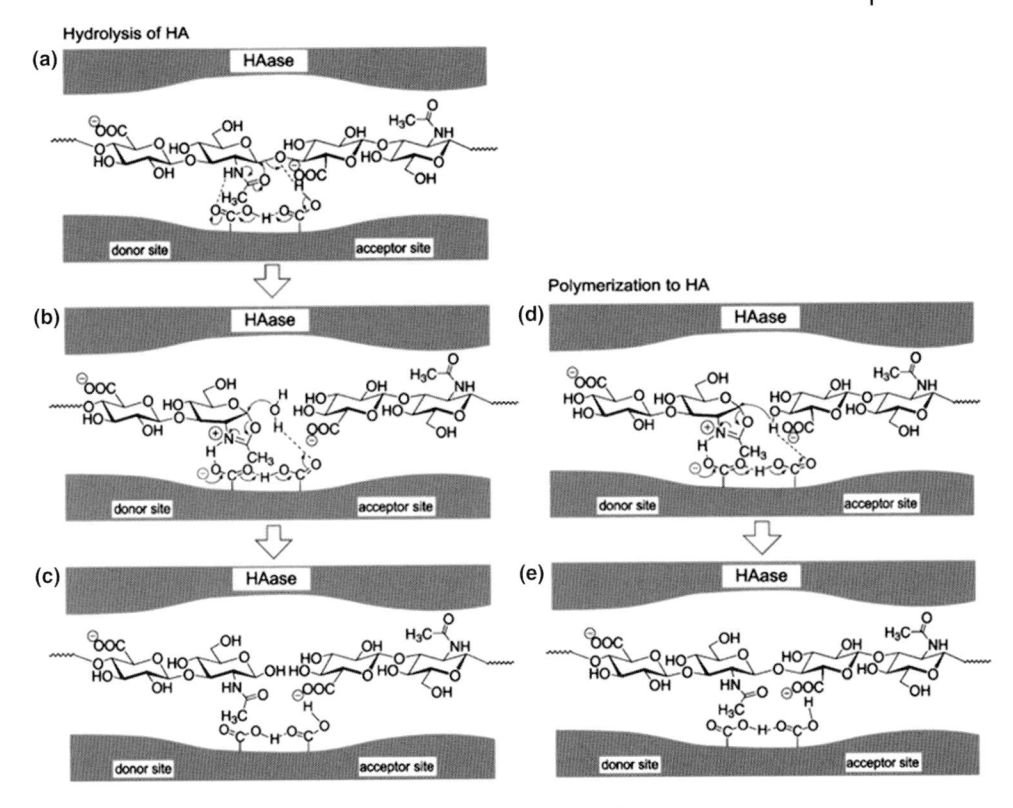

Figure 9.16 Postulated reaction mechanism catalyzed by HAase. (Reproduced with permission from [232]. Copyright © (2006) Wiley-VCH Verlag).

The proposed mechanism for enzymatic hydrolysis of hyaluronan involves nucleophilic attack of water at the anomeric carbon of the D-GlcNAc moiety *via* an oxazolinium ion intermediate (see Figure 9.16b). A water molecule can nucleophilically attack the oxazolinium anomeric carbon atom to open the oxazolinium ring, resulting in the formation of the hydrolysis products of a shortened HA molecule (see Figure 9.16c).

The oxazoline monomer is regarded as a transition state analog of the monomer, which is recognized and activated at the enzyme's donor site *via* protonation of the oxazoline ring nitrogen. The 4-hydroxyl group of GlcA in another molecule of the monomer located in the acceptor binding site then attacks the activated oxazolinium moiety to form a β-(1-4)-linkage (see Figure 9.16d,e). Repetition of this process gives rise to artificial HA of relatively high MW and perfectly controlled structure.

Wang *et al.* could show that this polymerization also proceeds under endo-β-N-acetylglucosaminidase catalysis [252].

Kobayashi and coworkers also showed that hyaluronidase is able to catalyze the enzymatic polymerization of chondroitin [229], chondroitin sulfate [230], their

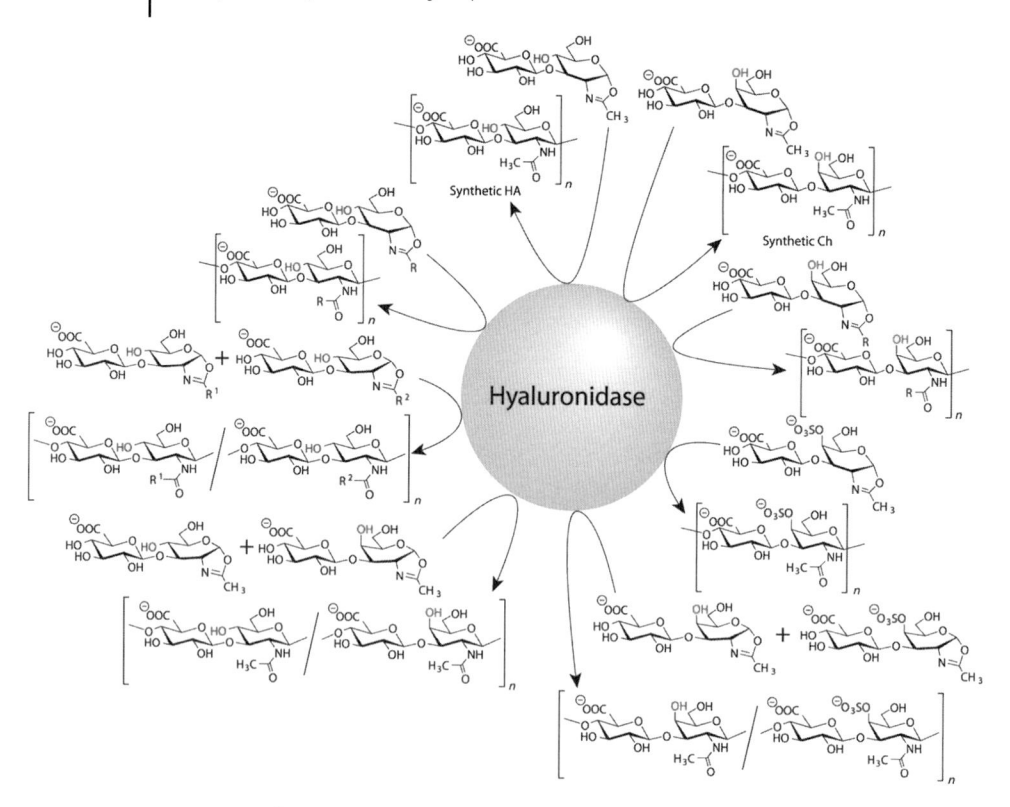

Figure 9.17 Illustration of the range of natural and non-natural glycosaminoglycans synthesizes using HAase. (Reproduced with permission from [232]. Copyright © (2006) Wiley-VCH Verlag).

derivatives [229, 231] and even of non-natural GAGs [232]. The range of polysaccharides they were able to synthesize with this enzyme is illustrated in Figure 9.17.

Considering the importance of GAGs as already discussed in Section 9.2.5 it is obvious that these findings will have a large impact on the field of biomaterials.

9.3.3
Glycosynthases

Another interesting approach is the use of glycosynthases, that are classified as mutated retaining glycosidases where the active-site carboxylate nucleophile has been replaced by a non-nucleophilic amino acid side chain (such as serine or alanine). Glycosynthases have proven to be very valuable biocatalysts for the synthesis of oligosaccharides and some excellent reviews appeared in the last years [3, 4, 213, 253–264].

Mutation of the catalytic nucleophile disables the enzyme as a hydrolase because no glycosyl–enzyme intermediate can be formed. However, an activated glycosyl donor with an anomeric configuration opposed to that of the donor substrate in the wild-type reaction (i.e., an α-glycosyl fluoride for a β-glycosidase) mimics the glycosyl–enzyme intermediate and is therefore able to react with an acceptor.

These mutants are effectively converted into inverting glycosidases which use opposite anomeric-configured donor substrates compared to their wild type glycosidases.

The cavity created in the active site – by mutation of the carboxylate residue acting as nucleophile in the wild-type enzyme by a smaller residue – allows binding of the glycosyl fluoride with the opposite anomeric configuration. As with the wild-type enzyme, transglycosylation is kinetically favored but the transglycosylation reaction is now irreversible because of the lack of the catalytic nucleophile; the product is no longer hydrolyzed and accumulates to give high transglycosylation yields.

The field of glycosynthases was pioneered by Withers et al. (glycosynthase by engineering the β-glycosidase from Agrobacterium sp. by mutating the nucleophile residue E358 to a non-nucleophilic residue [265]); Planas et al. (endo-acting glycosynthase by site-directed mutation of the nucleophile E134 of the retaining 1,3–1,4-β-glucanase from Bacillus licheniformis [266]); and Moracci et al. (glycosynthase that uses an activated glycosyl species as substrate but rescues the transglycosylation activity of the nonhydrolyzing glycosidase mutant with sodium formate as an external nucleophile [267]).

Glycosynthases from glycosidases belonging to more than a dozen glycoside hydrolase families have been discovered so far including, glycosynthases from cellulase [268, 269] amylosucrase [270], β-glucuronidase [271], exo-β-oligoxylanase [272, 273], 1,2-α-l-fucosidase [274] etc.

Glycosynthases have so far mainly been used to synthesize oligosaccharides but some interesting examples of their use for the synthesis of polysaccharides have appeared recently.

With glycosynthases mutants of cellulose, the successful synthesis of low molecular mass cellulose II [268], functionalized cellooligosaccharides [275], cellodextrins [276], xyloglycans oligomers and xylo-glucooligosaccharides [277, 278], and glycosylated flavonoids [279] has been reported.

The enzymatic polymerization of α-xylobiosyl fluoride using glycosynthases mutants of various xylanases resulted in β-(1-4)-xylans [280–282]. β-(1-3)-Glucans were synthesized by mutants of β-(1-3)-glucanase using α-laminaribiosyl fluoride as monomer [283]. Using the same monomer and a series of other oligosaccharide fluoride monomers various artificial (1-3)–(1-4) mixed-linked β-glucans were synthesized using glycosynthases mutants of β-(1-3)–(1-4)-glucanase [266, 284–287].

9.4
Conclusion

Polysaccharides are important biobased materials with applications spanning the whole range of cheap commodity plastics to advanced medical applications. The major limitations of the usage of polysaccharides in these applications is the difficulty in producing large quantities of well-defined molecules efficiently, as already mentioned in the introduction.

The examples reviewed in this chapter prove that enzymatic polymerizations using glycosyltransferases and glycosidases are powerful techniques for synthesizing various well-defined polysaccharides ranging from natural saccharides such as cellulose, amylose, amylopectin etc. to non-natural hybrid polysaccharides.

In contrast to many of the enzymes used to synthesize polyesters (see Chapter 4), vinyl polymers (see Chapter 6), polyphenols (see Chapter 7) etc. the enzymes reviewed in this chapter are mostly not commercially available.

In addition many of the glycosyltransferases and glycosidases used so far are not very stable. However, it can be expected that with biotechnological methods it will become possible to obtain enzymes with enhanced stability that can be used more frequently in polysaccharide synthesis. One example is for instance the new mutants of glycosidases seen in Section 9.3.3.

Furthermore, limitations such as low solubility of the synthesized polysaccharide versus stability of the enzyme will be overcome by continued research on polymerization techniques (such as biphasic systems, new solvent systems (see also Chapter 13).

Considering the polysaccharides discussed in this chapter there can be no doubt that new approaches to synthesize them via enzymatic polymerization will have a huge impact on the field.

References

1 Nagel, B., Dellweg, H., and Gierasch, L.M. (1992) *Pure Appl. Chem.*, **64**, 143.
2 Leloir, L.F. (1971) *Science*, **172**, 1299.
3 Davies, G.J., Charnock, S.J., and Henrissat, B. (2001) *Trends Glycosci. Glycotechnol.*, **13**, 105.
4 Jakeman, D.L., and Withers, S.G. (2002) *Trends Glycosci. Glycotechnol.*, **14**, 13.
5 Qian, X.P., Sujino, K., Palcic, M.M., and Ratcliffe, R.M. (2002) *J. Carbohydr. Chem.*, **21**, 911.
6 Song, J., Zhang, H.C., Li, L., Bi, Z.S., Chen, M., Wang, W., Yao, Q.J., Guo, H.J., Tian, M., Li, H.F., Yi, W., and Wang, P.G. (2006) *Curr. Org. Synth.*, **3**, 159.
7 Blixt, O., and Razi, N. (2008) Enzymatic Glycosylation by Transferases in *Glycoscience* (eds B. Fraser-Reis, K. Tatsuta, and J. Thiem), Springer, Berlin, Heidelberg, Germany, pp. 1361–1385.
8 Lairson, L.L., Henrissat, B., Davies, G.J., and Withers, S.G. (2008) *Annu. Rev. Biochem.*, **77**, 521.
9 Homann, A., and Seibel, J. (2009) *Appl. Microbiol. Biotechnol.*, **83**, 209.
10 Kanehisa, M., and Goto, S. (2000) *Nucl. Acids Res.*, **28**, 27.
11 Kanehisa, M., Goto, S., Hattori, M., Aoki-Kinoshita, K.F., Itoh, M., Kawashima, S., Katayama, T., Araki, M., and Hirakawa, M. (2006) *Nucl. Acids Res.*, **34**, D354.
12 Greene, L.H., Lewis, T.E., Addou, S., Cuff, A., Dallman, T., Dibley, M., Redfern, O., Pearl, F., Nambudiry, R., Reid, A., Sillitoe, I., Yeats, C., Thornton, J.M., and Orengo, C.A. (2007) *Nucl. Acids Res.*, **35**, D291.
13 Cantarel, B.L., Coutinho, P.M., Rancurel, C., Bernard, T., Lombard, V., and Henrissat, B. (2009) *Nucl. Acids Res.*, **37**, D233.

14 Chang, A., Scheer, M., Grote, A., Schomburg, I., and Schomburg, D. (2009) *Nucl. Acids Res.*, **37**, D588.

15 Kanehisa, M., Goto, S., Furumichi, M., Tanabe, M., and Hirakawa, M. (2010) *Nucl. Acids Res.*, **38**, D355.

16 Ball, S.G., van de Wal, M., and Visser, R.G.F. (1998) *Trends Plant Sci.*, **3**, 462.

17 Guan, H.P., and Keeling, P.L. (1998) *Trends Glycosci. Glycotechnol.*, **10**, 307.

18 Ball, S.G., and Morell, M.K. (2003) *Annu. Rev. Plant Biol.*, **54**, 207.

19 Ugalde, J.E., Parodi, A.J., and Ugalde, R.A. (2003) *Proc. Natl. Acad. Sci. U. S. A.*, **100**, 10659.

20 Buléon, A., Colonna, P., Planchot, V., and Ball, S. (1998) *Int. J. Biol. Macromol.*, **23**, 85.

21 Iwanoff, L. (1902) *Ber. d. deutsch. bot. Ges.*, **20**, 366.

22 Zaleski, W. (1906) *Ber. d. deutsch. bot. Ges.*, **24**, 285.

23 Zaleski, W. (1911) *Ber. d. deutsch. bot. Ges.*, **29**, 146.

24 Suzuki, U., Yoshimura, K., and Takaishi, M. (1906) *Tokyo Kagaku Kaishi*, **27**, 1330.

25 Bodńar, J. (1925) *Biochem. Z.*, **165**, 1.

26 Cori, G.T., and Cori, C.F. (1936) *J. Biol. Chem.*, **116**, 119.

27 Cori, G.T., and Cori, C.F. (1936) *J. Biol. Chem.*, **116**, 129.

28 Cori, C.F., Colowick, S.P., and Cori, G.T. (1937) *J. Biol. Chem.*, **121**, 465.

29 Cori, C.F., and Cori, G.T. (1937) *Proc. Soc. Exp. Biol. Med.*, **36**, 119.

30 Kiessling, W. (1938) *Biochem. Z.*, **298**, 421.

31 Wolfrom, M.L., and Pletcher, D.E. (1941) *J. Am. Chem. Soc.*, **63**, 1050.

32 Palm, D., Klein, H.W., Schinzel, R., Buehner, M., and Helmreich, E.J.M. (1990) *Biochemistry*, **29**, 1099.

33 Schäfner, A., and Specht, H. (1938) *Naturwissenschaften*, **26**, 494.

34 Kiessling, W. (1939) *Naturwissenschaften*, **27**, 129.

35 Kiessling, W. (1939) *Biochem. Z.*, **302**, 50.

36 Cori, C.F., Schmidt, G., and Cori, G.T. (1939) *Science*, **89**, 464.

37 Hanes, C.S. (1940) *Proc. R. Soc. B*, **128**, 421.

38 Hanes, C.S. (1940) *Proc. R. Soc. B*, **129**, 174.

39 Ostern, P., and Holmes, E. (1939) *Nature*, **144**, 34.

40 Cori, G.T., Cori, C.F., and Schmidt, G. (1939) *J. Biol. Chem.*, **129**, 629.

41 Ostern, P., Herbert, D., and Holmes, E. (1939) *Biochem. J.*, **33**, 1858.

42 Green, D.E., and Stumpf, P.K. (1942) *J. Biol. Chem.*, **142**, 355.

43 Weibull, C., and Tiselius, A. (1945) *Arkiv för Kemi Mineralogi Och Geologi*, **19**, 1.

44 Whelan, W.J., and Bailey, J.M. (1954) *Biochem. J.*, **58**, 560.

45 Pfannemüller, B., and Burchard, W. (1969) *Makromol. Chem.*, **121**, 1.

46 Pfannemüller, B. (1975) *Naturwissenschaften*, **62**, 231.

47 Suganuma, T., Kitazono, J.I., Yoshinaga, K., Fujimoto, S., and Nagahama, T. (1991) *Carbohydr. Res.*, **217**, 213.

48 Ohdan, K., Fujii, K., Yanase, M., Takaha, T., and Kuriki, T. (2006) *Biocatal. Biotransformation*, **24**, 77.

49 Yanase, M., Takaha, T., and Kuriki, T. (2006) *J. Sci. Food Agric.*, **86**, 1631.

50 Ohdan, K., Fujii, K., Yanase, M., Takaha, T., and Kuriki, T. (2007) *J. Biotechnol.*, **127**, 496.

51 Izawa, H., Kaneko, Y., and Kadokawa, J. (2009) *J. Carbohydr. Chem.*, **28**, 179.

52 Emmerling, W.N., and Pfannemüller, B. (1978) *Makromol. Chem.*, **179**, 1627.

53 Emmerling, W.N., and Pfannemüller, B. (1981) *Starke*, **33**, 202.

54 Emmerling, W.N., and Pfannemüller, B. (1983) *Makromol. Chem.*, **184**, 1441.

55 Ziegast, G., and Pfannemüller, B. (1984) *Makromol. Chem.*, **185**, 1855.

56 Müller-Fahrnow, A., Hilgenfeld, R., Hesse, H., Saenger, W., and Pfannemüller, B. (1988) *Carbohydr. Res.*, **176**, 165.

57 Pfannemüller, B. (1988) *Starke*, **40**, 476.

58 Pfannemüller, B., and Kühn, I. (1988) *Makromol. Chem.*, **189**, 2433.

59 Taravel, F.R., and Pfannemüller, B. (1990) *Makromol. Chem.*, **191**, 3097.

60 Tuzov, I., Cramer, K., Pfannemüller, B., Kreutz, W., and Magonov, S.N. (1995) *Adv. Mater.*, **7**, 656.

61. Biermann, M., Schmid, K., and Schulz, P. (1993) *Starke*, **45**, 281.
62. von Rybinski, W., and Hill, K. (1998) *Angew. Chem. Int. Ed.*, **37**, 1328.
63. Hill, K., and Rhode, O. (1999) *Fett Lipid*, **101**, 25.
64. Ziegast, G., and Pfannemüller, B. (1987) *Carbohydr. Res.*, **160**, 185.
65. Niemann, C., Nuck, R., Pfannemüller, B., and Saenger, W. (1990) *Carbohydr. Res.*, **197**, 187.
66. Zhao, B., and Brittain, W.J. (2000) *Prog. Polym. Sci.*, **25**, 677.
67. Advincula, R., Zhou, Q.G., Park, M., Wang, S.G., Mays, J., Sakellariou, G., Pispas, S., and Hadjichristidis, N. (2002) *Langmuir*, **18**, 8672.
68. Edmondson, S., Osborne, V.L., and Huck, W.T.S. (2004) *Chem. Soc. Rev.*, **33**, 14.
69. Loos, K., von Braunmühl, V., Stadler, R., Landfester, K., and Spiess, H.W. (1997) *Macromol. Rapid Commun.*, **18**, 927.
70. Fiske, C.H., and Subbarow, Y. (1925) *J. Biol. Chem.*, **66**, 375.
71. Breitinger, H.-G. (2002) *Tetrahedron Lett.*, **43**, 6127.
72. Stupp, S.I. (1998) *Curr. Opin. Colloid In.*, **3**, 20.
73. Klok, H.A., and Lecommandoux, S. (2001) *Adv. Mater.*, **13**, 1217.
74. Lee, M., Cho, B.K., and Zin, W.C. (2001) *Chem. Rev.*, **101**, 3869.
75. Loos, K., and Munõz-Guerra, S. (2005) *Supramolecular Polymers*, vol. w, 2nd edn (ed. A. Ciferri) CRC Press, Boca Raton, p. 393.
76. Husemann, E., and Reinhardt, M. (1962) *Makromol. Chem.*, **57**, 109.
77. Husemann, E., and Reinhardt, M. (1962) *Makromol. Chem.*, **57**, 129.
78. Burchard, W., Kratz, I., and Pfannemüller, B. (1971) *Makromol. Chem.*, **150**, 63.
79. Bittiger, H., Husemann, E., and Pfannemüller, B. (1971) *Starke*, **23**, 113.
80. Andresz, H., Richter, G.C., and Pfannemüller, B. (1978) *Makromol. Chem.*, **179**, 301.
81. Emmerling, W., and Pfannemüller, B. (1978) *Chem. Ztg.*, **102**, 233.
82. Jonas, G., and Stadler, R. (1991) *Makromol. Chem. Rapid Commun.*, **12**, 625.
83. Jonas, G., and Stadler, R. (1994) *Acta Polymerica*, **45**, 14.
84. Haupt, M., Knaus, S., Rohr, T., and Gruber, H. (2000) *J. Macromol. Sci. Pure*, **37**, 323.
85. Loos, K., Jonas, G., and Stadler, R. (2001) *Macromol. Chem. Phys.*, **202**, 3210.
86. von Braunmühl, V., Jonas, G., and Stadler, R. (1995) *Macromolecules*, **28**, 17.
87. von Braunmühl, V., and Stadler, R. (1996) *Macromol. Symp.*, **103**, 141.
88. Kobayashi, K., Kamiya, S., and Enomoto, N. (1996) *Macromolecules*, **29**, 8670.
89. Wataoka, I., Urakawa, H., Kobayashi, K., Akaike, T., Schmidt, M., and Kajiwara, K. (1999) *Macromolecules*, **32**, 1816.
90. Kamiya, S., and Kobayashi, K. (1998) *Macromol. Chem. Phys.*, **199**, 1589.
91. Narumi, A., Kawasaki, K., Kaga, H., Satoh, T., Sugimoto, N., and Kakuchi, T. (2003) *Polym. Bull.*, **49**, 405.
92. Kadokawa, J., Nakamura, Y., Sasaki, Y., Kaneko, Y., and Nishikawa, T. (2008) *Polym. Bull.*, **60**, 57.
93. Sasaki, Y., Kaneko, Y., and Kadokawa, J. (2009) *Polym. Bull.*, **62**, 291.
94. Kaneko, Y., Matsuda, S., and Kadokawa, J. (2007) *Biomacromolecules*, **8**, 3959.
95. Matsuda, S., Kaneko, Y., and Kadokawa, J. (2007) *Macromol. Rapid Commun.*, **28**, 863.
96. Omagari, Y., Matsuda, S., Kaneko, Y., and Kadokawa, J. (2009) *Macromol. Biosci.*, **9**, 450.
97. Ziegast, G., and Pfannemüller, B. (1984) *Makromol. Chem. Rapid Commun.*, **5**, 363.
98. Ziegast, G., and Pfannemüller, B. (1984) *Makromol. Chem. Rapid Commun.*, **5**, 373.
99. Pfannemüller, B., Schmidt, M., Ziegast, G., and Matsuo, K. (1984) *Macromolecules*, **17**, 710.
100. Akiyoshi, K., Kohara, M., Ito, K., Kitamura, S., and Sunamoto, J. (1999) *Macromol. Rapid Commun.*, **20**, 112.

101 Akiyoshi, K., Maruichi, N., Kohara, M., and Kitamura, S. (2002) *Biomacromolecules*, **3**, 280.
102 Loos, K., and Stadler, R. (1997) *Macromolecules*, **30**, 7641.
103 Loos, K., and Müller, A.H.E. (2002) *Biomacromolecules*, **3**, 368.
104 Loos, K., Böker, A., Zettl, H., Zhang, A.F., Krausch, G., and Müller, A.H.E. (2005) *Macromolecules*, **38**, 873.
105 Bosker, W.T.E., Agoston, K., Stuart, M.A.C., Norde, W., Timmermans, J.W., and Slaghek, T.M. (2003) *Macromolecules*, **36**, 1982.
106 Haddleton, D.M., and Ohno, K. (2000) *Biomacromolecules*, **1**, 152.
107 Narumi, A., Miura, Y., Otsuka, I., Yamane, S., Kitajyo, Y., Satoh, T., Hirao, A., Kaneko, N., Kaga, H., and Kakuchi, T. (2006) *J. Polym. Sci. Pol. Chem.*, **44**, 4864.
108 Kadokawa, J.-I., Kaneko, Y., Nakaya, A., and Tagaya, H. (2001) *Macromolecules*, **34**, 6536.
109 Kadokawa, J., Kaneko, Y., Tagaya, H., and Chiba, K. (2001) *Chem. Commun.*, 449.
110 Kadokawa, J.-I., Kaneko, Y., Nagase, S., Takahashi, T., and Tagaya, H. (2002) *Chem. A Eur. J.*, **8**, 3321.
111 Kadokawa, J.-I., Nakaya, A., Kaneko, Y., and Tagaya, H. (2003) *Macromol. Chem. Phys.*, **204**, 1451.
112 Kaneko, Y., and Kadokawa, J.-I. (2005) *Chem. Rec.*, **5**, 36.
113 Kaneko, Y., Beppu, K., and Kadokawa, J.I. (2007) *Biomacromolecules*, **8**, 2983.
114 Kaneko, Y., Beppu, K., and Kadokawa, J.I. (2008) *Macromol. Chem. Phys.*, **209**, 1037.
115 Kaneko, Y., Saito, Y., Nakaya, A., Kadokawa, J.I., and Tagaya, H. (2008) *Macromolecules*, **41**, 5665.
116 Boyer, C., and Preiss, J. (1977) *Biochemistry*, **16**, 3693.
117 Lerner, L.R., and Krisman, C.R. (1996) *Cell. Mol. Biol.*, **42**, 599.
118 Takata, H.T., Okada, S., Takagi, M., and Imanaka, T. (1996) *J. Bacteriol.*, **178**, 1600.
119 Abad, M.C., Binderup, K., Rios-Steiner, J., Arni, R.K., Preiss, J., and Geiger, J.H. (2002) *J. Biol. Chem.*, **277**, 42164.
120 Palomo, M., Kralj, S., van der Maarel, M., and Dijkhuizen, L. (2009) *Appl. Environ. Microbiol.*, **75**, 1355.
121 Griffin, H.L., and Wu, Y.V. (1968) *Biochemistry*, **7**, 3063.
122 Griffin, H.L., and Wu, Y.V. (1971) *Biochemistry*, **10**, 4330.
123 Ponstein, A.S., Vos-Scheperkeuter, G.H., Feenstra, W.J., and Witholt, B. (1987) *Food Hydrocolloids*, **1**, 497.
124 Vos-Scheperkeuter, G.H., de Wit, J.G., Ponstein, A.S., Feenstra, W.J., and Witholt, B. (1989) *Plant Physiol.*, **90**, 75.
125 Ozbun, J.L., Hawker, J.S., Greenberg, E., Lammel, C., Preiss, J., and Lee, E.Y.C. (1973) *Plant Physiol.*, **51**, 1.
126 Pisigan, R.A., and del Rosario, E.J. (1976) *Phytochemistry*, **15**, 71.
127 Boyer, C.D., and Preiss, J. (1978) *Carbohydr. Res.*, **61**, 321.
128 Boyer, D., and Fisher, B. (1984) *Phytochemistry*, **23**, 737.
129 Manners, D.J. (1989) *Carbohydr. Polym.*, **11**, 87.
130 Boyer, C.D., and Preiss, J. (1981) *Plant Physiol.*, **67**, 1141.
131 Matters, G.L., and Boyer, C.D. (1981) *Phytochemistry*, **20**, 1805.
132 Ozbun, J.L., Hawker, J.S., and Preiss, J. (1972) *Biochem. J.*, **126**, 953.
133 Hawker, J.S., Ozbun, J.L., Ozaki, H., Greenberg, E., and Preiss, J. (1974) *Arch. Biochem. Biophys.*, **160**, 530.
134 Brown, B.I., and Brown, D.H. (1966) *Meth. Enzymol.*, **8**, 395.
135 Kawaguchi, K., Fox, J., Holmes, E., Boyer, C., and Preiss, J. (1978) *Arch. Biochem. Biophys.*, **190**, 385.
136 Larner, J. (1955) *Methods in Enzymology*, vol. 1, Academic Press, p. 222.
137 Haworth, W.N., Peat, S., and Bourne, E.J. (1944) *Nature*, **154**, 236.
138 Borovsky, D., Smith, E.E., and Whelan, W.J. (1976) *Eur. J. Biochem.*, **62**, 307.
139 Takata, H., Ohdan, K., Takaha, T., Kuriki, T., and Okada, S. (2003) *Jpn Soc. Appl. Glycosci.*, **50**, 15.
140 Andersson, L., Andersson, R., Andersson, R.E., Rydberg, U., Larsson, H., and Aman, P. (2002) *Carbohydr. Polym.*, **50**, 249.

141 Takeda, Y., Guan, H., and Preiss, J. (1993) *Carbohydr. Res.*, **240**, 253.
142 Guan, H., Kuriki, T., Sivak, M., and Preiss, J. (1995) *Plant Biol.*, **92**, 964.
143 Takata, H., Takaha, T., Kuriki, T., Okada, S., Takagi, M., and Imanaka, T. (1994) *Appl. Environ. Microbiol.*, **60**, 3096.
144 van der Maarel, M.J.E.C., Vos, A., Sanders, P., and Dijkhuizen, L. (2003) *Biocatal. Biotransfor.*, **21**, 199.
145 Kim, E.J., Ryu, S.I., Bae, H.A., Huong, N.T., and Lee, S.B. (2008) *Food Chem.*, **110**, 979.
146 Cori, G.T., and Cori, C.F. (1943) *J. Biol. Chem.*, **151**, 57.
147 Borovsky, D., Smith, E.E., and Whelan, W.J. (1975) *Eur. J. Biochem.*, **59**, 615.
148 Praznik, W., Rammesmayer, G., and Spies, T. (1992) *Carbohydr. Res.*, **227**, 171.
149 Fujii, K., Takata, H., Yanase, M., Terada, Y., Ohdan, K., Takaha, T., Okada, S., and Kuriki, T. (2003) Bioengineering and application of novel glucose polymer. *Biocatal. Biotransform.*, **21**, 167–172.
150 Hernandez, J.M., Gaborieau, M., Castignolles, P., Gidley, M.J., Myers, A.M., and Gilbert, R.G. (2008) *Biomacromolecules*, **9**, 954.
151 Kajiura, H., Kakutani, R., Akiyama, T., Takata, H., and Kuriki, T. (2008) *Biocatal. Biotransformation*, **26**, 133.
152 van der Vlist, J., Reixach, M.P., van der Maarel, M., Dijkhuizen, L., Schouten, A.J., and Loos, K. (2008) *Macromol. Rapid Commun.*, **29**, 1293.
153 Takata, H., Kajiura, H., Furuyashiki, T., Kakutani, R., and Kuriki, T. (2009) *Carbohydr. Res.*, **344**, 654.
154 Waldmann, H., Gygax, D., Bednarski, M.D., Shangraw, W.R., and Whitesides, G.M. (1986) *Carbohydr. Res.*, **157**, C4.
155 Dang, P.L., and Boyer, C.D. (1988) *Phytochemistry*, **27**, 1255.
156 Goldner, W., and Beevers, H. (1989) *Phytochemistry*, **28**, 1809.
157 Matsumoto, A., Nakajima, T., and Matsuda, K. (1990) *J. Biochem.*, **107**, 118.
158 Matsumoto, A., Nakajima, T., and Matsuda, K. (1990) *J. Biochem.*, **107**, 123.
159 Palomo Reixach, M., Kralj, S., van der Maarel, M., and Dijkhuizen, L. (2009) *Appl. Environ. Microbiol.*, **75**, 1355.
160 Potocki-Veronese, G., Putaux, J.L., Dupeyre, D., Albenne, C., Remaud-Simeon, M., Monsan, P., and Buleon, A. (2005) *Biomacromolecules*, **6**, 1000.
161 Putaux, J.L., Potocki-Veronese, G., Remaud-Simeon, M., and Buleon, A. (2006) *Biomacromolecules*, **7**, 1720.
162 Kralj, S., van Geel-Schutten, G.H., van der Maarel, M., and Dijkhuizen, L. (2004) *Microbiol. Sgm*, **150**, 2099.
163 Pijning, T., Vujicic-Zagar, A., Kralj, S., Eeuwema, W., Dijkhuizen, L., and Dijkstra, B.W. (2008) *Biocatal. Biotransformation*, **26**, 12.
164 van Leeuwen, S.S., Kralj, S., van Geel-Schutten, I.H., Gerwig, G.J., Dijkhuizen, L., and Kamerling, J.P. (2008) *Carbohydr. Res.*, **343**, 1237.
165 Gibson, G.R., and Roberfroid, M.B. (1995) *J. Nutr.*, **125**, 1401.
166 Hellmuth, H., Hillringhaus, L., Hobbel, S., Kralj, S., Dijkhuizen, L., and Seibel, J. (2007) *ChemBioChem*, **8**, 273.
167 Hellmuth, H., Wittrock, S., Kralj, S., Dijkhuizen, L., Hofer, B., and Seibel, J. (2008) *Biochemistry*, **47**, 6678.
168 van Hijum, S., Bonting, K., van der Maarel, M., and Dijkhuizen, L. (2001) *FEMS Microbiol. Lett.*, **205**, 323.
169 Kaper, T., Talik, B., Ettema, T.J., Bos, H., van der Maarel, M., and Dijkhuizen, L. (2005) *Appl. Environ. Microbiol.*, **71**, 5098.
170 van der Maarel, M., Capron, I., Euverink, G.J.W., Bos, H.T., Kaper, T., Binnema, D.J., and Steeneken, P.A.M. (2005) *Starch-Starke*, **57**, 465.
171 Lee, K.Y., Kim, Y.R., Park, K.H., and Lee, H.G. (2006) *Carbohydr. Polym.*, **63**, 347.
172 Hansen, M.R., Blennow, A., Pedersen, S., Norgaard, L., and Engelsen, S.B. (2008) *Food Hydrocolloids*, **22**, 1551.
173 Oh, E.J., Choi, S.J., Lee, S.J., Kim, C.H., and Moon, T.W. (2008) *J. Food Sci.*, **73**, C158.
174 Hansen, M.R., Blennow, A., Pedersen, S., and Engelsen, S.B. (2009) *Carbohydr. Polym.*, **78**, 72.

175 Takaha, T., Yanase, M., Okada, S., and Smith, S.M. (1993) *J. Biol. Chem.*, **268**, 1391.
176 Terada, Y., Fujii, K., Takaha, T., and Okada, S. (1999) *Appl. Environ. Microbiol.*, **65**, 910.
177 Takaha, T., and Smith, S.M. (1999) *Biotechnology and Genetic Engineering Reviews*, vol. 16, Intercept Ltd Scientific, Technical & Medical Publishers, Andover, p. 257.
178 Takaha, T., Yanase, M., Takata, H., Okada, S., and Smith, S.M. (1996) *J. Biol. Chem.*, **271**, 2902.
179 Lapcik, L., Lapcik, L., De Smedt, S., Demeester, J., and Chabrecek, P. (1998) *Chem. Rev.*, **98**, 2663.
180 Cowman, M.K., and Matsuoka, S. (2005) *Carbohydr. Res.*, **340**, 791.
181 Hargittai, I., and Hargittai, M. (2008) *Struct. Chem.*, **19**, 697.
182 Mashimoto, M.S., Saegusa, H., Chiba, S., Kitagawa, H., and Myoshi, T. (1988) Production of hyaluronic acid, Vol. JP 63123392, Japan.
183 Akasaka, H., Seto, S., Yanagi, M., Fukushima, S., and Mitsui, T. (1988) *J. Soc. Cosmet. Chem. Jpn.*, **22**, 35.
184 Weigel, P.H., Hascall, V.C., and Tammi, M. (1997) *J. Biol. Chem.*, **272**, 13997.
185 DeAngelis, P.L. (2002) *Glycobiology*, **12**, 9R.
186 Weigel, P.H. (2002) *IUBMB Life*, **54**, 201.
187 Weigel, P.H., and DeAngelis, P.L. (2007) *J. Biol. Chem.*, **282**, 36777.
188 Markovitz, A., Cifonelli, J.A., and Dorfman, A. (1959) *J. Biol. Chem.*, **234**, 2343.
189 DeAngelis, P.L., Papaconstantinou, J., and Weigel, P.H. (1993) *J. Biol. Chem.*, **268**, 19181.
190 DeAngelis, P.L., Papaconstantinou, J., and Weigel, P.H. (1993) *J. Biol. Chem.*, **268**, 14568.
191 DeAngelis, P.L., and Achyuthan, A.M. (1996) *J. Biol. Chem.*, **271**, 23657.
192 Itano, N., and Kimata, K. (1996) *Biochem. Biophys. Res. Commun.*, **222**, 816.
193 Itano, N., and Kimata, K. (1996) *J. Biol. Chem.*, **271**, 9875.
194 Meyer, M.F., and Kreil, G. (1996) *Proc. Natl. Acad. Sci. U. S. A.*, **93**, 4543.
195 Shyjan, A.M., Heldin, P., Butcher, E.C., Yoshino, T., and Briskin, M.J. (1996) *J. Biol. Chem.*, **271**, 23395.
196 Spicer, A.P., Augustine, M.L., and McDonald, J.A. (1996) *J. Biol. Chem.*, **271**, 23400.
197 Watanabe, K., and Yamaguchi, Y. (1996) *J. Biol. Chem.*, **271**, 22945.
198 DeAngelis, P.L., Jing, W., Graves, M.V., Burbank, D.E., and VanEtten, J.L. (1997) *Science*, **278**, 1800.
199 Kumari, K., and Weigel, P.H. (1997) *J. Biol. Chem.*, **272**, 32539.
200 Spicer, A.P., Olson, J.S., and McDonald, J.A. (1997) *J. Biol. Chem.*, **272**, 8957.
201 DeAngelis, P.L., Jing, W., Drake, R.R., and Achyuthan, A.M. (1998) *J. Biol. Chem.*, **273**, 8454.
202 Spicer, A.P., and McDonald, J.A. (1998) *J. Biol. Chem.*, **273**, 1923.
203 Jong, A., Wu, C.H., Chen, H.M., Luo, F., Kwon-Chung, K.J., Chang, Y.C., LaMunyon, C.W., Plaas, A., and Huang, S.H. (2007) *Eukaryot. Cell*, **6**, 1486.
204 Blank, L.M., Hugenholtz, P., and Nielsen, L.K. (2008) *J. Mol. Evol.*, **67**, 13.
205 DeAngelis, P.L., Oatman, L.C., and Gay, D.F. (2003) *J. Biol. Chem.*, **278**, 35199.
206 Jing, W., and DeAngelis, P.L. (2004) *J. Biol. Chem.*, **279**, 42345.
207 Williams, K.J., Halkes, K.M., Kamerling, J.P., and DeAngelis, P.L. (2006) *J. Biol. Chem.*, **281**, 5391.
208 Jing, W., Haller, F.M., Almond, A., and DeAngelis, P.L. (2006) *Anal. Biochem.*, **355**, 183.
209 Bourquelot, E., and Bridel, M. (1913) *Ann. Chim. Phys.*, **29**, 145.
210 Bourquelot, E. (1914) *J. Pharm. Chim.*, **10**, 361.
211 Hehre, E.J. (2001) *Carbohydr. Res.*, **331**, 347.
212 Planas, A., and Faijes, M. (2002) *Afinidad*, **59**, 295.
213 Hamilton, C.J. (2004) *Nat. Prod. Rep.*, **21**, 365.
214 Rowan, A.S., and Hamilton, C.J. (2006) *Nat. Prod. Rep.*, **23**, 412.

215 Shaikh, F.A., and Withers, S.G. (2008) *Biochem. Cell Biol. Biochim. Biol. Cell.*, **86**, 169.
216 Homann, A., and Seibel, J. (2009) *Nat. Prod. Rep.*, **26**, 1555.
217 Kasche, V. (1986) *Enzyme Microb. Technol.*, **8**, 4.
218 Kobayashi, S., Kashiwa, K., Kawasaki, T., and Shoda, S. (1991) *J. Am. Chem. Soc.*, **113**, 3079.
219 Lee, J.H., Brown, R.M., Jr., Kuga, S., Shoda, S., and Kobayashi, S. (1994) *Proc. Natl. Acad. Sci. U. S. A.*, **91**, 7425.
220 Kobayashi, S., Hobson, L.J., Sakamoto, J., Kimura, S., Sugiyama, J., Imai, T., and Itoh, T. (2000) *Biomacromolecules*, **1**, 168.
221 Kobayashi, S., Hobson, L.J., Sakamoto, J., Kimura, S., Sugiyama, J., Imai, T., and Itoh, T. (2000) *Biomacromolecules*, **1**, 509.
222 Kobayashi, S., Shimada, J., Kashiwa, K., and Shoda, S. (1992) *Macromolecules*, **25**, 3237.
223 Kobayashi, S., Wen, X., and Shoda, S. (1996) *Macromolecules*, **29**, 2698.
224 Kobayashi, S., Kiyosada, T., and Shoda, S.I. (1996) *J. Am. Chem. Soc.*, **118**, 13113.
225 Sakamoto, J., Sugiyama, J., Kimura, S., Imai, T., Itoh, T., Watanabe, T., and Kobayashi, S. (2000) *Macromolecules*, **33**, 4155.
226 Sakamoto, J., Sugiyama, J., Kimura, S., Imai, T., Itoh, T., Watanabe, T., and Kobayashi, S. (2000) *Macromolecules*, **33**, 4982.
227 Ohmae, M., Sakaguchi, K., Kaneto, T., Fujikawa, S., and S. Kobayashi (2007) *ChemBioChem*, **8**, 1710.
228 Kobayashi, S., Morii, H., Itoh, R., Kimura, S., and Ohmae, M. (2001) *J. Am. Chem. Soc.*, **123**, 11825.
229 Kobayashi, S., Fujikawa, S., and Ohmae, M. (2003) *J. Am. Chem. Soc.*, **125**, 14357.
230 Fujikawa, S., Ohmae, M., and Kobayashi, S. (2005) *Biomacromolecules*, **6**, 2935.
231 Ochiai, H., Ohmae, M., Mori, T., and Kobayashi, S. (2005) *Biomacromolecules*, **6**, 1068.

232 Kobayashi, S., Ohmae, M., Ochiai, H., and Fujikawa, S.I. (2006) *Chem. A Eur. J.*, **12**, 5962.
233 Shoda, S., and Kobayashi, S. (1997) *Trends Polym. Sci.*, **5**, 109.
234 Shoda, S., Fujita, M., and Kobayashi, S. (1998) *Trends Glycosci. Glycotechnol.*, **10**, 279.
235 Kobayashi, S. (1999) *J. Polym. Sci. Part A Polym. Chem.*, **37**, 3041.
236 Kobayashi, S., Sakamoto, J., and Kimura, S. (2001) *Prog. Polym. Sci.*, **26**, 1525.
237 Kobayashi, S., Uyama, H., and Kimura, S. (2001) *Chem. Rev.*, **101**, 3793.
238 Kobayashi, S., and Uyama, H. (2003) *Macromol. Chem. Phys.*, **204**, 235.
239 Kobayashi, S., Fujikawa, S., Itoh, R., Morii, H., Ochiai, H., Mori, T., and Ohmae, M. (2005) *Polymer Biocatalysis and Biomaterials*, vol. 900 (eds H.N. Cheng and R.A. Gross), American Chemical Society, Washington, DC, USA, p. 217.
240 Kobayashi, S., Ohmae, M., Fujikawa, S., and Ochiai, H. (2005) *Macromol. Symp.*, **226**, 147.
241 Kobayashi, S., and Ohmae, M. (2006) *Enzyme-Catalyzed Synthesis of Polymers* (Advances in Polymer Science), Springer, Berlin, Heidelberg, Germany, vol. 194, p. 159.
242 Ohmae, M., Fujikawa, S., Ochiai, H., and Kobayashi, S. (2006) *J. Polym. Sci. Part A Polym. Chem.*, **44**, 5014.
243 Faijes, M., and Planas, A. (2007) *Carbohydr. Res.*, **342**, 1581.
244 Ohmae, M., Makino, A., and Kobayashi, S. (2007) *Macromol. Chem. Phys.*, **208**, 1447.
245 Kobayashi, S., and Makino, A. (2009) *Chem. Rev.*, **109**, 5288.
246 Kobayashi, S., Kawasaki, T., Obata, K., and Shoda, S. (1993) *Chem. Lett.*, 685.
247 Shoda, S., Kawasaki, T., Obata, K., and Kobayashi, S. (1993) *Carbohydr. Res.*, **249**, 127.
248 Shoda, S., Obata, K., Karthaus, O., and Kobayashi, S. (1993) *J. Chem. Soc. Chem. Commun.*, 1402.
249 Karthaus, O., Shoda, S., Takano, H., Obata, K., and Kobayashi, S. (1994) *J. Chem. Soc. Perkin Trans. I*, 1851.

250 Kobayashi, S. (2005) *J. Polym. Sci. Part A Polym. Chem.*, **43**, 693.

251 Egusa, S., Kitaoka, T., Goto, M., and Wariishi, H. (2007) *Angew. Chem. Int. Ed.*, **46**, 2063.

252 Ochiai, H., Huang, W., and Wang, L.X. (2009) *Carbohydr. Res.*, **344**, 592.

253 Ly, H.D., and Withers, S.G. (1999) *Annu. Rev. Biochem.*, **68**, 487.

254 Williams, S.J., and Withers, S.G. (2002) *Aust. J. Chem.*, **55**, 3.

255 Daines, A.M., Maltman, B.A., and Flitsch, S.L. (2004) *Curr. Opin. Chem. Biol.*, **8**, 106.

256 Hanson, S., Best, M., Bryan, M.C., and Wong, C.H. (2004) *Trends Biochem. Sci.*, **29**, 656.

257 Perugino, G., Trincone, A., Rossi, M., and Moracci, M. (2004) *Trends Biotechnol.*, **22**, 31.

258 Perugino, G., Cobucci-Ponzano, B., Rossi, M., and Moracci, M. (2005) *Adv. Synth. Catal.*, **347**, 941.

259 Hancock, S.M., Vaughan, M.D., and Withers, S.G. (2006) *Curr. Opin. Chem. Biol.*, **10**, 509.

260 Trincone, A., and Giordano, A. (2006) *Curr. Org. Chem.*, **10**, 1163.

261 Boltje, T.J., Buskas, T., and Boons, G.J. (2009) *Nat. Chem.*, **1**, 611.

262 Kelly, R.M., Dijkhuizen, L., and Leemhuis, H. (2009) *J. Biotechnol.*, **140**, 184.

263 Kitaoka, M., Honda, Y., Fushinobu, S., Hidaka, M., Katayama, T., and Yamamoto, K. (2009) *Trends Glycosci. Glycotechnol.*, **21**, 23.

264 Wang, L.X., and Huang, W. (2009) *Curr. Opin. Chem. Biol.*, **13**, 592.

265 Mackenzie, L.F., Wang, Q., Warren, R.A.J., and Withers, S.G. (1998) *J. Am. Chem. Soc.*, **120**, 5583.

266 Malet, C., and Planas, A. (1998) *FEBS Lett.*, **440**, 208.

267 Moracci, M., Trincone, A., Perugino, G., Ciaramella, M., and Rossi, M. (1998) *Biochemistry*, **37**, 17262.

268 Fort, S., Boyer, V., Greffe, L., Davies, G.J., Moroz, O., Christiansen, L., Schülein, M., Cottaz, S., and Driguez, H. (2000) *J. Am. Chem. Soc.*, **122**, 5429.

269 Ducros, V.M.A., Tarling, C.A., Zechel, D.L., Brzozowski, A.M., Frandsen, T.P., Von Ossowski, I., Schülein, M., Withers, S.G., and Davies, G.J. (2003) *Chem. Biol.*, **10**, 619.

270 Champion, E., André, I., Moulis, C., Boutet, J., Descroix, K., Morel, S., Monsan, P., Mulard, L.A., and Remaud-Siméon, M. (2009) *J. Am. Chem. Soc.*, **131**, 7379.

271 Müllegger, J., Chen, H.M., Chan, W.Y., Reid, S.P., Jahn, M., Warren, R.A.J., Salleh, H.M., and Withers, S.G. (2006) *ChemBioChem*, **7**, 1028.

272 Honda, Y., and Kitaoka, M. (2006) *J. Biol. Chem.*, **281**, 1426.

273 Honda, Y., Fushinobu, S., Hidaka, M., Wakagi, T., Shoun, H., Taniguchi, H., and Kitaoka, M. (2008) *Glycobiology*, **18**, 325.

274 Wada, J., Honda, Y., Nagae, M., Kato, R., Wakatsuki, S., Katayama, T., Taniguchi, H., Kumagai, H., Kitaoka, M., and Yamamoto, K. (2008) *FEBS Lett.*, **582**, 3739.

275 Boyer, V., Fort, S., Frandsen, T.P., Schülein, M., Cottaz, S., and Driguez, H. (2002) *Chem. A Eur. J.*, **8**, 1389.

276 Fort, S., Christiansen, L., Schülein, M., Cottaz, S., and Driguez, H. (2000) *Isr. J. Chem.*, **40**, 217.

277 Fauré, R., Saura-Valls, M., Brumer Iii, H., Planas, A., Cottaz, S., and Driguez, H. (2006) *J. Org. Chem.*, **71**, 5151.

278 Saura-Valls, M., Fauré, R., Ragàs, S., Piens, K., Brumer, H., Teeri, T.T., Cottaz, S., Driguez, H., and Planas, A. (2006) *Biochem. J.*, **395**, 99.

279 Yang, Z., Liang, G., Guo, Z., and Xu, B. (2007) *Angew. Chem. Int. Ed.*, **46**, 1.

280 Kim, Y.W., Fox, D.T., Hekmat, O., Kantner, T., McIntosh, L.P., Warren, R.A.J., and Withers, S.G. (2006) *Org. Biomol. Chem.*, **4**, 2025.

281 Sugimura, M., Nishimoto, M., and Kitaoka, M. (2006) *Biosci. Biotechnol. Biochem.*, **70**, 1210.

282 Ben-David, A., Bravman, T., Balazs, Y.S., Czjzek, M., Schomburg, D., Shoham, G., and Shoham, Y. (2007) *ChemBioChem*, **8**, 2145.

283 Hrmova, M., Imai, T., Rutten, S.J., Fairweather, J.K., Pelosi, L., Bulone, V., Driguez, H., and Fincher, G.B. (2002) *J. Biol. Chem.*, **277**, 30102.

284 Faijes, M., Fairweather, J.K., Driguez, H., and Planas, A. (2001) *Chem. A Eur. J.*, **7**, 4651.

285 Fairweather, J.K., Faijes, M., Driguez, H., and Planas, A. (2002) *ChemBioChem*, **3**, 1.

286 Faijes, M., Pérez, X., Pérez, O., and Planas, A. (2003) *Biochemistry*, **42**, 13304.

287 Faijes, M., Imai, T., Bulone, V., and Planas, A. (2004) *Biochem. J.*, **380**, 635.

10
Polymerases for Biosynthesis of Storage Compounds
Anna Bröker and Alexander Steinbüchel

10.1
Introduction

Biopolymers are either synthesized by template-dependent or template-independent enzymatic processes. For the synthesis of nucleic acids and proteins a template is required, whereas all other polymers are synthesized by template-independent processes. The templates for nucleic acids are desoxyribonucleic acids or ribonucleic acids depending on the type of nucleic acid synthesized. For proteins, the template is messenger ribonucleic acid (mRNA). This has different impacts on the structure and on the molecular weights (MWs) of the polymers. Although both nucleic acids and proteins are copolymers with each type consisting of 4 or 22 different constituents, respectively, the distribution of the constituents is absolutely defined by the matrix and is not random. Furthermore, each representative of the two polymers has a defined MW. Polymers synthesized in template-dependent processes are monodisperse. All this is different in polymers synthesized by template-independent processes: first of all, these polymers are polydisperse; secondly, if these polymers are copolymers, the distribution of the constituents is more or less fully random.

In this chapter, we focus on the synthesis of polyhydroxyalkanoic acids (PHA) and cyanophycin (cyanophycin granule polyperptide, CGP) and the key enzymes PHA synthase (PhaC) and cyanophycin synthetase (CphA), respectively. Both polymers are synthesized by template-independent processes. The issue of template dependency and template independency will be illustrated in more detail with polymers consisting of amino acids.

In general, mechanisms for the biosynthesis of polyamides can be divided into three different pathways, which mainly differ in the mode of activation of the monomers (adenylation or phosphorylation), the dependency on a template, and the enzyme apparatus. In comparison to the activation by phosphorylation, adenylation involves synthesis of a phosphodiester bond between the hydroxyl group of the carboxylic group of the amino acid and the α-phosphate group of adenosine triphosphate (ATP). Activation by phosphorylation has been proposed that is, for synthesis of the tripeptide glutathione (Gly-Glu-Cys) or transpeptidase, the

Biocatalysis in Polymer Chemistry. Edited by Katja Loos
Copyright © 2011 WILEY-VCH Verlag GmbH & Co. KGaA, Weinheim
ISBN: 978-3-527-32618-1

enzyme catalyzing the polymerization of two peptidoglycan precursors in bacterial cell wall biogenesis.

1. The best investigated mechanism, distributed ubiquitously in living matter, is the template-dependent ribosomal synthesis of proteins. Here the amino acids are activated by adenylation catalyzed by aminoacyl-tRNA synthetases [1].

2. The second mechanism is performed by nonribosomal peptide synthetases (NRPS) which are multienzyme complexes consisting of four domains [2, 3].

   The adenylation domain required for activation of the substrate at the expense of ATP, via the formation of an enzyme-stabilized aminoacyl adenylate. As the formed adenylate is not stable, the energy is further conserved by transfer of the peptide to:

   the thiolation domain giving rise to a thioester bond formed between the amino acid and a cysteine of the enzyme complex.

   The condensation domain is required for formation of peptide bonds between the itemized monomers, and

   the thioesterase domain catalyzes the release of the final product from the NRPS by cyclization to an amide or ester, or by hydrolysis to the free acid [3]. The transfer of the intermediates is mediated by the cofactor pantetheine. Numerous peptide antibiotics such as penicillin, bacitracin, and actinomycin are synthesized by NRPS (reviewed by [2]). These compounds are known to contain non-proteinogenic amino acids, D-amino acids, hydroxy acids, methylated or cyclic forms, and other unusual constituents; these modifications are catalyzed by the NRPS. As the enzyme complex itself functions as matrix, the resulting peptides have a strictly defined length which is in contrast to poly(amino acids).

3. The third mechanism is represented by nonmodular one-step peptide synthesis. Enzymes belonging to this group catalyze the biosynthesis of poly(amino acids). Naturally occurring poly(amino acids) comprise cyanophycin [multi-L-arginyl-poly-(L-aspartic acid); cyanophycin granule polypeptide, (CGP)], (poly-(ε-lysine) (PL), and poly-(γ-glutamate) (PGA). As a consequence of non-ribosomal biosynthesis these peptides reveal a polydisperse mass distribution.

   Further aspects distinguishing poly(amino acids) from proteins are the following:

   proteins consist of a mixture of 22 amino acids, whereas poly(amino acids) consist of one amino acid in the polymer backbone;

   their biosynthesis is not constrained by translational inhibitors such as chloramphenicol [4];

   the amide bonds formed between the monomers are not exclusively linked between the α-carboxylic and α-amino groups, as for proteins, but also between β- or γ-carboxylic groups or ε-amino groups.

Enzymes involved in the biosynthesis of PGA have only been partially characterized; investigations are rather difficult as the synthetases are membranous complexes. A synthetase complex of *Bacillus licheniformis* catalyzing the activation, racemization, and polymerization of L-glutamate to poly-D-glutamate was described by Troy in 1985 [5]. The enzyme catalyzing PL synthesis was only detected recently and is referred to as PL synthetase [6]. The latter was found to be a membranous protein with adenylation and thiolation domains characteristic of the NRPS but it had no traditional condensation or thioesterase domain [6]. For the synthesis of CGP only one enzyme is required, a soluble protein referred to as cyanophycin synthetase (CphA). Deletion of the chromosomally encoded *cphA* gene leads to complete inhibition of CGP synthesis by the respective mutant cells [7, 8]. Among the enzymes catalyzing poly(amino acid) synthesis, CphA is the best characterized.

10.2
Polyhydroxyalkanoate Synthases

10.2.1
Occurrence of Polyhydroxyalkanoate Synthases

Polyhydroxyalkanoates (PHAs) are synthesized by a vast variety of prokaryotes. It seems to be the most important storage compound for carbon and energy in bacteria. Apart from the prokaryotes, PHAs have never been detected as an insoluble storage material. There are only a few groups of prokaryotes which do not synthesize PHAs; one major group is the group of methanogenic archae bacteria; another group is the lactic acid bacteria. One could consider that it does not make much sense for these bacteria from a physiological perspective to synthesize PHAs and to accumulate them as storage compounds. However, recently poly(3-hydroxybutyrate) (PHB) biosynthesis and accumulation was successfully established in the lactic acid bacterium *Lactococcus lactis* [9], thereby demonstrating that these bacteria are obviously not generally impaired in the synthesis of these polyesters. PHB biosynthesis and accumulation in *L. lactis* is remarkable because it is a homofermentative lactic acid bacterium normally converting glucose solely to lactic acid via pyruvate.

All bacteria capable of synthesizing PHAs and accumulating these polyesters as storage compounds possess a PHA synthase. All these PhaCs share a certain degree of homology and they are normally easily recognized as such. The authors are not aware of a PHA-accumulating bacterium in which no homolog of the four related classes of PhaCs (see below) was found. Vice versa, lipases or other enzymes are *in vivo* not capable of synthesizing PHAs.

Poly(3-hydroxybutyrate) and a few other PHAs were in prokaryotic microorganisms and in eukaryotic micro-organisms as well as in higher organisms also found in complexes and/or associated with other molecules in the laboratory of Reusch [10]. In these cases PHAs do not occur as insoluble inclusions, have a much lower MW, and they contribute only marginally (about 0.2% of the cell dry matter) to the cell mass. These PHAs are only detectable if certain precautions are considered and if sensitive devices are employed. The enzyme for the synthesis of these PHAs and their functions remained unknown for 20 years after the discovery of these PHAs. *Escherichia coli* is such a bacterium which does not naturally accumulate PHAs as storage compounds but synthesizes these 'complexed' PHB. In this bacterium a PhaC activity has recently been assigned to the periplasmic protein YdcS, which is a component of a putative ABC transporter [11]. The enzyme responsible for the synthesis of these PHAs is not known, and consequently a phenotype of a respective mutant is also unknown.

10.2.2
Chemical Structures of Polyhydroxyalkanoates and their Variants

As the general term polyhydroxyalkanoates (PHAs) indicates, not only PHB but many different polyesters are synthesized by bacteria. Polyhydroxyalkanoates are polyesters in which the hydroxyl group and the carboxyl group of hydroxyalkanoic acids are linked via oxoester bonds. About 150 different hydroxyalkanoic acids have been found as constituents in bacterial PHAs [12]. These hydroxyalkanoic acids are distinguished by the position of the hydroxyl group in relation to the carboxyl group (from position 2 to 6), by the length of the alkyl-side chain (from 0 to 12 carbon atoms), by a large variety of substituents in the side chains, and by one additional methyl group at carbon atoms between the hydroxyl and the carboxyl groups [12]. The chemical structures of three well-studied PHAs are shown in Figure 10.1. If these hydroxyalkanoic acids possess a chiral carbon atom, as is the case with most constituents, than generally the *R*-stereoisomer was found. *S*-stereoisomers of hydroxyalkanoic acids have never been detected as PHA constituents. The MWs of

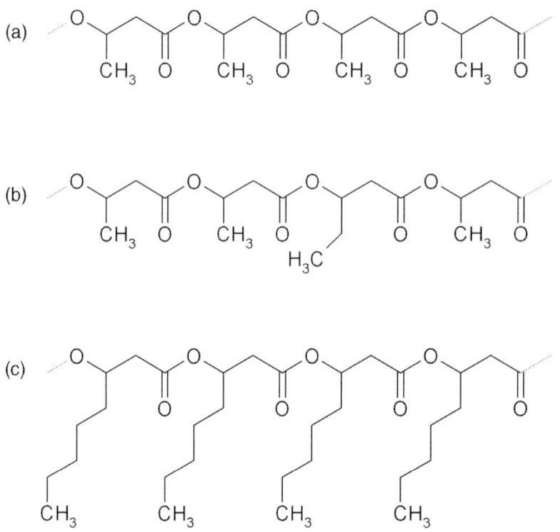

Figure 10.1 Chemical structures of (a) poly(3-hydroxybutyric acid), (b) poly(3-hydroxybutyric acid-*co*-3-hydroxyvaleric acid) and (c) poly(3-hydroxyoctanoic acid).

PHAs may range from about 80 kDa to several million kDa; they depend on the micro-organism, the cultivation conditions and the type of PHA accumulated.

Frequently more than one constituent occur in PHAs; homopolymers of the 150 constituents are rare. In the copolymers, which contain two, three or even more different constituents, the constituents are more or less randomly distributed in the chain. Only in a very few cases was evidence for 'blocky' structures obtained; however, true block copolyesters are not synthesized in bacteria.

Beside the well-known polyoxoesters, structurally related polythioesters (PTE) also occur. They are even synthesized by the same enzyme as the polyoxoesters, that is, by the PhaCs [13]. In PTEs the oxygen atom of the oxoester bond in the polymer backbone is replaced by a sulfur atom, thereby constituting thioester bonds. 3-Mercaptopropionic acid, 3-mercaptobutyric acid and 3-mercaptovaleric acid are currently known to occur in PTEs [14].

The large variety of PHAs, which is due to the many constituents that are incorporated, can be further diversified by chemical reactions or by irradiation. These measures chemically modify the side chains or cross-link polyester chains (for review see [15]).

10.2.3
Reaction Catalyzed by the Key Enzyme

PHA synthases catalyze the covalent linkage of an additional hydroxyalkanoic acid to the growing polyester chain. They do not use the free hydroxyalkanoic acid but essentially the hydroxyacyl-CoA thioesters. Coenzyme A is released during the attachment of the hydroxyacyl moiety to the growing polyester chain. The reaction is depicted in Figure 10.2. The enzyme requires a primer molecule and is obviously inactive if only the monomeric substrate is present.

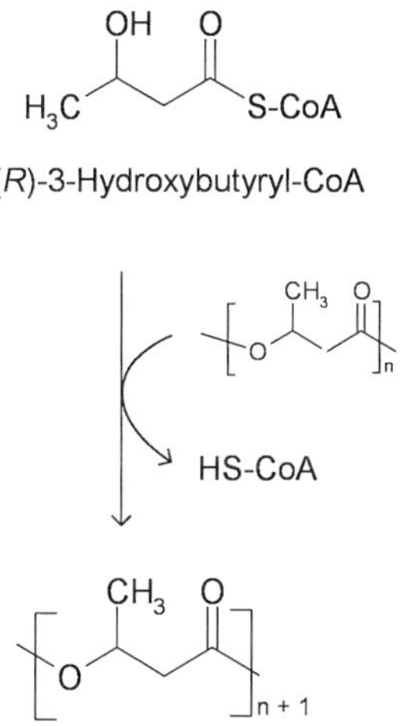

(R)-3-Hydroxybutyryl-CoA

HS-CoA

Figure 10.2 Reaction catalyzed by PHA synthase yielding poly(3-hydroxybutyric acid).

10.2.4
Assay of Enzyme Activity

The quantitative determination of PhaC activity and their substrate range *in vitro* is one of the obstacles in PHA research. One limiting factor is the limited availability of hydroxyacyl coenzyme A thioesters which have to be used as substrates to measure the polymerization reaction. Whereas 3-hydroxybutyryl-CoA is commercially available or can be relatively easily synthesized, all other hydroxyacyl-CoA esters are not commercially available and have to be synthesized by tedious methods [16].

A radiometric and a spectrometric assay have been developed to measure PhaC activity. The radiometric assay measures the incorporation of isotope-labeled hydroxyacyl moieties into the polyester, which is present from the beginning as primer [17]. [3-^{14}C]R-(-)-3-hydroxybutyryl-CoA or [^{3}H]-R,S-3-hydroxybutyryl-CoA or in principle any other CoA thioester of a radioactively-labeled hydroxyacyl moiety could be used as substrate. Only the radioactivity that is really incorporated into the insoluble polyester is measured. The time course of the assay, the need to synthesize the substrates and the high costs make the assay very inconvenient and it is hardly used anymore. A more convenient assay is the spectrometric assay which measures the release of coenzyme A during the polymerization reaction in presence of Ellmann's reagent 5,5'-dithiobis-(2-nitrobenzoic acid) (DTNB) yielding 5'-thio(2-nitrobenzoate) that absorbs at about 412 nm [16]. Here, the enzyme activity is measured directly without delay. However, it is not the formation of the polymeric product that is measured, but the release of coenzyme A, which can also be due to the hydrolytic cleavage of the substrate by another enzyme that does not have any PhaC activity, like a thioesterase. Nevertheless, this assay is now most frequently used due to its convenience.

Because of its easiness, in several studies a so-called *in vivo* enzyme activity assay was performed. This assay uses the amounts and also the composition of PHAs that are synthesized per time by whole cells during their cultivation as an indication for enzyme activity and PHA substrate specificity, respectively. Although such an assay allows the demonstration of the functionality of the PhaC protein, quantitative data can hardly be obtained because the conditions of the PhaC with regard to substrate concentration and the presence of other biological molecules, which affect the enzyme activity in a positive or negative way, can hardly be controlled.

10.2.5
Location of Enzyme and Granule Structure

The PhaC is soluble in the cytoplasm when the cells do not accumulate PHB. The enzyme becomes particle bound upon the onset of PHB biosynthesis [18]. According to the 'micelle' model of PHA granule formation and according to the mechanism of PHA formation (see below) the enzyme is covalently bound to the growing polyester chain. Therefore, the enzyme is found in most PHA-accumulating bacteria at the surface of PHA granules and in the PHA granule fraction (for review see [19]). Although the micelle model is the most favored model for granule formation, other models are also discussed (for review see [20]). Nor is it known why in some bacteria the observed distribution of the granules in the cytoplasm was not even, but localized at the cell poles [21].

Most of the proteins bound to the PHA granule surface of PHA_{SCL}-accumulating bacteria are small amphiphilic proteins, which are referred to as phasins. These phasins were found in *R. eutropha* and in other PHA_{SCL}-accumulating bacteria and are the major proteins in cells having accumulated PHAs, contributing to about 3 to 5% of the total protein, and they bind tightly to the granule surface probably covering most of the surface [22–24]. For the binding a short segment at the carboxy-terminal region of the phasin is responsible in *Rhodococcus ruber* [25]. By this, they stabilize the suspension of the PHA granules with its amorphous polyester content in the cytoplasm and allow the existence of several discrete PHB granules. In a phasin-negative mutant all granules coalesce to one single large granule [23]. The situation is, however, much more complex, since other proteins like lysozyme or heat shock proteins can, for example, unspecifically bind to the PHA granule surface [26, 27].

The granules of PHA_{MCL}-accumulating bacteria seem to be different and more complex. It has been studied in most detail in *Pseudomonas putida*. In pseudomonads two PhaCs are present, and there seem to be additional proteins involved. In addition, the organization of the genes relevant for PHA_{MCL} metabolism including the intracellular degradation of the polyester and for the structure of the granules is different than in other PHA accumulating bacteria and much more compact [28, 29].

10.2.6
Primary Structures of the Enzyme

According to size, protein composition and substrate specificity four different classes of PhaCs can be distinguished, to which most of the many known PhaCs can be allocated.

Whereas class-I and class-II PHA synthases consist of only one type of subunit exhibiting MWs in the range of about 60 to 65 kDa in most cases, class-III and class-IV PHA synthases consist of two different types of subunits each exhibiting a size of about 40 plus 40 kDa (class-III) or of about 40 plus 22 kDa (class-IV), respectively.

Other remarkable differences of these two classes refer to the substrate specificities of the enzymes. Whereas class-I PhaCs, like the enzyme of *Ralstonia eutropha*, are restricted to 3-, 4-, and 5-hydroxyalkanoic acids of short-carbon-chain length (SCL), class-II PhaCs like the enzyme from *Pseudomonas aeruginosa* are restricted to 3-hydroxyalkanoic acids of medium-carbon-chain length (MCL). The enzyme of *R. eutropha* accepts also 3-mercaptoalkanoate$_{SCL}$-CoA thioesters as substrates.

Class-III PhaCs have been detected in unoxygenic phototrophic bacteria like *Allochromatium vinosum* or *Thiocapsa pfennigii* and occur also in cyanobacteria and sulfate-reducing bacteria. PhaC and PhaE constitute the subunits of this PhaC class; without PhaE the enzyme is practically inactive. Most members are PHA$_{SCL}$ synthases; some have a broader substrate range like the enzyme from *T. pfennigii* which resembles that of *R. eutropha* regarding the position of the hydroxyl group and the capability of also incorporating mercaptoalkanoic acids, and incorporating in addition also some 3HA$_{MCL}$.

A Class-IV PHA synthase has been detected in *Bacillus megaterium*. PhaC and PhaR (the transcriptional regulator of phasin expression), the latter exhibiting only little sequence homology to PhaE but being essential for enzyme activity, constitute this PhaC. Rehm [19] concluded that PhaR, eventually like PhaE, may functionally replace the N-terminus of class-I PhaCs.

The three-dimensional (3D) structure of PHA synthases has not been directly determined and is unknown. To the best of our knowledge, crystals have not been obtained from any PhaC; therefore, X-ray diffraction analysis could not be done. However, employing the SWISS-MODEL protein threading algorithm and other threading algorithms, a model for the 3D structure of the enzyme of *A. vinosum* could be partially revealed [30]. These models suggest that PhaCs belong to the α/β-hydrolase superfamily [19].

10.2.7
Special Motifs and Essential Residues

Some PHA synthases like the enzymes from *R. eutropha*, *A. vinosum* and *P. aeruginosa* have been investigated in detail regarding the enzymatic mechanism. The overall amino acid identities of the many known PhaCs vary between 8 and 96% [19] and only eight amino acids are strictly conserved in all of them. All PhaCs, which have been identified as such possess a modified lipase box in which the serine occurring in lipases has been replaced by a cysteine in the PhaCs (GXCXG). If this cysteine (Cys319 in *R. eutropha* PhaC; Cys149 in *A. vinosum* PhaC) is exchanged by a serine in a mutated PhaC, the enzyme becomes inactive [31]. Those who are interested in foregoing information should examine, for example, the review of Rehm [19] and of others.

10.2.8
The Catalytic Mechanism of Polyhydroxyalkanoate Synthases

Above, it was mentioned that the PHA synthase proteins are covalently bound to the growing polyester chain. Two thiol groups are directly involved in the catalytic mechanism of PHA synthase: the first is acting as the loading site for a new hydroxyacyl moiety from hydroxyacyl-CoA; the other serves as an elongation site. The most likely candidate for one of these thiol groups is the strictly conserved cysteine of the lipase box. Since a second strictly conserved thiol group was not found and since also pantetheinylation of the protein could not be confirmed [32, 33], the second thiol group could be provided by the strictly conserved cysteine of a second subunit. Alternatively, a serine was proposed to fulfill the function of the second thiol group [19].

10.2.9
In Vitro Synthesis

Polyhydroxyalkanoates can be synthesized *in vitro* in cell-free systems with purified PHA synthase protein. This has been demonstrated several times not only with different PHA_{SCL} synthases [34–37] but also with PHA_{MCL} synthases [38]. Such studies were of course done for kinetic and mechanistic analysis. Studies going beyond this point are, however, hampered by the high costs and/or lacking availability of the hydroxyacyl-CoA esters or by the need to use CoA in stoichiometric amounts.

These problems were circumvented by two different *in vitro* approaches, in which purified enzymes were used.

(i) One approach was the recycling of CoA. This was achieved by adding to the two-component PhaCs from *A. vinosum* or *T. pfennigii*, the propionyl-CoA transferase from *Clostridium propionicum,* and a commercially available acetyl-CoA synthetase [39]. Thereby it was possible to start PHB biosynthesis from 3-hydroxybutyrate and acetate and to use CoA in only catalytic amounts. This also prevented the accumulation of CoA which obviously inhibited the reaction of the PhaC. The reaction was driven by cleavage of ATP. Although in this case ATP instead of CoA has to be used in stoichiometric amounts, this reaction is more affordable because ATP is much cheaper than CoA.

(ii) The other approach was to start directly from free hydroxyfatty acids and to employ the so-called BPEC-pathway [40]. The pathway consists of the butyrate kinase (Buk) and the phosphotransbutyrylase (Ptb) from *Clostridium acetobutylicum* and the two-component PHA synthase (PhaEC) from *T. pfennigii* or from another anoxygenic phototrophic bacterium. This short *in vitro* engineered pathway was later established in *E. coli* by expressing the respective genes (see below) [41].

None of these approaches can be used on a technical scale for commercial PHA production. However, *in vitro* synthesis of PHAs on a semipreparative scale is possible, allowing production of small amounts of novel PHAs also including constituents that cannot be provided by the metabolism or of a certain distribution of the comonomers, and sufficient amounts for determination of crystallinity, melting point etc. can be obtained.

10.2.10
Embedding in General Metabolism

Although the PhaC is the key enzyme of PHA biosynthesis, it is not directly linked to the central metabolism. The enzyme must be provided with hydroxyacyl-CoA thioesters as substrates and for this a link between the central metabolism and the PhaC must exist or must be established [42]. If this is not ensured, no PHAs will be synthesized and accumulated in the organisms although a PhaC is perfectly expressed. This seems to limit PHA production in some heterologous systems. In natural PHA biosynthesis pathways this is, for example, realized by a ß-ketothiolase plus an acetoacetyl-CoA reductase in most PHB synthesizing bacteria allowing synthesis of PHB from acetyl-CoA. In bacteria synthesizing PHA$_{MCL}$, for example, from carbohydrates like *Pseudomonas putida*, a 3-hydroxyacyl-ACP:CoA transferase (PhaG) links the fatty acid *de novo* biosynthesis and PHA$_{MCL}$ synthases [43]. Several enzymes are known to provide links between the fatty acid ß-oxidation pathway and PHA$_{MCL}$ synthases in bacteria capable of synthesizing PHA$_{MCL}$ from lipids and fatty acids like *P. oleovorans* and most other pseudomonads [42]. A few additional examples of such links are known.

The knowledge of these pathways is important when PHA biosynthesis pathways are established in bacteria or other organisms naturally not capable of synthesizing PHAs. In this case also non-natural links could be used. One example is the so-called BPEC-pathway which was originally engineered for *in vitro* PHA biosynthesis [40]. This non-natural pathway, which consisted of Buk and Ptb from *C. acetobutylicum* and PhaEC from an anoxygenic bacterium (for example *T. pfennigii*), constituted a novel pathway via a phosphorylated intermediate, and various added hydroxyfatty acids could be converted into the corresponding PHAs. This pathway was then also used for *in vivo* PHA biosynthesis in a recombinant *E. coli* strain after functional expression of the three enzymes [41]. It was not only shown that this pathway allowed the synthesis of various PHA homopolyesters but also of various PTEs homopolymers [14] which seem to be non-biodegradable (for a review see [44]).

Numerous other interesting and intriguing pathways have meanwhile been engineered. There is not sufficient space to describe all of them. Only lack of fantasy seems to limit the possibilities for such novel pathways. I would like to mention the efforts of the laboratory of Dr. Taguchi, who succeeded to establish a metabolic route enabling *in vivo* production of lactic acid containing polyesters [45, 46]. For this they expressed in a lactic acid overproducing mutant strain of *E. coli* a PHA synthase mutant (Ser325Thr, Gln481Lys) from *Pseudomonas* sp. 61-3 together with a ß-ketothiolase, a NADPH-dependent acetoacetyl-CoA reductase and a propionyl-CoA transferase. This recombinant *E. coli* strain produced a tercopolymer consisting of lactic acid, 3-hydroxybutyric acid and 3-hydroxyvaleric acid when cultivated in presence of glucose and propionic acid [45].

10.2.11
Biotechnological Relevance

Polyhydroxyalkanoates have been considered for a large variety of different applications for about 50 years when scientists from the company Grace detected that PHB has thermoplastic properties. Since it is biodegradable, it could be used as biodegradable packaging material and later, when composting became in many countries popular in civil engineering, it was considered as compostable material. Later also several other applications were considered, for example as resorbable material in pharmacy and medicine. More applications came up when the large flexibility of the chemical structures and thereby the material properties of PHAs that could be produced became evident. Nowadays also the aspect of renewable resources from which these polyesters can be produced became relevant.

Consequently several companies eagerly tried to establish large scale processes for the production of these PHAs. Beside ICI and companies related to it, Monsanto, DuPont and several companies in Asian countries were active in this field.

On the other hand the experience and knowledge revealed during studying PHAs prompted other companies to develop and establish chemical processes for the production of biodegradable polyesters and other biodegradable polymers demonstrating that synthetic polymers must not necessarily be persistent to degradation. Several companies developed, for example, polylactic acid with Natureworks™ from Cargill as the most important product. Ecoflex produced by BASF is another biodegradable synthetic polymer which is on the market since several years.

At present, Metabolix Inc (Cambridge, MA) and Tepha Inc (Lexington, MA) are most probably the most advanced companies in this area. Metabolix has build together with Archer Daniels Midland Company (ADM) a new plant in Clinton (Iowa) to produce about 50000 tons per year of various PHA_{SCL} for bulk applications. Tepha is focusing on the production of various sophisticated medical devices like sutures, cardiovascular tissues, meshes etc.

Not only are the various PHAs of interest for materials, technical applications and devices in industry, but also the PHA granules and the phasins as well as other granule-associated proteins (GAP). Soon after the discovery of the phasins it was demonstrated that segments of the proteins containing the PHA granule binding motif could be used to immobilize other proteins at the granules [25]. The PHA granules could be used as small beads in the below micrometer-range as matrix for protein purification and also as carrier for the distribution of proteins that could be immobilized to the granules for example by using the PHA-granule binding domain of GAPs which are fused the other medically or pharmacologically relevant proteins and also in drug delivery [47, 48].

10.3
Cyanophycin Synthetases

10.3.1
Occurrence of Cyanophycin Synthetases

Cyanophycin synthetases are widely distributed among bacteria [49] while no sequence similarities towards *cphA* genes could be detected in eukaryotes after *in silico* analysis of genome sequences [50]. Although thought to exist exclusively in cyanobacteria, Krehenbrink *et al.* revealed in 2002 [49] the presence of sequences with high similarity to cyanobacterial *cphA* genes in several heterotrophic bacteria. Putative *cphA* genes were found in *Acinetobacter baylyi*, *Bordetella bronchiseptica*, *B. parapertussis*, *B. pertussis*, *Desulfitobacterium hafniense*, *Clostridium botulinum*, and *Nitrosomonas europaea* [49]. Their functionality was ascertained for *A. baylyi* ADP1 and *D. hafniense* DCB-2 [49, 51]. The molecular size distribution of CGP isolated from cyanobacteria varies significantly from that isolated from other bacteria or recombinant eukaryotic cells. Cyanobacteria accumulate a CGP with a molecular mass distribution ranging from 25 kDa to up to 100 kDa [52–55], while other organisms synthesize a polymer with a maximal molecular mass of 40 kDa [56]. These deviations might be due to a differing enzyme-to-substrates ratio, an inadequate amount of substrates, the absence of specific catalytic factors, or other physiological divergences [57]. The function of the polymer is mainly referred to as nitrogen reserve material as the constituting amino acids aspartate and especially arginine exhibit a high content of nitrogen. Additionally, CGP may serve as carbon and energy reserve material [7, 55, 58]. Energy may be provided through conversion of arginine, obtained after hydrolysis of the polymer, to carbon dioxide, ammonia, ornithine, and ATP. This reaction involves the action of the enzymes arginine deiminase, ornithine carbamoyl transferase, and carbamoyl phosphokinase [59], and is referred to as the arginine deiminase or dihydrolase pathway. This pathway was detected in some cyanobacteria previously [60].

Due to low polymer and cell yields, and a complicated cultivation procedure in photobioreactors, cyanobacteria are not suitable for an efficient production of the polymer [61]. Therefore, several *cphA* genes were applied for heterologous expression in bacteria, yeasts, and plants [56, 62–64]. In bacteria maximal polymer contents of 46.0% (w/w per cell dry matter) were detected in *A. baylyi* ADP1 [65] and 40.0% in recombinant cells expressing *cphA* from *Synechocystis* sp. PCC 6308 [66]. Concerning transgenic yeasts, maximal contents of 23.2% were achieved using the yeast *Pichia pastoris* expressing *cphA* from *Synechocystis* sp. PCC 6308 [56], in transgenic plants up to 7.5% of the polymer were accumulated in potato tubers, and up to 6.9% were achieved in *Nicotiana tabacum* expressing *cphA* from *Thermosynechococcus elongatus*, respectively [62, 64].

10.3.2
Chemical Structure of Cyanophycin

The chemical structure of CGP, as depicted in Figure 10.3, consists of a poly(aspartic acid) backbone with arginine residues linked to the β-carboxyl group of each aspartate by the α-amino group, and was proposed by Simon and Weathers in 1976 [53]. Nuclear magnetic resonance (NMR) spectroscopy of the polymer confirmed the postulated structure and provides a suitable tool for detection and characterization of the polymer [67–69]. Suarez *et al.* [67] characterized the polymer from *Synechocystis* sp. PCC 6308 by ^{1}H, ^{13}C, and ^{15}N NMR spectroscopy. Steinle *et al.* [69] showed by ^{1}H NMR analysis that the primary structures of the soluble and the insoluble forms of the polymer (see below) are identical, thus presuming that both forms differ in their secondary or tertiary structure. Using circular dichroism and Raman spectroscopy Simon *et al.* [70] proposed a secondary structure of CGP comprising 50% β-sheet, 45% random coil, and 5% α-helix. Construction of molecular models showed that a β-sheet aspartate backbone with

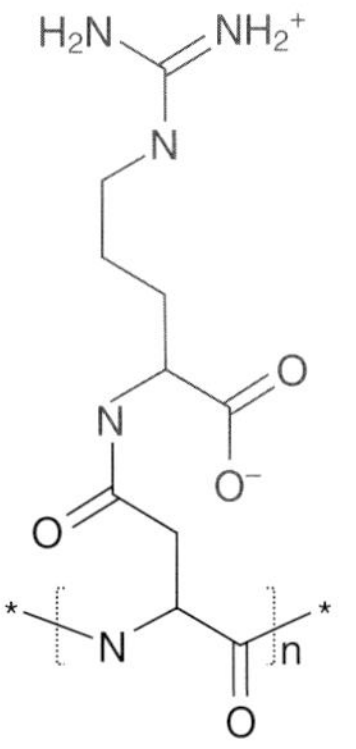

Figure 10.3 Chemical structure of cyanophycin.

the aspartate β-carboxyl groups peptide-bonded to the arginine residues can form a compact structure with no disallowed contacts [70]. The authors detected that the polymer exhibits the respective defined secondary structure in acidic solution and presumably under insoluble conditions, but not in alkaline medium. The tertiary structure of the polymer was not ascertained.

10.3.3
Variants of Cyanophycin

Several CGP variants, concerning the amino acid constituents, were detected *in vitro* as well as *in vivo* previously [69, 71–74]. Lysine is incorporated into the polymer to up to 18 mol% instead of arginine in recombinant organisms after expression of CphAs with a broad substrate range [49, 51, 56, 69, 75]. An incorporation of ornithine and citrulline, respectively, into CGP was only observed lately after expression of *cphA* from *Synechocystis* sp. PCC 6308 in strains of *Saccharomyces cerevisiae* exhibiting defects in arginine metabolism [69]. Cells with deleted ornithine carbamoylphosphate transferase accumulated a polymer consisting of up to 8.3 mol% ornithine, while cells with deleted argininosuccinate synthetase revealed a molar ratio of citrulline of 20 mol %. Both compounds were detected as constituents replacing arginine. The incorporation of glutamate instead of arginine, observed as a function of specific cultivation conditions such as nitrogen starvation in cells of *Synechocystis* sp. PCC 6308 [71] or *Synechococcus* sp. strain G2.1 [72], could not be corroborated by *in vitro* studies, and thus remain dubious.

Synthesis of CGP variants is of special interest for (i) the production of a variety of dipeptides (see below); (ii) environmentally friendly synthesis of a wider selection of bulk chemicals; and (iii) still unknown functions of new CGP being of biotechnical interest might be beneficial for industry. The *in vitro* synthesis of variants of the polymer is discussed below.

10.3.4
Reaction Catalyzed by the Key Enzyme

CphA catalyzes the ATP-dependent incorporation of aspartate and arginine onto a CGP primer. As determined *in vitro* by use of purified CphA from *Synechococcus* sp. MA19, CTP or GTP cannot act as substitutes for ATP during CGP biosynthesis [76]. Additionally, it was shown that activation of the amino acid substrates is performed by phosphorylation instead of adenylation as no AMP could be detected as reaction product [77]. The constituent amino acids are incorporated stepwise into the growing polymer, in the order aspartate followed by arginine [73]. The amino acid substrates are incorporated at the C terminus of the primer, which was shown by use of synthetic blocked CGP primers [73]: blocking the primer at the C terminus results in inhibition of the elongation while blocking the N terminus does not have an effect on the polymer elongation. Thus, the elongation is mediated by activation of the carboxylic group of the C-terminal amino acids which is in accordance with experiments performed with the structurally related enzymes D-alanine-D-alanine ligase [78], UDP-N-acetylmuramate:L-alanine ligase (MurC) [79], or UDP-N-acetylmuramoyl-L-alanine:D-glutamate (MurD) from *E. coli*, respectively [80, 81].

10.3.5
Assay of Enzyme Activity

Originally Simon described a radiometric enzyme assay employing L-[U^{14}C]-arginine for CphA in 1976. Therefore, CphA from the filamentous cyanobacterium *Anabaena cylindrica* was enriched 92-fold to investigate the basic properties of the enzyme [82]. The assay was modified by Ziegler et al. in 1998 [77] and further simplified by Aboulmagd et al. in 2000 [57]. For purification of the enzyme three chromatographic steps are essential comprising dye-ligand, size-exclusion, and ion-exchange [77]. For catalytic activity CphA requires not only the incorporating substrates aspartate and arginine, but also Mg^{2+}, K^+ and a sulfur compound such as β-mercaptoethanol at low concentrations. Additionally, ATP is required as energy source, and a CGP primer, consisting of at least three dipeptide units [(β-Asp-Arg)$_3$] is essential for biosynthesis of the polymer. An exceptional CphA is that from *T. elongatus* BP-1, as it can synthesize the polymer *de novo* independently of a CGP primer [83]. In other organisms, especially recombinant ones, provision of the primer creates a 'chicken and egg' problem. Presumably, small peptides occurring naturally in cells can substitute the CGP primer [57]. The purified CphA from the thermophilic cyanobacterium *Synechococcus* sp. MA19 accepted a modified CGP with reduced arginine content, α-arginyl aspartic acid dipeptide, or poly-α,β-DL-aspartic acid as primers [76]. A pH of 8.5 is the optimum pH value for CphA activity, although binding of CGP remains relatively constant between pH 9.0 and 6.3, and drops only significantly at pH values below 6 [74, 82, 84]. Maximum activity of CphA is obtained at 50 °C. With the exception of CphA from *Synechococcus* sp. MA19 staying active even after prolonged incubation at the respective temperature [76], prolonged incubation at 50 °C leads to inactivition of CphA after 30 min, whereas the enzyme stayed active at 28 °C [74].

10.3.6
Location of Enzyme–Granule Structure

Cell inclusions consisting of CGP had already been discovered in 1887 by microscopic analysis of cyanobacteria [85]. Koop *et al.* [86] made investigations on the localization of CGP in recombinant cells of R. eutropha by record of nitrogen distribution maps employing energy-filtering transmission electron microscopy. Biochemical investigations of the interaction of CphA to CGP inclusions have not yet been performed. In its natural hosts the polymer is stored as membraneless granules, but a soluble type of CGP was observed in recombinant yeasts and plants [56, 62–64, 69, 75], and in an *E. coli* strain expressing *cphA* from *D. hafniense* [51]. Physicochemical reasons leading to the formation of this type of CGP have not yet been elucidated and thus require further research [51, 63].

10.3.7
Kinetic Data of Wild Type Enzyme

Kinetic parameters, such as substrate affinity and specificity, specified using chromatographically purified CphAs were determined for the enzymes from the cyanobacteria *A. cylindrica* [82], *Synechococcus* sp. MA19 [54], *Synechocystis* sp. PCC 6803 [77], *A. variabilis* ATCC 29413 [73], *Synechocystis* sp. PCC 6308 [74] and from the heterotrophic bacterium *A. baylyi* ADP1 [84]. A high affinity towards the reactants L-aspartic acid, L-arginine, ATP, and cyanophycin was determined for CphA from *Synechocystis* sp. PCC 6308 (CphA$_{6308}$) with K_m-values of 450 µM, 49 µM, 200 µM, and 35 µg ml^{-1}, respectively [61]. Assays employing CphA from *A. baylyi* ADP1 (CphA$_{ADP1}$) emphasized the observation made from isolated CGP that it exhibits a narrow substrate range without significant affinity towards lysine [84]. K_m-values for L-arginine (47 µM) and L-aspartic acid (240 µM) determined for CphA$_{ADP1}$ were similar to those of known CphAs from cyanobacteria [84]. Additionally, the two different ATP-binding sites of the enzyme were characterized independently of each other with respect to their affinities for ATP. CGP-Asp was applied as primer in the reaction mixture to determine the K_m-value of the ATP-binding site involved in the incorporation of arginine. Interestingly, the ATP-binding site responsible for the addition of arginine exhibited a much higher affinity for ATP (38 µM) than those responsible for the addition of aspartic acid (210 mM). Experiments conducted to analyze the role of Mg^{2+}, demonstrated that binding of CphA$_{ADP1}$ to CGP-Arg is independent of Mg^{2+}, whereas binding to CGP-Asp requires the presence of Mg^{2+} to be effective [84]. Analysis of CphA from *Synechococcus* sp. strain MA19 demonstrated that α-arginyl aspartic acid dipeptide, citrulline, ornithine, arginine amide, agmatine, or norvaline could not replace arginine [76]. However, this enzyme could significantly incorporate canavanine instead of arginine and β-hydroxyaspartic acid instead of L-aspartic acid

[76]. Additionally, compounds such as α-arginyl aspartic acid dipeptide, modified cyanophycin containing less arginine, and poly-α,β-DL-aspartic acid were used as primers by the respective CphA [76].

10.3.8
Primary Structures and Essential Motifs of the Enzyme

Analysis of the primary structure of CphA revealed different binding motifs or domains in the enzyme [49, 73, 77, 87, 88]. Two distinct ATP binding motifs, which were shown to be involved in the incorporation of arginine or aspartate, respectively, were detected [73, 87]. Experimental proof was obtained by Berg [87] employing point-mutated CphAs from *A. variabilis* ATCC 29413 (CphA$_{29413}$). CphA$_{29413}$ harboring the point mutation in the N-terminal ATP binding site could not catalyze the addition of aspartate onto the CGP primer, whereas one molecule of arginine could still be attached. CphA$_{29413}$ harboring the point mutation in the C-terminal ATP binding site behaved vice versa, aspartate was still added but not arginine. From these experiments it was postulated that the enzyme possibly contains two distinct active sites, each being involved in binding one amino acid substrate [73].

This hypothesis could be corroborated by comparison of the amino acid sequences of CphA and homologous proteins which showed that CphA could be divided into N-terminal and C-terminal regions [73, 77, 88]. Several proteins exhibited sequence similarities to the N or the C terminus of CphA (Figure 10.4). The N-terminal regions of CphA show high similarities to a superfamily of ATP-dependent ligases [89]. These enzymes are characterized by activation of carboxylates for nucleophilic attack by phosphorylation with Mg^{2+}ATP. This group of enzymes contains the highly conserved ATP-binding motifs referred to as ATP-

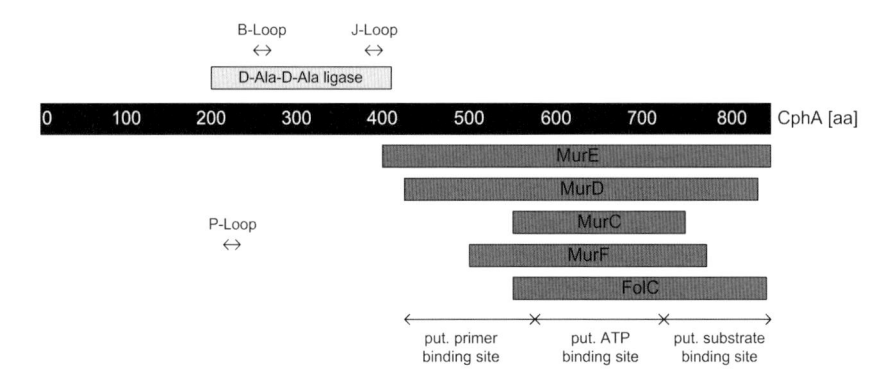

Figure 10.4 Proteins showing sequence similarities to CphA. Motifs conserved among the proteins are indicated as well as identified binding sites for ATP, primer and substrates [105]. D-Ala, D-Alanine; MurE, UDP-N-acetylmuramoyl-L-alanyl-D-glutamate:*meso*-diaminopimelate ligase; MurD, UDP-N-acetylmuramoyl-L-alanine:D-glutamate ligase; MurC, UDP-N-acetylmuramate:L-alanine ligase; MurF, UDP-N-acetylmuramoyl-tripeptide:D-alanyl-D-alanine ligase; FolC, folyl-poly-γ-glutamate synthetase-dihydrofolate synthetase.

grasp or B-loop and the so-called J-loop. These loops are flexible in an ATP-free state and become rigid in the ATP-bound state [90]. However, primary structures beyond these domains show rather low identities to the ones from CphA. Members of this group include, among others, D-alanine-D-alanine ligase [78], biotin carboxylase α-chain [91], glutathione synthetase [92], succinyl-CoA synthetase [93], and carbamoylphosphate synthetase [94] and are exemplary members of the ATP-grasp enzyme superfamily. In contrast, the C-terminal region of CphA, approximately starting at amino acid residue 400 [73, 77], shows high sequence similarities to the well investigated superfamily of murein ligases (MurC, MurD, MurE, MurF) and folyl-poly-γ-glutamate dihydrofolate synthetase (FolC) [79, 95–97]. A feature shared by all members of enzymes belonging to this superfamily is the similar catalytic function [97] involving (i) the formation of a peptide or amide bond through hydrolysis of ATP to ADP and P_i; (ii) the enzymatic catalysis via a similar mechanism involving the formation of acyl phosphate [98] and tetrahedral intermediates [99, 100]. All these enzymes exhibit the so-called 'P-loop' motif, the primary structure of which typically consists of a glycine-rich sequence followed by a conserved lysine and a threonine or serine [101]. The P-loop is involved in the binding of ATP; in MurD, the α-phosphate oxygens of ATP form hydrogen bonds with one of the conserved glycine residues, and the β-phosphate oxygens with the leucine and threonine residue of the P-loop [81]. For two CphAs the essentiality of Lys497, being part of the P-loop, was demonstrated by point mutation into Ala497 resulting in total enzyme activity loss [87, 88]. Besides the conserved P-loop motif this family of enzymes reveals additional strongly conserved regions which are putatively involved in the binding of other substrates than ATP [77, 79, 95]. For detection of putative substrate binding sites in CphA special regard was given to those of the murein ligases, as both enzymes operate via similar mechanisms and bind the same sort of substrates (reviewed by [88]). The murein ligases, MurC, MurD, MurE, and MurF, catalyze the subsequent addition of L-alanine, D-glutamate, meso-diaminopimelate or L-lysine, and D-alanyl-D-alanine, respectively. UDP-N-acetylmuramic acid functions as primer, and ATP is used for activation of the substrates by phosphorylation. In contrast to CphA, the 3D structures of the murein ligases were determined, and thus the exact binding sites for the respective substrates were elucidated [80, 81, 102–105]. Regions involved in the binding of the primer, ATP, and the amino acid substrates are indicated in Figure 10.4. As depicted, the ATP-binding region is overlapping in the primary structure of these proteins, and thus it is conceivable that also the remaining substrates are bound in the respective regions. Comparison of the primary structure between CphAs with wide substrate range with the once exhibiting a low substrate specificity (see above) did not reveal obvious differences in the putative amino acid binding sites [88].

10.3.9
Catalytic Cycle

During the catalytic cycle, the α-carboxylic group and the β-carboxylic group of aspartate are subsequently activated by phosphorylation consuming one molecule

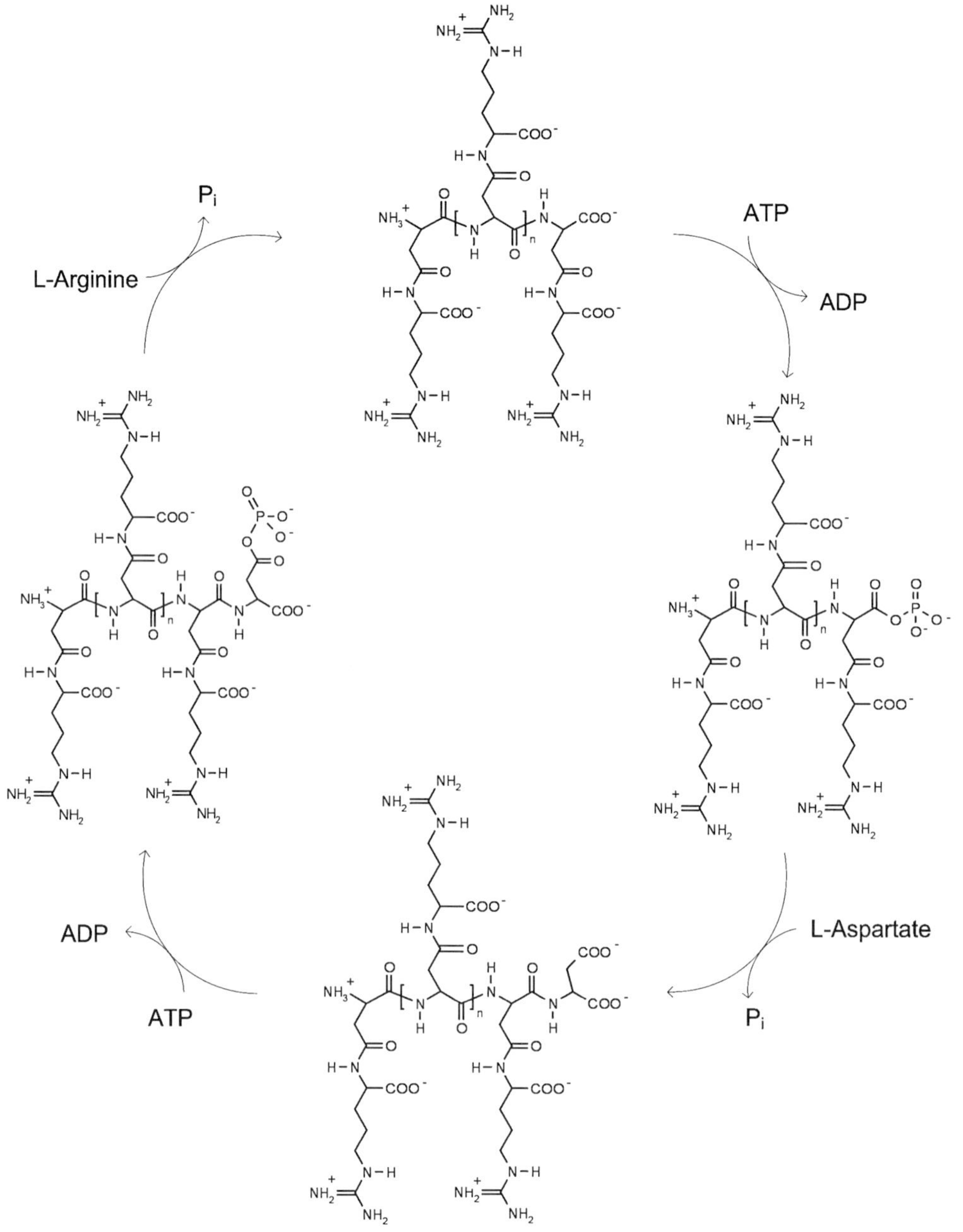

Figure 10.5 Postulated catalytic cycle of CphA (modified according to [73]).

of ATP for each step (Figure 10.5). As mentioned above, aspartate is bound to the CGP primer prior to arginine [87]. The catalytic cycle is divided into four steps (Figure 10.5) and incorporation of the substrates requires one molecule of ATP each as energy source. Catalysis starts by activation of the α-carboxylic group of the C-terminal aspartate residue with ATP which is performed by the attack of the γ-phosphoryl group of ATP to the oxygen of the respective carboxylic group forming an acylphosphate as intermediate. In the next step one molecule of aspartate is bound to the α-carboxylic group of the CGP primer by its α-amino group as a consequence of the nucleophilic attack of the amino group of the condensing amino acid, with elimination of phosphate and subsequent peptide bond formation. For incorporation of arginine, the β-carboxylic group is then activated by ATP as described for the binding of aspartate, again forming an acylphosphate as intermediate. This reaction is followed by binding of the α-amino group of arginine to the activated β-carboxylic group of aspartate. The existence of acylphosphate and tetrahedral intermediates has been proposed for the L-alanine- and D-glutamate-adding enzymes [79].

10.3.10
Mutant Variants of the Enzyme

Enzyme engineering is the method of choice to direct synthesis of products in a specific direction and to identify key residues involved in catalytic mechanisms [98, 106]. Mutagenesis of *cph*As is beneficial for three aims: (i) to obtain enzymes with enhanced specific activity or other useful features; (ii) to elucidate the function of specific amino acid residues during catalysis; or (iii) to generate CphAs with altered specificity for synthesis of novel CGPs. Several site-directed mutations were performed employing $CphA_{6308}$, $CphA_{29413}$, or CphA from *Nostoc ellipsosporum* NE1 ($CphA_{NE1}$), respectively. By truncation of $CphA_{6308}$ and $CphA_{NE1}$, enzymes with increased enzyme activities were obtained [56, 107]. Increase of enzyme activity as a result of C-terminal truncation is a frequently observed phenomenon also with other enzymes [108–110]. Deletion of 29 residues of $CphA_{NE1}$, normally constituting of 901 amino acids, resulted in a 2.2-fold increase of the CphA activity, while deletion of 57 residues resulted in a complete activity loss when compared to the wild type enzyme [107]. As $CphA_{6308}$ consists of 874 residues only, a stepwise truncation by one amino acid each was performed to investigate whether truncation leads to enhanced specific activity of the enzyme [56]. Due to the strong loss of activity of $CphA_{6308}\Delta2$ and $CphA_{6308}\Delta3$, it can be concluded that amino acid residues Ser872 and Ser873 play a significant role during catalysis; a structural role can be excluded as the protein was still soluble, detectable by immunological methods, and catalytically active [56]. Comparable experiments carried out by Hai *et al.* in 2009 [111] using $CphA_{NE1}$, proposed a crucial function of residues 867 to 870 for thermostability.

For further site-directed mutagenesis, determination of the 3D structure of CphA after crystallization would be convenient for precise identification of amino acid residues that constitute the substrate binding sites. Well-defined point mutations in the region encoding the substrate binding sites could enable generation of CphAs with altered substrate specificity. As crystallization of proteins is a time-consuming, cost-effective and complicated procedure, the generation of so-called homology models gained preferential interest in the past. Here the main bottleneck for analysis of CphA is the unavailability of a protein with significant sequence similarity to the entire CphA [88]. Among the enzymes involved in CGP metabolism, the 3D structure has only been resolved for the CGP-degrading protein cyanophycinase (CphB) [112].

A set of point-mutated CphAs from *Synechocystis* sp. PCC 6308 was constructed and analyzed previously (reviewed by [56, 88, 113]). While most point mutations (C59A, C133A, C218A, K261K, K497A, F692H, R731E, H748F, R777K, E800R, D802E, R805K, E835Q) led to decreased or even loss of enzyme activity, point mutation C595S led to a 1.6-fold increased specific activity and led to higher CGP accumulation after heterologous expression when compared to the wild type CphA$_{6308}$ [56]. However, the function of residue Cys595 during catalysis was not ascertained.

10.3.11
In Vitro Synthesis

As shown by *in vitro* assays and by analysis of the composition of accumulated polymer, specific CphAs exhibit differing substrate specificities. CphAs exhibiting a narrow substrate specificity, only accepting arginine and aspartate, include the ones from *A. baylyi* ADP1 and *N. ellipsosporum* NE1. CphAs with a wide substrate range resulting in incorporation of lysine after recombinant expression in *E. coli*, yeasts or plants comprise the cyanobacterial ones from *Anabaena variabilis* ATCC29413, *Synechocystis* sp. strains PCC6308 and PCC6803, *T. elongatus*, *Nostoc* sp. PCC7120, *Synechococcus* sp. MA19, and CphA from the heterotrophic bacterium *D. hafniense* DCB-2 [51, 54, 62, 73, 74, 77, 84]. Using the purified enzymes of *A. variabilis* ATCC29413 [73], and of *Synechocystis* sp. PCC 6308 [74], respectively, it was shown that other compounds than arginine or aspartate can be successfully incorporated into the polymer.

Additionally, several compounds were shown to inhibit the incorporation of (i) both amino acid substrates or (ii) arginine into the polymer [74]. Substrates belonging to the first group exhibited an equal effect on the incorporation of arginine and aspartate and comprise arginine methyl ester, argininamide, S-(2-aminoethyl) cysteine, β-hydroxy aspartic acid, aspartic acid β-methyl ester, norvaline, citrulline and asparagine. Compounds belonging to the second group like canavanine, lysine, agmatine, D-aspartic acid, L-glutamic acid and ornithine inhibited the incorporation of arginine to a greater extend than the incorporation of aspartic acid. Analysis of the proteinogenic amino acids like alanine, histidine, leucine, proline, tryptophan or glycine revealed no effect on the incorporation of arginine, thus assuming that these compounds are not recognized as substrates for CphA$_{6308}$ [74]. Additionally, CphA from *Synechococcus* MA19 incorporated β-hydroxyaspartic acid instead of aspartic acid and L-canavanine instead of L-arginine at a significant rate [76].

10.3.12
Embedding in General Metabolism

Synthesis of recombinant proteins requires additional synthesis of precursor substrates or provision of catalytically essential compounds which might lead, especially in recombinant organisms, to a certain so-called metabolic burden [114]. In case of CGP synthesis especially detraction of the amino acids aspartate and arginine from for example, protein biosynthesis plays an important role; these compounds thus need to be synthesized in higher amounts than under normal metabolic conditions. In contrast to organisms expressing *cphA* heterologously, cyanobacteria possess two differently regulated arginine biosynthesis pathways, enabling strict regulation of provision of arginine for (i) protein synthesis or (ii) CGP synthesis. As observed in several bacteria, especially sufficient provision of arginine plays a crucial factor for efficient CGP synthesis, and is much more important than provision of aspartate. This can be explained by the fact that the latter requires only one enzymatic step from an intermediate of the tricarboxylic acid cycle, whereas arginine is synthesized by eight successive steps starting from α-ketoglutarate [115]. Additionally, two molecules of ATP are detracted per dipeptide subunit of the polymer from the regular metabolism of the CGP-synthesizing host, which may lead to weaker growth due to use of the host cell's resources.

Intracellular biodegradation of CGP is mediated by cyanophycinase (CphB) (reviewed by [116]). Products of the degradation mechanism are usually β-dipeptides which are then split to free amino acids by intracellular dipeptidases. CphB shows a high specificity for CGP as a substrate and is inhibited by serine protease inhibitors. Its primary structure reveals a serine residue within a lipase box motif (Gly-Xaa-Ser-Xaa-Gly) which forms a catalytic triad together with a glutamic acid and a histidine residue.

10.3.13
Biotechnological Relevance

Putative biotechnical applications for CGP are several-fold (reviewed by [116]). However, until now fields of applications were only determined for degradation products of the polymer and not for CGP itself, but especially the new variants of the polyamide require attention as they may fulfill chemical functions interesting for industry. Synthesis of bulk chemicals first requires hydrolysis of CGP into its monomers [117, 118]. Besides arginine and aspartate themselves, other compounds such as ornithine, 1,4-butanediamine or acrylonitrile can be obtained [119]. Ornithine is obtained by conversion of arginine by arginase, while 1,4-butanediamine, also referred to as putrescine, is obtained after decarboxylation of ornithine and represents a precursor of nylon [119]. Acrylonitrile, which can be obtained from aspartate as precursor, is employed for the manufacture of plastics [120]. At present most nitrogen-containing compounds are obtained chemically from fossil resources; due to high energy consumption required by such processes, the establishment of environmentally friendly processes is attempted [117, 118]. β-Dipeptides, obtained after α-cleavage of the polymer by specific cyanophycinases are of special interest for pharmaceutical use [121]; they can be applied as dietary supplement for malnourished patients, or as medicament against specific diseases [122]. Poly(aspartic acid), obtained after hydrolytic removal of the arginine side chains, can replace poly(acrylic acid) [123, 124]. As the latter is, in contrast to PAA, not biodegradable, PAA might serve as substitute in various technical processes such as, for example, as dispersant, antiscalant, or water-softener. Currently PAA is produced by thermal condensation of the reactants derived from petrochemistry [125].

10.4
Conclusions

Due to the known environmental concerns, the search for environmentally-friendly substitutes for non-biodegradable plastics or chemicals is nowadays an object. Therefore, PHAs and CGP represent promising candidates. Production of PHAs has become more economically feasible due to intensive research in the last four decades: however, it is still competing with established processes for the well-known polymers which are currently used. The biochemistry of PHA metabolism is mostly known and well understood; some open questions remain for the PHA granule structure and the biogenesis of the granules. In particular the availability of an increasing number of genome sequences from PHA synthesizing bacteria help to complete our view of this metabolism. The genome sequence of the model organism R. eutropha strain H16 and its analysis have unraveled some interesting and unexpected information also regarding the PHA metabolism [126, 127]. Furthermore, proteome analyzes related to PHA metabolism were also carried out [128]. Answers to these questions are probably more of academic interest and will most likely have only little impact on the economics. Production of CGP requires further intensive research for establishment of an environmentally and economically feasible production chain. Essential initial steps using recombinant bacteria, yeasts, and plants were performed showing that the process of CGP production becomes feasible. Further research will focus on obtaining CGP variants to widen the applicability of the polymer, not only therefore further intensive research on the biochemistry and enzymology of CphA is necessary. Especially resolving the 3D structure of this unique enzyme would be beneficial to determine residues involved in direct binding of substrates.

References

1 Wintermeyer, W., and Rodnina, M.V. (2003) Ribosomal protein synthesis, in Biopolymers: Polyamides and Complex Proteinaceous Materials I, vol. 7 (eds S.R. Fahnenstock and A. Steinbüchel), Wiley-VCH Verlag GmbH, Weinheim, Germany, pp. 1–24.

2 Kleinkauf, H., and von Döhren, H. (1990) Nonribosomal biosynthesis of peptide antibiotics. *Eur. J. Biochem.*, **192**, 1–15.

3 Schwarzer, D., Finking, R., and Marahiel, M.A. (2003) Nonribosomal peptides: from genes to products. *Nat. Prod. Rep.*, **20**, 275–287.

4 Simon, R.D. (1973) The effect of chloramphenicol on the production of cyanophycin granule polypeptide in the blue-green alga *Anabaena cylindrica*. *Arch. Mikrobiol.*, **92**, 115–122.

5 Troy, F.A. (1985) Capsular poly-γ-D-glutamate synthesis in *Bacillus licheniformis*. *Methods Enzymol.*, **113**, 146–168.

6 Yamanaka, K., Maruyama, C., Takagi, H., and Hamano, Y. (2008) ε-Poly-L-lysine dispersity is controlled by a highly unusual nonribosomal peptide synthetase. *Nat. Chem. Biol.*, **4**, 766–772.

7 Li, H., Sherman, D.M., Bao, S., and Sherman, L.A. (2001) Pattern of cyanophycin accumulation in nitrogen-fixing and non-nitrogen-fixing cyanobacteria. *Arch. Microbiol.*, **176**, 9–18.

8 Ziegler, K., Stephan, D.P., Pistorius, E.K., Ruppel, H.G., and Lockau, W. (2001) A mutant of the cyanobacterium *Anabaena variabilis* ATCC 29413 lacking cyanophycin synthetase: growth properties and ultrastructural aspects. *FEMS Microbiol. Lett.*, **196**, 13–18.

9 Mifune, J., Grage, K., and Rehm, B.H.A. (2009) Production of functionalized biopolyester granules by recombinant *Lactococcus lactis*. *Appl. Environ. Microbiol.* **75**, 4668–4675.

10 Reusch, R.N., and Sadoff, H.L. (1988) Putative structure and function of a poly-beta-hydroxybutyrate/calcium polyphosphate channel in bacterial plasma membranes. *Proc. Natl. Acad. Sci. U. S. A.*, **85**, 4176–4180.

11 Dai, D., and Reusch, R.N. (2008) Poly-3-hydroxybutyrate synthase from the periplasm of *Escherichia coli*. *Biochem. Biophys. Res. Commun.*, **374**, 485–489.

12 Steinbüchel, A., and Valentin, H.E. (1995) Diversity of bacterial polyhydroxyalkanoic acids. *FEMS Microbiol. Lett.*, **128**, 219–228.

13 Lütke-Eversloh, T., Bergander, K., Luftmann, H., and Steinbüchel, A. (2001) Biosynthesis of a new class of biopolymer: bacterial synthesis of a sulfur containing polymer with thioester linkages. *Microbiology (SGM)*, **147**, 11–19.

14 Lütke-Eversloh, T., Fischer, A., Remminghorst, U., Kawada, J., Marchessault, R.H., Bögershausen, A., Kalwei, M., Eckert, H., Reichelt, R., Liu, S.-J., and Steinbüchel, A. (2002) Biosynthesis of novel thermoplastic polythioesters by engineered *Escherichia coli*. *Nature*, **1**, 236–240.

15 Hazer, B., and Steinbüchel, A. (2007) Increased diversification of polyhydroxyalkanoates by modification reactions for industrial and medical applications. *Appl. Microbiol. Biotechnol.*, **74**, 1–12.

16 Valentin, H.E., and Steinbüchel, A. (1994) Application of enzymatically synthesized short-chain-length hydroxy fatty acid coenzyme A thioesters for assay of polyhydroxyalkanoic acid synthesis. *Appl. Microbiol. Biotechnol.*, **40**, 699–709.

17 Schubert, P., Steinbüchel, A. and Schlegel, H.G. (1988) Cloning of the *Alcaligenes eutrophus* gene for synthesis of poly-ß-hydroxybutyric acid and synthesis of PHB in *Escherichia coli*. *J. Bacteriol.* **170**, 5837–5847.

18 Haywood, G.W., Anderson, A.J., and Dawes, E.A. (1989) The importance of PHB-synthase substrate specificity in polyhydroxyalkanoate synthesis by *Alcaligenes eutrophus*. *FEMS Microbiol. Lett.*, **57**, 1–6.

19 Rehm, B.H.A. (2003) Polyesters synthases: natural catalysts for plastics. *Biochem. J.*, **376**, 15–33.

20 Pötter, M., and Steinbüchel, A. (2006) Biogenesis and structure of polyhydroxyalkanoate granules, in *Inclusions in Prokaryotes, Volume 1 of Microbiology Monographs* (ed. M. Shively), Springer, Heidelberg, pp. 109–136.

21 Peters, V., Becher, D., and Rehm, B.H.A. (2007) The inherent property of polyhydroxyalkanoate synthase to form spherical PHA granules at the cell poles: the core region is required for polar localization. *J. Biotechnol.*, **132**, 238–245.

22 Pieper-Fürst, U., Madkour, M.H., Mayer, F., and Steinbüchel, A. (1994) Purification and characterization of a 14-kDa protein that is bound to the surface of polyhydroxyalkanoic acid granules in *Rhodococus ruber*. *J. Bacteriol.*, **176**, 4328–4337.

23 Wieczorek, R., Pries, A., Steinbüchel, A., and Mayer, F. (1995) Analysis of a 24-kDa protein associated with the polyhydroxyalkanoic acid granules in *Alcaligenes eutrophus*. *J. Bacteriol.*, **177**, 2425–2435.

24 Steinbüchel, A., Aerts, K., Babel, W., Föllner, C., Liebergersell, M., Madkour, M.H., Mayer, F., Pieper-Fürst, U., Pries, A., Valentin, H.E., and Wieczorek, R. (1995) Considerations on the structure and biochemistry of bacterial polyhydroxyalkanoic acid inclusions. *Can. J. Microbiol.*, **41** (Suppl. 1), 94–105.

25 Pieper-Fürst, U., Madkour, M.H., Mayer, F., and Steinbüchel, A. (1995) Identification of the region of a 14-kDa protein of *Rhodococcus rubber* that is responsible for the binding of this phasin to polyhydroxyalkanoic acid granules. *J. Bacteriol.*, **177**, 2513–2523.

26 Liebergesell, M., and Steinbüchel, A. (1996) New knowledge about the *pha*-locus and PHA granule-associated proteins in *Chromatium vinosum*. *Biotechnol. Lett.*, **18**, 719–724.

27 Tessmer, N., König, S., Reichelt, R., Pötter, M., and Steinbüchel, A. (2007) Heat shock protein HspA and other proteins mimicking the function of phasins *sensu strictu* in recombinant strains of *Escherichia coli* accumulating polythioesters or polyhydroxyalkanoates. *Microbiology (SGM)*, **153**, 366–374.

28 Huisman, G.W., Wonink, E., de Koning, G.J.M., Preusting, H., and Witholt, B. (1992) Synthesis of poly(3-hydroxyalkanoates) by mutant and recombinant *Pseudomonas* strains. *Appl. Microbiol. Biotechnol.*, **38**, 1–5.

29 de Eugenio, L.I., P. Garcia, Luengo, J.M., Sanz, J.M., Roman, J.S., Garcia, J.L., and Prieto, M.A. (2007) Biochemical evidence that *phaZ* gene encodes a specific intracellular medium chain length polyhydroxyalkanoate depolymerase in *Pseudomonas putida* KT2442. *J. Biol. Chem.*, **282**, 4951–4962.

30 Jia, Y., Kappock, T.J., Frick, T., Sinskey, A.J., and Stubbe, J. (2000) Lipases provide a new mechanistic model for polyhydroxybutyrate (PHB) synthases: characterization of the functional residues in *Chromatium vinosum* PHB synthase. *Biochemistry*, **39**, 3927–3936.

31 Rehm, B.H.A., Antonio, R.V., Spiekermann, P., Amara, A.A., and Steinbüchel, A. (2002) Molecular characterization of the poly(3-hydroxybutyrate) (PHB) synthase from *Ralstonia eutropha*: in vitro evolution, site-specific mutagenesis and development of a PHB synthase protein model. *Biochim. Biophys. Acta*, **1594**, 178–190.

32 Gerngross, T.U., Snell, K.D., Peoples, O.P., Sinskey, A.J., Csuhai, E., Masamune, S., and Stubbe, J. (1994) Overexpression and purification of the soluble polyhydroxyalkanoate synthase from *Alcaligenes eutrophus*: evidence for a required posttranslational modification for catalytic activity. *Biochemistry*, **33**, 9311–9320.

33 Hoppensack, A., Rehm, B.H.A., and Steinbüchel, A. (1999) Analysis of 4-phosphopantetheinylation of polyhydroxybutyrate synthase from *Ralstonia eutropha*: generation of beta-alanine auxotrophic Tn5 mutants and cloning of the *panD* gene region. *J. Bacteriol.*, **181**, 1429–1435.

34 Jossek, R., Reichelt, R., and Steinbüchel, A. (1998) *In vitro* biosynthesis of poly(3-hydroxybutyric acid) by using purified poly(hydroxyalkanoic acid) synthase of *Chromatium vinosum*. *Appl. Microbiol. Biotechnol.*, **49**, 258–266.

35 Gerngross, T.U., and Martin, D.P. (1995) Enzyme-catalyzed synthesis of poly[(R)-(-)-3-hydroxybutyrate] – formation of macroscopic granules in-vitro. *Proc. Natl. Acad. Sci. U. S. A.*, **92**, 6279–6283.

36 Lenz, R.W., Farcet, C., Dijkstra, P.J., Goodwin, S., and Zhang, S.M. (1999) Extracellular polymerization of 3-hydroxyalkanoate monomers with the polymerase of *Alcaligenes eutrophus*. *Int. J. Biol. Macromol.*, **25**, 55–60.

37 Su, L., Lenz, R.W., and Takagi, Y. (2000) Enzymatic polymerization of (R)-3-hydroxyalkanoates by a bacterial polymerase. *Macromolecules*, **33**, 229–231.

38 Qi, Q., Steinbüchel, A., and Rehm, B.H.A. (2000) *In vitro* synthesis of poly(3-hydroxydecanoate): purification of type II polyhydroxyalkanoate synthases PhaC1 and PhaC2 from Pseudomonas aeruginosa and development of an enzyme assay. *Appl. Microbiol. Biotechnol.*, **54**, 37–43.

39 Jossek, R., and Steinbüchel, A. (1998) *In vitro* synthesis of poly(3-hydroxybutyric acid) by using an enzymatic coenzyme A recycling system. *FEMS Microbiol. Lett.*, **168**, 319–324.

40 Liu, S.-J., and Steinbüchel, A. (2000) Exploitation of butyrate kinase and phosphotransbutyrylase from *Clostridium acetobutylicum* for the *in vitro* biosynthesis of poly(hydroxyalkanoic acid). *Appl. Microbiol. Biotechnol.*, **53**, 545–552.

41 Liu, S.-J., and Steinbüchel, A. (2000) A novel genetically engineered pathway for synthesis of poly(hydroxyalkanoic acids) in *Escherichia coli*. *Appl. Environ. Microbiol.*, **66**, 739–743.

42 Steinbüchel, A. (2001) Perspectives for biotechnological production and utilization of biopolymers: metabolic engineering of polyhydroxyalkanoate biosynthesis pathways as a successful example. *Macromol. Biosci.*, **1**, 1–24.

43 Rehm, B.H.A., Krüger, N., and Steinbüchel, A. (1998) A new metabolic link between fatty acid de novo synthesis and polyhydroxyalkanoic acid synthesis. The *phaG* gene from *Pseudomonas putida* KT2440 encodes a 3-hydroxyacyl-acyl carrier protein-coenzyme A transferase. *J. Biol. Chem.*, **273**, 24044–24051.

44 Steinbüchel, A. (2005) Non-biodegradable biopolymers from renewable resources: perspectives and impacts. *Curr. Opin. Biotechnol.*, **16**, 607–613.

45 Shozui, F., Matsumoto, K., Nakai, T., Yamada, M., and Taguchi, S. (2010) Biosynthesis of novel terpolymers poly(lactate-co-3-hydroxybutyrate-co-3-hydroxyvalerate)s in lactate-overproducing mutant *Escherichia coli* JW0885 by feeding propionate as a precursor of 3-hydroxyvalerate. *Appl. Microbiol. Biotechnol.*, **85**, 949–954.

46 Matsumoto, K., and Taguchi, S. (2010) Enzymatic and whole-cell synthesis of lactate-containing polyesters: toward the complete biological production of polylactate. *Appl. Microbiol. Biotechnol.*, **85**, 921–932.

47 Moldes, C.P., García, J.L., and Prieto, M.A. (2004) *In vivo* immobilization of the fusion proteins on bioplastics by the novel tag BioF. *Appl. Environ. Microbiol.*, **70**, 3205–3212.

48 Grage, K., Jahns, A.C., Parlane, N., Palanisamy, R., Rasiah, I.A., Atwood, J.A., and Rehm, B.H.A. (2009) Bacterial polyhydroxyalcanoate granules: biogenesis, structure, and poteintial use as nano-/micro-beads in biotechnological and biomedical applications. *Biomacromolecules*, **10**, 660–669.

49 Krehenbrink, M., Oppermann-Sanio, F.B., and Steinbüchel, A. (2002) Evaluation of non-cyanobacterial genome sequences for occurrence of genes encoding proteins homologous to cyanophycin synthetase and cloning of an active cyanophycin synthetase from *Acinetobacter* sp. strain DSM 587. *Arch. Microbiol.*, **177**, 371–380.

50 Füser, G., and Steinbüchel, A. (2007) Analysis of genome sequences for genes of cyanophycin metabolism: identifying putative cyanophycin metabolizing prokaryotes. *Macromol. Biosci.*, **7**, 278–296.

51 Ziegler, K., Deutzmann, R., and Lockau, W. (2002) Cyanophycin synthetase-like enzymes of non-cyanobacterial Eubacteria: characterization of the polymer produced by a recombinant synthetase of *Desulfitobacterium hafniense*. *Z. Naturforsch.*, **57**, 522–529.

52 Simon, R.D. (1971) Cyanophycin granules from the blue-green alga *Anabaena cylindrica*: a reserve material consisting of copolymers of aspartic acid and arginine. *Proc. Natl. Acad. Sci. U. S. A.*, **68**, 265–267.

53 Simon, R.D., and Weathers, P. (1976) Determination of the structure of the novel polypeptide containing aspartic acid and arginine which is found in cyanobacteria. *Biochim. Biophys. Acta*, **420**, 165–176.

54 Hai, T., Oppermann-Sanio, F.B., and Steinbüchel, A. (1999) Purification and characterization of cyanophycin and cyanophycin synthetase from the thermophylic *Synechococcus* sp. MA19. *FEMS Microbiol. Lett.*, **181**, 229–236.

55 Simon, R.D. (1973) Measurement of the cyanophycin granule polypeptide contained in the blue-green alga *Anabaena cylindrica*. *J. Bacteriol.*, **114**, 1213–1216.

56 Steinle, A., Witthoff, S., Krause, J.P., and Steinbüchel, A. (2010) Establishment of cyanophycin biosynthesis in *Pichia pastoris* and optimization by use of engineered cyanophycin synthetases. *Appl. Environ. Microbiol.*, **76**, 1062–1070. doi: 10.1128/AEM.01659-09

57 Aboulmagd, E., Oppermann-Sanio, F.B., and Steinbüchel, A. (2000) Molecular characterization of the cyanophycin synthetase from *Synechocystis* sp. strain PCC 6308. *Arch. Microbiol.*, **174**, 297–306.

58 Mackerras, A.H., de Chazal, N.M., and Smith, G.D. (1990) Transient accumulations of cyanophycin in *Anabaena cylindrica* and *Synechocystis* 6308. *J. Gen. Microbiol.*, **136**, 2057–2065.

59 Stanier, R.Y., and Cohen-Bazire, G. (1977) Phototrophic prokaryotes: the cyanobacteria. *Ann. Rev. Microbiol.*, **31**, 225–274.

60 Weathers, P.J., Chee, H.L., and Allen, M.M. (1978) Arginine catabolism in *Aphanocapsa* 6308. *Arch. Microbiol.*, **118**, 1–6.

61 Hai, T., Ahlers, H., Gorenflo, H., and Steinbüchel, A. (2000) Axenic cultivation of anoxygenic phototrophic bacteria, cyanobacteria, and microalgae in a new closed tubular glass photobioreactor. *Appl. Microbiol. Biotechnol.*, **53**, 383–389.

62 Hühns, M., Neumann, K., Hausmann, T., Ziegler, K., Klemke, F., Kahmann, U., Staiger, D., Lockau, W., Pistorius, E.K., and Broer, I. (2008) Plastid targeting strategies for cyanophycin synthetase to achieve high-level polymer accumulation in *Nicotiana tabacum*. *Plant Biotechnol. J.*, **6**, 321–336.

63 Steinle, A., Oppermann-Sanio, F.B., Reichelt, R., and Steinbüchel, A. (2008) Synthesis and accumulation of cyanophycin in transgenic strains of *Saccharomyces cerevisiae*. *Appl. Environ. Microbiol.*, **74**, 3410–3418.

64 Hühns, M., Neumann, K., Hausmann, T., Klemke, F., Lockau, W., Kahmann, U., Kopertekh, L., Staiger, D., Pistorius, E.K., Reuther, J., Waldvogel, E., Wohlleben, W., Effmert, M., Junghans, H., Neubauer, K., Kragl, U., Schmidt, K., Schmidtke, J., and Broer, I. (2009) Tuber-specific *cphA* expression to enhance cyanophycin production in potatoes. *Plant Biotechnol. J.*, **7**, 883–898.

65 Elbahloul, Y., Krehenbrink, M., Reichelt, R., and Steinbüchel, A. (2005) Physiological conditions conducive to high cyanophycin content in biomass of *Acinetobacter calcoaceticus* strain ADP1. *Appl. Environ. Microbiol.*, **71**, 858–866.

66 Voss, I., and Steinbüchel, A. (2006) Application of a KDPG-aldolase gene-dependent addiction system for enhanced production of cyanophycin in *Ralstonia eutropha* strain H16. *Metab. Eng.*, **8**, 66–78.

67 Suarez, C., Kohler, S.J., Allen, M.M., and Kolodny, N.H. (1999) NMR study of the metabolic ^{15}N isotopic enrichment of cyanophycin synthesized by the cyanobacterium

Synechocystis sp. strain PCC 6308. Biochim. Biophys. Acta, **1426**, 429–438.

68 Erickson, N.A., Kolodny, N.H., and Allen, M.M. (2001) A rapid and sensitive method for the analysis of cyanophycin. Biochim. Biophys. Acta, **1526**, 5–9.

69 Steinle, A., Bergander, K., and Steinbüchel, A. (2009) Metabolic engineering of Saccharomyces cerevisiae for production of novel cyanophycins with an extended range of constituent amino acids. Appl. Environ. Microbiol., **75**, 3437–3446.

70 Simon, R.D., Lawry, N.H., and Mc Lendon, G.L. (1980) Structural characterization of the cyanophycin granule polypeptide of Anabaena cylindrica by circular dichroism and Raman spectroscopy. Biochim. Biophys. Acta, **626**, 277–281.

71 Merritt, M.V., Sid, S.S., Mesh, L., and Allen, M.M. (1994) Variations in the amino acid composition of cyanophycin in the cyanobacterium Synechocystis sp. PCC 6308 as a function of growth conditions. Arch. Microbiol., **162**, 158–166.

72 Wingard, L.L., Miller, S.R., Sellker, J.M., Stenn, E., Allen, M.M., and Wood, A.M. (2002) Cyanophycin production in a phycoerythrin-containing marine Synechococcus strain of unusual phylogenetic affinity. Appl. Environ. Microbiol., **68**, 1772–1777.

73 Berg, H., Ziegler, K., Piotukh, K., Baier, K., Lockau, W., and Volkmer-Engert, R. (2000) Biosynthesis of the cyanobacterial reserve polymer multi-L-arginyl-poly-L-aspartic acid (cyanophycin). Mechanism of the cyanophycin synthetase reaction studied with synthetic primers. Eur. J. Biochem., **267**, 5561–5570.

74 Aboulmagd, E., Oppermann-Sanio, F.B., and Steinbüchel, A. (2001) Purification of Synechocystis sp. strain PCC 6308 cyanophycin synthetase and its characterization with respect to substrate and primer specificity. Appl. Environ. Microbiol., **67**, 2176–2182.

75 Neumann, K., Stephan, D.P., Ziegler, K., Hühns, M., Broer, I., Lockau, W., and Pistorius, E.K. (2005) Production of cyanophycin, a suitable source for the biodegradable polymer polyaspartate, in transgenic plants. Plant Biotechnol. J., **3**, 249–258.

76 Hai, T., Oppermann-Sanio, F.B., and Steinbüchel, A. (2002) Molecular characterization of a thermostable cyanophycin synthetase from the thermophylic cyanobacterium Synechococcus sp. MA19 and in vitro synthesis of cyanophycin and related polyamides. Appl. Environ. Microbiol., **68**, 93–101.

77 Ziegler, K., Diener, A., Herpin, C., Richter, R., Deutzmann, R., and Lockau, W. (1998) Molecular characterization of cyanophycin synthetase, the enzyme catalyzing the biosynthesis of the cyanobacterial reserve material multi-L-arginyl-poly-L-aspartate (cyanophycin). Eur. J. Biochem., **254**, 154–159.

78 Fan, C., Moews, P.C., Shi, Y., Walsh, C.T., and Knox, J.R. (1995) A common fold for peptide synthetases cleaving ATP to ADP: glutathion synthetase and D-alanine:D-alanine ligase of Escherichia coli. Proc. Natl. Acad. Sci. U. S. A., **92**, 1172–1176.

79 Bouhss, A., Mengin-Lecreulx, D., Blanot, D., van Heijenoort, J., and Parquet, C. (1997) Invariant amino acids in the Mur peptide synthetases of bacterial peptidoglycan synthesis and their modification by site-directed mutagenesis in the UDP-MurNAc:l-Alanine ligase from Escherichia coli. Biochemistry, **36**, 11556–11563.

80 Bertrand, J.A., Auger, G., Fanchon, E., Martin, L., Blanot, D., Le Beller, D., van Heijenoort, J., and Dideberg, O. (1999) Determination of the MurD mechanism through crystallographic analysis of enzyme complexes. J. Mol. Biol., **289**, 579–590.

81 Bertrand, J.A., Auger, G., Martin, L., Fanchon, E., Blanot, D., van Heijenoort, J., and Dideberg, O. (1997) Crystal structure of UDP-N-acetylmuramoyl-l-alanine:D-glutamate ligase from Escherichia coli. EMBO J., **16**, 3416–3425.

82 Simon, R.D. (1976) The biosynthesis of multi-l-arginyl-poly(L-aspartic acid) in the filamentous cyanobacterium *Anabaena cylindrica*. *Biochim. Biophys. Acta*, **422**, 407–418.

83 Arai, T., and Kino, K. (2008) A cyanophycin synthetase from *Thermosynechococcus elongatus* BP-1 catalyzes primer-independent cyanophycin synthesis. *Appl. Microbiol. Biotechnol.*, **81**, 69–78.

84 Krehenbrink, M., and Steinbüchel, A. (2004) Partial purification and characterization of a non-cyanobacterial cyanophycin synthetase from *Acinetobacter calcoaceticus* strain ADP1 with regard to substrate specificity, substrate affinity and binding to cyanophycin. *Microbiology*, **150**, 2599–2608.

85 Borzi, A. (1887) Le comunicazioni intracellulari delle *Nostochinee*. *Malpighia*, **1**, 28–74.

86 Koop, A., Voss, I., Thesing, A., Kohl, H., Reichelt, R., and Steinbüchel, A. (2007) Identification and localization of cyanophycin in bacteria cells via imaging of the nitrogen distribution using energy-filtering transmission electron microscopy. *Biomacromolecules*, **8**, 2675–2683.

87 Berg, H. (2003) Untersuchungen zu Funktion und Struktur der Cyanophycin-Synthetase von Anabaena variabilis ATCC 29413. Dissertation. Humboldt-Universität zu Berlin, Germany.

88 Steinle, A., and Steinbüchel, A. (2008) Cyanophycin synthetases, in *Protein Engineering Handbook* (eds S. Lutz and U.T. Bornscheuer), Wiley-VCH Verlag GmbH, Weinheim, Germany, pp. 829–848.

89 Galperin, M.Y., and Koonin, E.V. (1997) A diverse superfamily of enzymes with ATP-dependent carboxylate-amine/thiol ligase activity. *Protein Sci.*, **6**, 2639–2643.

90 Hibi, T., Nishioka, T., Kato, H., Tanizawa, K., Fukui, T., Katsube, Y., and Oda, J. (1996) Structure of the multifunctional loops in the nonclassical ATP-binding fold of glutathione synthetase. *Nat. Struct. Biol.*, **3**, 16–18.

91 Kondo, S., Nakajima, Y., Sugio, S., Yong-Biao, J., Sueda, S., and Kondo, H. (2004) Structur of the biotin carboxylase subunit of pyruvate cyarboxylase from *Aquifex aeolicus* at 2.2 Å resolution. *Acta Crystallogr. D*, **60**, 486–492.

92 Matsuda, K., Mizuguchi, K., Nishioka, T., Kato, H., Go, N., and Oda, J. (1996) Crystal structure of glutathione synthetase at optimal pH: domain architecture and structural similarity with other proteins. *Protein Eng.*, **9**, 1083–1092.

93 Joyce, M.A., Fraser, M.E., James, M.N.G., Bridger, W.A., and Wolodko, W.T. (2000) ADP-binding site of *Escherichia coli* succinyl-CoA synthetase revealed by X-ray crystallography. *Biochemistry*, **39**, 17–25.

94 Thoden, J.B., Holden, H.M., Wesenberg, G., Raushel, F.M., and Rayment, I. (1997) Structure of carbamoyl phosphate synthetase: a journey of 96 Å from substrate to product. *Biochemistry*, **36**, 6305–6316.

95 Eveland, S.S., Pompliano, D.L., and Anderson, M.S. (1997) Conditionally lethal *Escherichia coli* murein mutants contain point defects that map to regions conserved among murein and folyl poly-γ-glutamate ligases: identification of a ligase superfamily. *Biochemistry*, **36**, 6223–6229.

96 Sun, X., Bognar, A.L., Baker, E.N., and Smith, C.A. (1998) Structural homologies with ATP and folate-binding enzymes in the crystal structure of folylpolyglutamate synthetase. *Proc. Natl. Acad. Sci. U. S. A.*, **95**, 6647–6652.

97 Dementin, S., Bouhss, A., Auger, G., Parquet, C., Mengin-Lecreulx, D., Dideberg, O., van Heijenoort, J., and Blanot, D. (2001) Evidence of a functional requirement for a carbamoylated lysine residue in MurD, MurE and MurF synthetases as established by chemical rescue experiments. *Eur. J. Biochem.*, **268**, 5800–5807.

98 Bouhss, A., Dementin, S., Parquet, C., Mengin-Lecreulx, D., Bertrand, J.A., Le Beller, D., Dideberg, O., can

Heijenoort, J., and Blanot, D. (1999) Role of the ortholog and paralog amino acid invariants in the active site of the UDP-MurNac-L-alanine:D-glutamate ligase (MurD). *Biochemistry*, **38**, 12240–12247.

99 Tanner, M.E., Vaganay, S., van Heijenoort, J., and Blanot, D. (1996) Phosphinate inhibitors of the D-glutamic acid-adding enzyme of peptidoglycan biosynthesis. *J. Org. Chem.*, **61**, 1756–1760.

100 Gegnas, L.D., Waddell, S.T., Chabin, R.M., Reddy, S., and Wong, K.K. (1998) Inhibitors of the bacterial cell wall biosynthesis enzyme MurD. *Bioorg. Med. Chem. Lett.*, **8**, 1643–1648.

101 Saraste, M., Sibbald, P.R., and Wittinghofer, A. (1990) The P-Loop – a common motif in ATP- and GTP-binding proteins. *Trends Biochem. Sci.*, **15**, 430–434.

102 Yan, Y., Munshi, S., Li, Y., Pryor, K.A.D., Marsilio, F., and Leiting, B. (1999) Crystallization and preliminary X-ray analysis of the *Escherichia coli* UDP-MurNAc-tripeptide d-alanyl-d-alanine-adding enzyme (MurF). *Acta Crystallogr. D Biol. Crystallogr.*, **55**, 2033–2034.

103 Gordon, E., Flouret, B., Chantalat, L., van Heijenoort, J., Mengin-Lecreulx, D., and Dideberg, O. (2001) Crystal structure of UDP-N-acetylmuramoyl-L-alanyl-D-glutamate: *meso*-diaminopimelate ligase from *Escherichia coli*. *J. Biol. Chem.*, **276**, 10999–11006.

104 Yan, Y., Munshi, S., Leiting, B., Anderson, M.S., Chrzas, J., and Chen, Z. (2000) Crystal structure of *Escherichia coli* UDP-MurNAc-tripeptide D-alanyl-D-alanine-adding enzyme (MurF) at 2.3 Å resolution. *J. Mol. Biol.*, **304**, 435–445.

105 Mol, C.D., Brooun, A., Dougan, D.R.M., Hilgers, T., Tari, L.W., Wijnands, R.A., Knuth, M.W., McRee, D.E., and Swanson, R.V. (2003) Crystal structures of active fully assembled substrate- and product-bound complexes of UDP-*N*-acetyl-muramic acid:L-alanine-ligase (MurC) from *Haemophilus influenza*. *J. Bacteriol.* **185**, 4152–4162.

106 Bergendahl, V., Linne, U., and Marahiel, M.A. (2002) Mutational analysis of the C-domain in nonribosomal peptide synthesis. *Eur. J. Biochem.*, **269**, 620–629.

107 Hai, T., Frey, K.M., and Steinbüchel, A. (2006) Activation of cyanophycin synthetase of *Nostoc ellipsosporum* strain NE1 by truncation at the carboxy-terminal region. *Appl. Microbiol. Biotechnol.*, **72**, 7652–7660.

108 Amara, A.A., Steinbüchel, A., and Rehm, B.H.A. (2002) *In vivo* evolution of the Aeromonas punctata polyhydroxyalkanoate (PHA) synthase: isolation and characterization of modified PHA synthases with enhanced activity. *Appl. Microbiol. Biotechnol.*, **59**, 477–482.

109 Park, S.R., Cho, S.J., Kim, M.K., Ryu, S.K., Lim, W.J., An, C.L., Hong, S.Y., Kim, J.H., Kim, H., and Yun, H.D. (2002) Activity enhancement of Cel5Z from *Pectobacterium chrysanthemi* PY35 by removing C-terminal region. *Biochem. Biophys. Res. Commun.*, **22**, 425–430.

110 Lim, W.J., Hong, S.Y., An, C.L., Cho, K.M., Choi, B.R., Kim, Y.K., An, J.M., Kang, J.M., Lee, S.M., Cho, S.J., Kim, H., and Yun, H.D. (2005) Construction of minimum size cellulase (Cel5Z) *Pectobacterium chrysanthemi* PY35 by removal of the C-terminal region.

111 Hai, T., Lee, J.-S., Kim, T.-J., and Suh, J.-W. (2009) The role of the C-terminal region of cyanophycin synthetase from Nostoc ellipsosporum NE1 in its enzymatic activity and thermostability: a key function of Glu^{856}. *Biochim. Biophys. Acta*, **1794**, 42–49.

112 Law, A.M., Lai, S.W.S., Tavares, J., and Kimber, M.S. (2009) The structural basis of β-peptide-specific cleavage by the serine protease cyanophycinase. *J. Mol. Biol.*, **392**, 393–404.

113 Kroll, J., Steinle, A., Reichelt, R., Ewering, C., and Steinbüchel, A. (2009) Establishment of a novel anabolism-based addiction system with an artificially introduced mevalonate pathway: complete stabilization of plasmids as universal application in white biotechnology. *Metab. Eng.*, **11**, 168–177.

114 Glick, B.R. (1995) Metabolic load and heterologous gene expression. *Biotechnol. Adv.*, **13**, 247–261.

115 Diniz, C.S., Voss, I., and Steinbüchel, A. (2006) Optimization of cyanophycin production in recombinant strains of *Pseudomonas putida* and *Ralstonia eutropha* employing elementary mode analysis and statistical experimental design. *Biotechnol. Bioeng.*, **93**, 698–717.

116 Sallam, A., Steinle, A., and Steinbüchel, A. (2009) Cyanophycin: biosynthesis and applications, in *Microbial Production of Biopolymers and Polymer Precursors* (ed. B.H.A. Rehm), Caister Academic Press, Norfolk, UK, pp. 79–99.

117 Mooibroek, H., Oosterhuis, N., Giuseppin, M., Toonen, M., Franssen, H., Scott, E., Sanders, J., and Steinbüchel, A. (2007) Assessment of technological options and economical feasibility for cyanophycin biopolymer and high-value amino acid production. *Appl. Microbiol. Biotechnol.*, **77**, 257–267.

118 Sanders, J., Scott, E., Weusthuis, R., and Mooibroek, H. (2007) Bio-Refinery as the Bio-inspired process to bulk chemicals. *Macromol. Biosci.*, **7**, 105–117.

119 Scott, E., Peter, F., and Sanders, J. (2007) Biomass in the manufacture of industrial products – the use of proteins and amino acids. *Appl. Microbiol. Biotechnol.*, **75**, 751–762.

120 Désormeaux, A., Moreau, M.E., Lepage, Y., Chanard, J., and Adam, A. (2008) The effect of electronegativity and angiotensin-converting enzyme inhibition on the kinin-forming capacity of polyacrylonitrile dialysis membranes. *Biomaterials*, **29**, 1139–1146.

121 Sallam, A., Kast, A., Przybilla, S., Meiswinkel, T., and Steinbüchel, A. (2009) Process for biotechnological production of β-dipeptides from cyanophycin at technical scale and its optimization. *Appl. Environ. Microbiol.*, **75**, 29–38.

122 Sallam, A., and Steinbüchel, A. (2009) Cyanophycin-degrading bacteria in digestive tracts of mammals, birds and fish and consequences for possible applications of cyanophycin and its dipeptides in nutrition and therapy. *J. Appl. Microbiol.*, **107**, 474–484.

123 Joentgen, W., Groth, T., Steinbüchel, A., Hai, T., and Oppermann, F.B. (1998) Polyaspartic acid homopolymers and copolymers, biotechnical production and use thereof. Patent application WO 98/39090.

124 Conrad, U. (2005) Polymers from plants to develop biodegradable plastics. *Trends Plant Sci.*, **10**, 511–512.

125 Joentgen, W., Müller, N., Mitschker, A., and Schmidt, H. (2003) Polyaspartic acids, in *Biopolymers: Polyamides and Complex Proteinaceous Materials I*, vol. 7 (eds S.R. Fahnenstock and A. Steinbüchel), Wiley-VCH Verlag GmbH, Weinheim, Germany, pp. 175–199.

126 Pohlmann, A., Fricke, W.F., Reinecke, F., Kusian, B., Liesegang, H., Cramm, R., Eitinger, T., Ewering, C., Pötter, M., Schwartz, E., Strittmatter, A., Voß, I., Gottschalk, G., Steinbüchel, A., Friedrich, B., and Bowien, B. (2006) Hydrogen-based biotechnology: genome sequence of the bioplastic-producing 'Knallgas' bacterium *Ralstonia eutropha* H16. *Nat. Biotechnol.*, **24**, 1257–1262.

127 Reinecke, F., and Steinbüchel, A. (2009) *Ralstonia eutropha* strain H16 as model organism for PHA metabolism and for biotechnological production of technically interesting biopolymers. *J. Mol. Microbiol. Biotechnol.*, **16**, 91–108.

128 Raberg, M., Reinecke, F., Reichelt, R., Malkus, U., König, S., Pötter, M., Fricke, W.F., Pohlmann, A., Voigt, B., Hecker, M., Friedrich, B., Bowien, B., and Steinbüchel, A. (2008) *Ralstonia eutropha* H16 flagellation changes according to nutrient supply and state of poly(3-hydroxybutyrate) accumulation. *Appl. Environ. Microbiol.*, **74**, 4477–4490.

11
Chiral Polymers by Lipase Catalysis
Anja Palmans and Martijn Veld

11.1
Introduction

A wide variety of chemical catalysts is nowadays available to polymerize monomers into well-defined polymers and polymer architectures that are applicable in advanced materials for example, as biomedical applications and nanotechnology. However, synthetic polymers rarely possess well-defined stereochemistries in their backbones. This sharply contrasts with the polymers made by nature where perfect stereocontrol is the norm. An interesting exception is poly-L-lactide, a polyester that is used in a variety of biomedical applications [1]. By simply playing with the stereochemistry of the backbone, properties ranging from a semi-crystalline, high melting polymer (poly-L-lactide) to an amorphous high T_g polymer (poly-meso-lactide) have been achieved [2].

The synthetic synthesis of known chiral polymers mostly starts from optically pure monomers obtained form the chiral pool. The optically pure fermentation product L-lactic acid, for example, is the starting material for the synthesis of poly(L-lactide). However, converting a *racemic* or achiral monomer *quantitatively* into a homochiral polymer is less straightforward [3]. This is surprising considering the enormous potential of biocatalysis and tandem catalysis that has emerged in the past decades to prepare optically active intermediates [4].

Enzymes are perfectly equipped to convert substrates into products in high enantio-, regio-, or chemoselectivity, a property that is commonly used in industry to prepare optically active fine-chemical intermediates [5]. More specifically, lipases appeared as ideal catalysts as a result of their high enantioselectivity, broad substrate scope and stability. In addition, lipases are powerful catalysts for the preparation of polyesters, polycarbonates and even polyamides, as is reviewed in Chapters 4 and 5 of this book. Moreover, a variety of different polymer architectures such as block copolymers, graft copolymers etc have been prepared using lipases as the catalyst (see Chapter 12).

In this chapter we will summarize the recent achievements in the preparation of *chiral* polymers and *chiral* polymer architectures starting from both natural and non-natural monomers using lipase catalysis. First, the catalytic mechanism

Biocatalysis in Polymer Chemistry. Edited by Katja Loos
Copyright © 2011 WILEY-VCH Verlag GmbH & Co. KGaA, Weinheim
ISBN: 978-3-527-32618-1

of lipases and the origin of their high enantioselectivity are discussed. Then, the polymerization of optically pure monomers by lipase catalysis is discussed, followed by the enantioselective kinetic resolution polymerization of racemic monomers. The limitations of enantioselective polycondensations can be overcome by introducing a continuous substrate racemization step, referred to as dynamic kinetic resolution polymerization (DKRP). Finally, the use of chirality as a tool to influence material properties and the use of enantioselective lipase-catalyzed post-modification are discussed.

11.2
Reaction Mechanism and Enantioselectivity of Lipases

Lipases belong to the subclass of α/β-hydrolases and their structure and reaction mechanism are well understood (see also Chapter 14). The common enzyme fold is characterized by an α-helix that is connected with a sharp turn, referred to as the nucleophilic elbow, to the middle of an β-sheet array [6]. All lipases possess an identical catalytic triad consisting of an aspartate or glutamate, a histidine and a nucleophilic serine residue [6]. The latter residue is located at the nucleophilic elbow and is found in the middle of the highly conserved Gly-AA$_1$-Ser-AA$_2$-Gly sequence in which AA$_1$ and AA$_2$ can vary [7]. The histidine residue is spatially located at one side of the serine residue, whereas at the opposite of the serine a negative charge can be stabilized in the so-called oxyanion hole by a series of hydrogen bond interactions. Below the reaction mechanism of the class of α/β-hydrolases is briefly discussed using *Candida antarctica* lipase B (CALB) as a typical example.

The catalytic triad of CALB consists of Asp187, His224, and Ser105 while the oxyanion hole is formed by the backbone amide protons of Thr40 and Gln106 and the side chain of Thr40 [8]. First, a substrate reversibly complexes to the free enzyme (Figure 11.1 top left) thereby forming a Michaelis–Menten complex. After correct positioning of the substrate, a nucleophilic attack of Ser105 onto the substrate carbonyl group occurs and a first tetrahedral intermediate is formed (Figure 11.1 top right). In this tetrahedral intermediate, the negative charge on the former substrate carbonyl oxygen is stabilized by threefold hydrogen bonding interaction with the oxyanion hole, whereas the positive charge on His224 is stabilized by interaction with Asp187. Subsequently, proton transfer from His224 to the substrate alkyl oxygen happens and the alcohol part of the residue is liberated from the enzyme. As a result, a covalently bound acyl enzyme intermediate is formed at the end of the acylation step (Figure 11.1 bottom right). Subsequently, the acyl enzyme intermediate is deacylated by an incoming nucleophile R'NuH, which generally is water, an alcohol or an amine. A second tetrahedral intermediate is formed by attack of the nucleophile onto the acyl enzyme carbonyl group (Figure 11.1 bottom left). In this process, the proton is transferred from the nucleophile to the His224 residue and the positive and negative charges are effectively stabilized by Asp187 and the oxyanion hole, respectively. Then, the proton is

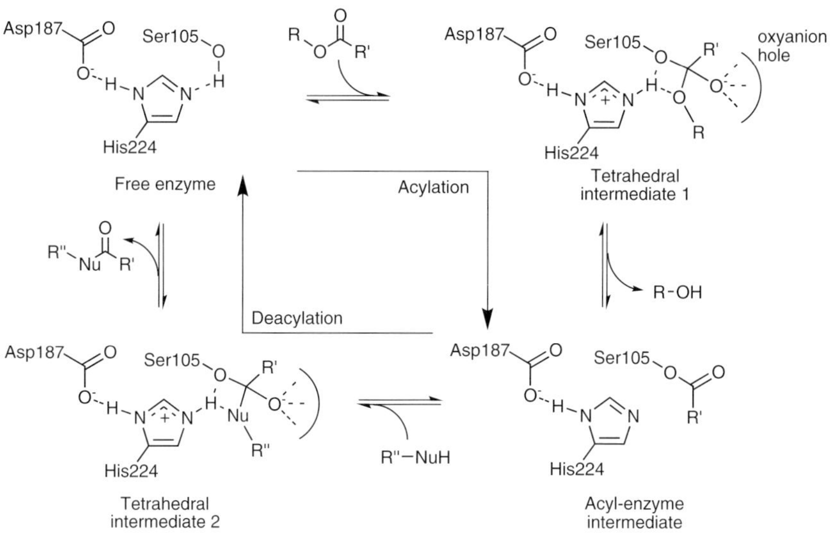

Figure 11.1 Catalytic mechanism of CALB showing an acylation and deacylation step and the formation of a covalently bound acyl-enzyme intermediate (bottom right) [9].

transferred from the His224 residue to the Ser105 alkyl oxygen while restoring the carbonyl bond of the bound substrate. As a result, a weakly bound enzyme–product complex is formed and the free enzyme species is regenerated after release of the reaction product.

Although the naturally occurring fatty acid ester substrates of lipases are achiral in nature, lipases can show excellent enantioselectivity in the deacylation step when a chiral center is present in the nucleophile. The chiral nature of the amino acids and the unique three-dimensional spatial organization of the catalytic residues favor the deacylation the acyl-enzyme intermediate by one of the nucleophile enantiomers. Especially when the chiral center is close to the nucleophilic site, the degree of enantioselectivity can be high. In these cases, the enantiopreference of lipases is well understood and depends on the relative size of the substituents at the chiral center. Generally speaking lipases prefer R-nucleophiles over S-nucleophiles [10]. Serine proteases, which also possess an α/β-hydrolase fold, have a mirror image arrangement of the catalytic triad compared to lipases and therefore they show opposite enantioselectivity compared to lipases [11]. The ratio of the reaction rates for the R- and S-enantiomer of a nucleophile (k_R/k_S) in the deacylation of an achiral acyl-enzyme intermediate is conveniently referred to as the enantiomeric ratio or E-ratio. If the differences in steric demands of the substituents at the chiral center are sufficiently large, high E-ratios can be observed, meaning that only the preferred enantiomer will react. This will result in products that show a high enantiomeric excess (ee). Chiral substrates in which the center of chirality is located more distant from the reactive functional groups can also be recognized by lipases. However, in these cases the preferred enantioselectivity is often lower and more difficult to predict *a priori* [12].

11.3
Lipase-catalyzed Synthesis and Polymerization of Optically Pure Monomers

One of the easiest ways to prepare chiral polymers is the polymerization of optically pure monomers. These monomers can be derived or isolated from the chiral pool, be synthesized from prochiral substrates using asymmetric catalysis, or be obtained by lipase-catalyzed resolution of a racemate followed by further synthetic manipulation.

Only few examples from literature report on the use of lipases for the preparation of optically pure, chiral monomers. Trollsås *et al.* [13] used Novozym® 435, which consists of CALB immobilized on a macroporous polyacrylic resin, as enantioselective catalyst in the kinetic resolution of 2-hydroxy-9-decene. This optically pure alcohol was subsequently transformed in many synthetic steps into bisacrylate monomer **3** (Figure 11.2a), which was subjected to radical polymerization. A second example shows the use of a *Candida rugosa* lipase (CRL)-catalyzed kinetic resolution of oxanorbornene **2** (Figure 11.2b) [14]. After isolation of the optically pure residual substrate *R*-**2** and structural modification, ring opening metathesis polymerization was performed. The kinetic resolution of α-methyl-β-propiolactone (**4**) by *Pseudomonas fluorescens* lipase (PFL) in diethyl ether containing MeOH gave *R*-**4** with 93% *ee* (Figure 11.2c) [15]. The isolated lactone was subsequently subjected to anionic polymerization. Racemic 6-methyl caprolactone (**6**) has also been subjected to a kinetic resolution procedure using 4-*t*-butylbenzylalcohol (Figure 11.2d) [16]. The residual *R*-lactone was isolated with 98.8% *ee*, while the ester product was enzymatically ring-closed again and afforded the *S*-**6** in 99.6% *ee*. A final example of lipase catalysis for the preparation of optically pure monomers is the *Chromobacterium viscosum* lipase (CVL)-catalyzed selective hydrolysis of racemic diester **7** (Figure 11.2e) [17]. The lipase showed the preferential hydrolysis of the ester at the most substituted phenyl ring of the *S*-enantiomer (92% *ee*). Chemical hydrolysis of the second hydroxy group afforded enantiopure diol **8**, which could be used for chemical polymerization.

The use of lipase-catalyzed kinetic resolution as key step in monomer synthesis is a powerful approach for the synthesis of optically-enriched chiral polymers. An alternative route towards the synthesis of chiral polymers consists of the enzymatic polymerization of optically pure monomers.

The enzymatic polymerization of optically pure D,D-, or L,L-lactide and meso D,L-lactide for the preparation of biocompatible and biodegradable poly(lactide) has received a lot of attention. Only the polymerization reactions of D,D-, and L,L-lactide will be discussed here as only these polymers are chiral in nature. Matsumura *et al.* reported the first lipase-catalyzed preparation of poly(L-lactide) (**10**) (Scheme 11.1) [18]. Polymerization reactions were performed in bulk at temperatures between 80 and 130 °C for 7 days using *Pseudomonas cepacia* lipase

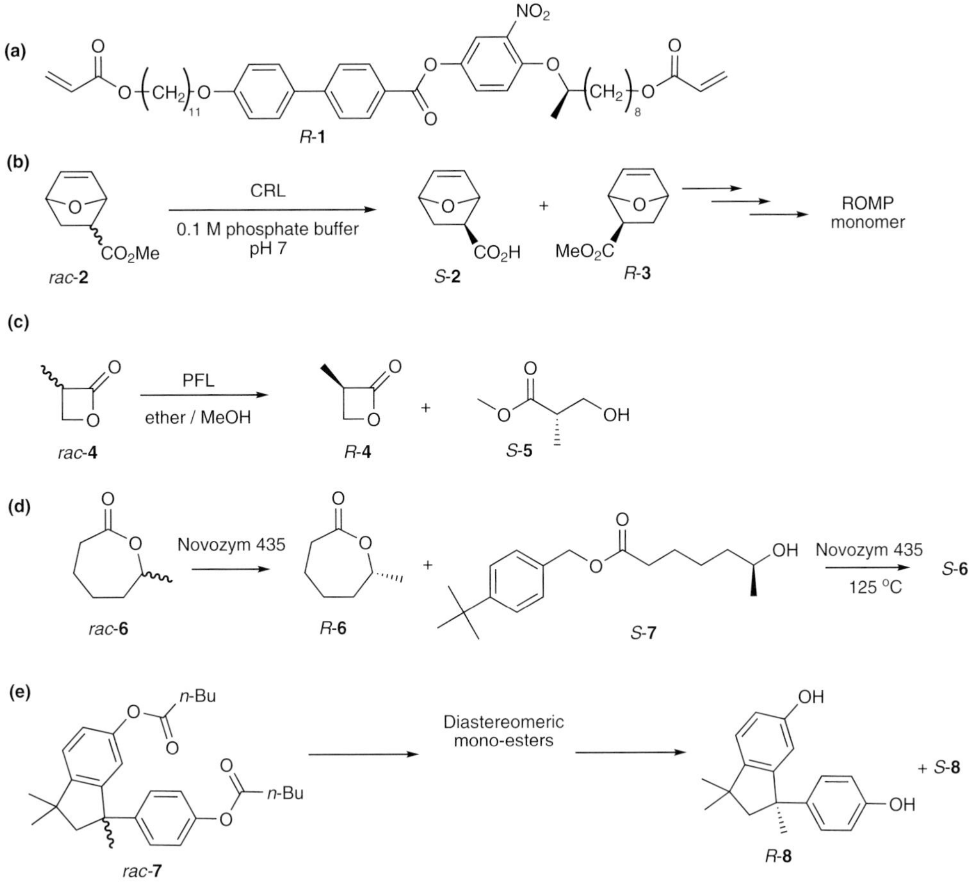

Figure 11.2 Examples of lipase-catalyzed kinetic resolution for the preparation of optically pure monomers that were subsequently chemically polymerized.

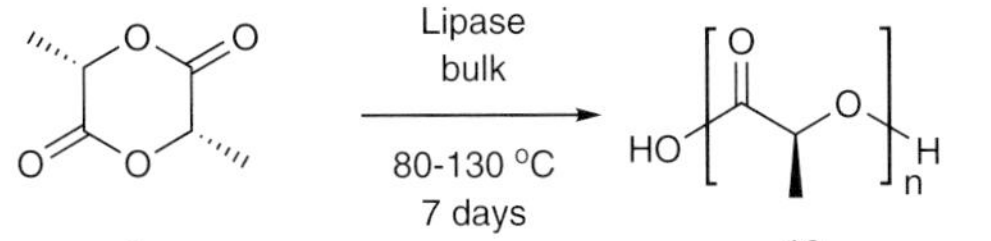

Scheme 11.1 Lipase-catalyzed synthesis of poly(L-lactide) (**10**).

(PCL) as catalyst [18]. Isolated polymer yields after precipitation based on weight varied between 5 and 10%, while M_w values generally were between 13 and 59 kg mol^{-1} having polydispersity index (PDI) values of 1.1–1.2.

Copolymerization of lactides catalyzed by porcine pancreatic lipase (PPL) with cyclic trimethylene carbonate to give random copolymers with M_w up to 21 kg mol^{-1} and a PDI of 1.3–1.4 has been shown as well [19]. While CALB was shown to be

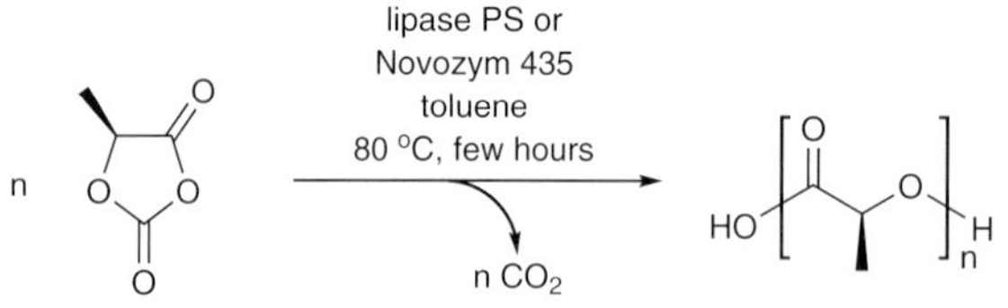

Scheme 11.2 Lipase PS or Novozym 435 initiated polymerization of L-lacOCA to give PLLA.

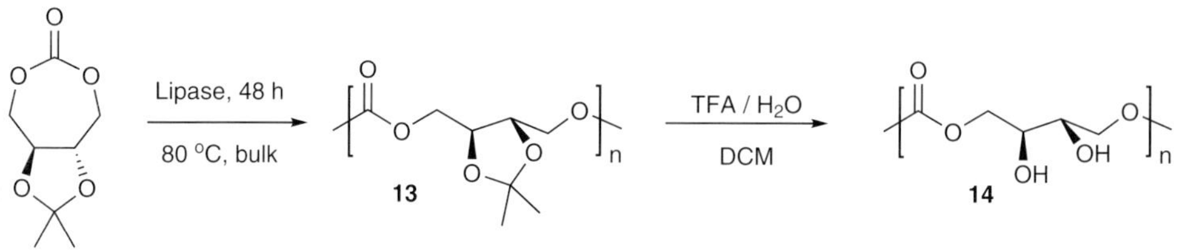

Scheme 11.3 Lipase-catalyzed polymerization of L-tartaric acid derive cyclic carbonate followed by deprotection of the ketal groups to give chiral, hydroxy functional polycarbonate.

inactive in the polymerization of L-lactide, the same enzyme gave low molecular weight polymers (M_n = 3.3 kg mol^{-1}, PDI = 1.2) after 3 days when using D-lactide as the substrate [20]. This difference in reactivity has been attributed to the R-configuration of the secondary alcohol obtained after ring opening of the D-lactide, whereas a non-propagating alcohol of the S-configuration is formed for L-lactide. Later on, the copolymerization of L-lactide with glycolide using both free and immobilized *Burkholderia cepacia* lipase (BCL) was investigated [21]. At reaction temperature of 100–130 °C, copolymers with small amounts of glycolyl repeat units were formed according to careful MALDI-TOF-MS analysis. For polymerization at 100 °C, M_w and PDI values were higher when using free enzyme (11.7 kg mol^{-1}, PDI = 1.8) compared to immobilized lipase (8.7 kg mol^{-1}, PDI = 1.5).

An alternative approach for the preparation of chiral poly(lactide) consists of the enzyme promoted ring opening of optically pure lactic acid derived O-carboxy anhydrides (lacOCA) (Scheme 11.2) [22]. Both PCL and Novozym 435 were able to induce the polymerization lacOCA, which was driven to completion by release of CO_2. Complete monomer conversion was observed within 4 h using 6 wt% of enzyme. Both enzymes gave molecular weight distributions with M_w values up to 38.4 kg mol^{-1} and a PDI as low as 1.2–1.3. The polymerization of pure L-lacOCA was faster than D-lacOCA and gave chiral PLLA and PDLA, respectively. When *rac*-lacOCA was used as monomer, no enantiopreference was observed and atactic PLA was obtained. Moreover, similarly low PDI values were observed for lacOCA polymerization using DMAP and an alcohol as initiator [23], suggesting that the polymerization process is only enzyme initiated and not enzyme-catalyzed.

Chiral, hydroxy functional polycarbonates were prepared by the enzymatic ring opening polymerization (ROP) of a seven-membered cyclic carbonate (**12**) derived from L-tartaric acid (Scheme 11.3) [24]. Among others Novozym 435, PFL, and

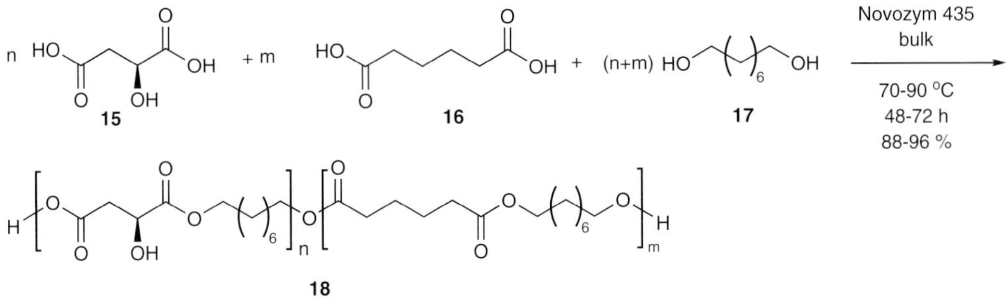

Scheme 11.4 Novozym 435-catalyzed terpolymerization of L-malic acid, adipic acid and octane-1,8-diol in bulk affording random terpolymers.

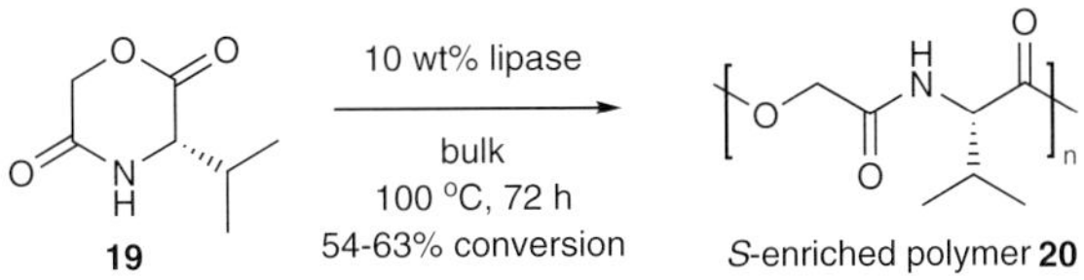

Scheme 11.5 Lipase-catalyzed ROP of (S)-3-isopropyl morpholino-2,5-dione giving enantiomerically enriched poly(depsipeptides).

PCL could polymerize this carbonate. Novozym 435 gave the highest molecular weight polymers (M_n = 5.5 kg mol^{-1}; PDI = 1.7) after 24 h. Deprotection of the ketal functional groups was possible without negatively affecting the molecular weight of the polymer to give hydroxy functional polymer **14** in good yield.

Another approach towards the synthesis of chiral, hydroxy functional polyesters consists of the Novozym 435-catalyzed terpolymerization of L-malic acid (**15**), adipic acid (**16**) and octane-1,8-diol (**17**) in bulk afforded copolyesters with chiral, pendant hydroxy groups (Scheme 11.4) [25]. According to ^{1}H-NMR, incorporation of the malic acid in the polymer was exclusively via the two carboxylic acid groups, showing a high degree of selectivity. Upon increasing the L-malic acid content from 0 to 20%, the molecular weight of the polymers dropped from 9.5 to 4.7 kg mol^{-1} while PDI values for all polymerizations were between 1.50 and 1.96.

The synthesis of chiral poly(depsipeptides), polymers with alternating amide and ester bonds, by lipase-catalyzed ring opening of 3-isopropyl morpholino-2,5-dione (**19**) was shown by Höcker and coworkers (Scheme 11.5) [26]. Various lipases were tested for the bulk polymerization of these heterocyclic monomers at temperatures of 100 °C or above. PPL and lipase type III from a pseudomonas species were shown to be effective catalysts. The isolated polymers showed M_n values of 3.5–17.5 kg mol^{-1}. The influence of reaction temperature, the amount of enzyme and the presence of water in the reaction medium were shown to be important factors on the high molecular weight fraction and were investigated in detail [26b]. Comparison of optical rotation values for polymers prepared by

Sn(Oct)$_2$-catalyzed ROP showed that in case of the lipase-catalyzed polymerization, significant amounts of racemization were observed [26]. The lipase-catalyzed ring opening of other 3-alkyl substituted morpholine-2,5-diones was shown as well [27].

A final example of the lipase-catalyzed polymerization of optically pure chiral monomers involves the polymerization of alditols with dicarboxylic acids [28]. These optically pure carbohydrates possess both primary and chiral secondary hydroxy groups and are acylated with a high degree of regioselectivity at the primary hydroxy groups in Novozym 435-catalyzed reactions. Polymerization of sorbitol with sebacic acid gave polymers with M_n up to 12 kg mol^{-1} while the regioselectivity according to ^{1}H-NMR measurements was excellent [28a]. A series of other alditols was used in the terpolymerization with adipic acid and octane-1,8-diol [28c]. Linear polymers were formed initially due to the high degrees of regioselectivity, whereas upon longer reaction times more hyperbranched polymers were obtained.

11.4
Kinetic Resolution Polymerization of Racemic Monomers

A more elegant approach towards lipase-catalyzed synthesis of chiral polymers is to take advantage of the intrinsic ability of lipases to discriminate between enatiomers. Chiral polymers can be formed when one of the two enantiomers of a chiral monomer preferentially reacts during the polymerization reaction. This is referred to as a kinetic resolution polymerization (KRP) and allows the optically active polymer to be directly procured from a racemic mixture, albeit in a maximum of 50% yield.

11.4.1
KRP of Linear Monomers

Early studies on KRP focused on the enantioselective condensation polymerization of an activated chiral diacyl donor (**21**) with achiral diols such as **22** (Scheme 11.6a) [29]. Lipases of different origin were tested and the formation of optically active trimers and pentamers was demonstrated for the dibromo monomer. The isolated oligomers were end capped with free hydroxy groups and showed small optical rotation values ($[\alpha]_D^{25}$ +4.5 and +4.3 for the trimer and pentamer, respectively). Similarly, the condensation polymerization of achiral diacyl donors **23** with diastereomeric mixtures of chiral diols (**24**) was shown as well, affording optically active oligomers (Scheme 11.6b) [29].

The use of the diastereomeric monomer mixtures complexes the analysis of the formed reaction products and hampers the formation of high molecular weight polymers due to end capping of the growing polymer chains with non-reactive end groups. Therefore, Wallace et al. investigated the use of a chiral diacyl donor lacking a *meso*-form (Scheme 11.7) [30]. Moreover, the C_2-symmetry present

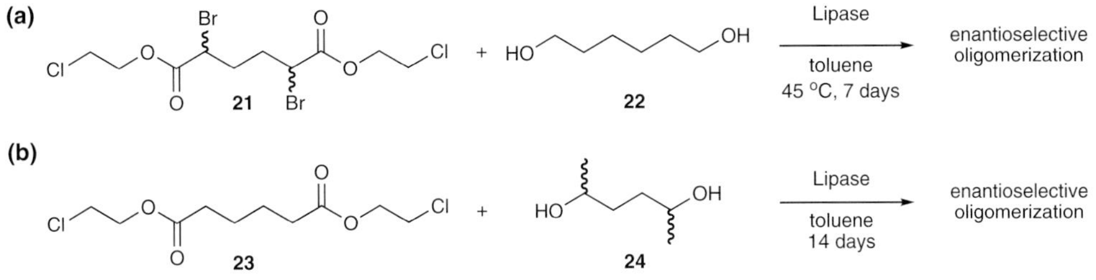

Scheme 11.6 First examples of enantioselective lipase-catalyzed oligomerization of (a) a diastereomeric chiral diacyl donor with a diol and (b) achiral diacyl donors with a diastereomeric chiral diol.

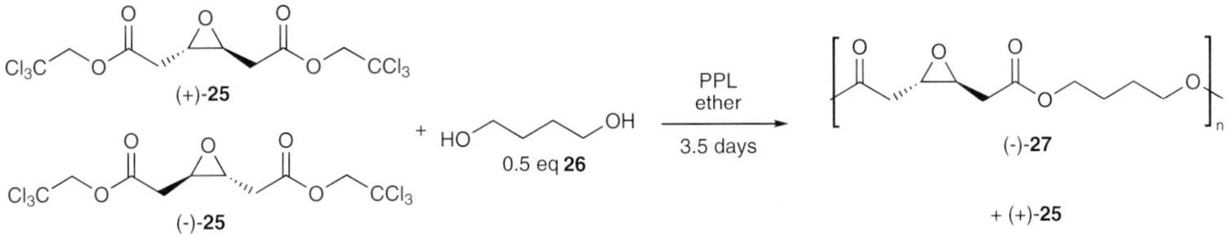

Scheme 11.7 Enantioselective porcine pancreatic lipase (PPL) catalyzed polymerization of chiral diacyl donor **25**.

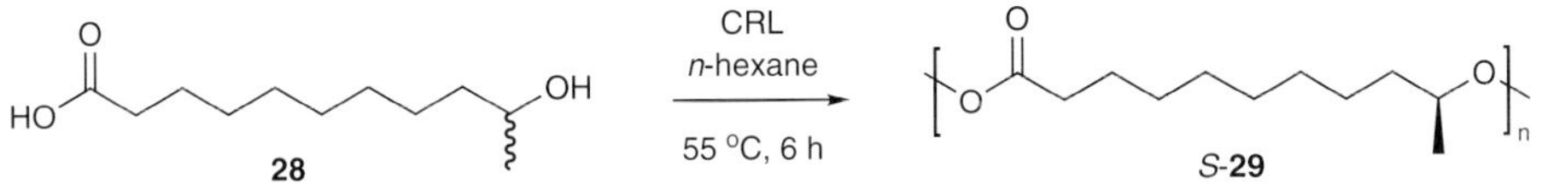

Scheme 11.8 Formation of S-enriched polyester **29** by CRL-catalyzed condensation polymerization of racemic 10-hydroxyundecanoic acid (**28**).

in the monomers ensured the same configuration at both acyl groups ensuring equal reactivity at both sides of the monomer. Stirring of racemic diacyl donor **25** with 0.5 eq of butane-1,4-diol (**26**) in a slurry of PPL indeed resulted in the highly enantioselective formation of polymer (-)-**27**. ^{1}H-NMR analysis of the residual monomer using a chiral shift reagent showed 95% *ee*. The ^{1}H-NMR spectrum of polymer (-)-**27** showed a regioregular structure and an M_n of 5.3 kg mol^{-1} (M_w = 7.9 kg mol^{-1} according to GPC using PS standards), while $[\alpha]_D$ was −13.4° (CHCl$_3$).

A final example involves the enantioselective polymerization of racemic 10-hydroxyundecanoic acid by CRL (Scheme 11.8) [31]. The polymerization reaction was stopped when 50% monomer conversion was reached and polymers with a molecular weight of 1 kg mol^{-1} (PDI = 1.3) were isolated. Subsequent ^{1}H-NMR analysis using the Mosher's acid derivatization procedure on the residual monomer and hydrolyzed monomer revealed an *ee* of 33 and 60%, respectively. Comparison with Mosher's esters of R-2-octanol and *rac*-2-octanol showed that the S-monomer was preferentially incorporated in the polymer.

11.4.2
KRP of Substituted Lactones

The lipase-catalyzed kinetic resolution of substituted lactones is known to be highly enantioselective in some cases [32, 33]. For example, hydrolysis of ω-alkyl substituted caprolactones, with the size of the alkyl group ranging from methyl to n-octyl, showed varying degrees of enantioselectivity and different enantiopreference depending on the substrate and enzyme used [33]. Additionally, the enantioselective ring opening of 3-, 4- and 5-methyl substituted ε-caprolactones was demonstrated using CAL [33]. Various degrees of enantioselectivity ($E = 13 - >100$) were found depending on the position of the substituent and except for 5-methyl-ε-caprolactone, a preference for ring opening of the S-enantiomer was found [33, 34]. No enantiopreference was found for 2-methyl-ε-caprolactone L [35].

If the hydroxy chain-end generated upon enantioselective ring opening of the lactone is able to deacylate another acyl-enzyme complex, enantioselective polymerization of the lactone is possible. The advantage of ROP over condensation polymerization is that no condensation product needs to be removed and as a result, higher molecular weight polymers can be formed. The ROP of α-methyl- and ω-methyl-substituted lactones of varying ring size to polymers with a M_n of 0.9–13 kg mol^{-1} (PDI 1.4–2.8) has indeed been described in literature using *Candida antarctica* lipase (CAL) or *Mucor miehei* lipase (MML) as catalyst [36]. However, no information on the degree of enantioselectivity of the polymerization was given. Partially fluorine-substituted lactones have been enantioselectively polymerized as well [37].

The two isomeric four-membered lactones with a methyl substituent (*rac*-4 and *rac*-31) have both been enantioselectively polymerized using lipase catalysis (Scheme 11.9). First of all, α-methyl-β-propiolactone (*rac*-4) was subjected to PFL-catalyzed ROP in toluene (Scheme 11.9a) [38]. After 144 h polymers were obtained ($M_n = 2.0$ kg mol^{-1}; PDI = 1.7) and a modest degree of enantioselectivity favoring reaction of the S-enantiomer was observed ($E = 4.0$). The use of 1,4-dioxane as solvent led to the formation of oligomeric products only. Secondly, β-butyrolactone (*rac*-31) was used as substrate for the bulk or solution polymerization in isooctane using a library of lipases obtained from CloneZyme (Scheme 11.9b) [39].

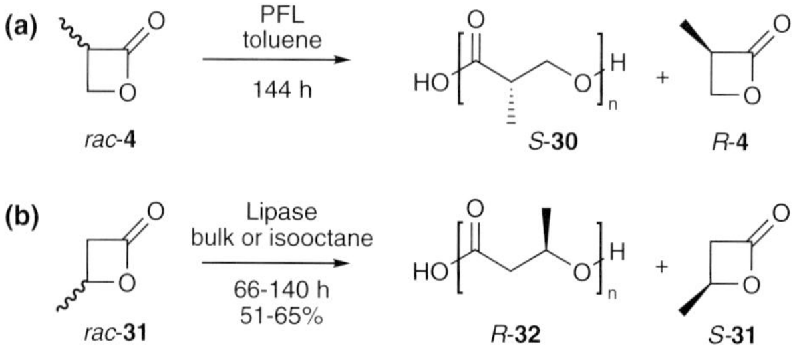

Scheme 11.9 Enantioselective ring opening polymerization of methyl substituted 4-membered resulting in the formation of enantiomerically enriched: (a) poly(S-2-methylpropionate) (*S*-30) and (b) poly(R-3-hydroxybutyrate) (*R*-32).

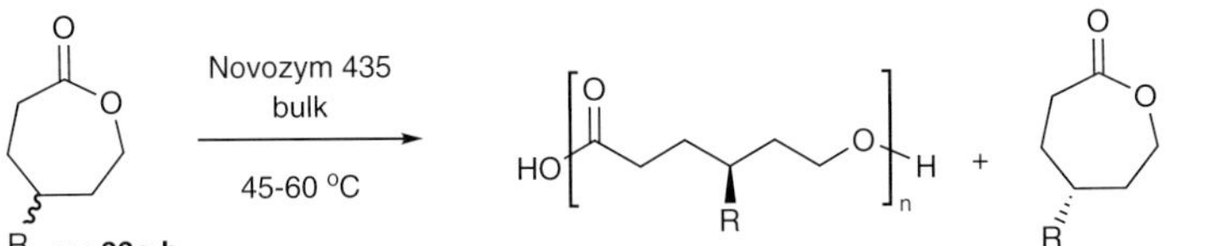

Scheme 11.10 Enantioselective Novozym 435-catalyzed ROP of 4-substituted caprolactone.

Enantioenriched polymers with ee_p values up to 37% and molecular weights in the order of 2.0–3.9 kg mol^{-1} were obtained. An identical sign for the optical rotation as for naturally occurring poly(3-hydroxybutyrate) was observed, showing the R-enantioselectivity of the polymerization reaction.

The enantioselective bulk polymerization of 4-methyl-ε-caprolactone (rac-31a) or 4-ethyl-ε-caprolactone (rac-31b) was demonstrated using Novozym 435 as catalyst (Scheme 11.10) [40]. Depending on the reaction temperature, polymers with a M_n between 5.3 (PDI = 1.5, T = 45 °C) and 8.1 kg mol^{-1} (PDI = 1.4, T = 60 °C) were obtained after 2.5 h [39a]. The enantiopurity of the polymers depended both on the temperature and reaction time: higher temperature and shorter reaction times were shown to give higher ee_p values. When the alkyl substituent was changed into n-propyl, opposite enantiopreference (E = 2.0 ± 0.1) was observed for the ROP and a relatively low molecular weight polymer (M_n = 1.7 kg mol^{-1}; PDI = 1.4) was obtained after 72 h reaction [39b].

The polymerization of 6-methyl-ε-caprolactone by Novozym 435 is impossible due to the preferential formation of a secondary alcohol with the S-configuration, which cannot propagate. Larger lactones with a methyl group at the ω-position, however, do show an R-preference for the enantioselective ring opening and could be efficiently polymerized [41]. In contrast, the Novozym 435-catalyzed polymerization of racemic ω-methyl-heptanolactone, ω-methyl-dodecanolactone and ω-methyl-pentadecanolactone gave polymers of R-configuration with M_n values in the range of 14.2–16.7 kg mol^{-1} (PDI = 1.2–2.3), while excellent ee_p values of 99% and higher were observed. The unreactive (S)-enantiomer of the ω-methyl lactones remained unreacted. The sudden change in enantiopreference from the S- to R-configuration coincides with the change in conformation around the ester bond from cisoid to transoid. The R-enantiopreference of the large transoid lactones perfectly matches the observed R-selectivity of open chain substrates that have an identical ester bond conformation, suggesting that the ester bond conformation has an important role on the enantiopreference of the lipase [40].

11.5
Dynamic Kinetic Resolution Polymerization of Racemic Monomers

While KRP affords optically-enriched polymers from racemic mixtures, the molecular weights of the polymers are usually low or the enantio-enrichment of the polymer is limited. In addition, the less reactive enantiomer remains unpolymerized and can be regarded as waste. To increase the monomer conversion and enantio-enrichment in the chiral polymers, researchers have started including racemization methods into the polymerizations with the aim of converting all monomer into a chiral polymer. This will be discussed below.

11.5.1
Dynamic Kinetic Resolutions in Organic Chemistry

The dynamic kinetic resolution (DKR) of secondary alcohols and amines (Scheme 11.11) is a prominent, industrially relevant, example of chemo-enzymatic chemistry in which a racemic mixture is converted into one enantiomer in essentially 100% yield and in high *ee*. This is in sharp contrast to enzyme-catalyzed kinetic resolutions that afford the desired end-product in a yield of at most 50%, while 50% of the starting material remains unreacted. In DKR processes, hydrolases are typically employed as the enantioselective acylation catalyst (which can be either *R* or *S* selective) while a concurrent racemization process racemizes the remaining substrate via an optically inactive intermediate. This ensures that all starting material is converted into the desired end-product. The importance of optically pure secondary alcohols and amines for the pharmaceutical industry triggered the development of a number of approaches that enable the racemization of *sec*-alcohols and amines via their corresponding ketones and imines, respectively [42].

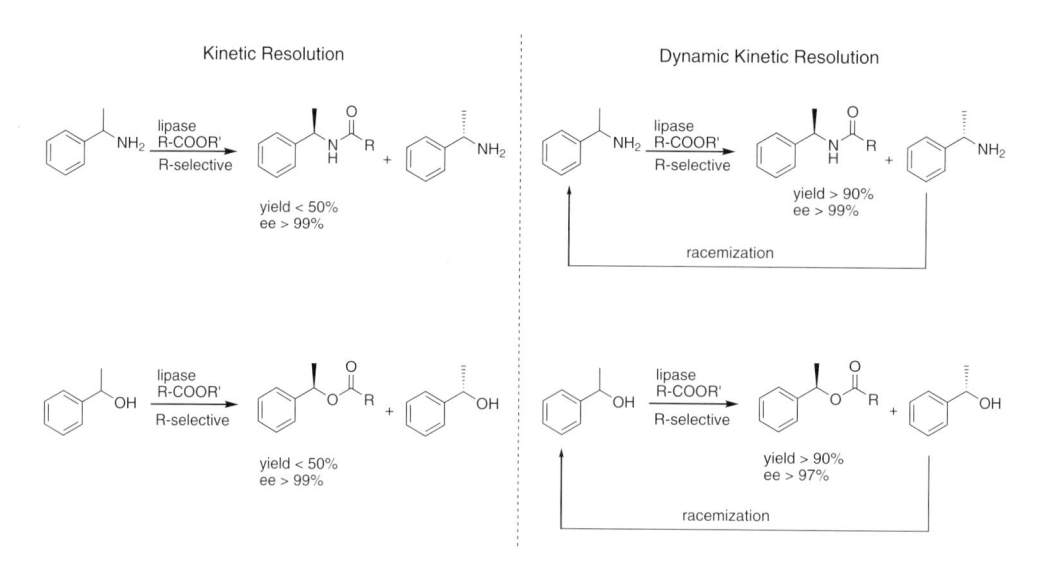

Scheme 11.11 Principle of kinetic resolution and dynamic kinetic resolution as applied for the synthesis of chiral secondary esters and amides.

The first examples of dynamic kinetic resolutions employing an enantioselective enzyme and a metal-based racemization catalyst were reported in 1996. Allyl acetates were quantitatively converted into enantio-enriched allyl alcohols using a Pd-based racemization catalyst and a lipase from *Pseudomonas fluorescens* while benzylic amines were converted into enantiopure amides in reasonable yields using Pd on carbon and a lipase from *Candida antarctica* [43, 44]. The quest for mild reaction conditions that are compatible with enzymes, short reaction times, and easy removal of the catalysts subsequently lead to the development of homogeneous and heterogeneous metal complexes and also to racemization mechanisms that do not depend on a metal catalyst, such as the thiyl radical mediated racemization of aliphatic amines [45, 46]. Moreover, hydrolases were screened that show either *S*- or *R*-selectivity [47]. As a result, the combination of enzyme-catalyzed kinetic resolution in combination with a suitable racemization process currently allows for the isolation of esters and amides in high yield and high *ee* and desired configuration. This combination of chemical and enzymatic catalysis is nowadays an industrially applied process leading to a variety of chiral end-products [48].

11.5.2
Extension of Dynamic Kinetic Resolutions to Polymer Chemistry

In principle, the DKR processes depicted in Scheme 11.11 are condensation reactions and therefore can be extended to bifunctional monomers which would allow the synthesis of chiral polyesters and polyamides. As a proof of principle, the well-known DKR of *(rac)*-1-phenylethanol was combined with the enzyme-catalyzed ring-opening polymerization (eROP) of ε-caprolactone (Scheme 11.12) by Howdle *et al.* [49] The eROP of ε-caprolactone was introduced in the 1990s by the groups of Kobayashi and Gross and is an interesting method of preparing well-defined poly(ε-caprolactone) [50]. Commercially available Novozym 435, lipase B of *Candida antarctica* immobilized on an acrylic resin, is the most frequently employed enzyme for eROP. This robust enzyme shows a broad substrate scope, high polymerization activity and high enantioselectivity for *(R)*-secondary alcohols and is compatible with a variety of metal catalysts in DKR conditions.

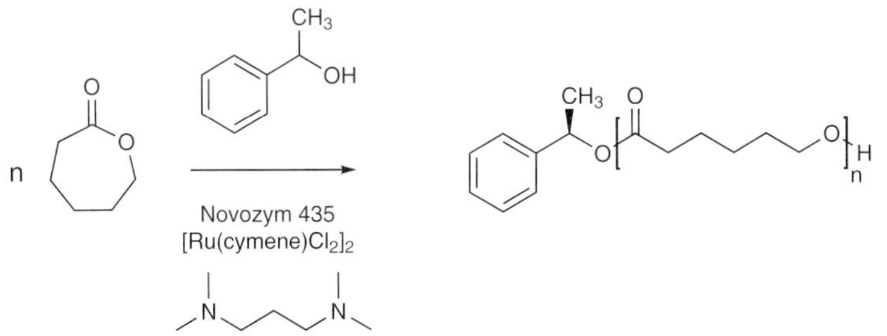

Scheme 11.12 DKR-eROP of *(rac)*-1-phenylethanol and ε-caprolactone using a Ru-based racemization catalyst and Novozym 435 as the acylation catalyst.

Polymerization of ε-caprolactone using *(rac)*-1-phenylethanol showed that around 90% of the *R*-enantiomer of 1-phenylethanol was incorporated in the polymer chain within 2 h of reaction. In contrast, the conversion of the *S*-enantiomer was negligible, even after 12 h. This showed that the enantioselectivity of Novozym 435 was not compromised in a polymerization reaction and that polymers with end-groups of defined chirality could be formed. The addition of a Ru-based racemization catalysts resulted in a 75% incorporation of *(R)*-1-phenylethanol into the polymer chain showing that DKR in combination with eROP indeed works. However, the rate of ε-caprolactone consumption was significantly reduced compared with the polymerization in the absence of the Ru-catalyst. This suggests that the Ru catalyst applied had a retardation effect on the enzymatic polymerization reaction.

11.5.3
Dynamic Kinetic Resolution Polymerizations

A straightforward extension of DKR to polymer chemistry is the use of diols and diesters (AA-BB monomers) or ester-alcohols (AB monomers) as substrates (Scheme 11.13, routes **a** and **b**). Such reactions have been referred to as dynamic kinetic resolution polymerizations (DKRP) and inevitably lead to the formation of oligomers/polymers because of the bifunctional nature of the reagents. The extension of DKR to polymer chemistry is not trivial since side reactions that are relatively unimportant in DKR (dehydrogenation, hydrolysis) have a major impact on the rate of polymerization and attainable chain lengths since the stoichiometry of the reactants is an important issue. As a result, the reaction conditions and catalyst combinations used in a typical DKR process will not *a priori* lead to chiral

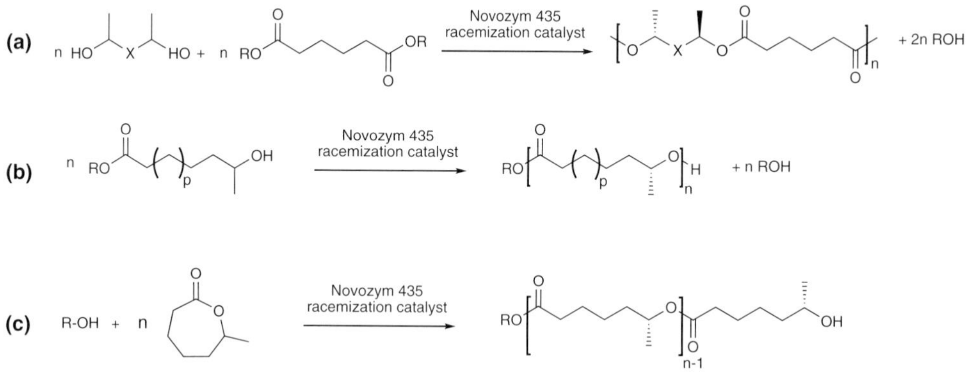

Scheme 11.13 Possible routes for the preparation of chiral polymers by DKRP. Route **a** uses AA and BB monomers to prepare chiral polymers from racemic/diasteromeric diols. Route **b** converts an enantiomer mixture of AB monomers to homochiral polymers. Route **c** is the enzymatic ring opening polymerization of ω-methylated lactones to homochiral polyesters.

Figure 11.3 (a) DKRP of 1,4-diol and dimethyl adipate (b) Chemical structure of Noyori type racemization catalyst **1** and Shvo's racemization catalyst **2**.

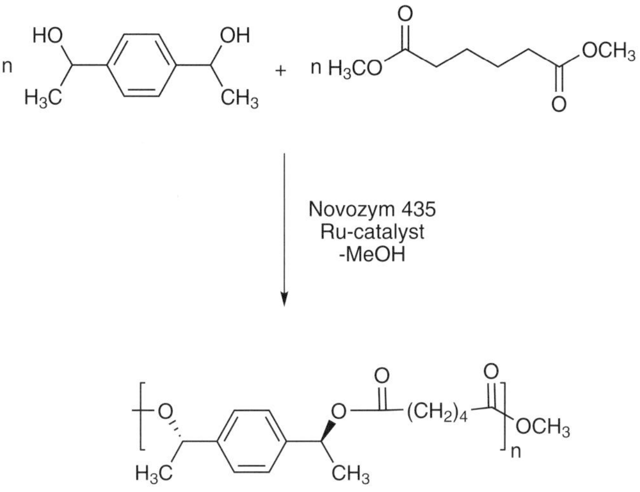

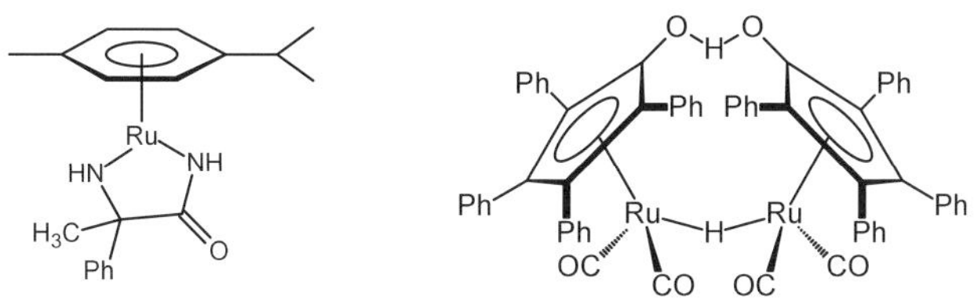

polymers from racemic or achiral monomers with good molecular weight (>10 kg mol^{-1}) and high *ee* (>95%).

Hilker et al. adapted the successful DKR of secondary alcohols developed at DSM[48] to prepare chiral polymers from α,α'-dimethyl-1,4-benzenedimethanol (1,4-diol) and dimethyl adipate (DMA) (Figure 11.3a) [51]. The applied catalytic system consisted of a Ru-Noyori type racemization catalyst **1** (Figure 11.3b) and Novozym 435. This catalyst combination tolerates a wide range of acyl donors, and it was anticipated that bifunctional monomers could result in the the formation of enantio-enriched polycondensates. Moreover, Novozym 435 is highly enantioselective in the transesterification of secondary benzylic alcohols (*E*-ratio is ca. 1 × 10^6) [52].

Before the start of the reaction the monomer mixture showed the expected diastereomer ratio of (*S,S*):(*R,R*):(*R,S*) of 1:1:2 of the 1,4-diol employed. After 30 hours of reaction the (*S,S*)-enantiomer almost completely disappeared while the ratio of *(R,R)* to *(R,S)*-monomer was ca. 3:1 (*R*:*S* ca. 7:1). At a hydroxy group conversion of 92% after 70 hours no further conversion was observed yielding a final ratio of *(R,R)* to *(R,S)* of 16:1 (*R*:*S* ca. 33:1). Unfortunately, the molecular weights of the polymer were moderate at best ($M_w = 3.4\,\text{kg}\,\text{mol}^{-1}$) and Novozym 435 had to be added every few hours to compensate for the activity loss of the lipase. This suggests that Ru-catalyst **1** and Novozym 435 are not fully compatible, but the origin of the incompatibility was not straightforward: a DKR of 1-phenylethanol and isopropylacetate under identical conditions gave the corresponding *(R)*-ester in 99% *ee* and >95% conversion within 48 h [48].

The catalytic system was significantly improved by van As *et al.* who changed the racemization catalyst from **1** to Shvo's catalyst **2** [53] (Figure 11.3b) and added a hydrogen donor 2,4-dimethyl-3-pentanol (DMP) to suppress dehydrogenation reactions [54]. DMP is a sterically hindered alcohol and is not accepted by Novozym 435 as a substrate. Although Shvo's catalyst **2** is a significantly slower racemization catalyst than **1**, it does not require the addition of K_2CO_3. This base, required to activate the precatalyst of **1**, contributed to the deactivation of Novozym 435 in DKRP conditions. For the 1,4-diol this resulted in polymers with *ee* of 94% and M_p of 8.3 kg mol^{-1} within 170 h. The 1,3-isomer, α,α'-dimethyl-1,3-benzenedimethanol (1,3-diol), showed slightly better results under similar conditions and had a higher solubility in toluene. Therefore, an optimization study was performed using 1,3-diol and employing diisopropyl adipate as acyl donor. The optimal conditions were 2 mol % **2**, 12 mg Novozym 435 per mmol alcohol group in the presence of 0.5 M DMP as the hydrogen donor. With these conditions, chiral polymers were obtained with peak molecular weights up to 15 kg mol^{-1}, *ee* values up to 99% and at most 1–3% ketone functional groups in ~120 h as a result of dehydrogenation.

Aliphatic secondary diols were also employed as the substrate, but DKRP of these diols did not lead to enantiopure polymers. At most, an *ee* of 46% was obtained with low molecular weights in the range of 3.3–3.7 kg mol^{-1}. The latter was attributed to the low of selectivity of Novozym 435 for these secondary diols as revealed by kinetic resolution experiments of 2,9-decandiol with vinyl acetate and Novozym 435. Apparently, the *S*-alcohol showed significant reactivity, decreasing the *ee* of the polymer.

In contrast, aliphatic AB type monomers encompassing a secondary hydroxy group and an ester moiety (Scheme 11.13, route **b**) did show high enantioselectivities in a Novozym 435-catalyzed transesterification reaction [55]. The *E*-ratio was high (*E*-ratio > 200) for all the monomers evaluated (see Scheme 11.14).

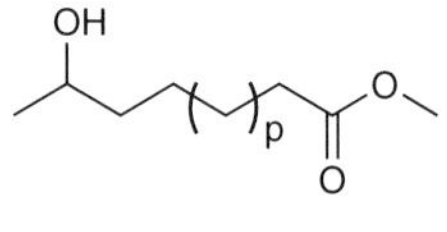

p = 1: Me-6HH
p = 2: Me-7HO
p = 3: Me-8HN
p = 8: Me-13HT

Scheme 11.14 Monomers studied in DKRP of AB monomers.

11.5 Dynamic Kinetic Resolution Polymerization of Racemic Monomers

An additional advantage of the use of AB monomers over AA-BB monomers is the lack of sensitivity on stoichiometric issues: polymerization of the monomers (Scheme 11.14) employing Novozym 435 and Shvo's catalyst **2** was slow (reaction time = 170 h) but did result in chiral polymers of good molecular weigth and *ee*. For example, DKRP of Me-7HO resulted in a polymer with an M_p of 16.3 kg mol^{-1} and an *ee* of 92% (determined after acid-catalyzed methanolysis).

A more challenging endeavor is the extention of the DKR of optically inactive, chiral diamines to polymer chemistry. The racemization of amines is much more difficult compared to that of secondary alcohols as a result of harsh conditions required and the risk of side products formed. The DKR of amines developed by Bäckvall was selected for evaluation of the preparation of chiral polyamides by DKRP [46f]. This DKR employed the more reactive Ru-cat **3** (Figure 11.4) for racemization of the amine and Novozym 435 as the enatioselective catalysts. Unfortunately, a large excess of acyl donor was employed in the original procedure and long reaction times were required for full conversion. Veld *et al.* optimized the DKR of amines for use with a single equivalent of acyl donor and improved the acylation and racemization rate while retaining high chemo- and enantiose-lectivity (Figure 11.4) [46h]. A major improvement was achieved by use of isopropyl 2-methoxyacetate as acyl donor, resulting in a significantly higher acylation rate of the amine substrates [56]. The racemization rate was improved by increasing the reaction temperature, while the chemoselectivity was simultaneously enhanced by applying reduced pressure and the addition of DMP as hydrogen donor.

The modified Bäckvall system was three times faster than the original system and gave similar yields and enantioselectivity for a series of amine and diamine substrates. Attempts to polymerize a chiral diamine with a diacyl donor using

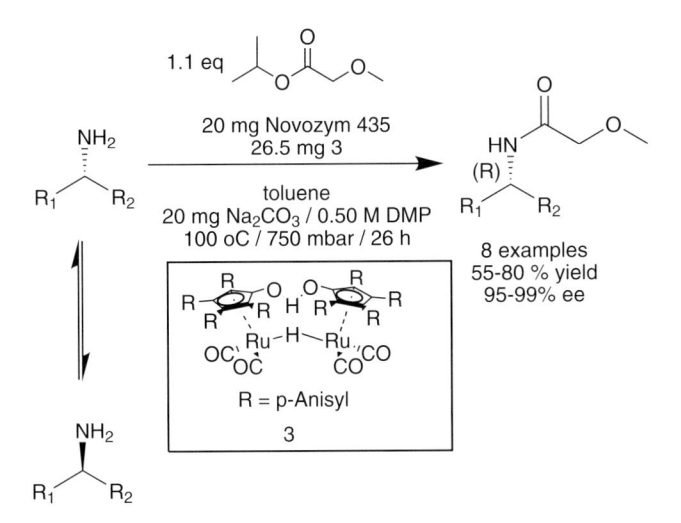

Figure 11.4 Modified Bäckvall system for the DKR of amine substrates.

the conditions of the optimized system did not result in the formation of a polymer although (insoluble) oligomers were obtained. As a result, the synthesis of chiral polyamides by DKRP also necessitates to address solubility issues.

11.5.4
Iterative Tandem Catalysis: Chiral Polymers from Racemic ω-Methylated Lactones

While the polymerization of optically inactive AA-BB and AB monomers under DKR conditions leads to chiral polyesters, these approaches always result in limited molecular weights since a condensation product is formed that needs to be effectively removed. A solution for this would be to use the eROP of lactones, where no condensation products are formed during polymerization. In principle, the eROP of lactones can lead to very high MW polyesters (>80 kg mol^{-1}) [57]. Addition of a methyl substituent at the ω-position of the lactone introduces a chiral center. Peeters et al. conducted a systematic study of substituted ε-caprolactones which revealed that monomers with a methyl at the 3-, 4-, or 5-position could be polymerized enantioselectively while a methyl at the 6-position (α-methyl-ε-caprolactone, 6-MeCL) could not [58]. The lack of reactivity of the latter monomer in a Novozym 435-catalyzed polymerization reaction was attributed to the formation of S-secondary alcohol end-groups. These cannot act as a nucleophile in the propagation reaction since the lipase-catalyzed transesterification is highly R-selective for secondary alcohols.

van As et al. proposed that by combining Novozym 435-catalyzed ring-opening of 6-MeCL with racemization of the terminal alcohol, it should be possible to polymerize 6-MeCL [59]. The method was termed iterative tandem catalysis (ITC) since chain propagation can only be achieved by a combination of two fundamentally different catalytic systems. Scheme 11.15 shows the different steps that are required to enable a successful ITC. First, ring-opening of (S)-6-MeCL, the preferred enantiomer of Novozym 435, will lead to a ring-opened product with an S-alcohol chain end. This product is unreactive and propagation does not occur. To enable polymerization of (S)-6-MeCL, racemization of the S-alcohol that is formed upon ring-opening is required, furnishing a reactive R-chain end. These can subsequently react with another molecule of (S)-6-MeCL. If the less reactive (R)-6-MeCL is incorporated (which will occur since the selectivity for lactones is moderate with an E-ratio of 12), a reactive R-chain end is obtained and propagation occurs instantly. In this way, both enantiomers of the monomer are consumed.

Racemization can be achieved with a variety of homogeneous catalysts. The Noyori type Ru-racemization catalyst **1** was first selected as a suitable candidate (Figure 11.3b). In fact, this was the first example in which DKR was combined with an enzyme-catalyzed polymerization reaction. It appeared, however, that polymerization with the Novozym 435/**1** catalytic system was problematic: only oligomers were obtained in two-step reactions because the catalysts were incompatible under the reaction conditions employed.

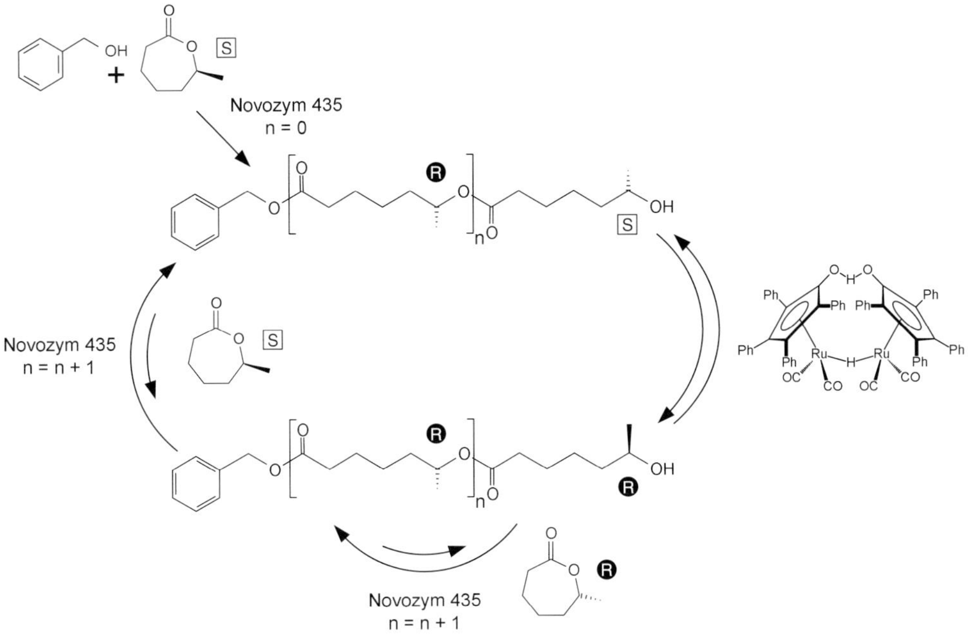

11.5 Dynamic Kinetic Resolution Polymerization of Racemic Monomers | 295

Scheme 11.15 ITC of 6-MeCL for the formation of chiral poly(6-MeCL).

A major improvement was achieved by van Buijtenen et al. when Shvo's catalyst 2 (Figure 11.3b) was introduced in combination with the addition of DMP to suppress dehydrogenation reactions [60]. Initial experiments were conducted with optically pure (S)-6-MeCL as the substrate. Although the polymerization was slow, the conversion of the lactone increased steadily and after 318 h full conversion was reached. After 5, 8, and 13 days of reaction, samples were evaluated by GPC which revealed a steady increase of the molecular weight over time, confirming that polymerization indeed takes place. Poly-(R)-6-MeCL with a promising ee of 86% and M_p of 8.2 kg mol^{-1} was obtained after work-up. The low rate of reaction compared to DKR (typically complete after 48 h with the Shvo catalyst) was attributed to the low concentration of the terminal alcohol as well as to the iterative nature of the system. Subsequently, racemic 6-MeCL was polymerized within 220 h with complete conversion of both enantiomers, yielding a high ee of 92% and a M_p of 9.4 kg mol^{-1}. Interestingly, poly((R)-6MeCL) appeared as a crystalline polymer with a melting point of 28 °C while its racemic analog did not crystallize. This shows that control of the stereochemistry in polyesters can lead to a significant improvement of the thermal properties. This is promising in view of the applications of chiral polymers since tuning the molecular weight of the monomer will allow for tunable thermal properties in the corresponding chiral polymers.

11.6
Tuning Polymer Properties with Chirality

In the previous section, chiral (co)polymers were prepared using a one-pot system; that is, all reactants and catalysts are present at the start of the reaction and both catalysts work simultaneously. This section will deal with the synthesis of chiral *block* copolymers (i) using a chiral polymer as the initiator or (ii) using different catalysts in sequence. Furthermore, some examples are highlighted in which the polymer properties are determined by the chirality of the polymer. The latter is introduced via lipase catalysis, before or after polymerization of the monomer.

11.6.1
Chiral Block Copolymers Using Enzymatic Catalysis

An interesting approach to chiral block copolymer synthesis was recently reported by Li et al. using poly[(R)-3-hydroxybutyrate] (PHB) as a macroinitiator for the eROP of ε-caprolactone using Novozym 435 (see Scheme 11.16) [61]. PHB is a microbial polyester, which can be produced in large quantities from renewable resources, and is optically pure. Incorporation in block copolymers could overcome limitations of the applicability of PHB in thermoplastic applications by improving its thermal properties. Microbial PHB comprises a primary and a secondary (chiral) hydroxy end-group. Interestingly, only the primary alcohol chain end intitiated the eROP of ε-caprolactone. Moreover, no randomization as a result of transesterification reactions was observed. This is thus an elegant example in which the enzymatic selectivity is used to distinguish between two hydroxyl groups in the enzymatic macroinitiation.

Peeters et al. combined the enantioselective eROP of 4-methyl-ε-caprolactone (4-MeCL) with controlled atom transfer radical polymerization (ATRP) of methyl methacrylate (Scheme 11.17) [58]. It was found that the addition of Ni(PPh$_3$)$_4$ inhibited Novozym 435 and at the same time catalyzed the ATRP reaction. On the other hand, Novozym 435 did not interfere with the ATRP of methyl methacrylate. As a result the sequence of the reaction was important: first the enantioselective eROP of 4-MeCL was conducted at low temperature to have a high enantioselectivity until the conversion was around 50%. Then, Ni(PPh$_3$)$_4$ and MMA were added and the temperature was raised to 80 °C to start up the ATRP reaction of MMA. After precipitation of the polymer to remove the unreacted (R)-4-MeCL, the chiral block copolymer (M_n = 11 kg mol^{-1} and 17 kg mol^{-1}) was

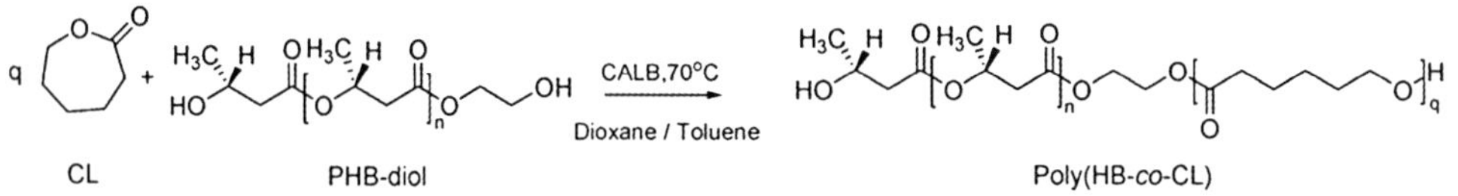

Scheme 11.16 eROP of ε-caprolactone with PHB-diol as the intiator.

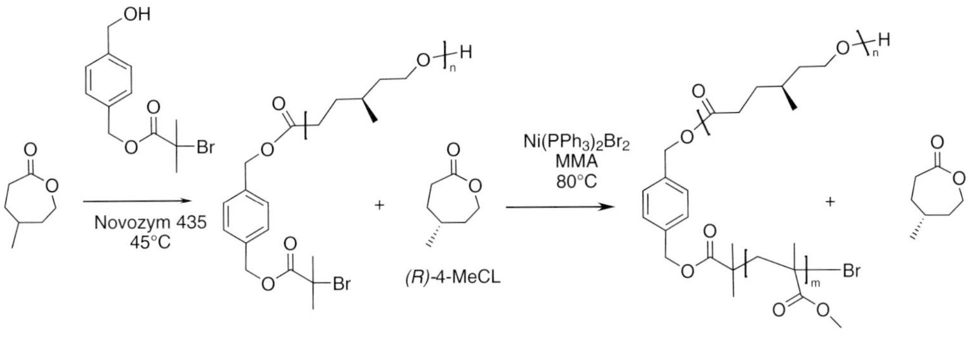

Scheme 11.17 Cascade approach to a chiral block copolymer combining enantioselective ROP of 4-methyl-ε-caprolactone and ATRP of methyl methacrylate.

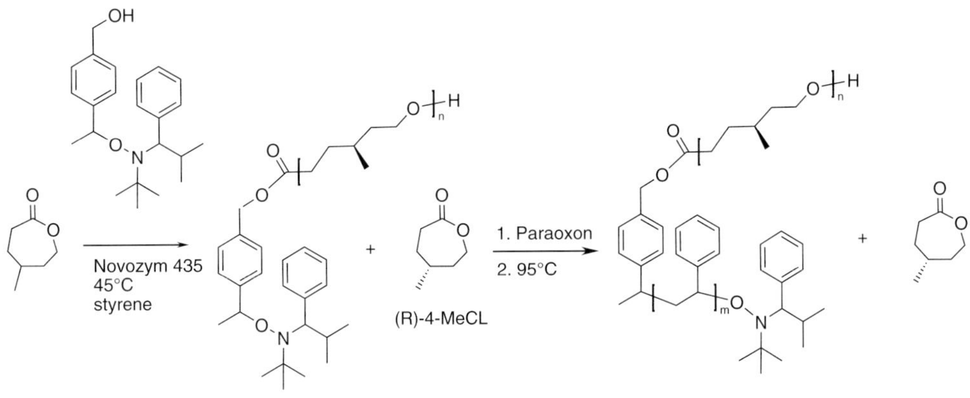

Scheme 11.18 One-pot enzymatic ROP and nitroxide mediated living free radical polymerization.

isolated as a solid compound showing two T_g's at −60 and 100 °C, indicative for phase separation between the two blocks.

A similar approach was followed with the eROP of 4-MeCL, followed by nitroxide mediated living free radical polymersation (NMP) of styrene using a bifunctional catalyst (Scheme 11.18) [62]. Styrene, the monomer for the NMP, was added already at the beginning since it proved to be a good solvent for the eROP of lactones. At low temperatures, no radical polymerization occurs thus the two polymerization mechanisms are thermally separated. When the eROP reached a conversion of 50%, a lipase inhibitor, paraoxon, was added to the reaction mixture to prevent further incorporation of the undesired enantiomer. Increasing the temperature to 95 °C started the nitroxide mediated LFRP, to afford block copolymers. After precipitation, the chiral block copolymers obtained showed two T_g's at −51 °C and 106 °C. The specific rotation $[\alpha]_D^{25}$ of the block copolymer was −2.6°.

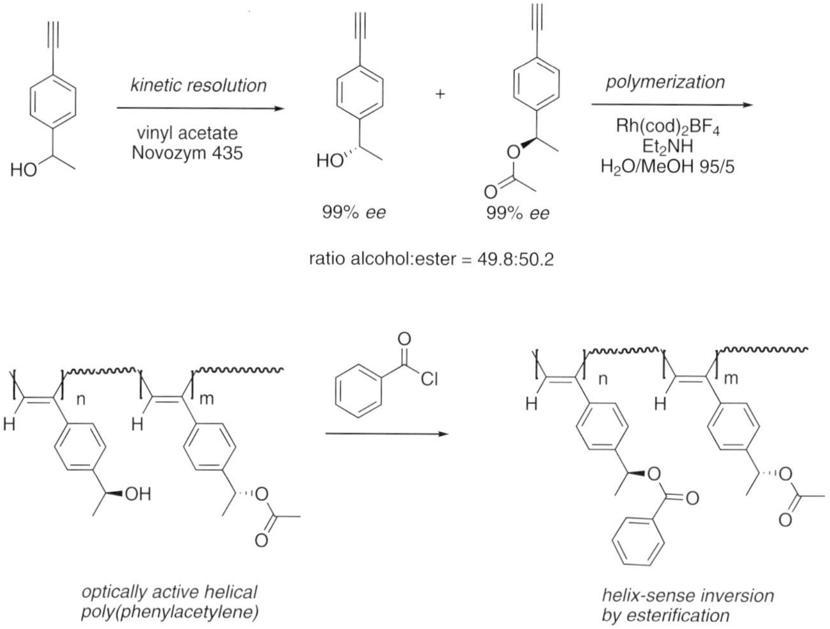

Scheme 11.19 Synthesis and polymerization of optically active, helical poly(acetylenes). Esterification of the chiral alcohols in the polymer reverses the helicity.

This is in good agreement with the optical rotation of −7.2° previously reported for poly((S)-4MeCL) [40a] with an *ee* of 90% considering that the block length ratio of chiral to non-chiral block in the poly((S)-4-MeCL-*co*-St) block copolymer is ca. 1:2.Control of Polymer Helicity by Lipase Catalysis

Yashima *et al.* recently showed an example where the polymer helicity was controlled by enzymatic enantioselective acylation of the monomers [63]. Optically active phenylacetylenes containing hydroxy or ester groups were obtained by the kinetic resolution of the corresponding racemic hydroxyfunctional phenylacetylene (see Scheme 11.19). Polymerization of the phenylacetylenes afforded an optically active poly(phenylacetylene) with a high molecular weight (M_n = 89 kg mol^{-1}; PDI = 2.0) and with a preferred helical sense, as evidenced by circular dichroism spectroscopy. Poly(acetylenes) are dynamic helical polymers composed of interconvertible left- and right-handed helical conformations separated by helical reversals. The optical activity introduced by the small excess of chiral monomers biases one of the helical confomations. Upon reaction of the chiral alcohols in the polymer using an acid chloride or an isocyanate, the helix sense was inverted. Although helix inversions have been reported previously for poly(phenylacetylenes) by temperature, solvent and chiral additives, this is the first example of helicity inversion by chemical modification using achiral compounds.

11.6.2
Enantioselective Acylation and Deacylation on Polymer Backbones

A particularly elegant approach to tune polymer properties with chirality was recently shown by Duxbury et al. [64] With the aid of two alcohol dehydrogenases that show opposite enantioselectivities in the reduction of ketones (ADH-LB and ADH-T), the two enantiomers of p-vinylphenylethanol were obtained in excellent yield and ee. Copolymers of these monomers with styrene using free radical polymerization afforded random block copolymers with compositions ranging from 100% R- to 100% S-alcohol in the side chains (Scheme 11.20). The backbones were designed to contain a total of approximately 45% of the secondary alcohol monomer (either R, S, or a mixture). The polymers showed number-average molecular weight $M_n = 5.0$–$6.0 \, \text{kg mol}^{-1}$ and polydispersities between 1.7 and 2.1. From differential scanning calorimetry (DSC) analysis a similar glass transition temperature T_g of approximately 115 °C was found for all polymers, irrespective of their chiral composition, suggesting that the enantiomeric composition has no effect on their thermal properties. The optical rotation of the polymers increased linearly from −20 to +20°. Novozym 435 was used to catalyze the esterification of the alcohol groups on the polymer backbone with vinyl acetate in toluene (Scheme 11.20). When a backbone containing 100% S-secondary alcohol groups (styrene copolymer containing approximately 45% alcohol monomer) was used for the enzymatic grafting of vinyl acetate, no reaction was observed over a period of 24 h. In contrast, when a backbone containing 100% R-secondary alcohol groups was used, the enzymatic esterification of vinyl acetate occurred from 75% of the alcohol groups within 24 h. This is the first example where chiral information stored in a polymer chain can be 'read out' by an enantioselective enzymatic reaction.

In a related approach, copolymers of styrene and a styrene derivative containing two pendant ester bonds were prepared with free radical polymerization (Scheme 11.21) [65]. Transesterification reactions were conducted with Novozym 435 as the

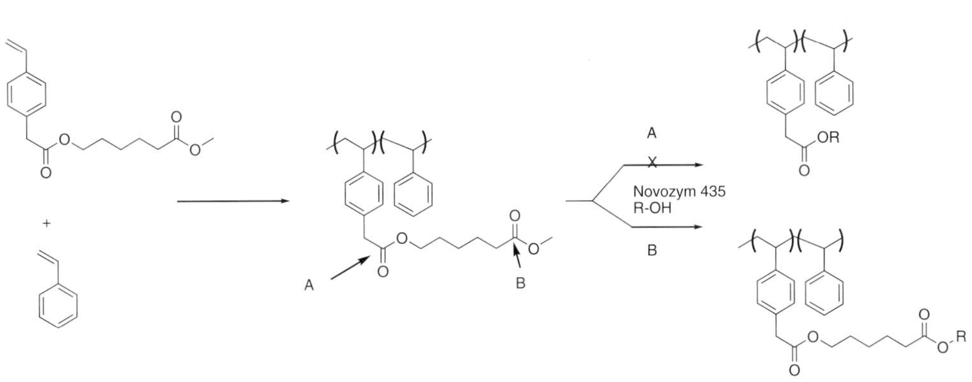

Scheme 11.20 Copolymers of the chiral monomers with styrene used for subsequent R-selective enzymatic grafting with vinyl acetate.

Scheme 11.21 Synthesis of a polymer with two ester groups for enzymatic transesterification and its modification by enzymatic transesterification.

catalyst and benzyl alcohol or (rac)-1-phenylethanol as the nucleophile. Interestingly, the ester bond closest to the polymer backbone (A, see Scheme 11.21) remained unaffected while ester bond B reacted in up to 98% to the corresponding benzyl ester. The transesterification was not only highly chemoselective but also enantioselective. Conversion of (rac)-1-phenylethanol in the transesterification reaction amounted to a maximum conversion of 47.9% of the R-alcohol and only at the ester position B.

11.6.3
Chiral Particles by Combining eROP and Living Free Radical Polymerization

Novozym 435-catalyzed KRP of racemic 4-MeCL was applied by Xiao et al. to prepare chiral crosslinked particle. By letting the conversion in the KRP of 4-MeCL proceed to 50% or 91%, poly((S)- and poly((rac)-4-MeCL), respectively, were obtained. The R-enantiomer of 4-MeCL could be recovered from the polymerization mixture that afforded poly((S)-4-MeCL), and was subsequently polymerized with Novozym 435 to poly((R)-4-MeCL) (Scheme 11.22). GPC analysis confirmed that molecular weights of the polymers were almost identical (M_n around $3 \, \text{kg mol}^{-1}$, PDI around 1.5 after precipitation) and optical rotation measurements revealed opposite signs for poly((S)- and poly((R)-4-MeCL) while

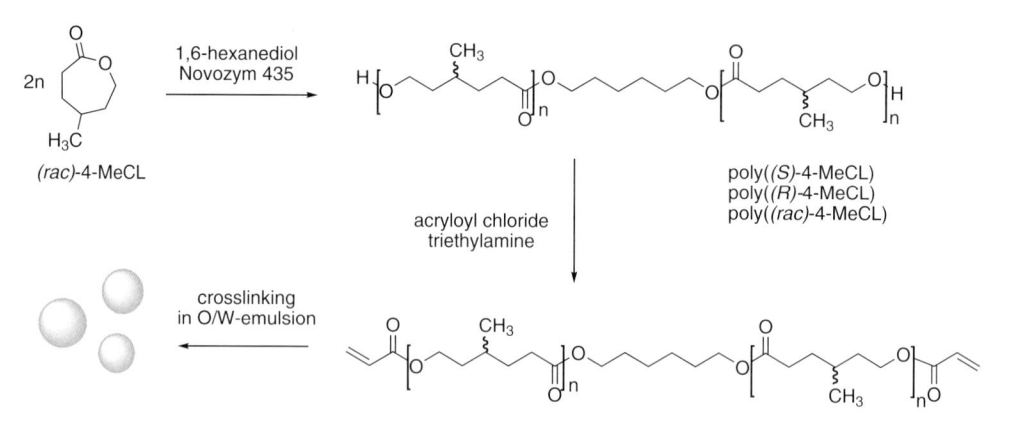

Scheme 11.22 Synthesis of chiral polymer precursors for the formation of chiral crosslinked particles.

poly((rac)-4-MeCL showed an optical rotation of 0°. Moreover, the polymers were fully amorphous with similar T_g's. Preliminary degradation experiments with *Candida antarctica* lipase B showed that the degradation rate of the polymers was dependent on the chirality in the polymer backbone.

After acrylation, cross-linked chiral polymer microspheres were synthesized by a modified oil-in-water (O/W) emulsion photopolymerization. The chiral particles showed a uniform spherical shape with a diameter ranging from a few microns to 80 microns. Amorphous chiral microspheres open the possibility to study the in vitro and in vivo enzymatic degradation as a function of chirality, without the influence of crystallinity. Moreover, they are promising as materials to conduct chiral separations of enantiomers.

11.7
Conclusions and Outlook

This review summarizes the recent results in the preparation of well-defined chiral polymers from optically inactive monomers. To date, optically active polycondensates based on non-natural monomers are still a curiosity in polymer chemistry. Expanding the catalytic toolbox in polymer chemistry by adopting methods from chemo-enzymatic synthesis may enable easy access to chiral polymers and allow the exploration of the added value of chirality in materials. Moreover, chemo-enzymatic approaches have the potential to further enhance macromolecular complexity and hence allow to access new materials with applications envisaged in nanomaterials and biomedical materials.

To date, the limited use of the enantioselectivity of biocatalysts in polymerization conditions and the lengthy synthetic procedures required to prepare optically pure monomers have hampered full exploitation of chemo-enzymatic approaches in polymer chemistry. However, a combined multidisciplinary effort at the interface of biocatalysis, polymer chemistry and organic catalysis, will allow to convert methods well-established in organic chemistry such as tandem catalysis, to the field of polymer chemistry. Undoubtedly, in the near future the exploitation of the selectivity of enzymes and the advantages of chemo-enzymatic approaches in a wide variety of polymerization chemistries will be recognized. This may lead to a paradigm shift in polymer chemistry and allow a higher level of structural complexity in macromolecules, reminiscent to those found in Nature.

References

1. Ikada, Y., and Tsuji, H. (2000) *Macromol. Rapid Commun.*, **21**, 117.
2. Lou, X., Detrembleur, C., and Jérôme, R. (2003) *Macromol. Rapid Commun.*, **24**, 161.
3. Okamoto, Y., and Nakano, T. (1994) *Chem. Rev.*, **94**, 349.
4. (a) Wasilke, J.C., Obrey, S.J., Baker, R.T., and Bazan, G.C. (2005) *Chem. Rev.*, **105**, 1001; (b) Bruggink, A., Schoevaart, R., and Kieboom, T. (2003) *Org. Process Res. Dev.*, **7**, 622.
5. (a) Schulze, B., and Wubbolts, M.G. (1999) *Curr. Opin. Biotechnol.*, **10**, 609;

(b) Gotor, V. (2002) *Org. Process Res. Dev.*, **6**, 420; (c) Jaeger, K.-E., and Eggert, T. (2002) *Curr. Opin. Biotechnol.*, **13**, 390.

6 Nardini, M., and Dijkstra, B.W. (1999) *Curr. Opin. Struct. Biol.*, **9**, 732.

7 Derewenda, Z.S., and Sharp, A.M. (1993) *Trends Biochem. Sci.*, **18**, 20.

8 Uppenberg, J., Hansen, M.T., Patkar, S., and Jones, T.A. (1994) *Structure*, **2**, 293.

9 Kirk, O., and Christensen, M.W. (2002) *Org. Process Res. Dev.*, **6**, 446.

10 Kazlauskas, R.J., Weissfloch, A.N.E., Rappaport, A.T., and Cuccia, L.A. (1991) *J. Org. Chem.*, **56**, 2656.

11 Mugford, P.F., Wagner, U.G., Jiang, Y., Faber, K., and Kazlauskas, R.J. (2008) *Angew. Chem. Int. Ed.*, **47**, 8782.

12 Hedenstrom, E., Nguyen, B.V., and Silks, L.A. (2002) *Tetrahedron Asymmetry*, **13**, 835.

13 Trollsås, M., Orrenius, C., Sahlen, F., Gedde, U.W., Norin, T., Hult, A., Hermann, D., Rudquist, P., Komitov, L., Lagerwall, S.T., and Lindstrom, J. (1996) *J. Am. Chem. Soc.*, **118**, 8542.

14 Schueller, C.M., Manning, D.D., and Kiessling, L.L. (1996) *Tetrahedron Lett.*, **37**, 8853.

15 Xu, J., Gross, R.A., Kaplan, D.L., and Swift, G. (1996) *Macromolecules*, **29**, 4582.

16 van As, B.A.C., Chan, D.K., Kivit, P.J.J., Palmans, A.R.A., and Meijer, E.W. (2007) *Tetrahedron Asymmetry*, **18**, 787.

17 Zhang, M., and Kazlauskas, R. (1999) *J. Org. Chem.*, **64**, 7498.

18 (a) Matsumura, S., Mabuchi, K., and Toshima, K. (1997) *Macromol. Rapid. Commun.*, **18**, 477; (b) Matsumura, S., Mabuchi, K., and Toshima, K. (1998) *Macromol. Symp.*, **130**, 285.

19 Matsumura, S., Tsukada, K., and Toshima, K. (1999) *Int. J. Biol. Macromol.*, **25**, 161.

20 Hans, M., Keul, H., and Moeller, M. (2009) *Macromol. Biosci.*, **9**, 239.

21 Huijser, S., Staal, B.B.P., Huang, J., Duchateau, R., and Koning, C.E. (2006) *Biomacromolecules*, **7**, 2465.

22 Bonduelle, C., Martin-Vaca, B., and Bourissou, D. (2009) *Biomacromolecules*, **10**, 3069.

23 Thillaye du Boullay, O., Marchal, E., Martin-Vaca, B., Cossio, F.P., and Bourissou, D. (2006) *J. Am. Chem. Soc.*, **120**, 16442.

24 Wu, R., Al-Azemi, T.F., and Bisht, K.S. (2008) *Biomacromolecules*, **9**, 2921.

25 Li, G.J., Yao, D.H., and Zong, M.H. (2008) *Eur. Polym. J.*, **44**, 1123.

26 (a) Feng, Y., Knüfermann, J., Klee, D., and Höcker, H. (1999) *Macromol. Rapid. Commun.*, **20**, 88; (b) Feng, Y., Knüfermann, J., Klee, D., and Höcker, H. (1999) *Macromol. Chem. Phys.*, **200**, 1506.

27 (a) Feng, Y., Klee, D., Keul, H., and Höcker, H. (2000) *Macromol. Chem. Phys.*, **201**, 2670.; (b) Feng, Y., Klee, D., and Höcker, H. (2001) *Macromol. Biosci.*, **1**, 66.

28 (a) Uyama, H., Klegraf, E., Wada, S., and Kobayashi, S. (2000) *Chem. Lett.*, 800; (b) Fu, H.Y., Kulshrestha, A.S., Gao, W., Gross, R.A., Baiardo, M., and Scandola, M. (2003) *Macromolecules*, **36**, 9804; (c) Hu, J., Gao, W., Kulshrestha, A., and Gross, R.A. (2006) *Macromolecules*, **39** (20), 6789.

29 Margolin, A.L., Crenne, J.Y., and Klibanov, A.M. (1987) *Tetrahedron Lett.*, **28**, 1607.

30 Wallace, J.S., and Morrow, C.J. (1989) *J. Polym. Sci. Part A Polym. Chem.*, **27**, 2553.

31 O'Hagan, D., and Parker, A.H. (1998) *Polym. Bull.*, **41**, 519.

32 Fellous, R., Lizzani-Cuvelier, L., Loiseau, M.A., and Sassy, E. (1994) *Tetrahedron Asymmetry*, **5**, 343.

33 Shioji, K., Matsuo, A., Okuma, K., Nakamura, K., and Ohno, A. (2000) *Tetrahedron Lett.*, **41**, 8799.

34 Kondaveti, L., Al-Azemi, T.F., and Bisht, K.S. (2002) *Tetrahedron Asymmetry*, **13**, 129.

35 Küllmer, K., Kikuchi, H., Uyama, H., and Kobayashi, S. (1998) *Macromol. Rapid. Commun*, **19**, 127.

36 Kikuchi, H., Uyama, H., and Kobayashi, S. (2002) *Polym. J.*, **34**, 835.

37 Runge, M., O'Hagan, D., and Haufe, G. (2000) *J. Polym. Sci. Part A Polym. Chem.*, **38**, 2004.

38 Svirkin, Y.Y., Xu, J., Gross, R.A., Kaplan, D.L., and Swift, G. (1996) *Macromolecules*, **29**, 4591.

39 Xie, W.H., Li, J., Chen, D.P., and Wang, P.G. (1997) *Macromolecules*, **30**, 6997.

40 (a) Al-Azemi, T.F., Kondaveti, L., and Bisht, K.S. (2002) *Macromolecules*, **35**, 3380; (b) Peeters, J.W., van Leeuwen, O., Palmans, A.R.A., and Meijer, E.W. (2005) *Macromolecules*, **38**, 5587.

41 van Buijtenen, J., van As, B.A.C., Verbruggen, M., Roumen, L., Vekemans, J.A.J.M., Pieterse, K., Hilbers, P.A.J., Hulshof, L.A., Palmans, A.R.A., and Meijer, E.W. (2007) *J. Am. Chem. Soc.*, **129**, 7393.

42 (a) Pellissier, H. (2008) *Tetrahedron*, **64**, 1563; (b) Ahn, Y., Ko, S.-B., Kim, M.-J., and Park, J. (2008) *Coord. Chem. Rev.*, **252**, 647; (c) Pàmies, O., and Bäckvall, J.-E. (2003) *Chem. Rev.*, **103**, 3247; (d) Huerta, F.F., Minidis, A.B.E., and Bäckvall, J.-E. (2001) *Chem. Soc. Rev.*, **30**, 321.

43 Allen, J.V., and Williams, J.M.J. (1996) *Tetrahedron Lett.*, **37**, 1859.

44 Reetz, M.T., and Schimossek, K. (1996) *Chimia*, **50**, 668.

45 For the DKR of secondary alcohols, see: (a) Larsson, A.L.E., Persson, B.A., and Bäckvall, J.-E. (1997) *Angew. Chem. Int. Ed.*, **36**, 1211; (b) Persson, B.A., Larsson, A.L.E., Ray, M.L., and Bäckvall, J.-E. (1999) *J. Am. Chem. Soc.*, **121**, 1645; (c) Koh, J.H., Jung, H.M., Kim, M.-J., and Park, J. (1999) *Tetrahedron Lett.*, **40**, 6281; (d) Choi, J.H., Kim, Y.K., Nam, S.H., Shin, S.T., Kim, M.-J., and Park, J. (2002) *Angew. Chem. Int. Ed.*, **41**, 2373; (e) Verzijl, G.K.M., De Vries, J.G., and Broxterman, Q.B. (2005) *Tetrahedron Assymetry*, **16**, 1603; (f) Berkessel, A., Luisa Sebastian-Ibarz, M., and Mueller, T.N. (2006) *Angew. Chem. Int. Ed.*, **45**, 6567.

46 For the DKR of amines, see: (a) Choi, Y.K., Kim, M.J., Ahn, Y., and Kim, M.J. (2001) *Org. Lett.*, **3**, 4099; (b) Parvulescu, A., De Vros, D., and Jacobs, P. (2005) *Chem. Commun.*, 5307; (c) Kim, V., Kim, W.H., Han, K., Choi, Y.K., and Park, J. (2007) *Org. Lett.*, **9**, 1157; (d) Gastaldi, S., Escoubet, S., Vanthuyne, N., Gil, G., and Bertrand, M.P. (2007) *Org. Lett.*, **9**, 837; (e) Stirling, M., Blacker, J., and Page, M.I. (2007) *Tetrahedron Lett.*, **48**, 1247; (f) Paetzold, J., and Bäckvall, J.-E. (2005) *J. Am. Chem. Soc.*, **127**, 17620; (g) Pàmies, O., Ell, A.H., Samec, J.S.M., Hermanns, N., and Bäckvall, J.-E. (2002) *Tetrahedron Lett.*, **43**, 4699; (h) Veld, M.A.J., Hult, K., Palmans, A.R.A., and Meijer, E.W. (2007) *Eur. J. Org. Chem.*, **32**, 5416.

47 Kim, M.-J., Chung, Y.I., Choi, Y.K., Lee, H.K., and Park, J. (2003) *J. Am. Chem. Soc.*, **125**, 11494.

48 Verzijl, G.K.M., De Vries, J.G., and Broxterman, Q.B. (2001) PCT Int. Appl. WO2001090396.

49 Zhou, J., Wang, W., Thurecht, K.J., Villarroya, S., and Howdle, S.M. (2006) *Macromolecules*, **39**, 7302.

50 For reviews on this topic, see: (a) Gross, R.A., Kumar, A., and Kalra, B. (2001) *Chem. Rev.*, **101**, 2097; (b) Kobayashi, S., Uyama, H., and Kimura, S. (2001) *Chem. Rev.*, **101**, 3793; (c) Varma, I.K., Albertsson, A.-C., Rajkhowa, R., and Srivastava, R.K. (2005) *Prog. Polym. Sci.*, **30**, 949; (d) Matsumura, S. (2006) *Adv. Polym. Sci.*, **194**, 95; (e) Kobayashi, S. (2009) *Macromol. Rapid Commun.*, **30**, 237.

51 Hilker, I., Rabani, G., Verzijl, G.K.M., Palmans, A.R.A., and Heise, A. (2006) *Angew. Chem. Int. Ed.*, **45**, 2130.

52 Magnusson, A.O., Takwa, M., Hamburg, A., and Hult, K. (2005) *Angew. Chem. Int. Ed.*, **44**, 2.

53 Karvembu, R., Phabharakan, R., and Natarajan, K. (2005) *Coord. Chem. Rev.*, **249**, 911.

54 van As, B.A.C., van Buijtenen, J., Mes, T., Palmans, A.R.A., and Meijer, E.W. (2007) *Chemistry*, **13**, 8325.

55 Kanca, U., van Buijtenen, J., van As, B.A.C., Korevaar, P., Vekemans, J.A.J.M., Palmans, A.R.A., and Meijer, E.W. (2008) *J. Polym. Sci. Part A Polym. Chem.*, **46**, 2721.

56 Cammenberg, M., Hult, K., and Park, S. (2006) *ChemBioChem*, **7**, 1745.

57 (a) Focarete, M.L., Scandola, M., Kumar, A., and Gross, R.A. (2001) *J. Polym. Sci. Part B Polym. Phys.*, **39**, 1721; (b) Gazzano, M., Malta, V., Focarete, M.L., Scandola, M., and Gross, R.A. (2003) *J. Polym. Sci. Part B Polym. Phys.*, **41**, 1009; (c) Van der Mee, L., Helmich, F., De Bruijn, R., Vekemans, J.A.J.M., Palmans, A.R.A., and Meijer, E.W. (2006) *Macromolecules*, **39**, 5021.

58 Peeters, J.W., Palmans, A.R.A., Veld, M., Scheijen, F., Heise A., and Meijer, E.W. (2004) *Biomacromolecules*, **5**, 1862.

59 van As, B.A.C., van Buijtenen, J., Heise, A., Broxterman, Q.B., Verzijl, G.K.M., Palmans, A.R.A., and Meijer, E.W. (2005) *J. Am. Chem. Soc.*, **127**, 9964.

60 van Buijtenen, J., van As, B.A.C., Meuldijk, J., Palmans, A.R.A., Vekemans, J.A.J.M., Hulshof, L.A., and Meijer, E.W. (2006) *Chem. Commun.*, 3169.

61 Dai, S.Y., and Li, Z. (2008) *Biomacromolecules*, **9**, 1883.

62 van As, B.A.C., Thomassen, P., Kalra, B., Gross, R.A., Meijer, E.W., Palmans, A.R.A., and Heise, A. (2004) *Macromolecules*, **37**, 8973.

63 Koayashi, S., Morino, K., and Yashima, E. (2007) *Chem. Commun.*, 2351.

64 Duxbury, C.J., Hilker, I., de Wildeman, S.M.A., and Heise, A. (2007) *Angew. Chem. Int. Ed.*, **46**, 8452.

65 Padovani, M., Hilker, I., Duxbury, C.J., and Heise, A. (2008) *Macromolecules*, **41**, 2439.

12
Enzymes in the Synthesis of Block and Graft Copolymers

Steven Howdle and Andreas Heise

12.1
Introduction

The potential for exploiting enzymes in the synthesis of various synthetic polymers has been shown extensively in recent years [1]. The application of enzymes in polymer synthesis and transformation is attractive because of their ability to function under mild conditions with high enantio- and regioselectivity. An exceptionally successful class of enzymes in that respect is lipases and in particular *Candida antarctica* lipase B (CALB) immobilized on macroporous resin (Novozym® 435) [2] (see also Chapter 3); the most common enzyme used for polymerization. If not otherwise noted, the term enzyme in this Chapter will refer to Novozym 435 or CALB. The stability of lipases in organic media and their ability to promote transesterification and condensation reactions on a broad range of low and high molar mass substrates has been used in numerous examples of ring-opening polymerization (ROP) and polycondensation (see also Chapters 4, 5 and 11) [3]. Enzymatic polymerization can therefore be an interesting complementary, and in some cases alternative synthetic method in the toolbox of polymer chemists. However, its full exploitation in polymer science will require an understanding of how to realize not only simple linear homo- and copolymers, but also more complex polymer architectures. The design of complex macromolecular architectures has become a particular focus in polymer science. Since some of these architectures possess unique properties, which makes them interesting candidates for specialty applications in nanostructured and biomedical materials such as block copolymers. The unique features of enzymatic polymerization have the potential to further increase the diversity of polymeric materials if they can be fully integrated with other (chemical) polymerization techniques so as to achieve block copolymers. In particular the combination of the distinctive attributes of enzymatic polymerization with, for example controlled polymerization techniques, can allow the formation of polymer structures with new properties not available by chemical polymerization. The synthesis of block copolymers by using enzymatic polymerization for at least one block has been the focus of an increasing number of papers over the last few years. The goal of this line of research is

(i) to develop understanding of the scope and limitations of enzymatic ROP in the synthesis of complex structures and (ii) to make materials not available from chemical catalysis by taking advantage of the unique selectivity of enzymes.

12.2
Synthetic Strategies for Block Copolymer Synthesis Involving Enzymes

Conventional strategies of macroinitiation have also been applied to the enzymatic synthesis of block copolymer as shown in Scheme 12.1. This includes enzymatic polymerization from functional polymers (macroinitiators) and the enzymatic synthesis of macroinitiators followed by another polymerization technique. Mechanistically, lipase-catalyzed ROP and also polycondensation can best be described as an activated monomer mechanism in which the lipase activates the carbonyl group of a lactone or carboxylic acid derivative (enzyme activated monomer, EAM). Attack of a nucleophile–in many cases a hydroxyl compound–results in the chain extension of the polymer. This mechanism has the consequence that the enzymatic polymerization has no 'living' character and provides limited control over molecular weight and polymer end-groups as compared to controlled polymerisations. Any strategy for the synthesis of block copolymers must take this into account as it might affect the yield of the block copolymer product. Typical concerns are the amounts of homopolymer and cyclic polymers formed from the enzymatic reaction step. Moreover, the lipase does not distinguish between the ester group of a monomer and the ester group of a polyester chain. Activation of the latter results in an enzyme activated chain segment (EACS), which then can be attacked by a nucleophile. Consequently, any nucleophile–whether low molecular weight or polymeric–will be incorporated by both mechanisms, that is, initiation and transesterification.

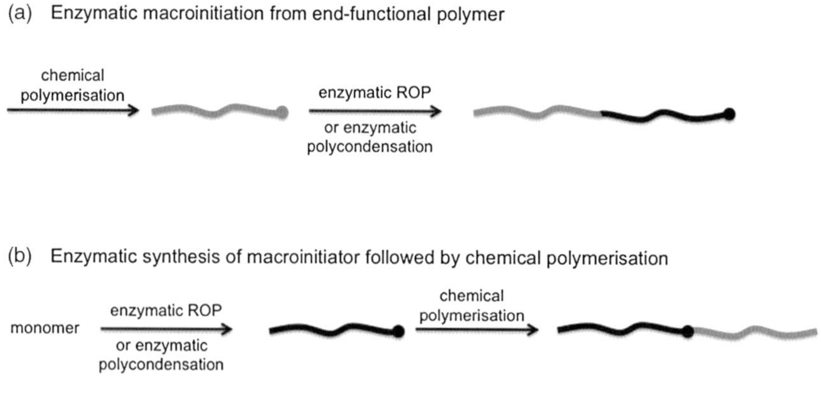

(a) Enzymatic macroinitiation from end-functional polymer

chemical polymerisation → enzymatic ROP or enzymatic polycondensation →

(b) Enzymatic synthesis of macroinitiator followed by chemical polymerisation

monomer → enzymatic ROP or enzymatic polycondensation → chemical polymerisation →

Scheme 12.1 Synthetic strategies reported for the synthesis of block copolymers involving enzymes.

Using the terms 'initiator' or 'macroinitiator' is mechanistically thus not completely correct for the enzymatic polymerization and these terms will only be used as descriptive terms in this chapter. Nevertheless, a number of very successful synthetic strategies have been published in the past and will be summarized in this chapter.

12.2.1
Enzymatic Polymerization from Functional Polymers (Macroinitiation)

The synthesis of block copolymers by enzymatic macroinitiation is mainly reported from chemically obtained end-functionalized polymers (see also Chapter 9). One example of enzymatic macroinitiation was published by Gross and Hillmyer using anionically synthesized hydroxy functional polybutadiene of various molecular weights (2600–19 000 g mol^{-1}) [4]. In a systematic study the authors investigated the efficiency of the macroinitiation of caprolactone (CL) and pentadecalactone (PDL) by Novozym 435 as a function of the polybutadiene macroinitiator. The reaction profile showed that polybutadiene consumption steadily increased with the reaction time according to incorporation by initiation and transesterification (Figure 12.1). It was found that all polybutadiene macroinitiators formed block copolymers irrespective of their molecular weight, with initiation efficiency of >80% and the presence of water-initiated chains was less than 30%. Moreover, the presence of cyclic polyesters was confirmed by ^{1}H-NMR spectroscopy.

Poly(ethylene glycol) (PEG)-based di- and triblock copolymers were prepared in a similar approach from the corresponding hydroxyl-functionalized PEG macroinitiators. Kaplan conducted a detailed mechanistic and kinetic study of the interplay between monomer conversion, chain initiation and chain propagation

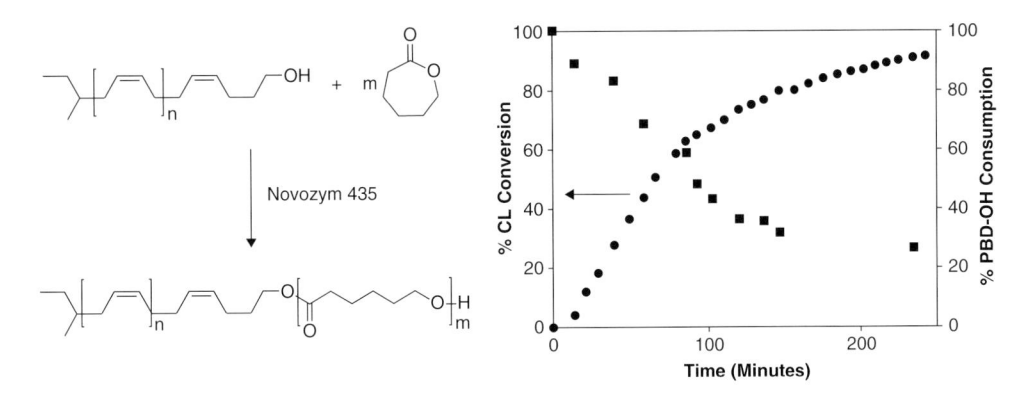

Figure 12.1 Synthesis of poly(butadiene-b-caprolactone) from hydroxyl functional polybutadiene. The plot shows the consumption of the macroinitiator (■) and the CL conversion (●) as a function of time (Reproduced with permission from [4]. Copyright (2002) American Chemical Society).

of the enzymatic ROP of CL in the presence of mono-hydroxy functional methoxy-PEG (MPEG) [5]. The results confirmed the dynamic character of the reaction and the presence of a product mixture containing unreacted MPEG, water-initiated PCL, cyclic PCL and the corresponding block copolymer. At otherwise identical reaction conditions, the amount of reacted MPEG was concentration dependent, being 35% at higher concentrations and up to 84% at lower concentrations for the same reaction time. It was suggested that PCL chains exclusively initiated by MPEG would be obtained at 'infinite reaction times'. In a more materials oriented study Feng He investigated the enzymatic ROP of CL in the presence of mono- and dihydroxyl PEG (4600 g mol^{-1}) [6]. It was found that the reactions in toluene produced higher block copolymer yields than the reactions in bulk and that the block copolymer composition was closer to the monomer feed ratio for the triblock copolymers. The semicrystalline polymers were solution-cast and enzymatic degradation experiments with *Pseudomonas* lipase revealed that the introduction of PEG did not significantly alter the degradation behavior. As part of a broader study on the enzymatic ROP from various alcohols Albertsson also reported the initiation from PEG diols (1000 and 2000 g mol^{-1}) with CL and 1,5-dioxepan-2-one (DXO) [7]. A correlation between the monomer feed ratio and the block copolymer molecular weights was found. Moreover, PCL diol was used as an initiator for DXO and within a 2-hour reaction time, gave rise to block copolymers.

Other amphiphilic block copolymers were obtained by porcine pancrease lipase initiation of 5-benzyloxytrimethylene carbonate (BTMC) from mono-hydroxyl poly(vinyl pyrolidone) (PVP-OH) [8]. Kaihara reported the enzymatic ROP of trimethylene carbonate from a copolymer of PEG and a cyclic acetal with two terminal hydroxyl groups [9]. The latter is a degradable segment and the resulting amphiphilic triblock copolymer was used to form micellar structures for which pH-dependent drug release was successfully shown.

A very interesting block copolymer synthesis was recently reported by Zhi Li using poly[(R)-3-hydroxybutyrate] (PHB) as a macroinitiator for enzymatic ROP of CL [10]. PHB is a microbial polyester, which can be produced in large quantities from renewable resources. Incorporation in block copolymers could overcome limitations of the applicability of PHB in thermoplastic applications by improving its thermal properties. Microbial PHB has primary and secondary (chiral) hydroxyl endgroups and the authors were able to demonstrate that PCL chain extension only occurred from the primary alcohols with no randomization (Figure 12.2). This is a very elegant example in which the enzymatic selectivity is used to distinguish between two hydroxyl groups in the enzymatic macroinitiation.

While all previous examples employ enzymatic ROP there are two reports on the block copolymer synthesis using enzymatic polycondensation. The first was published by Gross and Scandola and describes the synthesis and solid-state properties of polyesteramides with poly(dimethylsiloxane) (PDMS) blocks (Figure 12.3) [11]. The polycondensation was carried out with various

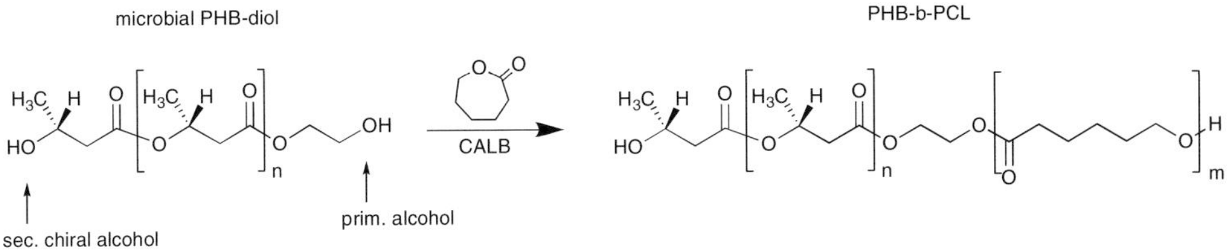

Figure 12.2 Enantioselective enzyme-catalyzed ring-opening polymerization of caprolactone from PHB-diol [10].

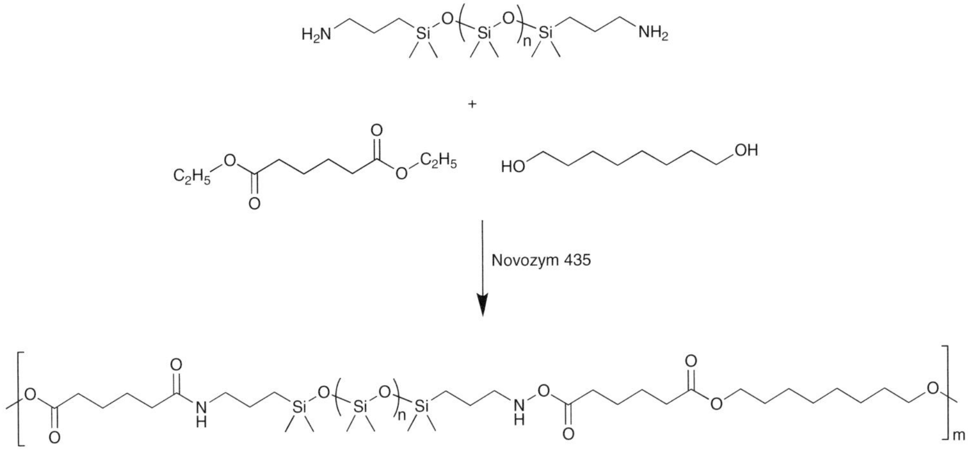

Figure 12.3 Poly(dimethyl siloxane) polyester amide block copolymers by enzymatic polycondensation of (diaminopropyl)polydimethylsiloxanes, diethyl adipate, and 1,8-octanediol [11].

ratios of dimethyl adipate, octanediol and diamine-functionalized PDMS (M_w = 875). Depending on the composition of the polymer, the physical properties varied from hard solid to sticky materials. The study shows that these versatile materials can be obtained under mild enzymatic conditions circumventing the usual harsh chemical reaction conditions. The second report by Kumar describes the synthesis of multi-block copolymers of poly(tetrahydrofuran) (PTHF) and PEG [12]. In a two-step procedure, first the hydroxyl groups of PTHF diol (M_n = 1000 g mol^{-1}) were enzymatically reacted with diethyl maloate. In the second step an enzymatic polycondensation with PEG diol of various molecular weights was carried out to molecular weights of 13 000–18 000 g mol^{-1}. Further experiments suggested that these amphiphilic polymers form micelles in water.

12.2.2
Enzymatic Synthesis of Macroinitiators Followed by Chemical Polymerization

The strategy of macroinitiaition from an enzymatically synthesized polymer is the most common way of producing block copolymers. Mostly controlled polymerization techniques like atom transfer radical polymerization ATRP, nitroxide-mediated polymerization (NMP) or reversible addition fragmentation chain-transfer (RAFT) were applied as the non-enzymatic polymerization techniques. They offer the advantage of high initiation efficiency and control over the molecular weight, which generally results in well-defined block copolymers. This method requires the end-functionalisation of the enzymatic block with an initiator moiety capable of selectively initiating the subsequent polymerization technique. Two main strategies have been applied to achieve this: (i) The enzymatic block is end-functionalized with an initiator end-group in a separate reaction step after the enzymatic polymerization. (ii) The initiator is incorporated during the enzymatic polymerization. The latter approach demands a so-called bifunctional or dual initiator (Figure 12.4).

12.2.2.1 Dual Initiator Approach

Several dual initiator structures have been reported and are summarized in Table 12.1. They have the common structural feature of a functional group capable of initiating the enzymatic polymerization (hydroxyl) and a group for the second polymerization. This approach has the advantage that polymer analogous end-group modification is avoided. On the other hand, for a high block copolymer yield it is important to ensure a high degree of incorporation of the dual initiator into the enzymatic block. Every non-functionalized block will be inactive in the macroinitiation and thus remain in the product as a homopolymer. Considering that nucleophiles like the applied dual initiators are incorporated by initiation and transesterification, controlling the end-group fidelity in this process becomes critical.

The initial contribution to the development of such dual processes has been made by Heise and Palmans, who reported the combination of enzymatic ROP and ATRP [13]. The authors applied dual initiator 1 and showed that during the course of the enzymatic ROP of CL the initiator is consumed to about 95%

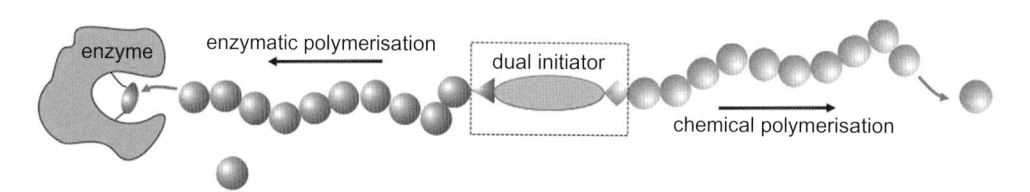

Figure 12.4 Dual or bifunctional initiator approach to block copolymers combining an enzymatic with a non-enzymatic chemical polymerization.

12.2 Synthetic Strategies for Block Copolymer Synthesis Involving Enzymes

Table 12.1 Initiators and monomer used in dual initiator approach to block copolymers. CL: caprolactone; 4MCL: 4-methyl caprolactone; MMA: methyl methacrylate; GMA: glycidyl methacrylate; FOMA: perfluorooctyl methacrylate; 10-HA: 10-hydroxydecanoic acid.

Dual initiator	Enzymatic monomer	Chemical monomer (technique)	References
1	CL	Styrene, MMA (ATRP)	[13, 14]
2	4MCL	MMA (ATRP)	[14, 15]
3	CL	FOMA, MMA (ATRP)	[14, 16–19]
4	CL, 10-HA	Styrene, GMA (ATRP)	[20–23]
5	CL, 4MCL	Styrene (NMP)	[24]
6	CL	Styrene (RAFT)	[25]
7	CL	Styrene (FRP)	[26]

(3 hours reaction time) and incorporated into the PCL. Subsequent macroinitiation of styrene from the ATRP initiator-functionalized PCL yielded the block copolymer. In a follow-up paper the same authors further investigated the influence of the reaction conditions and the initiator structure on the chemoenzymatic synthesis of block copolymer by this combination of polymerization techniques [14]. By applying a careful drying method, the amount of water-initiated PCL was reduced to <5%. Special attention was given to the comparison of the efficiency of the dual initiator in the enzymatic ROP of CL by comparison of initiators 1, 2 and 3. It was found that the latter two are incorporated relatively fast (>90% at

15 min. reaction time), while initiator 1 was incorporated at a very slow rate (40% at 60 min. reaction time). This suggests that 1 is a poor initiator and is predominantly incorporated by transesterification. This dependence of initiation process on the dual initiator structure was further confirmed by an investigation employing liquid chromatography under critical conditions (LCCC), which allows the separation of the formed PCL chains based on end-group functionality [27]. The obtained data show that all processes initially start with a high concentration of water-initiated PCL. As the reaction proceeds, the concentration of water-initiated polymers decreases, while the concentration of initiator-functionalized chains increases (Figure 12.5). The rate of the formation of the PCL end-capped with 1 is much slower than for 3, which means that it is much less feasible to produce a high concentration of ATRP macroinitiators with this dual initiator at short reaction times.

Moreover, Heise and Palmans investigated the possibility of conducting both polymerizations as a one-pot cascade reaction in the presence of the dual initiator, ATRP catalyst, CL and methacrylates [16]. The study revealed that certain ATRP catalysts have a strong inhibiting effect on the enzyme. A nickel catalyst, for example, completely inhibited the enzyme, while certain copper catalysts had no

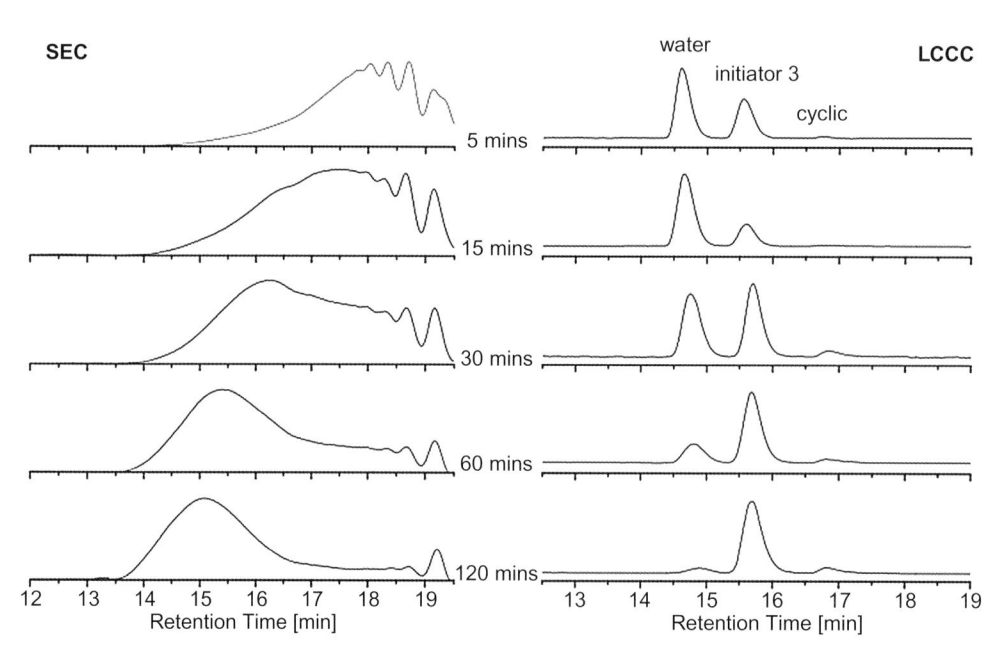

Figure 12.5 Comparison of size exclusion chromatography (SEC) and liquid chromatography under critical conditions (LCCC) for the bulk enzymatic polymerization of caprolactone in the presence of bifunctional initiator 3 as a function of time. LCCC shows the evolution of polymers end-capped (initiated) with water, initiator 3 and cyclic polymer. (Reproduced with permission from [27]. Copyright (2008) American Chemical Society).

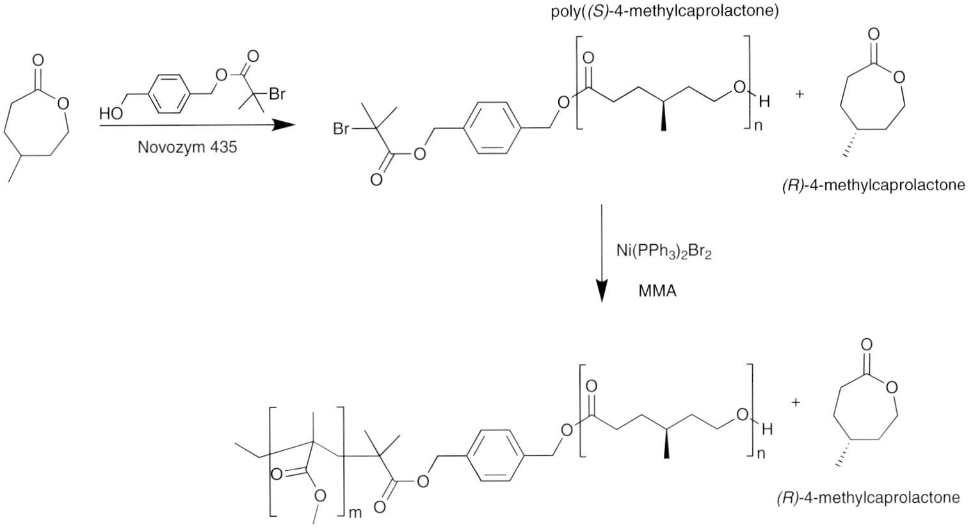

Figure 12.6 Chiral block copolymers combining enantioselective enzymatic ROP with ATRP.

or little effect on the enzyme. When the ATRP catalyst was added after the enzymatic process, high block copolymer yields were obtained, while homopolymer impurities were observed when all components were present from the beginning.

By taking advantage of nickel inhibition and the stereoselectivity of the enzyme, chiral block copolymers were obtained by this method from methyl-substituted CL such as 4-methyl caprolactone (MCL) and initiator 2 (Figure 12.6) [15]. The polymerization of racemic MCL showed good enantioselectivity and produced a chiral macroinitiator with ATRP end-group by polymerizing only the (R)-MCL. ATRP macroinitiation was then started by simply adding the nickel ATRP catalyst and MMA: methyl methacrylate to the reaction mixture, which produced the chiral block copolymer poly(MMA-b-(R)MCL). In this reaction nickel has a dual role: besides its catalytic function in the ATRP it also inhibits the enzyme and prevents the slow racemization of the polyester block by transesterification with the remaining (S)-MCL.

Howdle reported that a one-pot, simultaneous synthesis of block copolymers by enzymatic ROP and ATRP employing initiator 3, CL and MMA is possible in supercritical CO_2 (scCO_2) [17]. scCO_2 is a unique solvent because it combines gas-like and liquid-like properties; most importantly in this case it plasticizes and liquefies polymers very effectively, allowing enhanced mass transport which contributes to more efficient polymerization, particularly important for a supported solid phase enzyme catalyst (see also Chapter 13) [28]. The authors also show that the CL acts as a scCO_2 co-solvent which was crucial to allow the radical polymerization to remain homogeneous and controlled [18]. The unique ability

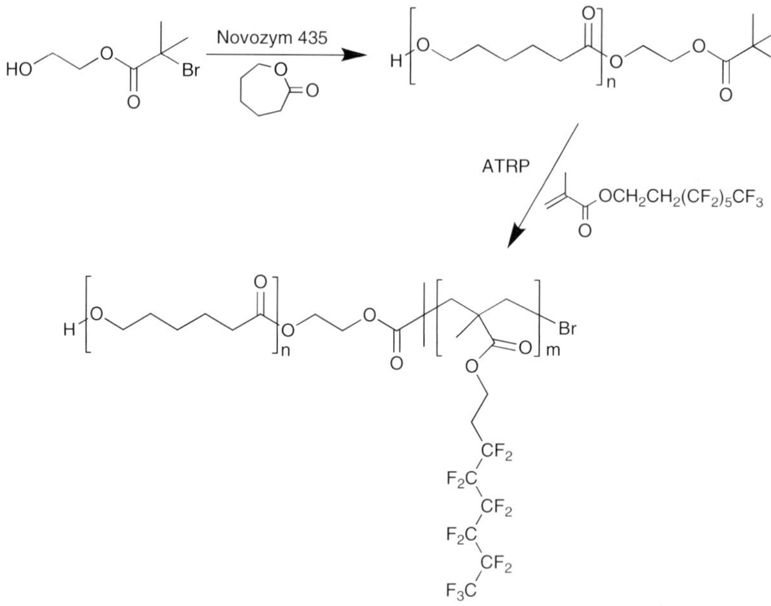

Figure 12.7 Two-step chemoenzymatic polymerization of PFOMA-b-PCL block copolymer [19].

of scCO$_2$ to solubilize highly fluorinated species was used by extending this methodology to the synthesis of novel copolymers consisting of a semifluorinated block of poly(1H, 1H, 2H, 2H-perfluorooctyl methacrylate) (FOMA) and PCL. Block copolymers were successfully synthesized by a two step process based on the sequential monomer addition (Figure 12.7). Parallel experiments in conventional solvents did not yield any block copolymers due to the limited solubility of FOMA in these solvents. The chemoenzymatic synthesis of block copolymers in scCO$_2$ has recently been reviewed [29].

Jingyuan Wang also applied ATRP for the chemoenzymatic synthesis of block copolymers. In a first series of publications the group reported the successful synthesis of a block copolymer comprising PCL and polystyrene blocks from the commercially available dual initiator 4 [20, 21]. This concept was then further applied for the chemoenzymatic synthesis of amphiphilic block copolymers by macroinitiation of glycidyl methacrylate (GMA) from the ATRP functional PCL [30]. This procedure yielded well-defined block copolymers, which formed micelles in aqueous solution (Figure 12.8). Jingyuan Wang was the first who applied the dual enzyme/ATRP initiator concept to an enzymatic polycondensation of 10-hydroxydecanoic acid [22]. The incorporation of the initiator was confirmed by ^{1}H-NMR spectroscopy and the subsequent ATRP of styrene produced the block copolymer. This concept was than extended to the ATRP of GMA and the formation of vesicles from the corresponding block copolymer [23].

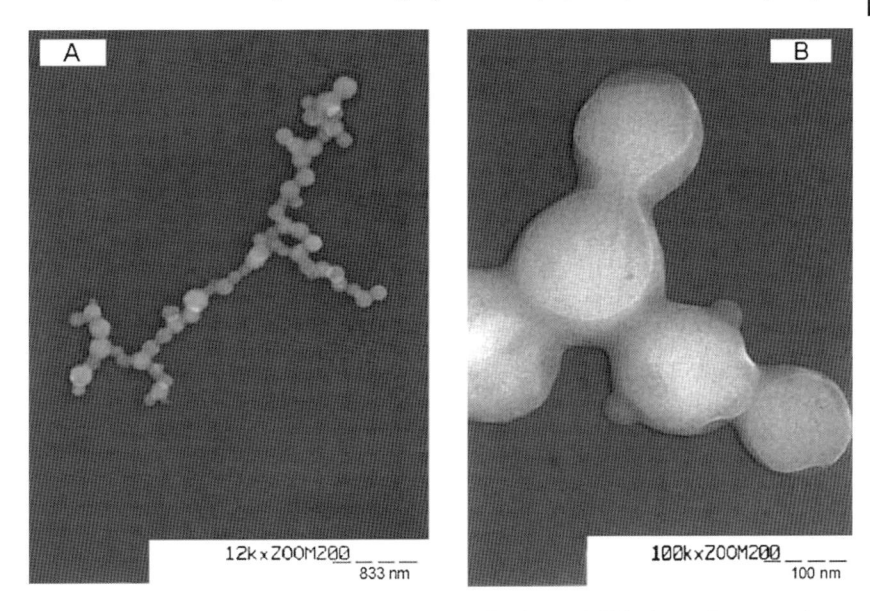

Figure 12.8 TEM micrographs of chemoenzymatically obtained diblock copolymer PCL-b-PGMA micelles sprayed from THF/H$_2$O with 2%, by volume, of THF. (Reproduced with permission from [30]. Copyright (2008) John Wiley & Sons Inc.).

Apart from ATRP the concept of dual initiation was also combined with other (controlled) polymerization techniques. Nitroxide-mediated living free radical polymerization is one example reported by Heise and Palmans, which has the advantage that no further metal catalyst is required [24]. Employing initiator 5 a PCL macroinitiator was obtained and subsequent polymerization of styrene produced a block copolymer. With this system it was for the first time possible to successfully conduct a one-pot chemoenzymatic cascade polymerization from a mixture containing 5, CL and styrene. Since the activation temperature of NMP is around 100 °C, no radical polymerization will occur at the reaction temperature of the enzymatic ROP (60 °C). The two reactions could thus be thermally separated by first carrying out the enzymatic polymerization at low temperature and then raise the temperature to around 100 °C to initiate the NMP. Moreover, it was shown that this approach is compatible with the stereoselective polymerization of MCL for the synthesis of chiral block copolymers (Figure 12.9) (see also Chapter 11).

Similarly successful was the combination of enzymatic ROP with RAFT in scCO$_2$ reported by Howdle [25]. Employing the RAFT agent 6 a block copolymer comprising PCL and polystyrene was obtained and the block structure confirmed by gradient polymer elution chromatography (GPEC). Like NMP, RAFT has the advantage that it does not require a metal catalyst and coupled with the green solvent CO$_2$ this process is environmentally desirable (see Chapter 13).

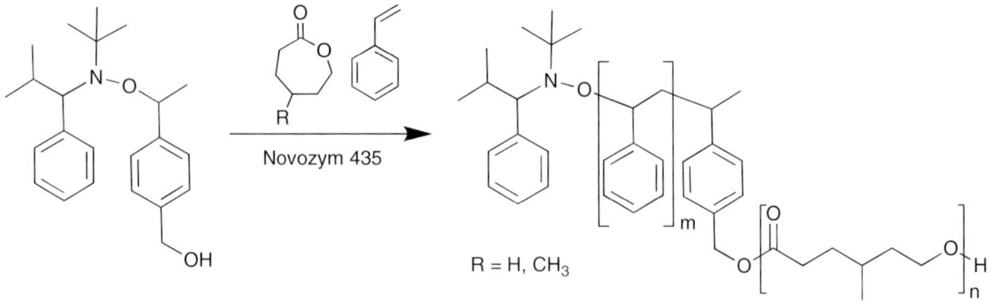

Figure 12.9 Block copolymers by one-pot enzymatic ROP and nitroxide-mediated living free-radical cascade polymerization [24].

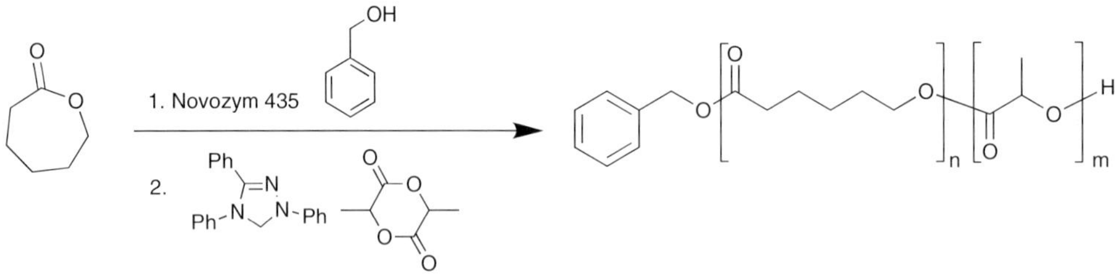

Figure 12.10 Block copolymers by combination of enymatic ROP and carbene-catalyzed ROP [31].

Ritter reported a radical chain transfer agent as a dual initiator 7 [26]. The first step builds on the fact that hydroxyl groups are much better nucleophiles in enzymatic ROP than thiols. Due to the chemoselectivity of the enzyme, PCL with predominantly thiol end-groups were obtained, which were subsequently used as a macroinitiator for styrene. The authors further report that the reaction yield can be further increased by microwave irradiation. While thiols provide less control over the radical polymerization than RAFT agents, the subsequent radical polymerization successfully lead to the synthesis of poly(CL-b-styrene).

Very recently Heise and Dubois reported the first example of the combination of enzymatic and chemical carbene-catalyzed ROP for the formation of block copolymers (Figure 12.10) [31]. In their approach they took advantage of the fact that CALB has a high catalytic activity for lactones but not for lactides (LLA), while it is visa versa for carbenes. In the applied synthetic strategy the enzymatic polymerization of CL was conducted first. Addition of lactide and carbene to the reaction mixture lead to the macroinitiation directly from the hydroxyl end-group of PCL allowing the syhthesis of PCL/PLLA block copolymers.

12.2.2.2 Modification of Enzymatic Blocks to Form Macroinitiators

The second strategy for the chemoenzymatic synthesis of block copolymers from enzymatic macroinitiators employs an individual modification step of the enzymatic block with an initiator for the chemical polymerization. This strategy has the advantage that it does not require a high incorporation rate of the dual initiator. On the other hand, quantitative end-functionalisation is becoming more difficult with increasing molecular weight of the polymer. Initial work following this approach was carried out by Jingyuan Wang who reported the quantitative functionalisation of enzymatically obtained PCL with α-bromopropionyl bromide [32]. The subsequent ATRP of styrene yielded the block copolymer. Based on this successful strategy also GMA was polymerized from the PCL macroinitiator to yield amphiphilic bock copolymers, which assemble into nanoscale micelles [33]. Pentablock copolymers where also obtained by combining this approach with the enzymatic macroinitiation of CL from PEG diol (Figure 12.11) [34]. The obtained PCL diol was modified with α-bromopropionyl to yield a telechelic ATRP macroinitiator. After ATRP of styrene, the obtained PS-*b*-PCL-*b*-PEG-*b*-PCL-*b*-PS revealed interesting self-assembly behavior in water forming spherical micelles, rod-like micelles, vesicles, lamellae and large compound micelles depending on the block copolymer composition and concentration (Figure 12.11). The same author also extended this concept to the synthesis of

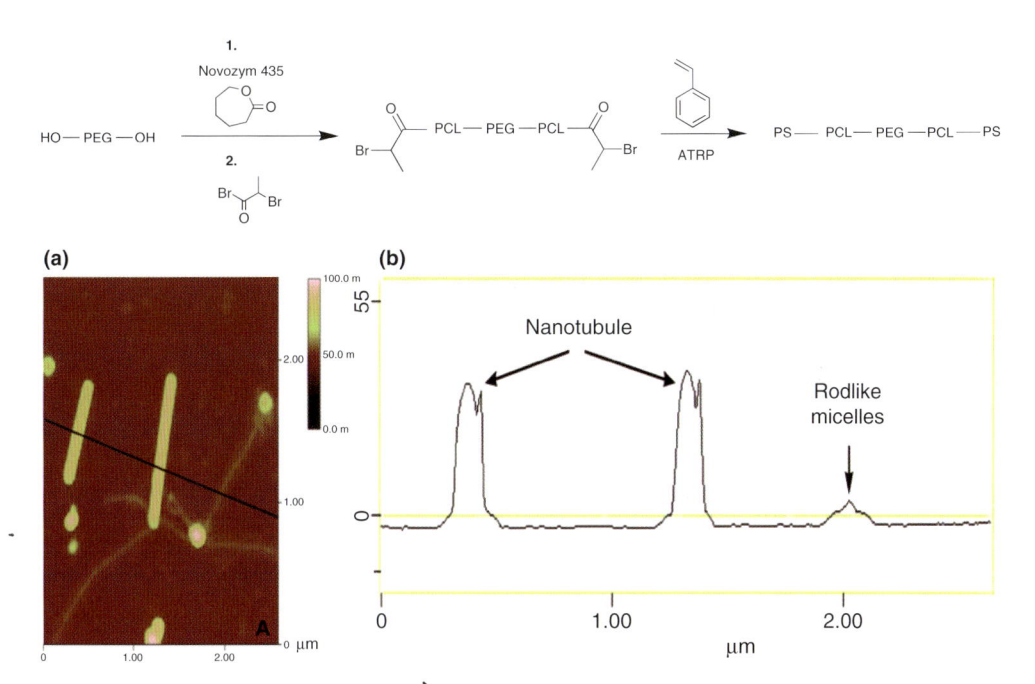

Figure 12.11 Synthesis of the pentablock copolymer PSt-*b*-PCL-*b*-PEG-*b*-PCL-*b*-PSt by enzymatic ROP and ATRP and AFM height image (a), and cross-sectional analysis (b) of the nanoscale aggregates of the block copolymer after exposure to high vacuum at an initial copolymer concentration of 1 mg ml^{-1}. (Reproduced with permission from [34]. Copyright (2008) American Chemical Society).

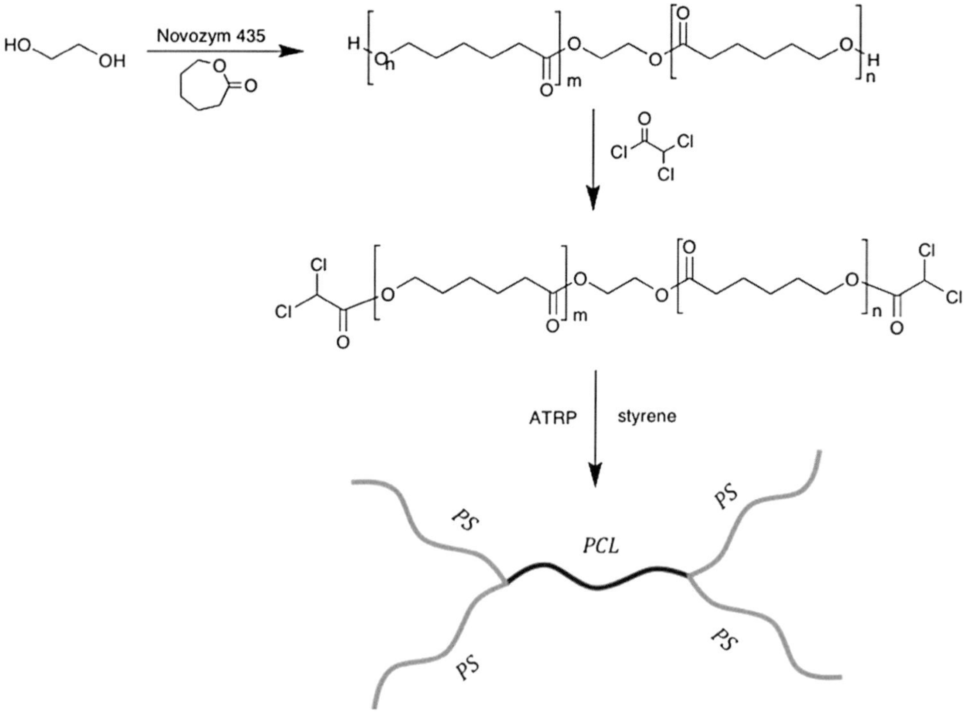

Figure 12.12 Synthesis route to the H-shape block copolymers by enzymatic ROP and ATRP [38].

hydroxyl functional polymer macroinitiators by enzymatic polycondensation of 10-decanoic acid [35].

A series of interesting block copolymer architectures has also been prepared by Jingyuan Wang. In a first paper the synthesis of H-shaped triblock copolymers was demonstrated from enzymatically obtained PCL diol after end-functionalisation with a difunctional ATRP initiator (Figure 12.12) [36]. This allowed growth of two polystyrene chains from each end of the telechelic PCL. When methanol instead of glycol was used as the initiator in the initial enzymatic CL polymerization, a PCL with one hydroxyl end-group was obtained. Functionalisation of this end-group with the difunctional ATRP initiator and subsequent ATRP of styrene resulted in Y-shaped polymers [37]. Again, by the polymerization of GMA from this polymer PCL precursor, amphiphilic Y-shaped block copolymers capable of micellization were realized [38].

A synthetic approach to poly(ester-urethanes) was recently published by Zhi Li [39]. Glycerol was employed as a trifunctional initiator in the enzymatic ROP of CL. The three arm PCL triol was then reacted with methylene-diphenyl diisocyanate (MDI) and hexanediol to yield a three arm PCL-based poly(ester-urethane) with shape memory properties.

12.3
Enzymatic Synthesis of Graft Copolymers

Graft copolymers were prepared by both classical strategies, that is, from enzymatically obtained macromonomers by subsequent chemical polymerization and by enzymatic grafting from hydroxyl functional polymers. Kalra *et al.* studied the synthesis of PPDL graft copolymers from macromonomers, which were obtained by the enzymatic ROP of pentadecalactone (PDL) from hydroxyethyl methacrylate (HEMA) and poly(ethylene glycol) methacrylate (PEGMA) [40]. Subsequently graft copolymers were obtained by free radical polymerization of the macromonomers. A similar approach was published by Srivastava for the HEMA-initiated enzymatic ROP of CL and subsequent free radical polymerization [41].

The enzymatic grafting from a polymer with pendant hydroxy groups was mainly studied by three groups using different polymer backbones and reaction conditions. Hans *et al.* employed well-defined polyglycidols and compared the chemical and enzymatic ROP of CL from linear and star-shaped polyglycidol obtained from anionic ROP [42]. It was found that the zinc-catalyzed ROP of CL from the multifunctional polyglycidol resulted in a quantitative initiation efficiency, while enzymatic ROP PCL was only grafted to 15–20% of the hydroxyl groups. This leads to different polymer architectures, that is, the fully grafted polymer resembles a core-shell structure with hydrophilic polyether core and hydrophobic PCL shell. The enzymatically-obtained partly grafted polymer, on the other hand, has a hydrophilic polyether head with a hydrophobic PCL tail. The difference in the initiation efficiency is due to the different polymerization mechanisms. While the chemical ROP is end-group activated and controlled, the enzymatic ROP is monomer activated. In the latter case the steric constrains in the nucleophilic attack of the hydroxyl groups on the EAM prevents the reactions of all hydroxyl groups. Similar results were obtained in a study by Duxbury *et al.* of the enzymatic ROP from polystyrene containing 10% of 4-vinylbenzyl alcohol [43]. A grafting efficiency of PCL of 50–60% was reported. Moreover, the results suggested that the grafting action is a combination of 'grafting from' and 'grafting onto' by transesterification of PCL.

The incomplete enzymatic grafting from polyalcohols opens opportunities to synthesize unique structures as was shown in several example by Hans *et al.* For example, heterografted molecular bottle brushes were synthesized starting from PCL grafted polyglycidol (50% grafting efficiency) [44]. The PCL graft ends were then selectively reacted with vinyl acetate and the hydroxy groups and the polyglycidol backbone chemically converted to ATRP initiator groups. Subsequent initiation of MMA and *n*-butyl acrylate yielded well-defined heterografted polymers. Alternatively the remaining hydroxyl groups on the polyglycidol backbone were used to initiate chemically-catalyzed ROP of lactide to produce a heterografted polymer comprising PCL and PLA grafts [45]. When the end-groups of the PCL were capped with acrylates, the resulting materials could be formulated into UV crosslinked microspheres.

Enzymatic synthesis of graft copolymers in supercritical CO_2 was reported by Villarroya *et al.* from a poly(MMA-co-HEMA) random copolymer by the combination of ATRP and enzymatic ROP of CL [46]. Similar to the results obtained in conventional solvents not all hydroxyl groups participated in the grafting reaction (33%), which was thought to be due to steric effects. In another study the authors confirmed two reasons for the limitation in the grafting efficiency: (i) The poly(MMA-co-HEMA) probably had a blocky structure due to the different reactivity ratios of the monomers in the radical polymerization. (ii) The hydroxy groups are too close to the polymer backbone. Consequently the grafting efficiency was significantly improved when highly randomized poly(MMA-co-HEMA) from starved-feed polymerization was used (80%) and when a PEG spacer was introduced between the polymer backbone and the hydroxyl group (100%) [47].

12.4
Summary and Outlook

Enzymatic polymerization has established itself as a useful technique in the synthetic toolbox of polymer chemists. The versatility of this technique for homo-and copolymerization has been known for some time. With the increasing number of reports on the synthesis of more complex structures like block copolymers, its versatility further increased. While not a controlled polymerization technique itself, by clever reaction design and integration with other polymerization techniques like controlled radical polymerization, well-defined polymer structures were realized. Early studies were mainly focused at a proof of principle of the compatibility of enzymatic polymerization with other techniques. In recent studies a shift of focus towards novel block copolymers can be observed. Specific unique attributes of the enzyme are applied to develop materials and processes difficult to achieve otherwise. Examples are the synthesis of chiral block copolymers (stereoselectivity), the selective modification of polymer end-groups using enzymatic chemoselectivity and even the synthesis of degradable/fluorinated block copolymers in a fully green process in $scCO_2$. It can be expected that this trend will continue and we will see more examples in which the specific strengths of the enzyme (selectivity) lead to unique materials.

References

1 Gross, R.A., Kumar, A., and Kalra, B. (2001) *Chem. Rev.*, **101**, 2097.
2 Kirk, O., and Christensen, M.W. (2002) *Org. Process Res. Dev.*, **6**, 446.
3 Kobayashi, S. (2009) *Macromol. Rapid Commun.*, **30**, 237.
4 Kumar, A., Gross, R.A., Wang, Y.B., and Hillmyer, M.A. (2002) *Macromolecules*, **35**, 7606.
5 Panova, A.A., and Kaplan, D.L. (2003) *Biotechnol. Bioeng.*, **84**, 103.

6. He, F., Li, S.M., Vert, M., and Zhuo, R.X. (2003) *Polymer*, **44**, 5145.
7. Srivastava, R.K., and Albertsson, A.C. (2006) *Macromolecules*, **39**, 46.
8. Wang, Y.X., He, F., and Zhuo, R.X. (2006) *Chin. Chem. Lett.*, **17**, 239.
9. Kaihara, S., Fisher, J.P., and Matsumura, S. (2009) *Macromol. Biosci.*, **9**, 613.
10. Dai, S., and Li, Z. (2008) *Biomacromolecules*, **9**, 1883.
11. Sharma, B., Azim, A., Azim, H., Gross, R.A., Zini, E., Focarete, L., and Scandola, M. (2007) *Macromolecules*, **40**, 7919.
12. Niu, L., Nagarajan, R., Guan, F., Samuelson, L.A., and Kumar, J. (2006) *J. Macromol. Sci. Pure Appl. Chem.*, **43**, 1975.
13. Meyer, U., Palmans, A.R.A., Loontjens, T., and Heise, A. (2002) *Macromolecules*, **35**, 2873.
14. de Geus, M., Peeters, J., Wolffs, M., Hermans, T., Palmans, A.R.A., Koning, C.E., and Heise, A. (2005) *Macromolecules*, **38**, 4220.
15. Peeters, J., Palmans, A.R.A., Veld, M., Scheijen, F., Heise, A., and Meijer, E.W. (2004) *Biomacromolecules*, **5**, 1862.
16. De Geus, M., Schormans, L., Palmans, A.A., Koning, C.E., and Heise, A. (2006) *J. Polym. Sci. A Polym. Chem.*, **44**, 4290.
17. Duxbury, C.J., Wang, W.X., de Geus, M., Heise, A., and Howdle, S.M. (2005) *J. Am. Chem. Soc.*, **127**, 2384.
18. Zhou, J., Villarroya, S., Wang, W., Wyatt, M.F., Duxbury, C.J., Thurecht, K.J., and Howdle, S.M. (2006) *Macromolecules*, **39**, 5352.
19. Villarroya, S., Zhou, J.X., Duxbury, C.J., Heise, A., and Howdle, S.M. (2006) *Macromolecules*, **39**, 633.
20. Sha, K., Li, D.S., Wang, S.W., Qin, L., and Wang, J.Y. (2005) *Polym. Bull.*, **55**, 349.
21. Sha, K., Li, D., Li, Y., Ai, P., Wang, W., Xu, Y., Liu, X., Wu, M., Wang, S., Zhang, B., and Wang, J. (2006) *Polymer*, **47**, 4292.
22. Sha, K., Li, D.S., Li, Y.P., Ai, P., Liu, X.T., Wang, W., Xu, Y.X., Wang, S.W., Wu, M.Z., Zhang, B., and Wang, J.Y. (2006) *J. Polym. Sci. A Polym. Chem.*, **44**, 3393.
23. Li, M., Wang, W., Pan, S., Zhang, B., and Wang, J. (2008) *Polym. Int.*, **57**, 1377.
24. van As, B.A.C., Thomassen, P., Kalra, B., Gross, R.A., Meijer, E.W., Palmans, A.R.A., and Heise, A. (2004) *Macromolecules*, **37**, 8973.
25. Thurecht, K.J., Gregory, A.M., Villarroya, S., Zhou, J., Heise, A., and Howdle, S.M. (2006) *Chem. Commun.*, 4383.
26. Kerep, P., and Ritter, H. (2007) *Macromol. Rapid Commun.*, **28**, 759.
27. de Geus, M., Peters, R., Koning, C.E., and Heise, A. (2008) *Biomacromolecules*, **9**, 752.
28. Woods, H.M., Silva, M.M.C.G., Nouvel, C., Shakesheff, K.M., and Howdle, S.M. (2004) *J. Mater. Chem.*, **14**, 1663.
29. Villarroya, S., Thurecht, K.J., Heise, A., and Howdle, S.M. (2007) *Chem. Commun.*, 3805.
30. Sha, K., Li, D., Li, Y., Liu, X., Wang, S., and Wang, J. (2008) *Polym. Int.*, **57**, 211.
31. Xiao, Y., Coulembier, O., Koning, C.E., Heise, A., and Dubois, P. (2009) *Chem. Commun.*, 2472.
32. Sha, K., Qin, L., Li, D.S., Liu, X.T., and Wang, J.Y. (2005) *Polym. Bull.*, **54**, 1.
33. Sha, K., Li, D., Li, Y., Liu, X., Wang, S., Guan, J., and Wang, J. (2007) *J. Polym. Sci. A Polym. Chem.*, **45**, 5037.
34. Sha, K., Li, D., Li, Y., Zhang, B., and Wang, J. (2008) *Macromolecules*, **41**, 361.
35. Li, D., Sha, K., Li, Y., Ai, P., Liu, X., Wang, W., and Wang, J. (2008) *Polym. Int.*, **57**, 571.
36. Zhang, B., Li, Y., Xu, Y., Wang, S., Ma, L., and Wang, J. (2008) *Polym. Bull.*, **60**, 733.
37. Zhang, B., Li, Y., Wang, W., Chen, L., Wang, S., and Wang, J. (2009) *Polym. Bull.*, **62**, 643.
38. Zhang, B., Li, Y., Sun, J., Wang, S., Zhao, Y., and Wu, Z. (2009) *Polym. Int.*, **58**, 752.
39. Xue, L., Dai, S., and Li, Z. (2009) *Macromolecules*, **42**, 964.
40. Kalra, B., Kumar, A., Gross, R.A., Baiardo, M., and Scandola, M. (2004) *Macromolecules*, **37**, 1243.
41. Srivastava, R.K., Kumar, K., Varma, I.K., and Albertsson, A. (2007) *Eur. Polym. J.*, **43**, 808.

42 Hans, M., Gasteier, P., Keul, H., and Moeller, M. (2006) *Macromolecules*, **39**, 3184.

43 Duxbury, C.J., Cummins, D., and Heise, A. (2007) *Macromol. Rapid Commun.*, **28**, 235.

44 Hans, M., Keul, H., Heise, A., and Moeller, M. (2007) *Macromolecules*, **40**, 8872.

45 Hans, M., Xiao, Y., Keul, H., Heise, A., and Moeller, M. (2009) *Macromol. Chem. Phys.*, **210**, 736.

46 Villarroya, S., Zhou, J., Thurecht, K.J., and Howdle, S.M. (2006) *Macromolecules*, **39**, 9080.

47 Villarroya, S., Dudek, K., Zhou, J., Irvine, D.J., and Howdle, S.M. (2008) *J. Mater. Chem.*, **18**, 989.

13
Biocatalytic Polymerization in Exotic Solvents

Kristofer J. Thurecht and Silvia Villarroya

Biocatalytic polymerization has evolved as an increasingly attractive approach for the synthesis of both bulk commodity plastics, as well as more specialized polymers. While enzymes have been used specifically because they are highly efficient and often impart unique properties on the polymer, (for example, due to their enantio- or chemoselectivity) the 'green' or environmentally friendly nature of these catalysts has recently become increasingly important. As such, the broad concept of biocatalytic polymerization has been extended to more exotic solvents. More often than not, these exotic solvents are employed in an attempt to maximize the 'green' aspect of the process and decrease the impact of environmentally damaging solvents. Indeed, the combination of environmentally friendly catalysts with 'green' solvents surely represents a unique and increasingly more desirable approach towards polymer synthesis. With this in mind, the key driver for many investigators is to make biocatalytic polymerization in exotic or 'green' solvents industrially competitive with conventional processes.

Numerous enzymes have been employed as polymerization catalysts in these exotic solvents. Table 13.1 outlines the enzymes discussed in this chapter, along with the application and the acronym or abbreviation that will be used throughout this report.

This chapter explores the application of biocatalytic polymerization in 'exotic' solvents. These solvents are often termed 'unconventional', in that they would not generally be considered as a polymerization media. However, their use over the previous decade has dramatically increased due to the international push for cleaner, greener reaction pathways in an effort to reduce volatile organic compounds (VOCs). The first solvents to be discussed (and by far the most fully investigated in the literature) are supercritical fluids. Within this field, supercritical CO_2 has been the most highly reported solvent. The second solvent class is ionic liquids. These have become increasingly popular over the last five years. Biphasic solvents will then be described and their application to biocatalytic polymerization. This section will be limited to biphasic solvents that are more unusual and, apart from a brief mention, will not encompass the broad field of emulsion polymerization in water. Finally, the use of fluorous solvents will be described. In all cases, the physical properties of the solvent imparts interesting,

Biocatalysis in Polymer Chemistry. Edited by Katja Loos
Copyright © 2011 WILEY-VCH Verlag GmbH & Co. KGaA, Weinheim
ISBN: 978-3-527-32618-1

Table 13.1 List of enzymes described in this report.

Enzyme	Application	Acronym/Tradename
Candida antarctica lipase B	Esterification, ROP	CALB; Novozym 435 (Immobilized on acrylic resin); Accurel (Immobilized on polypropylene); QDE 2-3-4 (immobilized on polypropylene/silica)
Porcine pancreatic lipase	Esterification	PPL
Pseudomonas cepacia lipase	Depolymerization	PCL
Mucor miehei lipase	Depolymerization/ ROP	MML
Horseradish peroxidase	Free radical polymerization	HRP
Soybean peroxidase	Free radical polymerization	SBP

and often vastly improved, properties on the final polymer compared with analogous reactions in more conventional solvents.

13.1
Supercritical Fluids

Supercritical fluids have received increasing attention over the previous two to three decades as a real alternative solvent for polymer synthesis [1, 2]. By far the most investigated of the supercritical solvents is CO_2. This has been partly driven by the need to develop new technologies in cleaner, greener solvents; supercritical CO_2 is nontoxic, nonflammable, relatively inert, and has easily accessible critical points. However, the interest in supercritical CO_2 as a solvent stems far from just environmental reasons. Indeed, many physical properties of the solvent can be easily tuned by subtle variation of the temperature and pressure–particularly in the region close to the critical point ($P_c = 7.38$ MPa and $T_c = 31.1\,°C$ for CO_2) [3]. This trait of supercritical fluids has important implications for polymer synthesis and dictates the solubility of the polymer in the solvent, or the solubility of the solvent in the polymer. Figure 13.1 shows the supercritical regime in a typical phase diagram. In addition, a series of photographs are presented showing the transition from the biphasic liquid/gas phase to the monophasic supercritical fluid.

Supercritical CO_2 has been quite extensively used as a solvent in biocatalysis [4–11]. Numerous reports exist describing improved rates and yields that can be

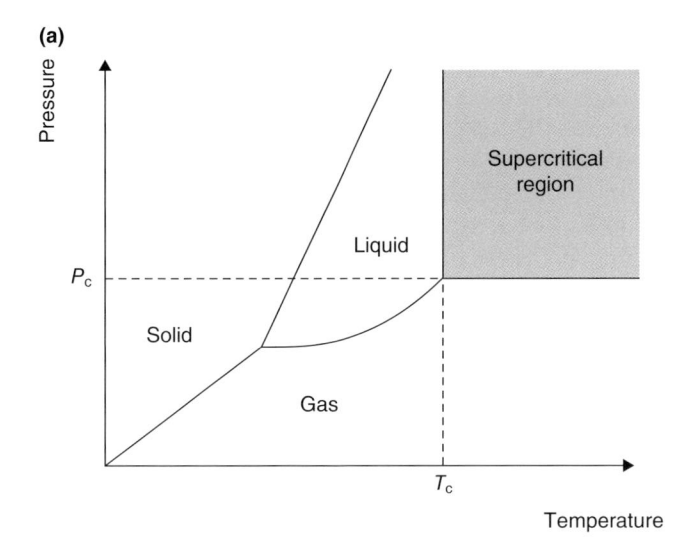

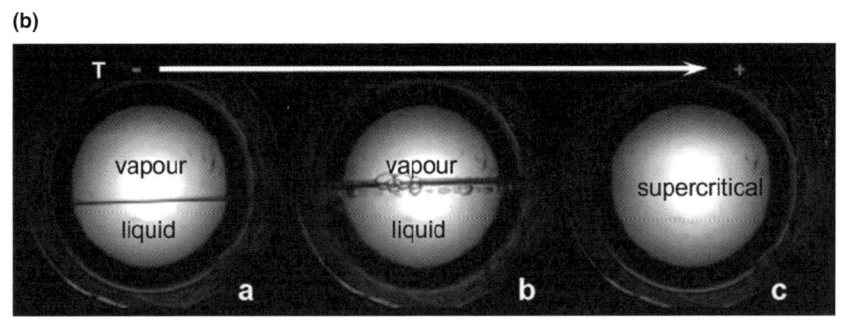

Figure 13.1 Phase diagram showing supercritical region (a) and series of photographs showing the phase transition from biphasic to monophasic upon transition into the supercritical regime in a high-pressure view cell.

obtained when supercritical CO_2 is used as a solvent rather than organic solvents or solvent-free systems [12, 13]. For example, the esterification of benzyl alcohol and butyl acetate by CALB was shown to be higher in supercritical CO_2 than in toluene or hexane [14]. In addition to improvements in conversion and reaction rates, there have been extensive reports showing enhanced enantioselectivity in supercritical CO_2 [15–18]. In most cases, this enhancement is attributed to the favorable properties of the fluid; low viscosity, high diffusivity. Furthermore, it has been reported that the formation of carbamate on the surface of the enzyme can actually improve enzymatic efficiency [15, 16, 19]. As with all reaction media, publications also exist in which supercritical CO_2 has been shown to be a poorer solvent than organic solvents and where the formation of carbamate is deleterious to the enzyme activity (rather than having a favorable effect) [20–22].

Supercritical fluids were first used as a solvent for biocatalytic polymerization by Russell and coworkers over a decade ago [23]. Since then, numerous advances in both the high pressure reactor design, as well as in enzyme preparation (e.g., immobilization by solid support [24], crosslinked enzyme aggregate [25] or crosslinked enzyme crystals [26] etc. (see also Chapters 2 and 3), have led to the preparation of high molecular weight polymer with some degree of control over the products. By far the most studied reaction systems are condensation or ring-opening polymerizations (ROPs). This is because lipases such as CALB have been shown to be exceptionally effective catalysts for these polymerization reactions. In addition, commercial interest in catalysis has lead to many enzymes being commercially available in immobilized and active forms. For example, CALB immobilized on an acrylic resin can be purchased from Novozymes under the tradename Novozym® 435 (see also Chapters 3 and 4). Indeed, Novozym 435 has become the preferred catalytic substrate for many researchers conducting polymerization reactions involving lipases. In the case of supercritical fluids, Novozym 435 shows an exceptionally high tolerance to the pressures required for supercritical solvents. The various classes of reactions that have been undertaken in supercritical fluids will be discussed in more detail in the following sections.

13.1.1
Lipase-catalyzed Homopolymerizations

Russell and coworkers [23] conducted an investigation into the condensation reaction between 1,4-butanediol and bis(2,2,2-trichloroethyl) adipate. The authors showed that low-dispersity polymers could be achieved in supercritical fluoroform at 50 °C and 372 bar using PCL. Typically, the polymers had a molecular weight less than 1500 Da. This was the first record of biocatalytic polymerization in a supercritical fluid and opened the way for biocatalytic polymerization in these solvents. Almost all polymerization reactions in supercritical fluids to follow were conducted in supercritical CO_2.

Approximately five years later, Matsuyama and coworkers [27] used a water-in-oil emulsion method to coat a lipase with aerosol OT – a common surfactant. The surfactant-coated enzyme was then used in supercritical CO_2 as a polymerization catalyst for various lactones including ε-caprolactone, 11-undecanolactone and 15-pentadecalactone. The authors showed that the rates of polymerization were increased using the surfactant-stabilized lipase (compared with the corresponding native enzyme) and by using supercritical CO_2 (compared with organic solvents). The authors also reported very narrow molecular weight distributions (1.08) when supercritical CO_2 was used as solvent. Russell and coworkers [28] formed fluorinated polyesters in supercritical CO_2 using CALB by the reaction between octafluorooctadiol and divinyladipate. These polymers had molecular weights of around 8 kDa and were formed at 50 °C and 200 bar in a batch reactor for 24 hours. Similarly, Kobayashi and coworkers reacted glycerol with a polyanhydride at 60–80 °C at 140 bar for 24 hours to yield the polyester [29].

Following these initial reports, numerous publications showed that enzymatic ring-opening polymerization (eROP) of lactones could be successfully achieved in supercritical CO_2. Kobayashi and coworkers [30] investigated the effects of parameters such as monomer and enzyme concentration as well as the effect of CO_2 pressure on the polymer product. They also showed that reaction of the monomers ε-caprolactone and 12-dodecanolide at 60 °C for 120 hours afforded a random copolymer. The 'randomness' of the copolymer was investigated by ^{13}C NMR and the copolymer diads suggested that the polymer was highly random. This was attributed to the occurrence of extensive transesterification during the reaction. Gross and coworkers [31] conducted an investigation into the effect of catalyst immobilization, temperature and pressure on the eROP of ε-caprolactone. A number of interesting points from this investigation were observed. Firstly, the temperature of reaction played an extremely important part in the effectiveness of the polymerization. The apparent activation rate of the enzyme was maximum at 80 °C, and the enzyme was observed to be fully de-activated if incubated at 140 °C. Interestingly, the pressure of the reactions was shown to have little effect on the polymerization parameters. Indeed, at constant temperature, little variation in the apparent activation of the enzyme was observed over the pressure range of 1 to 18 MPa – even at subcritical pressures. This was an interesting observation which highlighted two important points: (i) Pressure had little effect on the physical properties of the enzyme itself; and (ii) the effect of density, which changes drastically when passing through the critical point of the solvent, has little affect on the enzyme activation and more importantly, the polymerization products. This is perhaps surprising when one considers that eROP in supercritical CO_2 occurs via a precipitation polymerization mechanism and the increase in density of the solvent (and concomitant increase in solubilizing power) with pressure plays little role in the polymerization rate. Gross and coworkers further showed that the type of media on which the enzyme was immobilized had a dramatic effect on the polymerization rate. It is well known that immobilization of CALB on a matrix leads to enhancement of the catalyst, and the degree of enhancement depended on the hydrophobicity of the matrix. Thus, it was shown that by using more hydrophobic matrices such as Accurel (polypropylene-based matrix) and QDE 2-3-4 (polypropylene/silica-based matrix), higher rates of polymerization were observed. Indeed, the time to reach 50% conversion in supercritical CO_2 was 1.2 and 1.6 times faster than that for Novozym 435 (acrylate-based matrix) using Accurel and QDE 2-3-4, respectively. In a subsequent paper, Gross and coworkers investigated the effect of numerous matrices on the efficiency of physically immobilized CALB for eROP in various solvents [32].

Howdle and coworkers [33] showed that relatively high molecular weight polycaprolactone could be synthesized in supercritical CO_2 (M_n = 12–37 kDa), but with lower polydispersity (1.4–1.6) and higher yield (95–98%) than in conventional organic solvents. Furthermore, when Novozym 435 was used as catalyst in supercritical CO_2, the immobilized enzyme could be cleaned and recycled to allow repeated use, while unreacted monomer and low molecular weight oligomers were removed from the reaction mixture by venting of the CO_2. This allowed reasonable control over the molecular weight over repeated reaction cycles and pointed towards a possible continuous polymerization strategy employing supercritical CO_2 and enzymes for the synthesis of ε-caprolactone.

Numerous other enzymatic polymerizations have been conducted in supercritical CO_2 and we have included most of these in the following sections.

13.1.2
Lipase-catalyzed Depolymerization (Degradation)

A further application of lipases is the potential to degrade ester linkages in the presence of water. This has been demonstrated in supercritical CO_2 by a number of authors. Kobayashi and coworkers [34] showed that when polycaprolactone was exposed to CALB in supercritical CO_2, oligomeric caprolactone resulted. MALDI-TOF MS was used to show that the product was a mixture of linear and cyclic material. The authors also showed that when 5 vol% acetone was added, oligomers of molecular weight <500 Da resulted. At about the same time, Matsumura and coworkers [35, 36] showed that high molecular weight polycaprolactone (~110 kDa) could be degraded into low molecular weight oligomers in supercritical CO_2 in the presence of small amounts of water and CALB. The authors showed that the cyclic dimer was created in >90% yield. Importantly, this dimer was then re-polymerized in the presence of supercritical CO_2 and CALB, to yield high molecular weight polymer (~30 kDa). This was a significant step towards development of a sustainable chemical recycling mechanism for polycaprolactone. Later, Matsumura extended this approach to the degradation of polycarbonates and polylactic acid [37].

Recently, in an attempt to make this technique more commercially applicable, Matsumura *et al.* took this methodology a step further with the development of a continuous-flow reactor for polyester degradation [38]. In this approach, the enzyme (typically CALB) was immobilized onto a resin and packed into a column. The polymer (polyester, polycarbonate or polylactic acid) was dissolved in an organic solvent (typically toluene) and co-continuously fed into the supercritical CO_2 stream which passed through the enzyme-packed column. Typical reactions were performed at temperatures as low as 40 °C. The authors showed that compared with organic solvent, the use of supercritical CO_2 lead to faster reaction times, the ability to conduct the reaction at lower temperatures and the ability to use much higher concentration of polymer solution while still achieving full degradation upon passing through the enzyme-packed column. This was attributed to the high diffusivity, low viscosity and overall better mass-transfer properties of the supercritical media. In addition, at 40 °C, the lifetime of the column was maintained for greater than six months. A schematic of the continuous flow system is shown in Figure 13.2.

Various approaches have also been shown for the degradation of naturally-occurring polymers (see also Chapter 16). While this list is extensive, one particular approach worth mentioning is the degradation of biomass to glycerol products

backpressure regulator

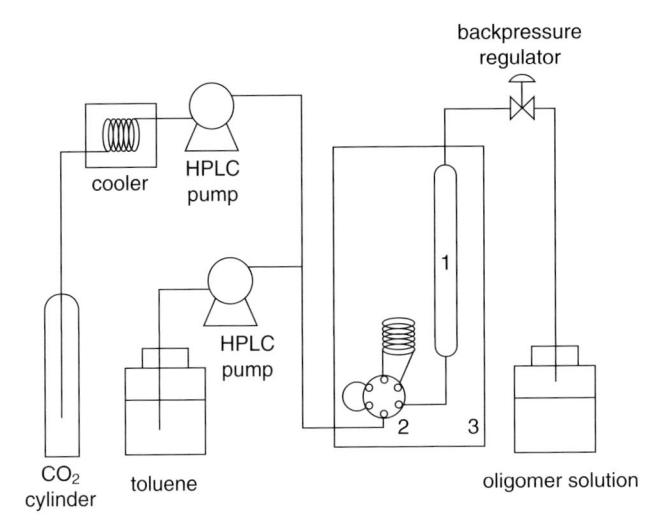

cooler
HPLC pump
HPLC pump
1
2   3
CO₂ cylinder
toluene
oligomer solution

(1) Immobilized lipase packed column
(2) Rheodine injector
(3) Column oven

Figure 13.2 Conceptual scheme of the degradation equipment used by Matsumura and coworkers to degrade polyesters, polycarbonates and polylactic acid to their oligomeric precursors. (Reproduced with permission from [38]. Copyright © (2006) IOP Science).

(see also Chapter 1). This has included the degradation of corn starch [39], cotton bolls [40], sunflower oil [41] and cellulose materials [42–44]. Table 13.2, while being by no means exhaustive, summarizes the various enzyme and biomass systems that have been studied. Note that supercritical H_2O has also been used in some cases. Additional examples of biomass hydrolysis in supercritical fluids can be found within the references given in Table 13.2.

13.1.3
Combination of Polymerization Mechanisms: Polymerization from Bifunctional Initiators

Recent advances in polymer synthetic chemistry have allowed the development of elegant and more complicated architectural polymers. This has been driven predominantly by the development of various controlled polymerization methodologies, particularly in the area of free radical polymerization [45–49]. This has equipped the polymer chemist with a rich and abundant synthetic toolbox. In general, these architectural polymers are based on the principle of being able to sequentially add different polymeric blocks with defined molecular weight into a single polymer chain [50]. The synthesis of block copolymers is particularly suited to the combination of two different polymerization techniques. This can be quite easily achieved by the use of a bifunctional initiator and is an elegant synthetic

Table 13.2 Enzymatic degradation of biomass in supercritical CO_2.

Product	Enzyme and Form	Conditions	Ref.
Corn starch	α-amylase (*B. licheniformis*)	Supercritical CO_2; 35 °C, 101 bar	[39]
Cellulose	Cellulase (*Humicola insolens*)	Supercritical CO_2; Immobilized on ceramic membrane; 45 °C, 101 bar	[42]
Cellulose (Avicel)	Cellulase	Supercritical CO_2; 46 °C, 137 bar, 2w/w water	[43]
Cotton boll fibers	Cellulase (*Trichoderma cirude*)	Supercritical CO_2; 50 °C, 160 bar	[40]
Biomass	Cellulase	Supercritical H_2O; 380 °C, 250–330 bar	[44]
Sunflower oil	Lipolase 100T (*Aspergillus niger*)	Supercritical CO_2; 40–50 °C, 100–200 bar	[41]

approach to prepare such block copolymers through the combination of mechanistically incompatible monomers (see also Chapter 12). Recently, Du Prez published an extensive review with a complete overview of the state-of-the-art on such heterofunctional initiators for polymerization [51]. Other reviews describe bi- and polyfunctional initiators for free radical polymerization [52, 53]. Perhaps the biggest advantage of bifunctional (or indeed, multifunctional initiators) is the ability to combine vastly different polymerization techniques without the need for intermediate transformation and protection steps. In this section, the various methodologies employed for the synthesis of functional polymers is described. The polymer architectures described range from classical block copolymers to graft copolymers and in all cases includes enzymatically-catalyzed polymerization in supercritical CO_2.

Numerous examples of block copolymers formed in supercritical CO_2 via the bifunctional initiator approach have been reported [54]. Perhaps the most common approach is to incorporate eROP with free-radical polymerization – the general scheme for this methodology is shown in Figure 13.3. Howdle *et al.* [55] was the first to report the synthesis of a block copolymer by the bifunctional initiator approach in supercritical CO_2 and showed the simultaneous eROP of ε-caprolactone with controlled free radical polymerization of methyl methacrylate by atom transfer radical polymerization (ATRP) – at this time simultaneous eROP and ATRP had not been reported in any media. The bifunctional initiator incorporated both a primary hydroxyl group (as an initiation site for eROP of ε-caprolactone) and a bromine moiety (for initiation of ATRP). Howdle showed that

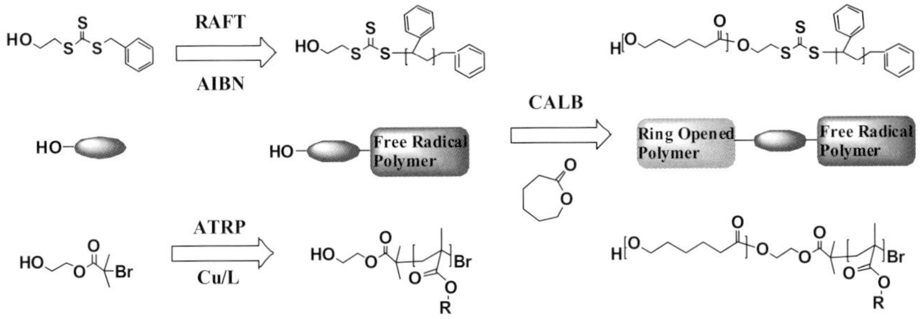

Figure 13.3 Overview showing the use of bifunctional initiators for block-copolymer synthesis. Two examples using either reversible addition fragmentation chain transfer polymerization or atom transfer radical polymerization combined with eROP are shown.

eROP, followed by ATRP of methyl methacrylate successfully formed a block copolymer. Perhaps more importantly, was the ability to simultaneously polymerize both the ε-caprolactone and methyl methacrylate using the two disparate mechanisms. In all the systems discussed in this section, the most important aspect of characterization is the proof of block copolymer formation (rather than a simple blend of two different monomers – this is generally indistinguishable by standard NMR techniques). Thus, Howdle et al showed that exposure of the polyester-block-polymethacrylate polymers to acid hydrolysis to degrade the polyester block and subsequent analysis by GPC led to a definite shift to lower molecular weight in the GPC curve. In this initial report, polymer blocks up to 10 kDa for PMMA and 30 kDa for ε-caprolactone were reported. While the PMMA block generally had a narrow polydispersity, the polycaprolactone block was often quite broad due to the competing transesterification and hydrolysis mechanism. Further analysis of the mechanism and subsequent block copolymers were reported by Howdle et al. in a follow-up publication [56]. Subsequent block copolymers incorporating a fluorinated block (1H,1H,2H,2H-perfluorooctyl methacrylate by ATRP) and polycaprolactone followed, synthesized using the same methodology [57]. This was an important publication as it showed that highly fluorinated block copolymers could be synthesized by using the unusual solubility of these polymers in supercritical CO_2.

The bifunctional 'initiator' approach using reversible addition fragmentation chain-transfer polymerization (RAFT) as the free-radical controlling mechanism was soon to follow and block copolymers of styrene and caprolactone ensued [58]. In this case, a trithiocarbonate species having a terminal primary hydroxyl group provided the dual 'initiation' (Figure 13.3). The resultant polymer was terminated with a trithiocarbonate; reduction of the trithiocarbonate to a thiol allows synthesis of α-hydroxyl-ω-thiol polymers which are of particular interest in biopolymer applications.

In addition to standard block copolymer synthesis, Howdle and coworkers [59] showed that eROP could occur in supercritical CO_2 from initiation sites along a polymer backbone. A random copolymer of methyl methacrylate and hydroxyethyl methacylate led to polymer that was decorated with hydroxyl side-groups. These acted as initiation sites for enzymatic polymerization of ε-caprolactone using Novozym 435. The subsequent polymer exhibited ~30–40% grafting efficiency, without extensive protection/deprotection steps – this is difficult to achieve using standard chemical catalysts. The authors related this observation to steric barriers imposed by hydroxyl groups on the polymer backbone in close proximity to each other. A schematic of the grafting approach is shown in Figure 13.4. Further investigation [60] showed that by increasing the chain length of the hydroxyl side-chain (i.e., using a polyethyleneglycol methacrylate (M_n 526 g mol^{-1}), the increased distance from the polymer back-bone lead to grafting efficiencies of ~100%. This was attributed to the ability of the enzyme to interact freely with all initiation sites.

The use of the dual initiator strategy has opened up a whole new area of block copolymer synthesis using enzymes and over the previous five years many reactions and systems have been extensively reported. In general, the previous reports have shown that the use of supercritical CO_2 has allowed the two reaction mechanisms of eROP and free radical polymerization to occur simultaneously to yield well-defined block and graft copolymers.

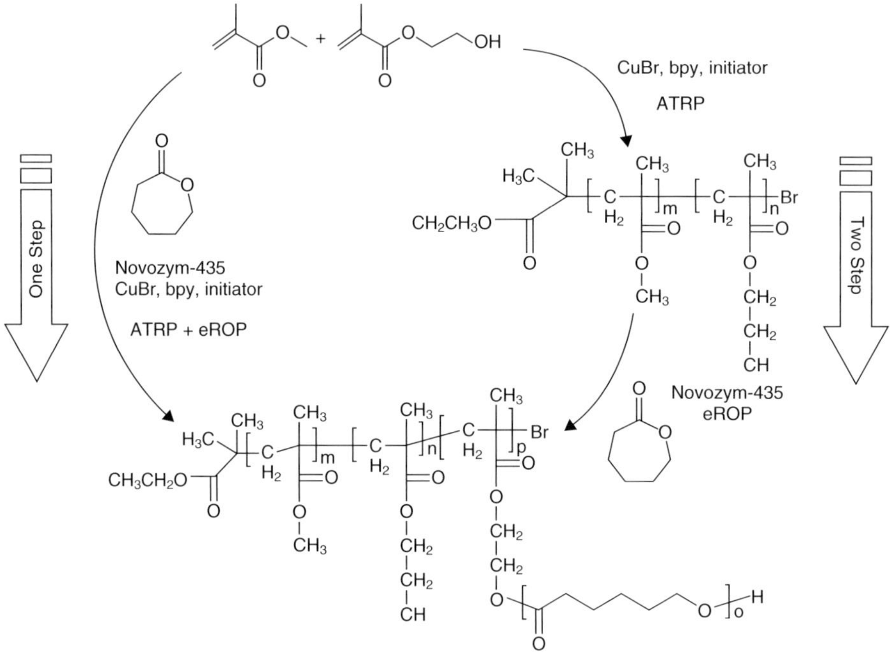

Figure 13.4 Schematic showing the steps required for synthesis of graft polymers by eROP and ATRP. Generally, the grafting efficiency was limited to 40% when short side-chain hydroxyl groups were used for initiation (e.g., HEMA).

13.1.4
Free Radical Polymerization Using Enzymatic Initiators

While most enzymatically-catalyzed polymerizations in supercritical CO_2 have been conducted by solution or precipitation polymerization mechanisms, the ability to perform emulsion-like polymerization has been extensively reported for nonenzymatic initiation [61–68].

Holmes et al. reported on enzymatic catalysis within the water pool of a reverse micelle formed in supercritical CO_2 [69]. The water-in-CO_2 (w/c) microemulsions have been shown to solubilize enzymes, particularly by using perfluoropolyether type surfactants, providing a unique medium for enzyme-catalyzed biotransformations [65]. Horseradish peroxidase (HRP) is well known to catalyze the oxidation of organic substances (phenols and anilines) (see Chapters 7 and 8) using hydrogen peroxide (H_2O_2) and/or alkyl peroxide [70]. The mechanism is generally accepted to involve the production of free radicals. The potential of HRP to catalyze free radical polymerization of vinyl monomers was first reported by Derango et al. [71] and since then others have also exploited the enzymatic route reporting HRP-mediated free radical polymerization of acrylamide in water [72, 73], and in the presence of surfactants (see also Chapter 6) [74].

Howdle et al. was the first to report on the enzymatic emulsion polymerization of water-soluble acrylamide monomers in a w/c microemulsion [75]. Acrylamide is one of the few monomers that show limited solubility in supercritical CO_2, and hence was particularly suited to inverse emulsion polymerization [61]. In this report, Howdle et al. showed that the yields and molecular weights obtained via polymerization in CO_2 are comparable to those generated in a conventional polymerization in an alkane/water medium. Indeed, polymer molecular weights up to 900 kDa were observed when a surfactant was not used and molecular weights around 300 kDa when a perfluropolyether was used. The generally accepted mechanism for this reaction is shown in Figure 13.5.

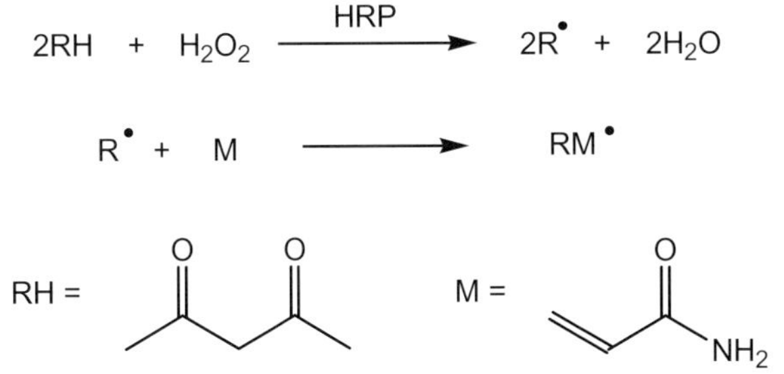

$$2RH + H_2O_2 \xrightarrow{HRP} 2R^\bullet + 2H_2O$$

$$R^\bullet + M \longrightarrow RM^\bullet$$

Figure 13.5 Reaction scheme showing the enzymatic production of free radicals by Horseradish peroxidase. These radicals then initiate polymerization in the presence of monomer. (Reproduced with permission from [75]. Copyright © (2008) Royal Society of Chemistry).

Numerous interesting observations evolved from this work. First, the pH at which HRP works effectively has been shown to be between 6 and 6.5 [76]. However, the pH within the water pool of a w/c inverse emulsion has been reported as ~3 [6], due to formation of carbonic acid in the aqueous core. Despite such a low pH being present, the emulsion polymerization worked effectively in supercritical CO_2 and the enzyme was still active. The second point of interest concerned elucidation of the HRP catalytic mechanism. By increasing the concentration of H_2O_2 (see Figure 13.5), lower yields were observed – this was attributed to poisoning of the catalyst as has been reported in the literature. This also showed that the peroxide concentration did not affect the molecular weight that was observed. On the other hand, the concentration of pentadione directly affected the molecular weight, indicating that the pentadione acted as the initiator in this system as proposed (Figure 13.5).

This report has been the only publication involving enzymatically-catalyzed free radical polymerization in supercritical CO_2, perhaps due to the difficulty in adequately matching monomer or polymer compatible systems with the solvent.

13.2
Biocatalytic Polymerization in Ionic Liquids

In recent years, room temperature ionic liquids (RTILs) have gained increasing attention as a possible alternative solvent to organic counterparts [77]. This class of solvent has been branded a 'green' alternative, mainly as a result of the almost zero vapor pressure of these compounds. As a result, the nonvolatility means that there is little environmental impact from solvent evaporation. On the other hand, many RTILs have been shown to be toxic and the very properties that make them attractive as alternative solvents (i.e., nonvolatility) means that disposal of the RTIL is difficult. Nonetheless, despite continuing debate on the 'green' credentials of the solvent, they have exhibited useful and interesting properties as a solvent for catalytic reactions. The solubility and activity of enzymes in RTILs has been previously reported [11, 78, 79]. Here, we will discuss some recent reports involving enzymatic polymerization in RTILs. A list of the RTILs used in this report and their structures are given in Table 13.3.

13.2.1
Free Radical Polymerization

Plant peroxidases (in particular SBP and HRP) have been shown to effectively catalyze polymerization of phenolic monomers (see also Chapter 7). However, the low solubility of monomer, and in particular oligomer or polymer, in aqueous and organic systems means that only low yields of oligomeric products are realized. Indeed, in aqueous systems predominantly dimers and trimers precipitate out of

Table 13.3 List of RTILs used as solvent for enzymatic polymerization.

Ionic Liquid	Abbreviation	Structure
1-butyl-3-methylimidazolium tetrafluoroborate	BMIM BF_4	
1-butyl-3-methylimidazolium hexafluorophosphate	BMIM PF_6	
1-butyl-3-methylimidazolium bistriflamide	BMIM NTf_2	
1-butyl-3-methylimidazolium tetrachloroferrate	BMIM $FeCl_4$	
1-butyl-3-methylimidazolium dicyanamide	BMIM DCA	
1-octyl-3-methylimidazolium dicyanamide	OMIM DCA	
1-butyl-1-methylpyrrolidinium tetrafluoroborate	BMPy BF_4	
1-butyl-1-methylpyrrolidinium dicyanamide	BMPy DCA	

solution [80, 81]. Unfortunately, when polar aprotic solvents are used to assist with monomer or polymer solubility issues, enzyme activity is compromised and is generally very poor [80, 82, 83].

Dordick and coworkers [84] have shown that polyphenols can be successfully polymerized using ionic liquids as solvent and certain peroxidases as catalyst. They reasoned that RTILs would provide a threefold advantage over conventional solvents. First, the RTIL provides a nonaqueous environment that should show good solubility of the monomer and polymer; second, the catalyst activity should

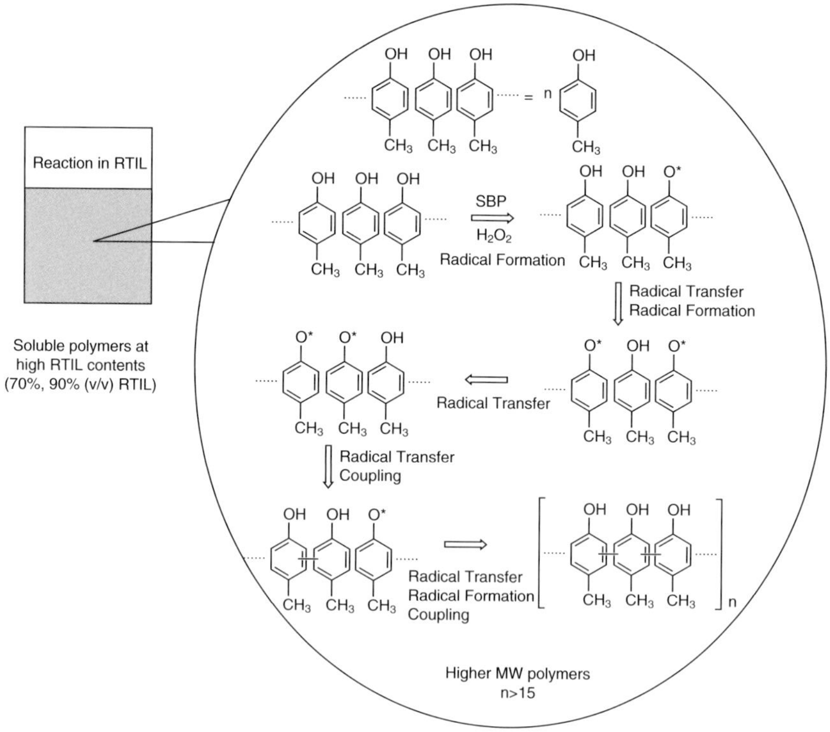

Figure 13.6 Schematic showing the free radical polymerization of phenols as catalyzed by FRP in RTILs. (Reproduced with permission from [84]. Copyright © (2009) Elsevier).

be maintained; and third, RTILs offer a greener approach due to their almost negligible vapor pressure. The authors based their study on two ionic liquids having different cations; BMIM BF_4 and BMPy BF_4. The concentration of ionic liquid in aqueous buffer solution was varied from 10–90% and it was shown that higher molecular weight polymers could be synthesized when the ionic liquid concentration was higher. Additionally, some level of control over the molecular weight was achieved by varying the buffer concentration. The successful polymerization of various phenol monomers was reported when catalyzed by either SBP or HRP in RTIL:buffer solutions. Figure 13.6 shows the proposed mechanism for polymerization in the RTIL.

The molecular weights of the polymers were characterized using gel permeation chromatography and MALDI-TOF MS with molecular weights up to 1700 Da and polydispersities between 1.1–1.3 being reported. Additionally, when buffer solution was added as a cosolvent, higher molecular weights were achieved (Mw up to 2900), but the polydispersity of the resultant polymer was quite high (1.5–1.75). The overall success of this reaction has been attributed to the solubility of the polymer and growing radical in the RTIL.

13.2.2
Lipase-catalyzed Polymerization in Ionic Liquids

Salunkhe and coworkers [85] reported the lipase (PCL)-catalyzed polycondensation reaction between diethylene octane-1,8-dicarboxylate and 1,4-butanediol in BMIM PF_6. The reaction scheme is presented in Figure 13.7.

In this first report in RTILs, the authors showed that catalysis was very slow and after seven days, a low molecular weight polymer of 2 kDa was recovered. When the reaction temperature was increased to 60 °C and conducted under vacuum, the molecular weight could be increased to around 4.5 kDa. The authors ascribed the low molecular weight to precipitation of the polymer product due to poor solubility in the solvent.

Uyama and coworkers [86] conducted the first ROP in ionic liquids using enzymatic catalysis. In this publication, the authors showed that ROP of ε-caprolactone occurred successfully in BMIM PF_6. However, after three days the molecular weights were less than 1000 Da and at least seven days were required to achieve molecular weights greater than 4 kDa. Uyama also investigated the polycondensation reaction between diethyl adipate or diethyl sebacate with 1,4-butanediol. However, molecular weights greater than 1.5 kDa were not achieved in either BMIM PF_6 or BMIM BF_4.

It was not until Heise and coworkers [87] also investigated the enzymatic polymerization by both a ring-opening and polycondensation mechanism in three different ionic liquids that improved products were observed. BMIM PF_6, BMIM BF_4 and BMIM NTf_2 were investigated as solvent using Novozym 435 as catalyst. The reaction scheme showing Heise's experiments are shown in Figure 13.8.

For the ROP of ε-caprolactone, both BMIM PF_6 and BMIM BF_4 result in precipitation of the polymer due to poor solubility in the solvent. Only for BMIM NTf_2 was the polymer soluble to high conversion and molecular weights between 7–9.5 kDa were obtained. The authors related this to bulk-like reaction conditions

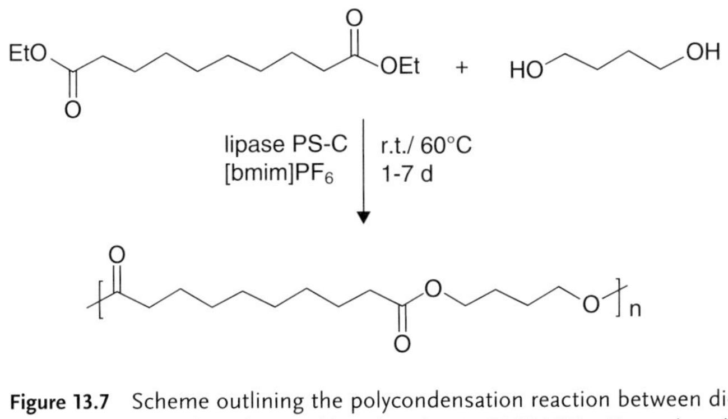

Figure 13.7 Scheme outlining the polycondensation reaction between diethylene octane-1,8-dicarboxylate and 1,4-butanediol undertaken in BMIM PF_6. (Reproduced with permission from [85]. Copyright © (2003) Elsevier).

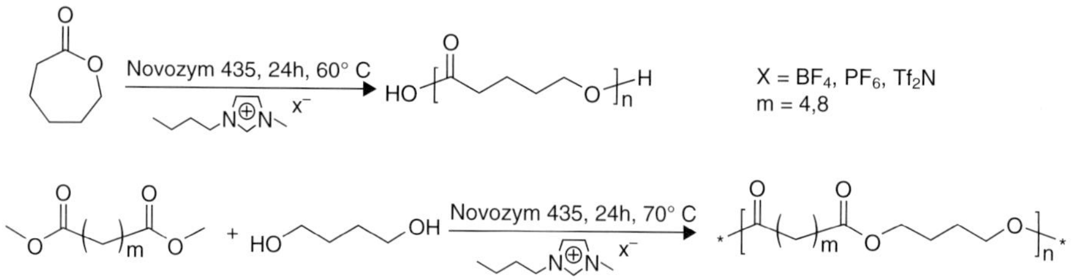

Figure 13.8 Reactions performed by Heise and coworkers in various ionic liquids. (Reproduced with permission from [87]. Copyright © (2006) Elsevier.)

Table 13.4 Solubility data for polylactone in various RTILs (Source: [88]).

Ionic liquid	Miscibility with water	Equilibrium solubility (g l^{-1})
BMIM[Tf$_2$N]	Immiscible	21
BMIM[FeCl$_4$]	Immiscible	17
BMIM[DCA]	Miscible	12
BMPYR[DCA]	Miscible	1
OMIM[DCA]	Miscible	8
BMIM[PF$_6$]	Immiscible	<0.01
BMIM[Br]	Miscible	<0.01

for the former and solvent-like conditions for the latter. In the case of polycondensation reaction between dimethyl adipate or dimethyl sebacate with butane diol, the reaction was performed in open vials at high temperatures very close to the boiling point of the condensate (70 °C). This was possible due to the low volatility of the ionic liquid solvent and allowed facile removal of the by-product. In this case, molecular weights of around 5 kDa were achieved. This approach uses the unique properties of ionic liquids (i.e., low vapor pressure; high boiling point) and highlights the advantages that these solvents could have over conventional organic solvents and the possibility for their application in industry.

Recently, Srienc and coworkers conducted an extensive investigation into the ROP of various lactones in RTILs, and thus their results warrant more detailed discussion [88]. Initially, the solubility of the polyesters in the RTILs was investigated. In general, the highest solubility was observed for RTILs with the largest anion. A summary of the solubility data for polyhydroxy butylvalerate in various RTILs is shown in Table 13.4.

The transesterification of ethyl valerate to methyl valerate was used as a model reaction to investigate the efficiency of the enzymes in each RTIL and to gain an understanding of the stability of these enzymes in the liquid. In general, the highest activity was recorded for those ionic liquids that were water immiscible. CALB (either immobilized or free enzyme) showed the highest activity of all

Table 13.5 Characteristics of various polylactone polymers formed in RTIL (Source: [88]).

Monomer	M_n wet	M_n dry	Comments
β-propiolactone	8000	11 900	1% CALB to monomer
β-butyrolactone	none	Oligomer	Rate limited by addition to polymer chain
γ-butyrolactone	none	Oligomer	Rate limited by addition to polymer chain
δ-valerolactone	2100	2 500	
ε-caprolactone	6300	9 700	Higher mwt also achieved at higher temperature / 1% CALB cf monomer
3-hydroxybutyric acid	none	13 000	Copolymer or P3HP and P3HB / 6% P3HB

enzymes studied in all RTILs in this study. As a result of this screening, BMIM NTf$_2$ was found to show the highest solubility of the polymer and the highest activity of the enzyme and hence was used as solvent for subsequent polymerization reactions in the report. The polymerization of a series of cyclic lactones in BMIM NTf$_2$ using CALB as catalyst was investigated. An overview of the results is presented in Table 13.5.

Degree of polymerization up to 180 was observed for β-propiolactone, and modest molecular weights were observed for ε-caprolactone and δ-valerolactone polymerizations. While polymerization was generally successfully achieved for most monomers, the authors did not report on any polydispersities and hence a comparison with other solvents is not available. Nonetheless, Srienc and coworkers showed that successful eROP is a function of both RTIL and the activity of the specific enzyme in the RTIL.

13.3
Enzymatic Polymerization under Biphasic Conditions

Polymerization under biphasic conditions offers a number of advantages over conventional solution or bulk polymerization. The ability to easily separate catalyst from reactants or products by way of immobilization in a second, immiscible phase provides a means of rapid and facile purification and separation. Traditional methods of biphasic polymerization include emulsion or suspension-type systems. While these are of huge importance both academically and commercially, they do not fit into the realm of 'exotic' solvents and so will not be described in depth in this section.

13.3.1
Ionic Liquid-Supported Catalyst

The extremely low vapor pressure of ionic liquids (the very parameter that gives them credibility as environmentally friendly solvents) also has some disadvantages for polymer synthesis. Generally, product recovery from an ionic liquid requires extraction with organic solvent. This was originally overcome by Blanchard *et al.* who showed that CO_2 exhibited very high solubility in 1-butyl-3-methyl imidazolium-based ionic liquids [89]. In contrast, the ionic liquids showed very low solubility in supercritical CO_2. Thus, naphthalene was solubilized in RTIL, then extracted using supercritical CO_2 to yield pure naphthalene and ionic liquid. Following this report, the use of biphasic IL-CO_2 systems has been extensively studied and the joint properties of the solvents investigated. Generally, the IL is used as a standard solvent for the reaction (or indeed, as a dispersant for the enzyme) and CO_2 is used to recover the product. A good example of this was reported by Lozano *et al.* [90] who showed that ionic liquids supported on a polymeric monolith provided a unique immobilization platform for biocatalysis in supercritical CO_2. Here, the insolubility of the polymer was advantageous for a successful solid support. CALB was immobilized in imidazolium-based ionic liquids that were embedded in the polymeric support (divinyl benzene-based or ethyleneglycol dimethacrylate-based). A continuous flow of CO_2 containing citronellol and vinyl propionate led to the synthesis of citronellyl propionate. A schematic is shown in Figure 13.9.

Unfortunately, most polymers are insoluble in supercritical CO_2, and hence extraction from the ionic liquid by this method is difficult. However, if CO_2-soluble polymers were synthesized (for example, fluoropolymers and polysiloxanes), then this method has the potential to be a very useful approach. Moreover, supercritical fluid-swollen ionic liquids offer a new solvent system that combines the viscosity-lowering properties of the supercritical fluid with the good solubilizing properties of the ionic liquid and may be a hybrid 'exotic' solvent of the future.

Mecerreyes *et al.* [91] used ionic liquid-supported HRP in water to conduct enzyme-catalyzed free radical polymerization. In this reaction procedure, the HRP was immobilized in BMIM NTf_2 by simple dissolution. This was then added to an aqueous solution of aniline, dodecylbenzenesulfonic acid (DBSA) and H_2O_2 at pH 4.3. Under these conditions, polymerization occurred immediately (as evidenced by the green color of the solution), but yield was low due to precipitation and association of polyaniline at the surface of the ionic liquid. To overcome this problem, a less hydrophobic ionic liquid, BMIM PF_6 was used. In this case, the solution separated into two liquid phases after ~0.5 hours; one contained the polymer in water and the other was ionic liquid-immobilized HRP. The IL-HRP could be recycled and polymerization successfully repeated up to five times. The resulting polymer had similar electric conductivity properties to conventionally prepared polymer. A schematic of this process is presented in Figure 13.10.

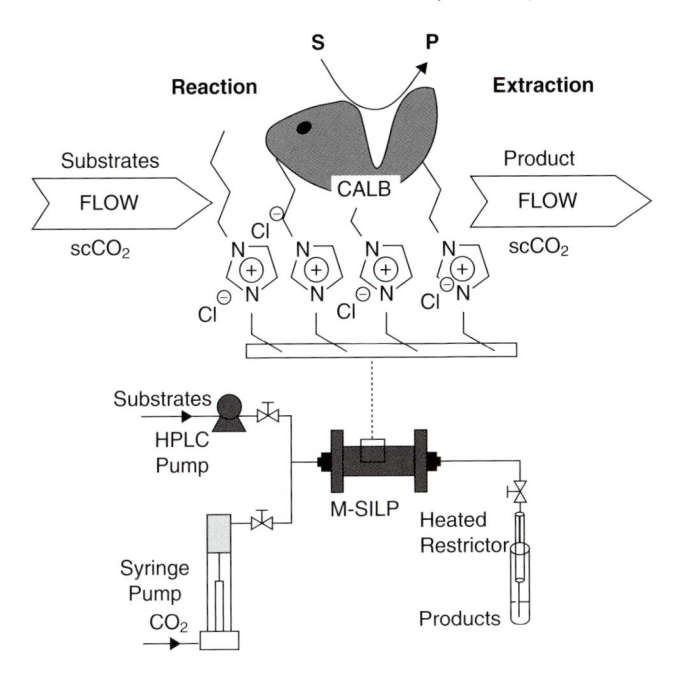

Figure 13.9 Schematic showing continuous flow biocatalytic process using ionic liquid-supported catalyst in supercritical CO_2. (Reproduced with permission from [90]. Copyright © (2007) Wiley-VCH Verlag).

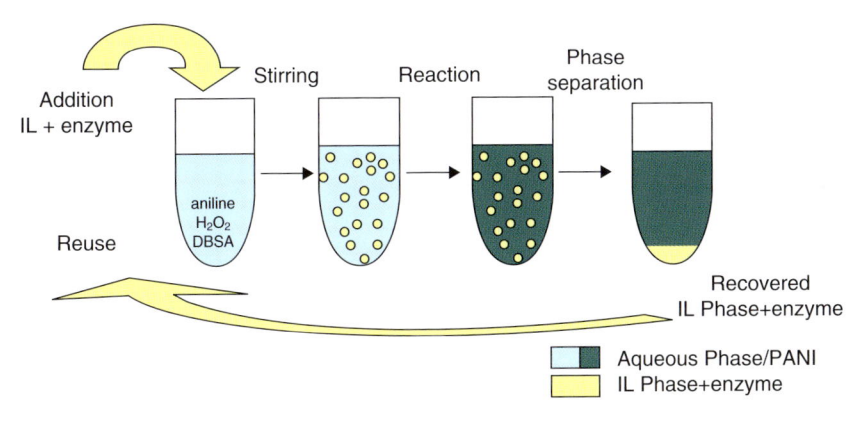

Figure 13.10 Schematic showing the biocatalytic biphasic polymerization in water and RTIL. (Reproduced with permission from [91]. Copyright © (2006) American Chemical Society).

13.3.2
Biphasic Polymerization of Polyphenols

Kaplan *et al.* [92] has demonstrated that HRP-mediated polymerization of polyphenols can be carried out in solution, emulsions or biphasic mixtures of buffer solution and isooctane. In the case of biphasic mixtures, polymers with molecular weights around 2 kDa were achieved with polydispersities about 2.0. These polymerizations were undertaken in an isooctane/buffer biphasic medium. The authors showed that higher molecular weight polymer and greater control over the polydispersity could be achieved if the reaction was carried out in a reverse micelle stabilized by a surfactant in water.

13.3.3
Fluorous Biphasic Polymerization

Fluorous biphasic solvents are an emerging class of extremely important reaction media that, while no reports of polymerization in such a media is presented, are still worthy of mention in this section. First described in detail in 1994 [93], the concept of a biphasic mixture at low temperature (for example, a mixture of a fluorinated solvent containing a catalyst and an organic solvent containing the substrate) which becomes miscible at higher temperature is indeed elegant. The reaction involving the substrate and catalyst can occur at the higher temperature, and then facile separation of the two components is facilitated by simply lowering the temperature following reaction. While the application of such a system to biocatalytic polymerization has not yet been realized, biocatalysis itself has been reported and a nice review has recently been published [94]. This concept definitely has potential as a future 'exotic' solvent for biocatalytic polymerization.

13.4
Other 'Exotic' Media for Biocatalytic Polymerization

Recently, Barzana *et al.* [95] proposed the use of fluorinated solvents for the eROP of ε-caprolactone. 1,1,1,2-tetrafluoroethane, commercially known as R-134a, was proposed as it is nontoxic, noninflammable, requires relatively low pressure to become liquid and importantly, it does not deplete the ozone layer. The viscosity of R-134a is similar to that of supercritical CO_2 (at 200 bar, 33 °C) and so the advantageous diffusivity properties of supercritical CO_2 should be transferred to reactions using R-134a. In addition, the fluoro-solvent has a surface tension much lower than most common organic solvents and a dipole moment and dielectric constant similar to dichloromethane. As such, this is expected to enhance the solubility of polar compounds and growing polymer chains.

Extensive studies on the polymerization of ε-caprolactone in this solvent showed that maximum yield and molecular weight was achieved at 2.0 MPa and 65 °C.

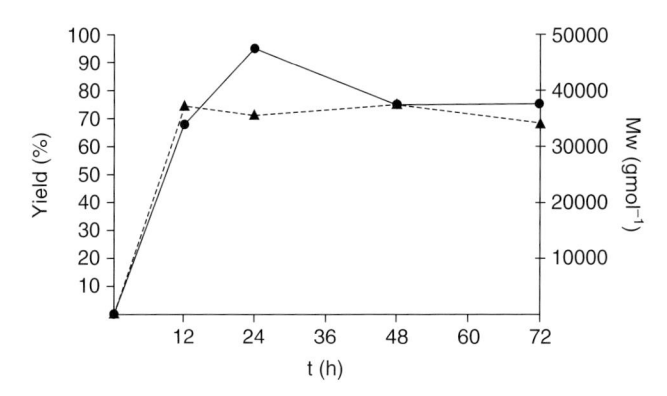

Figure 13.11 Plot of molecular weight and yield vs time for the synthesis of polycaprolactone in R-134a. (Reproduced with permission from [95]. Copyright © (2007) American Chemical Society).

Variations in temperature above and below 65 °C showed decreased yields. A maximum molecular weight of 37 600 Da with a PDI of 1.6 was achieved with a yield of up to 95%. Figure 13.11 shows a plot of yield and molecular weight for a typical reaction at 2.0 MPa and 65 °C.

Polymer exhibiting similar molecular weights and polydispersities to that formed in supercritical CO_2 were observed at order of magnitude lower pressures.

13.5
Conclusion

The use of exotic media for biocatalytic polymerization has ranged from the extensive, and often fundamental, studies using supercritical CO_2, to more exploratory and recent reports for media such as ionic liquids and fluoro-solvents. In all cases, however, intriguing results have lead to further investigation. As increasing pressure is exerted upon scientists in both academia and industry alike to develop and commercialize 'greener' reaction systems, it is expected that biocatalysis and reactions in these and other exotic solvents will continue to be of considerable interest into the future, as they have been over the previous decades.

References

1 Woods, H.M., Silva, M.M.C.G., Nouvel, C., Shakesheff, K.M., and Howdle, S.M. (2004) Materials processing in supercritical carbon dioxide: surfactants, polymers and biomaterials. *J. Mater. Chem.*, **14** (11), 1663–1678.

2 DeSimone, J.M., Maury, E.E., Menceloglu, Y.Z., McClain, J.B.,

Romack, T.J., and Combes, J.R. (1994) Dispersion polymerizations in supercritical carbon dioxide. *Science*, **265** (5170), 356–359.

3 Jessop, P.G., and W. Leitner (eds) (1999) *Chemical Synthesis Using Supercritical Fluids*, Wiley-VCH Verlag GmbH, Weinheim, Germany.

4 Mesiano, A.J., Beckman, E.J., and Russell, A.J. (1999) Supercritical biocatalysis. *Chem. Rev.*, **99** (2), 623–633.

5 Knez, Z., Habulin, M., and Primozic, M. (2005) Enzymatic reactions in dense gases. *Biochem. Eng. J.*, **27** (2), 120–126.

6 Kamat, S.V., Beckman, E.J., and Russell, A.J. (1995) Enzyme activity in supercritical fluids. *Crit. Rev. Biotechnol.*, **15** (1), 41–71.

7 Fontes, N., Almeida, M.C., and Barreiros, S. (2001) Biotransformations in supercritical fluids. *Methods Biotechnol.*, **15**, (Enzymes in Nonaqueous Solvents), 565–573.

8 Matsuda, T., Harada, T., and Nakamura, K. (2005) Biocatalysis in supercritical CO_2. *Curr. Org. Chem.*, **9** (3), 299–315.

9 Nakamura, K. (1990) Biochemical reactions in supercritical fluids. *Trends Biotechnol.*, **8** (10), 288–292.

10 Mori, T., and Okahata, Y. (2002) Control of enzymes reactions in supercritical fluids. *Biol. Ind.*, **19** (12), 14–21.

11 Garcia, S., Lourenco, N.M.T., Lousa, D., Sequeira, A.F., Mimoso, P., Cabral, J.M.S., Afonso, C.A.M., and Barreiros, S. (2004) A comparative study of biocatalysis in non-conventional solvents: ionic liquids, supercritical fluids and organic media. *Green Chem.*, **6** (9), 466–470.

12 Knez, Z., and Habulin, M. (1992) Lipase-catalyzed esterification in supercritical carbon dioxide. *Prog. Biotechnol.*, **8**, (Biocatalysis in Non-Conventional Media), 401–406.

13 Knez, Z., and Habulin, M. (1994) Lipase catalyzed esterification at high pressure. *Biocatalysis*, **9** (1–4), 115–121.

14 Tewari, Y.B., Ihara, T., Phinney, K.W., and Mayhew, M.P. (2004) A thermodynamic study of the lipase-catalyzed transesterification of benzyl alcohol and butyl acetate in supercritical carbon dioxide media. *J. Mol. Catal. B Enzym.*, **30** (3–4), 131–136.

15 Ikushima, Y., Saito, N., Arai, M., and Blanch, H.W. (1995) Activation of a lipase triggered by interactions with supercritical carbon dioxide in the near-critical region. *J. Phys. Chem.*, **99** (22), 8941–8944.

16 Ikushima, Y. (1997) Supercritical fluids: an interesting medium for chemical and biochemical processes. *Adv. Colloid Interface Sci.*, **71–72**, 259–280.

17 Catoni, E., Cernia, E., and Palocci, C. (1996) Different aspects of 'solvent engineering' in lipase biocatalyzed esterifications. *J. Mol. Catal. A Chem.*, **105** (1–2), 79–86.

18 Rantakyla, M., Alkio, M., and Aaltonen, O. (1996) Stereospecific hydrolysis of 3-(4-methoxyphenyl)glycidic ester in supercritical carbon dioxide by immobilized lipase. *Biotechnol. Lett.*, **18** (9), 1089–1094.

19 Mase, N., Sako, T., Horikawa, Y., and Takabe, K. (2003) Novel strategic lipase-catalyzed asymmetrization of 1,3-propanediacetate in supercritical carbon dioxide. *Tetrahedron Lett.*, **44** (28), 5175–5178.

20 Kamat, S., Critchley, G., Beckman, E.J., and Russell, A.J. (1995) Biocatalytic synthesis of acrylates in organic solvents and supercritical fluids: III. Does carbon dioxide covalently modify enzymes? *Biotechnol. Bioeng.*, **46** (6), 610–620.

21 Habulin, M., and Knez, Z. (2001) Activity and stability of lipases from different sources in supercritical carbon dioxide and near-critical propane. *J. Chem. Technol. Biotechnol.*, **76** (12), 1260–1266.

22 Habulin, M., and Knez, Z. (2001) Pressure stability of lipases and their use in different systems. *Acta Chim. Slov.*, **48** (4), 521–532.

23 Chaudhary, A., Beckman, E.J., and Russell, A.J. (1995) Rational control of polymer molecular weight and dispersity during enzyme-catalyzed polyester synthesis in supercritical fluids. *J. Am. Chem. Soc.*, **117** (13), 3728–3733.

24 Matsuda, T., Watanabe, K., Harada, T., Nakamura, K., Arita, Y., Misumi, Y.,

Ichikawa, S., and Ikariya, T. (2004) High-efficiency and minimum-waste continuous kinetic resolution of racemic alcohols by using lipase in supercritical carbon dioxide. *Chem. Commun. (Camb.)*, **20**, 2286–2287.

25 Matsuda, T., Tsuji, K., Kamitanaka, T., Harada, T., Nakamura, K., and Ikariya, T. (2005) Rate enhancement of lipase-catalyzed reaction in supercritical carbon dioxide. *Chem. Lett.*, **34** (8), 1102–1103.

26 Dijkstra, Z.J., Weyten, H., Willems, L., and Keurentjes, J.T.F. (2006) The effect of water concentration on the activity and stability of CLECs in supercritical CO_2 in continuous operation. *J. Mol. Catal. B Enzym.*, **39** (1–4), 112–116.

27 Mishima, K., and Matsuyama, K. (2000) Ring-opening polymerization of lactones using surfactant-coated enzymes in supercritical carbon dioxide. *Chorinkai Saishin Gijutsu*, **4**, 77–79.

28 Mesiano, A.J., Enick, R.M., Beckman, E.J., and Russell, A.J. (2001) The phase behavior of fluorinated diols, divinyl adipate and a fluorinated polyester in supercritical carbon dioxide. *Fluid Phase Equilibria*, **178** (1–2), 169–177.

29 Uyama, H., Wada, S., Fukui, T., and Kobayashi, S. (2003) Lipase-catalyzed synthesis of polyesters from anhydride derivatives involving dehydration. *Biochem. Eng. J.*, **16** (2), 145–152.

30 Takamoto, T., Uyama, H., and Kobayashi, S. (2001) Lipase-catalyzed synthesis of aliphatic polyesters in supercritical carbon dioxide. e-Polymers [online computer file]. Paper No 4.

31 Nakaoki, T., Kitoh, M., and Gross, R.A. (2005) Enzymatic ring opening polymerization of ĺμ-caprolactone in supercritical CO2. *ACS Symp. Ser.*, **900**, (Polymer Biocatalysis and Biomaterials), 393–404.

32 Nakaoki, T., Mei, Y., Miller, L.M., Kumar, A., Kalra, B., Miller, M.E., Kirk, O., Christensen, M., and Gross, R.A. (2005) Candida antarctica lipase B catalyzed polymerization of lactones: effects of immobilization matrices on polymerization kinetics and molecular weight. *Ind. Biotechnol.*, **1** (2), 126–134.

33 Loeker, F.C., Duxbury, C.J., Kumar, R., Gao, W., Gross, R.A., and Howdle, S.M. (2004) Enzyme-catalyzed ring-opening polymerization of ĺμ-caprolactone in supercritical carbon dioxide. *Macromolecules*, **37** (7), 2450–2453.

34 Takamoto, T., Uyama, H., and Kobayashi, S. (2001) Lipase-catalyzed degradation of polyester in supercritical carbon dioxide. *Macromol. Biosci.*, **1** (5), 215–218.

35 Matsumura, S., Ebata, H., Kondo, R., and Toshima, K. (2001) Organic solvent-free enzymatic transformation of poly(ĺμ-caprolactone) into repolymerizable oligomers in supercritical carbon dioxide. *Macromol. Rapid Commun.*, **22** (16), 1325–1329.

36 Matsumura, S. (2003) Depolymerization and Polymerization of polyester and polycarbonate in supercritical fluid, Keio University, Japan, Jpn. Kokai Tokkyo Koho JP2003079388; JP4171823.

37 Matsumura, S., and Osanai, Y. (2005) Method of Continuously depolymerizing polyesters, polycarbonates, or polylactic acid with supercritical fluid and apparatus for continuous depolymerization, Keio University, Japan, WO 2005026245; JP 2005082710.

38 Osanai, Y., Toshima, K., and Matsumura, S. (2006) Enzymatic transformation of aliphatic polyesters into cyclic oligomers using enzyme packed column under continuous flow of supercritical carbon dioxide with toluene. *Sci. Technol. Adv. Mater.*, **7** (2), 202–208.

39 Lee, H.S., Ryu, Y.W., and Kim, C. (1994) Hydrolysis of starch by ĺ±-amylase and glucoamylase in supercritical carbon dioxide. *J. Microbiol. Biotechnol.*, **4** (3), 230–232.

40 Muratov, G., Seo, K.-W., and Kim, C. (2005) Application of supercritical carbon dioxide to the bioconversion of cotton fibers. *J. Ind. Eng. Chem.*, **11** (1), 42–46.

41 Primozic, M., Habulin, M., and Knez, Z. (2005) Modeling of kinetics for the enzymatic hydrolysis of sunflower oil in a high-pressure reactor. *J. Am. Oil Chem. Soc.*, **82** (8), 543–547.

42 Park, C.Y., Ryu, Y.W., and Kim, C. (2001) Kinetics and rate of enzymic hydrolysis of cellulose in supercritical carbon dioxide. *Korean J. Chem. Eng.*, **18** (4), 475–478.

43 Zheng, Y., and Tsao, G.T. (1996) Avicel hydrolysis by cellulase enzyme in supercritical CO2. *Biotechnol. Lett.*, **18** (4), 451–454.

44 Habulin, M., Primozic, M., and Knez, Z. (2005) Enzymatic reactions in high-pressure membrane reactors. *Ind. Eng. Chem. Res.*, **44** (25), 9619–9625.

45 Kato, M., Kamigaito, M., Sawamoto, M., and Higashimura, T. (1995) Polymerization of methyl methacrylate with the carbon tetrachloride/dichlorotris- (triphenylphosphine) ruthenium(II)/methylaluminum bis(2,6-di-tert-butylphenoxide) initiating system: possibility of living radical polymerization. *Macromolecules*, **28** (5), 1721–1723.

46 Wang, J.-S., and Matyjaszewski, K. (1995) Controlled/'living' radical polymerization. halogen atom transfer radical polymerization promoted by a Cu(I)/Cu(II) redox process. *Macromolecules*, **28** (23), 7901–7910.

47 Chiefari, J., Chong, Y.K., Ercole, F., Krstina, J., Jeffery, J., Le, T.P.T., Mayadunne, R.T.A., Meijs, G.F., Moad, C.L., Moad, G., Rizzardo, E., and Thang, S.H. (1998) Living free-radical polymerization by reversible addition-fragmentation chain transfer: the RAFT process. *Macromolecules*, **31** (16), 5559–5562.

48 Moad, G., Rizzardo, E., and Solomon, D.H. (1982) Selectivity of the reaction of free radicals with styrene. *Macromolecules*, **15** (3), 909–914.

49 Georges, M.K., Veregin, R.P.N., Kazmaier, P.M., and Hamer, G.K. (1993) Narrow molecular weight resins by a free-radical polymerization process. *Macromolecules*, **26** (11), 2987–2988.

50 Hadjichristidis, N. (2002) *Block Copolymers: Synthetic Strategies, Physical Properties, and Applications*, John Wiley & Sons, Inc., Hoboken, USA.

51 Bernaerts, K.V., and Du Prez, F.E. (2006) Dual/heterofunctional initiators for the combination of mechanistically distinct polymerization techniques. *Prog. Polym. Sci.*, **31** (8), 671–722.

52 Simionescu, C.I., Comanita, E., Pastravanu, M., and Dumitriu, S. (1986) Progress in the field of bi- and polyfunctional free-radical polymerization initiators. *Prog. Polym. Sci.*, **12** (1–2), 1–109.

53 Jerome, R. (2002) Contribution of unsymmetrical difunctional initiators/monomers to the macromolecular engineering. *Macromol. Symp.*, **177**, (Synthesis of Defined Polymer Architectures), 43–59.

54 Villarroya, S., Thurecht, K.J., Heise, A., and Howdle, S.M. (2007) Supercritical CO_2: an effective medium for the chemo-enzymatic synthesis of block copolymers? *Chem. Commun. (Camb.)*, **37**, 3805–3813.

55 Duxbury Christopher, J., Wang, W., de Geus, M., Heise, A., and Howdle Steven, M. (2005) Can block copolymers be synthesized by a single-step chemoenzymatic route in supercritical carbon dioxide? *J. Am. Chem. Soc.*, **127** (8), 2384–2385.

56 Zhou, J., Villarroya, S., Wang, W., Wyatt, M.F., Duxbury, C.J., Thurecht, K.J., and Howdle, S.M. (2006) One-Step Chemoenzymatic synthesis of poly(e-caprolactone-block-methyl methacrylate) in supercritical CO_2. *Macromolecules*, **39** (16), 5352–5358.

57 Villarroya, S., Zhou, J., Duxbury, C.J., Heise, A., and Howdle, S.M. (2006) Synthesis of semifluorinated block copolymers containing poly(Iµ-caprolactone) by the combination of ATRP and enzymatic ROP in $scCO_2$. *Macromolecules*, **39** (2), 633–640.

58 Thurecht, K.J., Gregory, A.M., Villarroya, S., Zhou, J., Heise, A., and Howdle, S.M. (2006) Simultaneous enzymatic ring opening polymerization and RAFT-mediated polymerization in supercritical CO2. *Chem. Commun. (Camb.)*, **42**, 4383–4385.

59 Villarroya, S., Zhou, J., Thurecht, K.J., and Howdle, S.M. (2006) Synthesis of graft copolymers by the combination of ATRP and enzymatic ROP in $scCO_2$. *Macromolecules*, **39** (26), 9080–9086.

60 Villarroya, S., Dudek, K., Zhou, J., Irvine, D.J., and Howdle, S.M. (2008) Grafting polymers by enzymatic ring opening polymerization-maximizing the grafting efficiency. *J. Mater. Chem.*, **18** (9), 989–997.

61 Adamsky, F.A., and Beckman, E.J. (1994) Inverse emulsion polymerization of acrylamide in supercritical carbon dioxide. *Macromolecules*, **27** (1), 312–314.

62 Beckman, E.J. (2005) Inverse emulsion polymerization in carbon dioxide, In: *Supercritical Carbon Dioxide*, (eds M.F. Kemmere and T. Meyer) Polymer Reaction Engineering, Wiley-VCH Verlag GmbH, Weinheim, Germany, pp 139–156.

63 Tan, B., Lee, J.-Y., and Cooper, A.I. (2006) Ionic hydrocarbon surfactants for emulsification and dispersion polymerization in supercritical CO_2. *Macromolecules*, **39** (22), 7471–7473.

64 Hile, D.D., and Pishko, M.V. (2001) Emulsion copolymerization of D,L-lactide and glycolide in supercritical carbon dioxide. *J. Polym. Sci. [A1]*, **39** (4), 562–570.

65 Johnston, K.P., Harrison, K.L., Clarke, M.J., Howdle, S.M., Heitz, M.P., Bright, F.V., Carlier, C., and Randolph, T.W. (1996) Water-in-carbon dioxide microemulsions: an environment for hydrophiles including proteins. *Science*, **271** (5249), 624–626.

66 Johnston, K.P., Jacobson, G.B., Lee, C.T., Meredith, C., Da Rocha, S.R.P., Yates, M.Z., DeGrazia, J., and Randolph, T.W. (1999) Microemulsions, emulsions and latexes. *Chemical Synthesis Using Supercritical Fluids*, Wiley-VCH Verlag GmbH, Weinheim, Germany, 127–146.

67 Ohde, H., Wai, C.M., and Rodriguez, J.M. (2007) The synthesis of polyacrylamide nanoparticles in supercritical carbon dioxide. *Colloid Polym. Sci.*, **285** (4), 475–478.

68 Ye, W., and DeSimone, J.M. (2005) Emulsion polymerization of N-ethylacrylamide in supercritical carbon dioxide. *Macromolecules*, **38** (6), 2180–2190.

69 Holmes, J.D., Steytler, D.C., Rees, G.D., and Robinson, B.H. (1998) Bioconversions in a water-in-CO2 microemulsion. *Langmuir*, **14** (22), 6371–6376.

70 Barman, T.E. (1985) *Enzyme Handbook*, vol. 1, Springer-Verlag, New York, p. 234.

71 Derango, R.A., Chiang, L.C., Dowbenko, R., and Lasch, J.G. (1992) Enzyme-mediated polymerization of acrylic monomers. *Biotechnol. Tech.*, **6** (6), 523–526.

72 Emery, O., Lalot, T., Brigodiot, M., and Marechal, E. (1997) Free-radical polymerization of acrylamide by horseradish peroxidase-mediated initiation. *J. Polym. Sci. [A1]*, **35** (15), 3331–3333.

73 Teixeira, D., Lalot, T., Brigodiot, M., and Marechal, E. (1999) 1,2-diketones as key compounds in free-radical polymerization by enzyme-mediated initiation. *Macromolecules*, **32** (1), 70–72.

74 Kalra, B., and Gross, R.A. (2002) HRP-mediated polymerizations of acrylamide and sodium acrylate. *Green Chem.*, **4** (2), 174–178.

75 Villarroya, S., Thurecht, K.J., and Howdle, S.M. (2008) HRP-mediated inverse emulsion polymerisation of acrylamide in supercritical carbon dioxide. *Green Chem.*, **10** (8), 863–867.

76 Schomberg, D.S., Salzman, M., and Stephan, D. (1993) *Enzyme Handbook* vol. 7, Springer-Verlag, Berlin Heidelberg, pp. 1–6.

77 Wasserscheid, P., and T. Welton (eds) (2008) *Ionic Liquids in Synthesis*, 2nd edn, Wiley-VCH Verlag GmbH, Weinheim, Germany.

78 Madeira Lau, R., Sorgedrager, M.J., Carrea, G., Van Rantwijk, F., Secundo, F., and Sheldon, R.A. (2004) Dissolution of Candida antarctica lipase B in ionic liquids: effects on structure and activity. *Green Chem.*, **6** (9), 483–487.

79 van Rantwijk, F., Madeira Lau, R., and Sheldon, R.A. (2003) Biocatalytic transformations in ionic liquids. *Trends Biotechnol.*, **21** (3), 131–138.

80 Dordick, J.S., Marletta, M.A., and Klibanov, A.M. (1986) Peroxidases depolymerize lignin in organic media but not in water. *Proc. Natl. Acad. Sci. USA*, **83** (17), 6255–6257.

81 Akkara, J.A., Senecal, K.J., and Kaplan, D.L. (1991) Synthesis and characterization of polymers produced by horseradish peroxidase in dioxane. *J. Polym. Sci. [A1]*, **29** (11), 1561–1574.

82 Serdakowski, A.L., Munir, I.Z., and Dordick, J.S. (2006) Dramatic solvent and hydration effects on the transition state of soybean peroxidase. *J. Am. Chem. Soc.*, **128** (44), 14272–14273.

83 Ayyagari, M.S., Marx, K.A., Tripathy, S.K., Akkara, J.A., and Kaplan, D.L. (1995) Controlled free-radical polymerization of phenol derivatives by enzyme-catalyzed reactions in organic solvents. *Macromolecules*, **28** (15), 5192–5197.

84 Eker, B., Zagorevski, D., Zhu, G., Linhardt, R.J., and Dordick, J.S. (2009) Enzymatic polymerization of phenols in room-temperature ionic liquids. *J. Mol. Catal. B Enzym.*, **59** (1–3), 177–184.

85 Nara, S.J., Harjani, J.R., Salunkhe, M.M., Mane, A.T., and Wadgaonkar, P.P. (2003) Lipase-catalyzed polyester synthesis in 1-butyl-3-methylimidazolium hexafluorophosphate ionic liquid. *Tetrahedron Lett.*, **44** (7), 1371–1373.

86 Uyama, H., Takamoto, T., and Kobayashi, S. (2002) Enzymatic synthesis of polyesters in ionic liquids. *Polym. J.*, **34** (2), 94–96.

87 Marcilla, R., de Geus, M., Mecerreyes, D., Duxbury, C.J., Koning, C.E., and Heise, A. (2006) Enzymatic polyester synthesis in ionic liquids. *Eur. Polym. J.*, **42** (6), 1215–1221.

88 Gorke, J.T., Okrasa, K., Louwagie, A., Kazlauskas Romas, J., and Srienc, F. (2007) Enzymatic synthesis of poly(hydroxyalkanoates) in ionic liquids. *J. Biotechnol.*, **132** (3), 306–313.

89 Blanchard, L.A. (2001) Ionic liquid – carbon dioxide systems: phase behavior, solubilities and extraction, PhD Thesis Univ. of Notre Dame, Notre Dame, IN, USA, p. 157.

90 Lozano, P., Garcia-Verdugo, E., Piamtongkam, R., Karbass, N., De Diego, T., Burguete, M.I., Luis, S.V., and Iborra, J.L. (2007) Bioreactors based on monolith-supported ionic liquid phase for enzyme catalysis in supercritical carbon dioxide. *Adv. Synth. Catal.*, **349** (7), 1077–1084.

91 Rumbau, V., Marcilla, R., Ochoteco, E., Pomposo, J.A., and Mecerreyes, D. (2006) Ionic liquid immobilized enzyme for biocatalytic synthesis of conducting polyaniline. *Macromolecules*, **39** (25), 8547–8549.

92 Akkara, J.A., Kaplan, D.L., and Ayyagari, M. (2000) Process to control the molecular weight and polydispersity of substituted polyphenols and polyaromatic amines by enzymatic synthesis in organic solvents, microemulsions, and biphasic systems, United States Dept. of the Army, USA US 6096859; US 20010003774; US 6362314.

93 Horvath, I.T., and Rabai, J. (1994) Facile catalyst separation without water: fluorous biphase hydroformylation of olefins. *Science*, **266** (5182), 72–75.

94 Hobbs, H.R., and Thomas, N.R. (2007) Biocatalysis in supercritical fluids, in fluorous solvents, and under solvent-free conditions. *Chem. Rev.*, **107** (6), 2786–2820.

95 Garcia-Arrazola, R., Gimeno, M., and Barzana, E. (2007) Use of liquid 1,1,1,2-tetrafluoroethane as solvent media for enzyme-catalyzed ring-opening polymerization of lactones. *Macromolecules*, **40** (12), 4119–4120.

14
Molecular Modeling Approach to Enzymatic Polymerization
Gregor Fels and Iris Baum

14.1
Introduction

Molecular modeling combines a broad range of computational methods for studying molecules and their interactions. The output ranges from simple displays of molecules or the animations of a molecular interplay to fairly exact predictions of how molecules interact with each other [1]. The simplest application of molecular modeling is the visualization of a single molecule as calculated by a molecular modeling program. Figure 14.1 exemplarily shows a model of a lipase complexed with a molecule of Tween [2]. Figure 14.1a displays the protein as it can be extracted from the protein database (PDB) [3] with the only visual enhancement that the Tween has been highlighted in green for easy recognition. Figure 14.1b shows the same protein, however, now focusing on the ligand to demonstrate how it is embedded in the binding site of the protein. The front of the binding site (yellow surface) is transparent to allow the Tween molecule to be seen completely. The protein chain is simplified in a ribbon rendering of the backbone with α-helices colored in red, while β-sheets are shown in yellow. All amino acid sidechains are omitted in this figure to reduce the complexity of the graphics. All figures displayed in this article have been produced using the MOE software from Chemical Computing Group [4].

Databases exist that provide us with coordinates of molecules determined by X-ray crystallography. These are the PDB database [3] for proteins and the Cambridge structural database (CSD) [5] for small molecules. Structures of molecules that have not yet been deposited in these databases have to be predicted and this is where modeling programs come into play. Software tools allow the calculation of structures on three levels of accuracy, depending on the number of atoms involved. With small molecule an *ab initio* process can provide molecular details including electronic and energetic aspects, while in the case of medium sized molecules, in order to stay within acceptable computing time, semiempirical procedures are employed to find approximate solution to the Schrödinger equation treating only the valence electrons with quantum mechanics. If even larger systems like proteins are involved, a structural prediction can merely be be treated with *ab initio* techniques like Hartree–Fock or density function theory (DFT) procedures. The more correct way is the so called combined QM/MM approach, in which quantum mechanical (QM) procedures are applied to a subset of the protein involved in the reaction of interest, typically with up to 60–80 heavy atoms, while classical molecular mechanics (MM) is used for the rest of the system taking additional care for the boundary region between the QM and the MM part.

Molecular modeling, of course, has merely a predictive value. It can never prove or disprove a given mechanism but it can suggest plausible structures, reactions and path ways that eventually should be proven by experiments. Both requirements for molecular modeling, the methods of computational chemistry as well as the hardware aspects of computing, are improving year by year enabling the computational chemist more and more to correctly predict molecular processes and to explain experimental results.

14.2
Enzymatic Polymerization

Enzymes from four of the six enzyme classes have so far been used for enzymatic polymerization. These are oxireductases, transferases, hydrolases, ligases and the products achieved by help of these enzymes range from polyphenols (see Chapter 7) and polyalanines (see Chapter 8), via polysaccharides (see Chapter 9), polyesters and polycarbonates (see Chapter 4), to poly(amino acid)s (see Chapter 5) [7]. While there are, for instance, a great many reports on enzymatic polyester formation, there are, surprisingly, only very few publications on enzyme-catalyzed synthesis of polyamides (see also Chapter 5). Gu *et al.* [8] describe a polycondensation reaction using adipic, malonic and fumaric acid as diacid component and diethylenetriamine or triethylenetetramine as diamines. Remarkably, conventional diamines like 1,6-hexamethylenediamine have not been employed, despite the fact that polyester formation from for instance 1,6-dihydroxyhexane and dimethyl adipate are well known [7f]. Schwab *et al.* [9] report an enzymatic route to unbranched poly(β-alanine) using CALB immobilized as Novozym® 435 starting from unsubstituted β-lactam (2-azetidinone). Ring-opening of substituted β-lactams with lipase have been reported before [10] but not with the aim of polymerization.

None of the enzymatic polymerizations have as yet been looked at with computational chemistry techniques. This is all the more surprising as the literature of the last five to ten years provides us with numerous examples of particularly computational simulations of enzyme-catalyzed ester and amide formation and cleavage. This prompted us to look into these enzymatic reactions with particular emphasis on polyamide formation and in this article we outline our initial results of a computational chemistry approach towards the mechanism of CALB-catalyzed polyamide formation.

14.3
Candida antarctica Lipase B—Characterization of a Versatile Biocatalyst

Of the various enzymes available for enzyme catalyzed organic reactions the Novozym 435 preparation of CALB has extensively been used in biocatalysis reactions, using a broad range of substrates, often combined with outstanding chemo-, regio-, and/or enantioselectivity (see also Chapters 3–5, 11 and 12) [10d, 11]. The enzyme consists of 317 amino acids with a catalytic triad comprised of Ser_{105}, His_{224}, and Asp_{187}. It is noteworthy that His_{224} is the only histidine present in this enzyme. The crystal structure of CALB was solved in 1994 by Uppenberg *et al.* [2], and it was found that the enzyme belongs to the globular α/β-type protein class composed of seven central β strands that are flanked on both sides by ten α helices. CALB therefore belongs to the family of α/β-hydrolase fold proteins, and the short $\alpha 5$-helix (defined as residues 142–146, red in Figure 14.3a) was pointed out as a putative lid. In contrast to other lipases, CALB seems to be the only member of the family in which the fold is open, leaving the active site (blue mesh in Figure 14.3a) with the catalytic triad easily accessible to the solvent. Nevertheless, helix $\alpha 5$ together with helix $\alpha 10$ (residues 268–287, yellow in Figure 14.3a) have been identified as highly mobile regions in the CALB structure, with a relative motion between them [12]. In addition to the catalytic triad two further amino acids are of great mechanistic importance for the enzyme reactions. These are Gly_{106} and Thr_{40} which are involved in what is called an oxyanion hole. They can donate backbone hydrogens and hydroxyl hydrogen, respectively, for hydrogen bonding to the oxygen atom of a carbonyl group when bound to the catalytic

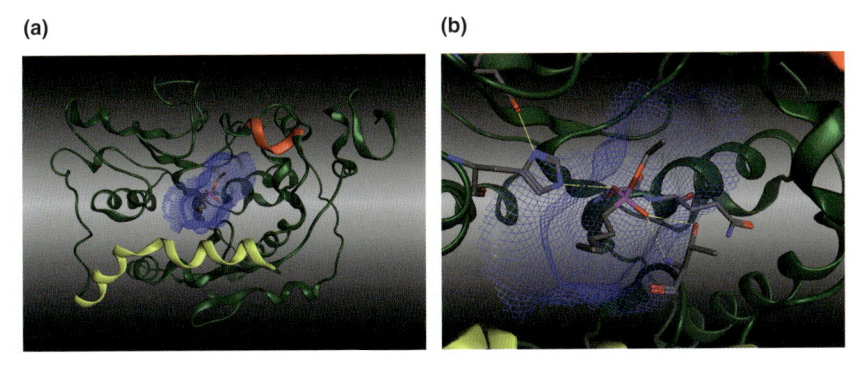

Figure 14.3 Representation of *Candida antarctica* B complexed with a covalently bound phosphonate ligand. (a) Ribbon rendering of the protein with the putative lid (red) in the open configuration and the ligand binding site (blue mesh). The mobile regions are shown in red ($\alpha 5$-helix) and yellow ($\alpha 10$-helix). (b) Close up view of the ligand binding site with the phosphonate ligand bound to Ser_{105}. The negatively charged P–O oxygen is stabilized by hydrogen bonding (yellow lines) to Thr_{40} and Gln_{106} (right side). His_{224} shows hydrogen bonds to Ser_{105}-oxygen and to Asp_{187} (left side). [6].

serine. This preferential binding is responsible for much of the catalytic efficiency and the stereoselectivity of the enzyme and it is of utmost importance for the enzymatic polymerization as described later on in this article.

Besides the crystal structure of the native enzyme, the PDB database [3] provides two further CALB crystal structures elucidated by Uppenberg et al. [6], which characterize complexes of the enzyme with small ligands. Of these structures the PDB code 1LBT describes the complex with a molecule of the detergent Tween 80 embedded in the active site while the PDB code 1LBS stands for the crystal structure of phosphonate inhibitor covalently bound to a Ser_{105} (Figure 14.3b). These structures can be used as a starting point for molecular modeling investigations.

14.4
Lipase Catalyzed Alcoholysis and Aminolysis of Esters

Lipases are usually active in aqueous systems as well as in organic solvents which makes them applicable not only for hydrolytic reactions but also for esterifications and transesterifications or for the formation of amides. These enzymes are known to accept a wide variety of substrates quite often with high enantioselectivity which makes them useful for chiral organic substrates. Numerous molecular modeling studies during the last decade describe these phenomena and provide us today with a clear picture of the mechanisms behind the enzymatic reaction particularly of CALB. Vicente Gotor, Karl Hult, and Romas J. Kazlauskas are three of the most cited names in this regard who have contributed fantastic work in this field.

First attempts to predict the selectivity of enzymes are dated back to 1964 when Prelog described an empirically determined rule for the addition of hydrogen to ketones by the yeast *Culvaria lunata* [13]. In 1991 Kazlauskas published the hydrolysis of acetates of secondary alcohols by Pancreatic cholesterol esterase, *Pseudomonas cepacia* and *Candida rugosa* and formulated the widely applicable Kazlauskas rule according to which esters of secondary alcohols with a specific substitution pattern of large (L) and medium (M) substituents are cleaved faster than the corresponding enantiomer [14].

However, due to the diversity of the active sites in different enzymes there is no universally valid rule for all hydrolases but rather individual enzymes

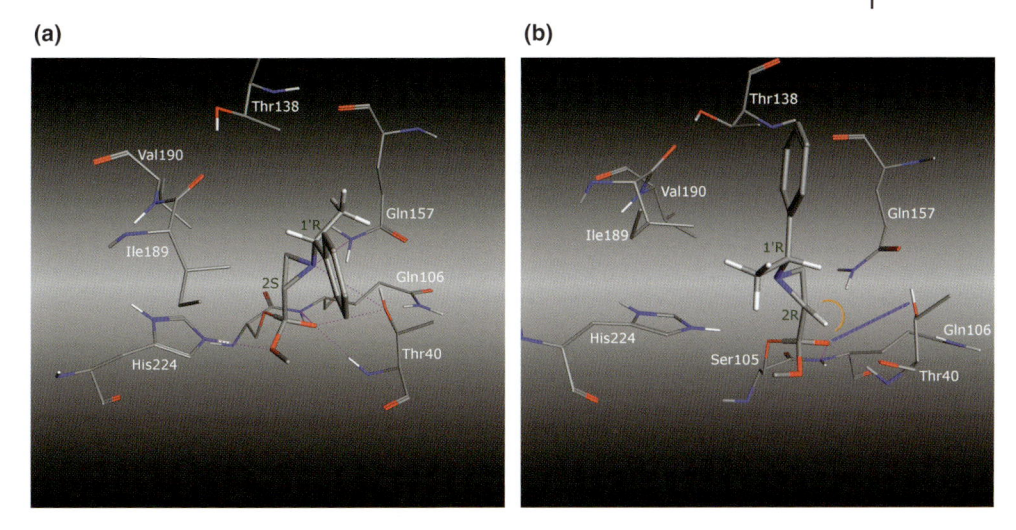

Figure 14.4 Tetrahedral intermediates of CALB aziridine-carboxylate complexes. The slow reacting (2R,1'R)-diastereomer (b) exhibits an umbrella-like inversion of substituents compared with the fast reacting (2S,1'R)-diastereomer (a). The hydrogen at the 2R stereocenter interferes with the hydroxyl group of Thr40 (orange), and one of the three hydrogen bonds of the oxyanion hole is lost (blue line) [11a].

control stereospecific ester or amide cleavage on a specific structural basis. A recent example from the Kazlauskas group for instance explains the enantiopreference of CALB towards the ammoniolysis of aziridine-2-carboxylates [11a] which is an extension of the classical Kazlauskas rule (Figure 14.4). While in this example the fast reacting (2S, 1'R)-diastereomer of the serine bound ligand perfectly fits to the active site for aminolysis (Figure 14.4a), the slow reacting (2R, 1'R)-diastereomer, can only form the tetrahedral intermediate at all after an umbrella like inversion of the substituents at the ring-stereocenter. This results in an unfavorable interaction with Thr_{40} and in the loss of one hydrogen bond of the oxyanion hole which in turn is responsible for the reduced reaction rate.

A rather different sterospecificity has recently been described by the Hult group for the esterification of pentan-2-ol with propionic acid methylester [11d]. It is well known that the water content in the enzyme – and in particular in the binding site – has a pronounced effect on both reaction rate and enantioselectivity. The presence of too much of water would prevent transesterification, while too low a water content destabilizes the enzyme, resulting in decreased reactivity. In this specific example the enantioselectivity is explained by the influence of a thermodynamic water molecule on CALB (Figure 14.5). While in the (R)-intermediate the propyl group of the chiral center is oriented towards the entrance of the pocket (Figure 14.5a) in the (S)-intermediate it points directly towards the stereospecificity pocket (Thr_{42} and Ser_{47}) that contains the decisive water molecule

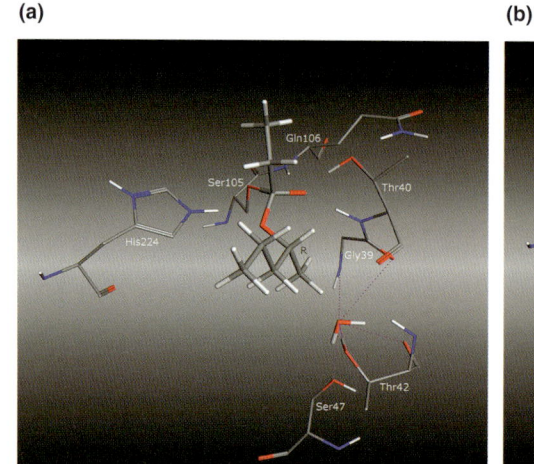

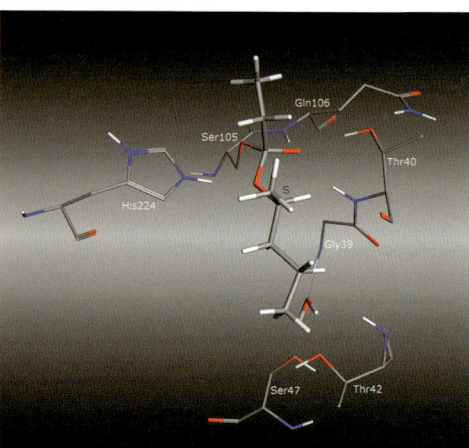

Figure 14.5 Models of (a) the (R)- and (b) the (S)-tetrahedral intermediates of 2-pentylpropanoic acid with a water molecule introduced into the stereospecificity pocket (Thr$_{24}$ andSer$_{47}$). The water molecule inside this pocket forms five hydrogen bonds in presence of the (R)-enantiomer while only three hydrogen bonds can be formed in the case of the (S)-enantiomer which thermodynamically favors the (R)-enantiomer [11d].

(Figure 14.5b) thereby reducing the number of hydrogen bonding of this water from five in the (R)-enantiomer to only three in the (S)-enantiomer, which indicates a better fit for the water molecule in the (R)-enantiomer and an unfavorable enzyme conformation necessary for the (S)-enantiomer to react.

Finally, a third example should serve further to demonstrate the diversity of stereoselective reactions provided by CALB. In a recent publication the Gotor group suggests the formation of zwitterionic species, resulting from the direct His-unassisted attack of an amine onto the carbonyl group of the acyl–enzyme, as the most plausible intermediate for the CALB-catalyzed aminolysis [11f]. This proposal differs slightly from the commonly accepted serine-mediated mechanism, where removal of the proton from the amine occurs simultaneously with the nucleophile attack to the acyl–enzyme complex. Subsequently, His-assisted deprotonation of the resulting ammonium group takes place, and a molecule of water in some cases is necessary to facilitate the transfer of the proton to the catalytic histidine. Proton transfer-aided pathways have been previously proposed for CALB-catalyzed reactions in which modeling predicted that a molecule of water is necessary to transfer the proton from the nucleophile to the catalytic histidine [10c, 15]. Stereoselectivity is achieved in this example by better binding of the substrate, together with a molecule of water in the (1R,2S)-2-cyclohexanamide intermediate, justifying the enantiopreference exhibited by CALB towards this substrate (Figure 14.6).

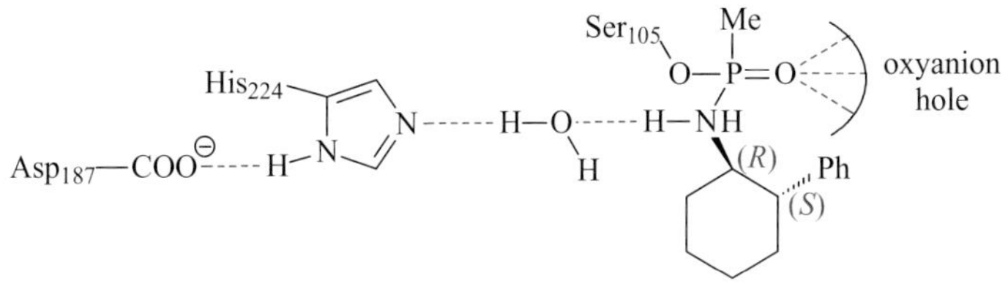

Figure 14.6 Geometry of the productive reaction intermediate for the acetylation of 2-phenylcyclohexanamines catalyzed by CALB [11f].

14.5
Lipase-Catalyzed Polyester Formation

The catalytic reaction of lipases follow the so called ping-pong bi-bi mechanism, a double displacement mechanism. This is a special multisubstrate reaction in which, for a two-substrate, two-product (i.e., bi-bi) system, an enzyme reacts with one substrate to form a product and a modified enzyme, the latter then reacting with a second substrate to form a second, final product, and regenerating the original enzyme (ping-pong).

In an attempt to apply this mechanism to a polyester formation and to investigate the single steps involved on a molecular basis one can for instance start with the CALB–phosphonate complex 1LBS [6] as the initial structure, as it is structurally close to the intermediate structure of an acylated Ser_{105} that certainly has to be passed through during an enzymatic polymerization process. In accordance with the accepted mechanism for serine hydrolases the enzymatic process consists of two mechanistically important steps. These are, in the case of for instance, a potential enzymatic esterification of adipic acid with 1,6-hexanediol:

- acylation of Ser_{105} by the diacid (steps a to c in Figure 14.7) which proceeds via a tetrahedral intermediate structure (Figure 14.7b)

- deacylation of Ser_{105} by attack of the diol (steps d to f in Figure 14.7) again via a tetrahedral intermediate structure (Figure 14.7e) regenerating the free Ser_{105} to be ready for the next elongation step.

It should be noted that the orientation of the molecules in the active site as shown in Figure 14.7 is not the result of an exact computational simulation but is rather meant to give an idea of the spatial arrangement of the catalytic triad, the growing polymer chain and the monomeric building blocks involved. A reliable simulation of this reaction still remains to be computed.

14.6
CALB -Catalyzed Polymerization of β-Lactam

While there are numerous molecular modeling studies that describe alcoholysis or aminolysis of acylated serine hydrolases the literature does not yet provide us

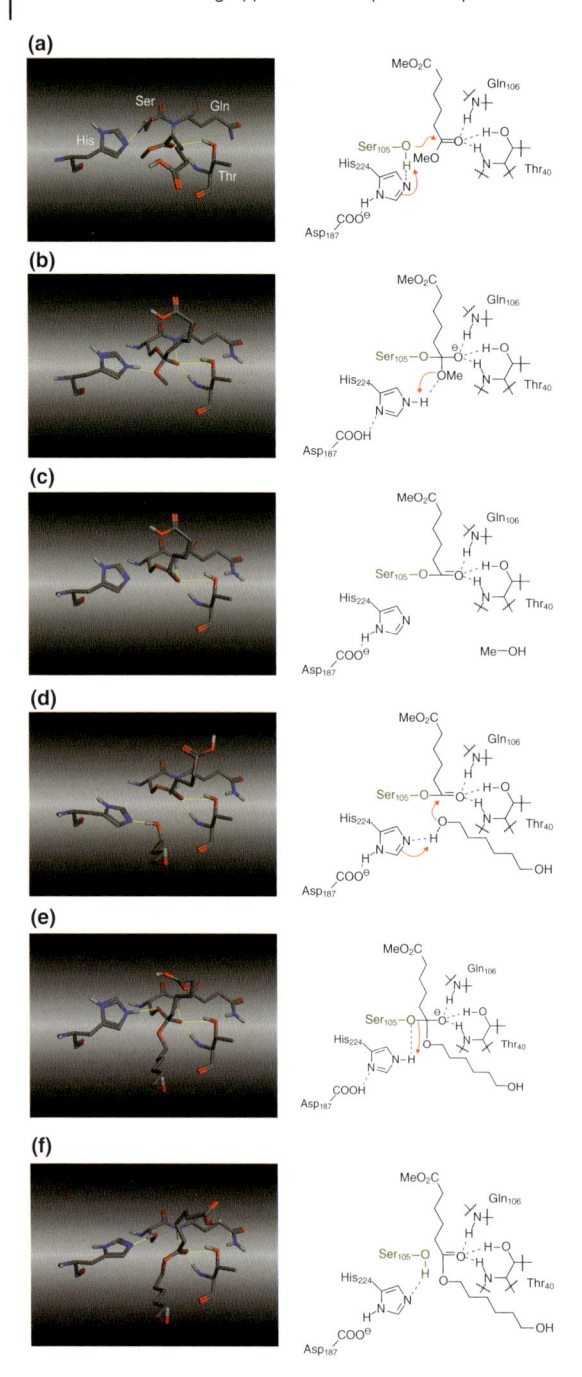

Figure 14.7 Screenshots of a hypothetical CALB-catalyzed esterification with acylation of serine by adipic acid (steps a to c) and deacylation by 1,6-hexanediol (steps d to f). Yellow lines depict hydrogen bondings.

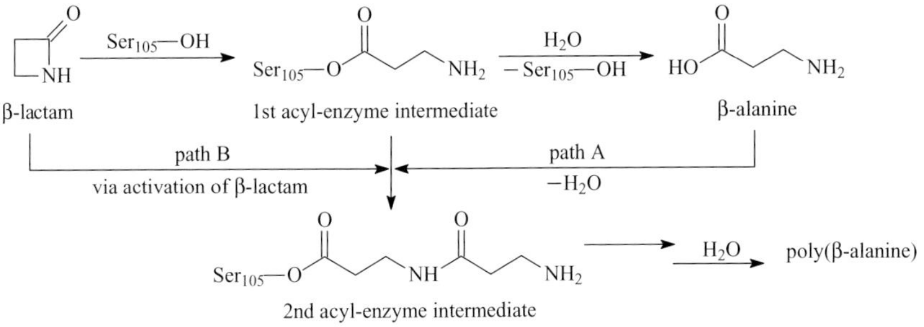

Figure 14.8 Putative mechanism of CALB-catalyzed polymerization of β-lactam adopted from the assumed polymerization mechanism for β-lactone [7h].

with a computational investigation of an enzymatic polyester or polyamide formation. In order to better understand the molecular basis of this processes and in particular to understand the CALB-catalyzed polyamide formation we have computationally investigated the enzymatic conversion of β-lactam to poly(β-alanine), an experiment that recently has been described by Schwab et al. [9].

If we apply the generally assumed mechanism of the CALB-catalyzed ring-opening polymerization of lactones [7h] to the polymerization of β-lactam the first step would be formation of a first acyl–enzyme intermediate as a consequence of the ring-opening of β-lactam by Ser_{105} assisted by His_{224} and Asp_{187} as depicted in Figure 14.8.

The first acyl–enzyme intermediate could potentially be hydrolyzed by a water molecule inherently present in the CALB active site to yield β-alanine. With sufficient β-alanine produced, a competitive reaction could occur between water and β-alanine which eventually would give rise to the formation of the second acyl–enzyme intermediate, elongated by one chain segment (path A in Figure 14.8), and which ultimately would yield the polymerized product by subsequently following this route. During this cycle, condensation of a β-alanine and the acyl–enzyme intermediate liberates the water molecule consumed during the release of β-alanine from Ser_{105}. Keeping in mind that the reaction is carried out in an organic solvent using a dried preparation of immobilized CALB (Novozym 435) (see also Chapter 3), the water balance therefore is neutral for the polymerization process, however, one water molecule per liberated polymer chain has to be contributed by the enzyme to release the final polymer into the solution.

In contrast to this mechanistic concept, Schwab et al. [9] surprisingly found, that β-alanine itself is not a substrate for the CALB-catalyzed polyamide formation. This can be understood in the light of a recent publication by Hollmann et al. [16] who could demonstrate that organic acids having a pK_a value of <4.8 irreversibly inhibit CALB, presumably by protonation of His_{224}, which in turn prevents the deprotonation of Ser_{105} necessary as part of the catalytic process. β-Alanine has a first pK_a of 3.6 [17] which clearly is below the critical value. Therefore, if β-alanine is generated in larger amounts is will reduce the overall enzyme activity by deactivation. Furthermore, Schwab et al. [4] only achieved low molecular weights in their β-lactam polymerization with a maximum chain length of 18 monomers and an average chain length of 8 repeating units. This is rather low compared with polyester formation from lactones [7e]. The polymerization process can be seen as a permanent competition between the elongation of the acyl–enzyme complex and the release of the polymer by attack of a water molecule. It was also found that β-alanine ethylester cannot be used as elongating monomer for the polymerization [9], which can be explained by the fact that the molecule is rather basic, having a pK_a of 9.25 [18], and is presumably protonated on its way to the active site, which prohibits the attack at the acylated serine. In this respect it is noteworthy that diethylenetriamine can be employed as nucleophilic component in a CALB-catalyzed polyamide formation with diethyladipate while the corresponding 1,5-diaminopentane inhibits polymerization [19]. This result can be rationalized by the comparatively low pK_a of 4.2 for the secondary amine while the two primary amines have a pK_a of 9.1 and 9.8, respectively [20].

As a consequence, a poly(β-alanine) can only be formed by a ring-opening polymerization, in which the chain elongation proceeds directly by reaction of β-lactam with an acyl–enzyme intermediate. Because of the low nucleophilicity of the lactam nitrogen, this process requires activation of β-lactam by a water molecule. On this basis we have proposed the catalytic cycle for a CALB-catalyzed polymerization of β-lactam depicted in Figure 14.9. The mechanism described is in accordance with experimental data and is assisted by in-depth computational calculation of the reactions involved. It contains two starting steps (I and II), five steps for the chain elongation (III–VII), and is completed by the release of the polymer (VIII).

Docking experiments with β-lactam into the empty binding pocket of CALB embeds the monomer in the so-called alcohol side of the active site, where it forms weak hydrogen bonds to Ser_{105}, His_{224} and Thr_{40} stabilizing the lactam in this position. Ser_{105}-O is 3.2 Å away from the carbonyl-C of the β-lactam monomer which is sufficiently close to permit an attack at the lactam carbonyl carbon. This process corresponds to the generally accepted initial step of hydrolysis by serine proteases which enables the nucleophilic attack of Ser_{105}-O at the carbonyl carbon of the β-lactam. As a result a first tetrahedral intermediate TI1 is formed (Figure 14.10a and step I in Figure 14.9).

We used a covalent docking procedure to study the intermediate structure TI1 by manually connecting the β-lactam carbonyl group to the Ser_{105}-O which generates a chiral center at the former carbonyl carbon with an (R)-configuration. There is no strong stereochemical control of this reaction as an (S)-configuration can also be achieved in the docking process, albeit QM/MM calculations show that the reverse reaction is favored over the (S)-enantiomer. The negatively charged oxygen of the former carbonyl group in TI1 is stabilized by two hydrogen bonds to Gly_{106} and Thr_{40} of the oxyanion hole (distances 2.6 Å and 2.7 Å, respectively). The β-lactam ring structure is no longer planar but shows a slight puckering of about 8°.

The next step is a ring-opening to yield the first acyl–enzyme complex (C in Figure 14.9) which requires transfer of the hydrogen from His_{224} to the lactam-N.

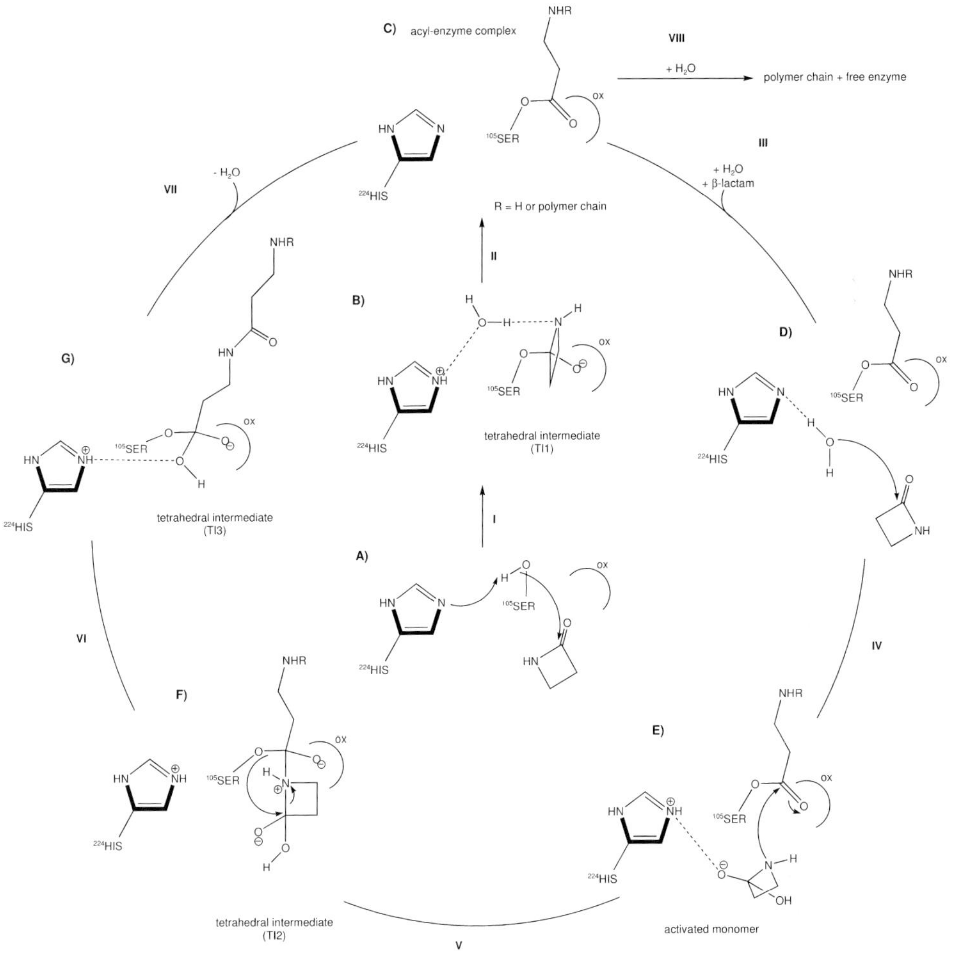

Figure 14.9 Catalytic cycle of CALB-mediated polymerization of β-lactam.

The proton has to be transmitted to the *cis*-configured nitrogen of the former β-lactam which exposes its lone pair toward the His$_{224}$ proton. In principle the position of the proton is arbitrary as it can rapidly invert between a *cis*- and a *trans*-configuration. In the *cis*-configuration, however, the distance of the hetero atoms involved is 3.8 Å and, therefore, too large for a direct hydrogen transfer. A distance of 2.6–3.2 Å is generally accepted as appropriate for hydrogen bonds. Here, a structurally conserved water molecule in the binding site, which is found in the crystal structures of native CALB (1TCA and 1TCB), is ideally positioned to assist the proton shuttle from His$_{224}$ to the lactam-N. Such a water-mediated mechanism has also been described for other CALB-catalyzed reactions [10c, 11f, 15]. As a result, a catalytically productive conformation can be generated by

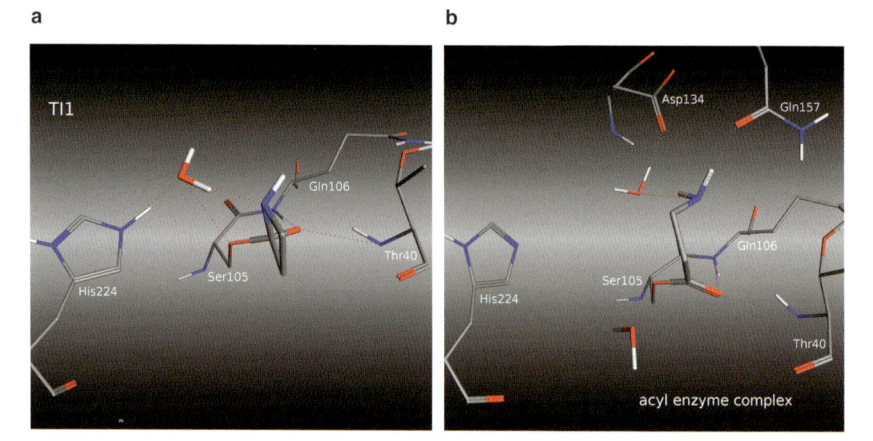

Figure 14.10 (a) Result of covalent docking and QM/MM geometry optimization after the attack of serine oxygen at the β-lactam carbonyl yielding the first tetrahedral intermediate TI1 (step I). (b) Ring-opening of TI1 results in a serine bound β-amino acyl side-chain (acyl–enzyme complex) located in the acyl pocket of the binding site (step II).

covalent docking with subsequent QM/MM minimization in which the conserved water molecule is included in the QM sphere, besides Ser_{105}, His_{224}, and the serine-bound ligand (Figure 14.10a). The two new hydrogen bonds show a distance of 2.6 Å between protonated His_{224} nitrogen and the H_2O oxygen, and 2.8 Å between the water and the lactam-N, respectively.

QM/MM calculations indeed reveal that the proton is transferred from the water molecule to the ring nitrogen, while at the same time the oxygen of the water molecule takes up the His_{224} proton and the ring-opening of the former β-lactam ring occurs. The conformation of the acyl chain at Ser_{105} relaxes from the all-*cis* conformation and the QM/MM optimization drives the β-aminoethyl chain over to the acyl side of the binding pocket (Figure 14.10b). The carbonyl oxygen of the acyl chain is still stabilized in the oxyanion hole with hydogen bonds of distances of ≤2.8 Å. The conformation gains further stabilization from hydrogen bonding of the terminal amino group to Gln_{157} (distance 2.9 Å) and to a water molecule (distance 3.0 Å). While the β-lactam monomer has entered the binding site from the so-called alcohol side, the acyl side-chain at Ser_{105} after ring-opening now ends up in the acyl side of the active site, which empties the alcohol side and makes it ready for the next monomer to come in. At this point the enzyme is primed for the polymerization process and can enter the catalytic cycle to add the next monomer and to elongate the chain. Structure C in Figure 14.9 (Figure 14.10b) is therefore the starting point for the polymerization process. It also resembles the intermediate structure that is always passed through after chain elongation, and it is finally the exit structure from which the polymer can be liberated from Ser_{105} by reaction with water. As mentioned earlier, for each chain termination one molecule of water has to be supplied by the enzyme. Obviously

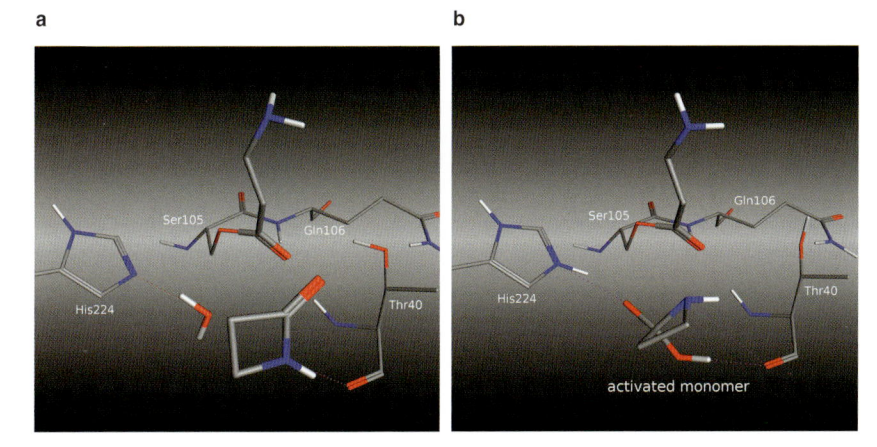

Figure 14.11 (a) Docking of a second β-lactam monomer into the acylated enzyme in the presence of a crystallographic water molecule (step III), and (b) activation of the β-lactam by the hydroxyl group of this water molecule (step IV).

this is not an obstacle to the CALB-catalyzed polymerization as for instance polyesters can be enzymatically produced with high molecular weights when using lactones as starting material [7e]. There seems to be enough water around in the immobilized CALB enzyme, even when dried preparations of Novozym 435 are used in dry solvent.

As mentioned earlier neither β-alanine nor the β-alanine ester can be used for chain elongation and they rather slow down the polyamide formation from β-lactam when simultaneously employed in enzymatic polymerizations. As a consequence, the β-lactam molecule is not only the initiating structure but also the elongating building block.

Because of the low electron density of the β-lactam nitrogen, the monomer has to be activated by water to be used in chain elongation. This can be achieved by re-using the water molecule previously employed for the proton shuttle. Docking studies of this activated β-lactam into the acyl–enzyme complex with a protonated His_{224} yield a structure as depicted in Figure 14.11a in which the activated former β-lactam is again stabilized by hydrogen bonds to His_{224} (2.5 Å) and to Thr_{40} (3.2 and 2.8 Å, respectively). In the resulting structure, the lone pair of the former amide nitrogen is ready to attack the acyl chain at Ser_{105}, although the distance between the nitrogen and the acyl carbonyl carbon is still 3.3 Å. The correct orientation at the nitrogen can be guaranteed as the NH proton can rapidly invert between a *cis* and *trans* configuration with respect to the hydroxyl group. The ring structure of the former β-lactam is slightly puckered with an angle of about 16°.

The nucelophilicity of the activated β-lactam should be high enough to attack the carbonyl carbon of the acyl–enzyme complex and it should yield the dimeric tetrahedral intermediate TI2 (Figure 14.12). We have modeled this structure in a covalent docking step, by connecting the nitrogen of the activated β-lactam

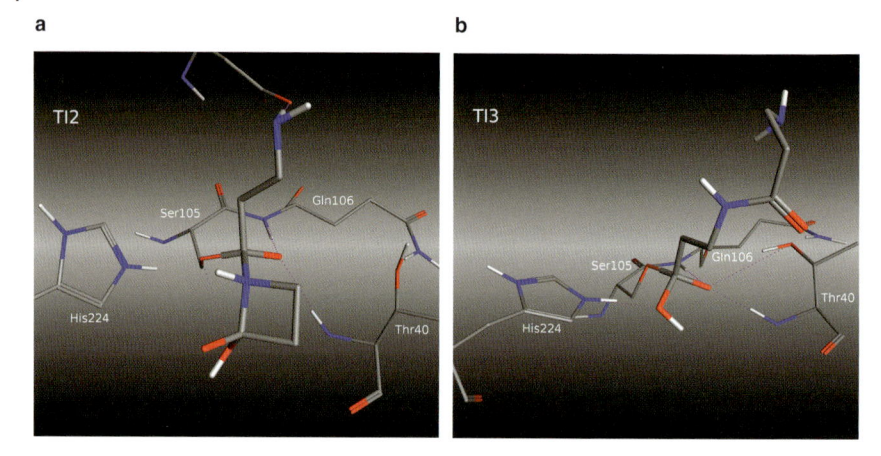

Figure 14.12 (a) Rearrangement step of the dimeric transition intermediate TI2 in which Ser_{105} jumps from the internal former carbonyl carbon to the terminal carboxyl carbon yielding the terminally bound tetrahedral intermediate TI3 (b, step VI).

monomer to the serine bound carbonyl carbon. In the resulting structures the negatively charged oxygen, as before, is stabilized by hydrogen bonding to the oxyanion hole (3.0 Å, 3.3 Å and 3.2 Å, respectively), and the negative charge is compensated by the positive charge of the attacking nitrogen. Additional stabilization is gained from hydrogen bonding between the hydroxy group of the attacking species and Ser-O (2.8 Å). The resulting molecule contains a chiral center with (S)-configuration at the former carbonyl carbon of the acyl–enzyme complex in which the absolute configuration is determined by the orientation of the attacking activated lactam molecule. The NH group of the former lactam is *cis* oriented with respect to the negatively charged carbonyl oxygen of the former β-lactam.

In the following steps QM/MM procedures have been employed to investigate the Ser_{105}-O transfer from the central to the terminal carbon of the dimeric β-alanine/β-lactam species. During the geometric and electronic optimization process, screen-shots of coordination files where collected at regular intervals to decide about the order of the successive steps involved. (Figure 14.13). This time-dependent analysis reveals that initially the former β-lactam ring is opened, yielding a structure in which the newly generated carboxyl group – that originates from the second β-lactam – is ideally positioned for an attack by the Ser_{105}-O, having a O–CO distance of 3.3 Å and an almost perpendicular orientation of the Ser_{105}-O to the carbonyl carbon. Chain elongation can then proceed by a rearrangement of Ser_{105}-O from the central to the terminal carbon, much like the mechanism of monomer insertion in transition metal-catalyzed polymerizations. As a result, the tetrahedral intermediate TI3 is formed (Figure 14.12b).

As an alternative to this mechanism one could also discuss a liberation of the dimeric β-alanine from Ser_{105} and regeneration of the native Ser_{105} by proton transfer from His_{224}. In contrast to the mechanism involving the rearrangement of Ser_{105}-O, a release of the dimer would generate a carboxyl group in the presence of an unprotonated His_{224} which then immediately could lead to protonation of the histidine and formation of a carboxylate, which would terminate the polymerization process. In the case where this deacylation does indeed occur, the β-alanine dimer could also potentially move out of the binding site into the solvent and would have to come back to elongate another serine bound acyl chain. Because of the orientation of the intermediate structure TI3 a direct re-binding of Ser_{105} to the carboxyl terminus of the dimer (starting from Figure 14.13b) appears to be more likely.

As explained above in our calculation we started from the assumption that β-alanine can neither initiate the polymerization nor can it be the building block for chain elongation. One could think of yet a third alternative in which a β-lactam is the initiating species for the polymerization which, however, is elongated by β-alanine produced from β-lactam and kept on hand, one at a time, inside the binding site. Such a potential procedure as outlined in Figure 14.13 is analogous to an interesterification mechanism as proposed by Li and Kanerva [11e], in which an alcohol generated by CALB-catalyzed ester cleavage is stored inside the active site and re-used for transesterification of another ester present in the reaction mixture.

If we apply this mechanism to the polymerization of β-lactam we would have to assume that the growing polymer chain is cleaved from Ser_{105} and temporarily stored in the binding site. A β-lactam then is ring-openend and liberated from the serine but would not leave the enzyme but rather also stays inside the binding site for further use. The growing polymer chain eventually is bound again to Ser_{105} and could finally be attacked by the β-lactam to initiate the next chain elongation (Figure 14.14).

The most plausible elongation mechanism, however, seems to be the transacylation as described in Figure 14.12, in which the growing polymer chain does not leave the serine during the elongation process (step IV in Figure 14.9). The

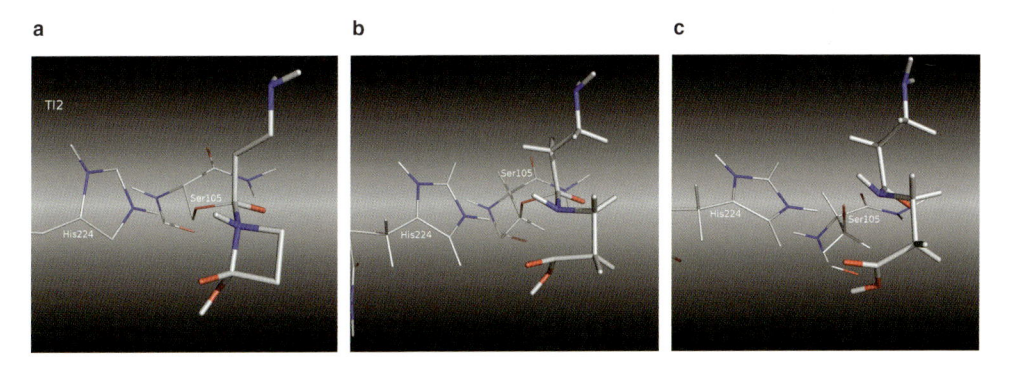

Figure 14.13 Three snapshots of step VI (Figure 14.9) depicted from a QM/MM calculated reaction sequence starting from the serine bound dimeric reaction intermediate TI2. (a) TI2 structure, (b) first ring-openend intermediate, (c) β-alanine dimer separated from the serine.

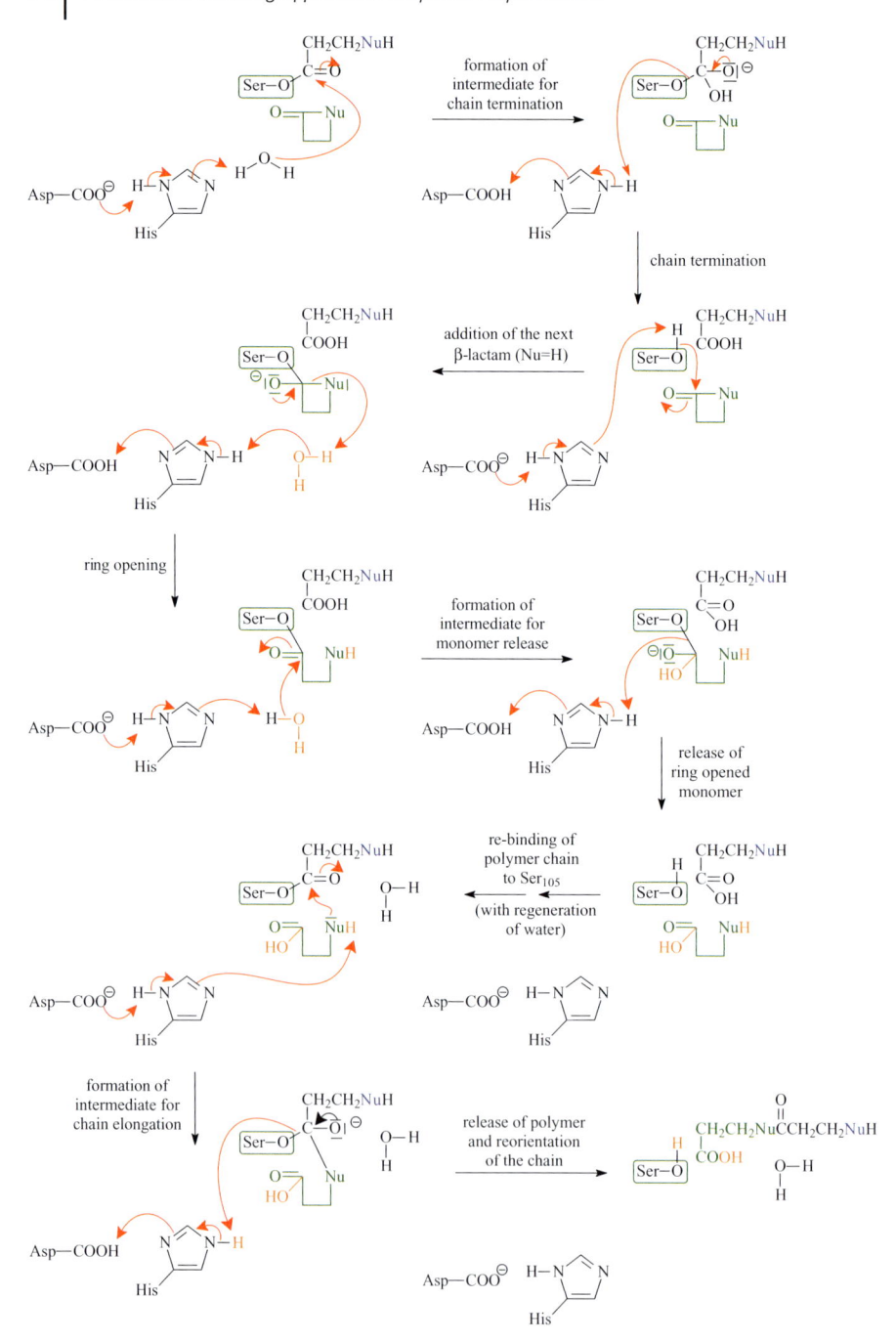

Figure 14.14 Chain elongation following a putative mechanism [11f] (Nu=NH or O).

corresponding transition intermediate TI3 (Figure 14.12b) is stabilized by the oxyanion hole (3.4 Å, 3.3 Å and 2.7 Å, respectively) and the OH group of TI3 can then be cleaved off after being protonated by the proton from His_{224} (2.7 Å between OH group and His_{224} nitrogen). At this point the water molecule that in step IV was used to activate the β-lactam monomer is regenerated and can be used again in the elongation step. Again, modeling of the corresponding acyl–enzyme intermediate by a covalent docking run results in an elongated analog of the β-aminoacyl chain of step II (see Figure 14.13c), which is stabilized by the oxyanion hole (3.1 Å, 3.2 Å and 2.6 Å, respectively) and an additional hydrogen bond between the carbonyl group of the amide bond and Thr_{40}-OH (2.7 Å).

This completes the catalytic cycle and the structure obtained can either react with another activated monomer (analogous to step V) to extend the growing polymer chain or the polymerization can be terminated by hydrolysis to yield the free polymer that could diffuse out into the solvent (step VIII).

14.7
General Remarks

The mechanism outlined for CALB-catalyzed polyamide formation should also apply very similarly to polyester formation from lactones and in general can be adapted to a polyamide/polyester formation from diacids and the corresponding diamine and diol, respectively. It should be noted that the proposed catalytic cycle is compatible with the polymerization pathway A of Figure 14.8 which so far is assumed for all the enzymatic polycondensation reactions in the literature [7]. Our investigation, to our knowledge, is the first attempt to explain the enzymatic polymerization process on a molecular level, and results in a unique mechanism in which the growing polymer never leaves the active site during chain elongation but rather stays bound to serine of the catalytic triad. Liberation of the growing chain would allow the polymer to leave the enzyme with the risk of never finding its way back in a correct orientation for chain elongation. Our proposed mechanism, however, still needs additional experimental verification and experiments along these lines are in progress in our laboratory together with the laboratory of Katja Loos at The University of Groningen, The Netherlands.

References

1 (a) Leach, A. (2001) *Molecular Modelling: Principles and Applications*, Prentice Hall; (b) Schlick, T. (2002) *Molecular Modeling and Simulation: An Interdisciplinary Guide*, Springer-Verlag, New York.

2 Uppenberg, J., Hansen, M.T., Patkar, S., and Jones, T.A. (1994) *Structure*, **2**, 293–308.

3 Berman, H.M., Westbrook, J., Feng, Z., Gilliland, G., Bhat T.N., Weissig, H., Shindyalov, I.N. and Bourne, P.E. (2000) RCSB Protein Data Bank, *Nucleic Acids Research*, **28**, 235–242, http://www.pdb.org (accessed 7 June 2010).

4 Molecular Operating Environment MOE, (2009) Chemical Computing

Group, Inc, Montreal, http://wwwchemcomp.com (accessed 7 June 2010).
5 Cambridge Structural Database (2010) The Cambridge Crystallographic Data Centre, http://www.ccdc.cam.ac.uk/products/csd/ (accessed 7 June 2010).
6 Uppenberg, J., Ohrner, N., Norin, M., Hult, K., Kleywegt, G.J., Patkar, S., Waagen, V., Anthonsen, T., and Jones, T.A. (1995) *Biochemistry*, **34**, 16838–16851.
7 Anderson, E.M., Karin, M., and Kirk, O. (1998) *Biocatal. Biotransformation*, **16**, 181–204; (b) Gotor-Fernandez, V., Busto, E., and Gotor, V. (2006) *Adv. Synth. Catal.*, **348**, 797–812; (c) Gross, R.A., Kalra, B., and Kumar, A. (2001) *Appl. Microbiol. Biotechnol.*, **55**, 655–660; (d) Gross, R.A., Kumar, A., and Kalra, B. (2001) *Chem. Rev.*, **101**, 2097–2124; (e) Kobayashi, S. (2006) *Macromol. Symp.*, **240**, 178–185; (f) Kobayashi, S. (2009) *Macromol. Rapid Commun.*, **30**, 237–266; (g) Kobayashi, S., Ritter, H., and Kaplan, D. (2007) *Enzyme Catalyzed Synthesis of Polymers (Advances in Polymer Science)*, Springer-Verlag, Berlin; (h) Matsumura, S. (2006) *Adv. Polym. Sci.*, **194**, 95–132.
8 Gu, Q.-M., Maslanka, W.W., and Cheng, H.N. (2008) *Polymer Biocatalysis and Biomaterials II*, vol. 999 (eds H.N. Cheng and R.A. Gross), ACS Symposium series, Ch 21, pp. 309–319.
9 Schwab, L.W., Kroon, R., Schouten, A.J., and Loos, K. (2008) *Macromol. Rapid Commun.*, **29**, 794–797.
10 (a) Forro, E., and Fulop, F. (2003) *Org. Lett.*, **5**, 1209–1212; (b) Li, X.G., Lahitie, M., and Kanerva, L.T. (2008) *Tetrahedron Asymmetry*, **19**, 1857–1861; (c) Park, S., Forro, E., Grewal, H., Fulop, F., and Kazlauskas, R.I. (2003) *Adv. Synth. Catal.*, **345**, 986–995; (d) Tasnadi, G., Forro, E., and Fülöp, F. (2007) *Tetrahedron Asymmetry*, **18**, 2841–2844.
11 (a) Park, J.H., Ha, H.J., Lee, W.K., Genereux-Vincent, T., and Kazlauskas, R.J. (2009) *Chembiochem*, **10**, 2213–2222; (b) Garcia-Urdiales, E., Rios-Lombardia, N., Mangas-Sanchez, J., Gotor-Fernandez, V., and Gotor, V. (2009) *Chembiochem*, **10**, 1830–1838; (c) Svedendahl, M., Carlqvist, P., Branneby, C., Allner, O., Frise, K., Berglund, P., and Brinck, T. (2008) *ChemBioChem*, **9**, 2443–2451; (d) Leonard, V., Fransson, L., Lamare, S., Hult, K., and Graber, M. (2007) *Chembiochem*, **8**, 662–667; (e) Li, X.G., and Kanerva, L.T. (2006) *Org. Lett.*, **8**, 5593–5596; (f) Gonzalez-Sabin, J., Lavandera, I., Rebolledo, F., and Gotor, V. (2006) *Tetrahedron Asymmetry*, **17**, 1264–1274.
12 Skjot, M., De Maria, L., Chatterjee, R., Svendsen, A., Patkar, S.A., Ostergaard, P.R., and Brask, J. (2009) *Chembiochem*, **10**, 520–527.
13 Prelog, V. (1964) *Pure Appl. Chem.*, **9**, 119–130.
14 Kazlauskas, R.J., Weissfloch, A.N.E., Rappaport, A.T., and Cuccia, L.A. (1991) *J. Org. Chem.*, **56**, 2656–2665.
15 Lavandera, I.W., Fernandez, S., Magdalena, J., Ferrero, M., Kazlauskas, R.J., and Gotor, V. (2005) *Chembiochem*, **6**, 1381–1390.
16 Hollmann, F., Grzebyk, P., Heinrichs, V., Doderer, K., and Thum, O. (2009) *J. Mol. Catal. B Enzym.*, **57**, 257–261.
17 Tarabek, P., Bonifacic, M., and Beckert, D. (2006) *J. Phys. Chem. A*, **110**, 7293–7302.
18 Williams, A., and Jencks, W.P. (1974) *J. Chem. Soc. Perkin Trans.* 2, 1753–1759.
19 Baum, I., Schwab, L., Loos, K., and Fels, G. (2010), Mechanistic insight into the enzymatic ring-opening polymerization of unsubstituted β-lactam catalyzed by Candida antarctica lipase B, *submitted for publication*.
20 Corradini, D., Bevilacqua, L., and Nicoletti, I. (2005),*Chromatographia*, **62**, 43–50.

15
Enzymatic Polymer Modification

Georg M. Guebitz

15.1
Introduction

Polymer-modifying enzymes are not new. The starch hydrolyzing diastase (now: amylase) was the first enzyme discovered in 1833 [1]. At the end of the nineteenth century it was found that proteases cause de-coagulation of proteins as well as hydrolysis [2]. Similarly, cellulases were discovered around this time [3]. Since then 'polymer hydrolases' have occupied by far the largest market share among enzymes. However, needless to say that many bulk enzyme applications such as in detergents or in the biomass-to-fuel area rather focus unspecific hydrolysis of biopolymers rather than on their specific modification. Nevertheless, in recent years enzymes have increasingly been employed in polymer processing, ranging from the lignocellulose to the food sector. For example, enzymatic crosslinking of pulp can reduce the amount of binders needed in the production of fiber-boards [4] while enzymatic crosslinking of proteins can upgrade a variety of food products [5]. Considering this wide range of enzyme applications in biopolymer modification and processing, only selected examples can be discussed here. In chapter 15.2 we will give a short overview on functionalization of natural polymers, discussing recent developments in enzymatic approaches and then move towards synthetic polymers.

15.2
Enzymatic Polymer Functionalization: From Natural to Synthetic Materials

For the production of phenolic resins, functional molecules have been enzymatically grafted onto lignin by using laccases (EC 1.10.3.2) (see also Chapter 7) [6]. Similarly, laccase was used to graft acrylamide onto lignin (see also Chapter 6) [7]. Generally, the potential of enzyme crosslinking to replace adhesive in various lignocellulose-based products such as MD-fiber boards has been intensively investigated [4, 8]. In recent approaches small molecules such as 4-hydroxy-3-methoxybenzylurea were added to enhance laccase-catalyzed crosslinking in particle board production. Apart from the replacement of adhesives, laccases have

Biocatalysis in Polymer Chemistry. Edited by Katja Loos
Copyright © 2011 WILEY-VCH Verlag GmbH & Co. KGaA, Weinheim
ISBN: 978-3-527-32618-1

also been used for the grafting of functional molecules onto lignocellulose-based materials [9]. These authors have also demonstrated that aromatic amines (i.e., tyramine) were covalently (4-O-5) bound to syringylglycerol-guaiacylether as a lignin model substrate. Numerous studies have focused on laccase treatment of pulp to improve paper properties such as strength. Again, addition of small phenolic molecules or peptides in laccase treatment has recently been shown to improve paper properties and/or impart novel functionalities such as antimicrobial behavior [10–12]. Flax is another fiber-based material where antimicrobial properties have been imparted with laccase-catalyzed grafting of phenolics [13]. Antimicrobial functionalization is equally important for protein-based fiber materials like wool, and has been achieved with tyrosinases. Tyrosinases have also shown potential for grafting chitosan with proteins while they are widely used for crosslinking and functionalization of proteins to upgrade a variety of food products [5].

Apart from tyrosinases, transglutaminases (EC 2.3.2.13) have been used to crosslink proteins to improve functional properties such as texture of food products [14]. In addition to food processing, transglutaminases have also been used for improving properties of protein-based fabrics such as wool leading to a higher tensile strength after chemical or protease pretreatment [15, 16]. Besides crosslinking, transglutaminases have been employed for grafting/coating of wool fabrics with silk sericin or keratin leading to increased bursting strength and softness and reduced felting shrinkage [17, 18]. In biomedical applications, transglutaminases have been used for tissue engineering [19, 20] or for the production of melt-extruded guides for peripheral nerve repair [21].

Apart from natural materials, oxidoreducates have been used to modify synthetic polymers. For example, using peroxidase, poly(4-hydroxystyrene) has been functionalized with aniline while poly(*p*-phenylene-2,6-benzobisthiazole) has been rendered more hydrophilic [22, 23]. Other authors have demonstrated that phenolics can be covalently bound to amino-functionalized polymers by using laccase resulting in increased fire resistance [13]A large number of scientific reports are available on enzymatic functionalization of poly(alkyleneterephthalate)s. Polyester fibers account for 73% of all synthetic fibers on the market with an annual production of approx. 27 million tons [24]. Similarly, polyamides and polyacrylonitriles have significant market shares. In contrast to natural polymers discussed above, hydrolases have shown higher potential for modification of these synthetic materials than oxidoreducates.

15.3
Surface Hydrolysis of Poly(alkyleneterephthalate)s

15.3 1
Enzymes and Processes

Limited surface hydrolysis of poly(alkyleneterephthalate)s (PAT), polyamides (PA) and polyacrylonitriles (PAN) by enzymes increases their hydrophilicity which is a key requirement for many applications, including gluing, painting, inking, anti-fogging, filtration, textile, electronics and biomedical [25].

Apart from many beneficial characteristics, PET is difficult to finish, and highly hydrophobic, builds up static charges, is unbreathable as fabrics, and shows poor adhesion and wetting properties due to the low surface energy. Thus, for many applications surface modification without compromising the bulk properties is required.

In coating PET, partial hydrolysis of the surface (e.g., introduction of carboxyl and hydroxyl groups) facilitates binding. Biocompatible/hemocompatible materials, antimicrobial surfaces and scaffolds for tissue engineering are obtained by coating PET with biomolecules/polymers [26]. An increased number of functional groups on the PET surface could enhance binding and reduce binder consumption in coating with PVC, a bulk application for the production of for example, truck tarpaulins [27]. Increased hydrophilicity of PET (i.e., a 15° lower contact angle) has been demonstrated to reduce bacterial adhesion and consequently infections of cardiovascular implants, such as artificial heart valve sewing rings and artificial blood vessels [28]. Similarly, in the production of flexible electronic devices (FEDs) such as displays or photovoltaic cells, surface hydrophilization is required for the attachment of functional layers [29]. Specific surface hydrolysis could also replace alkaline hydrolysis in the production of polymer brushes. Polymer brushes are produced by grafted polymer chains from surfaces and their design can regulate interactions with liquids, solids, particles, proteins, cells. Polymer brushes have been produced by polymerization of styrene from PET surfaces activated via alkaline hydrolysis. Using enzymatic surface activation the number of anchor points can be more specifically tuned [30]. PET is widely used in the textile industry with an annual production of 36 million tons [25, 31]. To reduce build up of static charges, improve moisture transport, and breathability and handle, alkaline treatment is conventionally used to increase hydrophilicity of PET-based textile materials. However, formation of pit-like structures results in high weight loss of up to 15% and leads to reduced fiber strength [24, 32, 33]. Strategies for PET surface modification include chemical hydrolysis, aminolysis, plasma-, UV-ozone-electrical discharge or corona treatments [34, 35].

PET hydrolases have been mainly recruited from the classes of lipases and cutinases, while some typical esterases and even proteases have also shown activity on PET. In screening experiments several authors have shown that PET hydrolyzing enzyme activities were inducible by addition of the plant polyester cutin [36, 37]. The plant cuticle contains waxes and insoluble cutin, consisting of oxygenated C_{16} and C_{18} fatty acids crosslinked by ester bonds [38]. Cutin is important for plant protection and its enzymatic degradation is one of the first steps in the infection of plants [39]. It has been suggested that cutin oligomers, resulting from cutin hydrolysis by small levels of constitutive cutinase activity, induce production of higher amounts of cutinases [40]. Fungal cutinases show both *exo*- and *endo*- esterase activity [41] and have first been purified and characterized from *Fusarium solani pisi* growing on cutin as a carbon source [42]. The most common components of the C_{16} family of monomeric hydrolysis products are 16-hydroxyhexadecanoic acid and 9,10,16-dihydroxyhexadecanoic acid. Usually a mixture of medium chain length positional isomers of dihydroxy acids is also present. The major members of C_{18} cutin monomers are 18-hydroxy-C_{18}-9-enoic acid, 18-hydroxy-C_{18}-9,12-dienoic acid, 18-hydroxy-9,10-epoxy-C_{18} acid, 18-hydroxy-9,10-epoxy-C_{18} acid, 9,12,18-trihydroxy-C_{18} acid and 9,10,18-trihydroxy-C_{18}-12-enoic acid [41]. Various fungal cutinases such as from *Aspergillus oryzae*, *A. niger* and *A. nomius* from *Humicola insolens*, and from *Furarium solani* and *F. oxysporium* have been reported to hydrolyze PET (Table 15.1). In addition, this behavior was found for the bacterial cutinases from *Streptomyces coelicolor* and *Thermobifida fusca* and *T. alba* (Table 15.1).

Various bacterial, fungal and plant lipases have been described to hydrolyze PET (Table 15.1). Lipases catalyze the hydrolysis of long chain water insoluble triglycerides and, unlike cutinase they are 'interfacially activated' in the presence of a water–lipid interface [63–65]. The active site of lipases is covered with a peptide segment called lid while upon opening the active site becomes accessible to the substrate. Consequently, it as been indicated that PET hydrolysis by lipase can be improved in the presence of detergents [55, 66]. Apart from typical lipases and cutinases, other esterases have been shown to hydrolyze PET. Nevertheless, it is not quite clear yet what constitutes a PET-hydrolase. On the one hand a comprehensive comparison of all reported enzymes on typical lipase and cutinase substrates in addition to PET is not available. On the other hand, apart from the active site architecture and specificities on water soluble substrates, the adsorption behavior onto polymers will also play a major role.

15.3.2
Mechanistic Aspects

In several studies the release of mono- and oligomeric reaction products from PET hydrolysis was investigated [36, 43, 49, 51, 52, 57, 60]. Interestingly, differences in the ratios of released molecules were found for the individual enzymes. A lipase from *T. lanuginosus* released higher amounts of mono(2-hydroxyethyl) terephthalate (MHET) than terephthalic acid (TA) whereas the amounts of TA and MHET were similar in case of a cutinase from *T. fusca*. Both enzymes additionally released small amounts of bis(2-hydroxyethyl) terephthalate (BHET) [24]. Interestingly, only TA and ethylene glycol (EG) were detected after alkaline treatment, indicating pure *exo*-type hydrolysis in contrast to the enzyme treatment [24]. Except for antipilling effects in detergents, only partial hydrolysis of the PET surface with a concomitant increase of hydrophilicity is required for most other applications without changing the bulk properties. Consequently, a release/solubilization of mono-/oligomers is not desired and therefore other parameters quantifying surface hydrolysis are important.

A variety of techniques including rising height and contact angle measurements, the drop dissipation test and tensiometry have been used to quantify increases in hydrophilicity [48, 49, 53, 57, 67]. For example, treatment of PET fabrics with a *T. fusca* cutinase and *T. lanuginosus* lipase lead to a wetting time of 120 s and 100 s, respectively, compared with 45 ± 2 min for the untreated material [24].

15.3 Surface Hydrolysis of Poly(alkyleneterephthalate)s

Table 15.1 Enzymatic modification of poly(alkylene terephthalates).

Fungal enzymes

Enzyme	Organism	Effect measured	Reference
Cutinase	Aspergillus oryzae	Water contact angle (WCA), K/S values after dying, moisture regain, weight loss of PET, release of oligomers (HPLC-UV)	[43]
	Aspergillus oryzae Aspergillus niger	WCA, K/S values after dying, surface energy determined by the Qwens-Wendt method	[44]
	Aspergillus nomius HS-1	Hydrolysis of ethylene glycol dibenzoate	[45]
	Penicillium citrinum	Release of oligomers, hydrophilicity	[36]
	Humicola insolens, Humicola sp.	Hydrophilicity	[46–48]
	Fusarium oxysporum Fusarium solani	Release of oligomers and terephthalic acid, hydrophilicity, XPS	[49–51]
Lipase	Candida antarctica, Candida sp.	XPS, release of oligomers	[50, 51]
	Thermomyces lanuginosus	Depilling assay, release of oligomers, hydrophilicity, XPS, Maldi-Tof, K/S values after dying, FTIR, surface derivatization	[24, 52–56]

Bacterial Enzymes

Enzyme	Organism	Effect measured	Reference
Cutinase	Thermobifida fusca	Release of oligomers, hydrophilicity, XPS, Maldi-Tof, K/S values after dying	[24, 37, 54, 55, 57]
	Thermobifida alba	Release of oligomers	[46]
Lipase	Burkholderia (formerly Pseudomonas cepacia)	Hydrolysis of oligomers	[58]
	Triticum aestivum	Hydrolysis of oligomers	[59]
Esterase	Pseudomonas spp. (Serin esterase)	Release of terephthalic acid, dye binding assay, hydrophilicity, depilling assay	[60]
	Bacillus sp. (Nitro-benzyl-esterases)	Hydrolysis of poly(ethylene terephthalate) oligomers	[61, 62]
Laccase	Streptomyces coelicolor	Hydrophilicity	[47]

However, besides enzymatic hydrolysis, the simple adsorption of enzyme protein can also increase the hydrophilicity of PET due to the hydrophilicity of the protein. Using X-ray photoelectron spectroscopy (XPS) analysis an increase of the nitrogen content of up to 7.2% due to adsorption of a lipase to PET was measured, while angle-resolved XPS confirmed the presence of a protein layer with thickness of 1.6–2.6 nm and 2.5–2.8 nm for cutinase from *F. solani* and lipase from *C. Antarctica*-treated PET, respectively [50]. Similarly, removal of PET oligomers was mostly attributed to adsorption of lipase from *Triticum aestivum* rather than to catalytic activity of this enzyme [59]. Consequently, complete removal of protein from PET is a prerequisite for the assignment of hydrophilicity effects to the catalytic action of enzymes. Therefore, washing procedures were developed such as involving ethanol extraction steps in addition to washing steps with detergent, sodium carbonate and deionized water [24]. After application of such procedures, complete absence of protein on the PET surface was confirmed by the absence of a nitrogen peak (binding energy 400 eV) in XPS analysis. Similarly, protease treatment was successfully used to remove PET-hydrolases from the surface [53]. Another alternative to avoid artifacts due to protein adsorption are control experiments with quantitative enzyme inhibitors such as mercury chloride [54].

Derivatization of carboxyl and hydroxyl groups resulting from enzymatic hydrolysis is another method to monitor changes in surface chemical composition. Carboxyl groups were esterified with a fluorescent alkyl bromide, 2-(bromonethyl)naphthalene (BrNP). Consequently, higher fluorescence intensity was measured for PET partially hydrolyzed with enzymes [53]. Similarly, derivatization especially with basic dyes was widely used to follow enzyme hydrolysis of PET [24, 54, 55, 57, 62]. Clear differences of color shades with increases of K/S (according to the Kubelka–Monk theory) by up to 200% were obtained [24].

Several recent studies focused on the investigation of the mechanism of enzymatic PET hydrolysis. All reports agree that polyesterases preferably attack the amorphous regions of polymers [37, 50, 53, 54, 68, 69]. For example, in a comparison of amorphous fibers with a degree of crystallinity of 5% and semi-crystalline fibers with a degree of crystallinity of 40%, clearly higher amounts of degradation products were measured for the amorphous fibers. A cutinase from *T. fusca* cutinase released up to 50-fold higher amounts of MHET and TA from amorphous fibers when compared with semi-crystalline fibers [24]. Similarly, a lipase enzyme displayed higher hydrolytic activity towards amorphous PET as shown by the decrease of the water contact angle (WCA) values [53]. In agreement with these results, spectral changes by FTIR-ATR analysis indicating increased crystallinity after enzymatic PET treatment was reported. Recently it has been shown that the addition of a plasticizer (*N*,*N*-diethylphenyl acetamide; DEPA) can enhance susceptibility of PET to enzymatic hydrolysis. This was indicated by a considerable increase in the amount of the hydrolysis products released from the PET fabric and from a semi-crystalline film in the presence of a *T. lanuginosus* lipase and a *T. fusca* cutinase [55].

Compared with PET, enzymatic hydrolysis of poly(trimethyleneterephthalate) (PTT) has been considerably less investigated. PTT was introduced under the

(a) NaOH treated   **(b)** Cutinase *T. fusca*   **(c)** Reference

Figure 15.1 Surface structure of PTT fibers (a) after hydrolysis with NaOH and (b) with the *T. fusca* cutinase leading to similar increase in surface hydrophilicity.(Reproduced with permission from [56]. Copyright © (2008) Elsevier).

trade names Corterra and Sorona and shows excellent properties such as the high elastic recovery and good dyeing ability. Additionally, one of the building blocks, namely 1,3-propanediol, is increasingly produced by microbial fermentation from renewable sources as substrates [54]. PTT oligomers and polymers (film, fabrics) were incubated with enzymes from *Thermomyces lanuginosus, Penicillium citrinum, Thermobifida fusca* and *Fusarium solani pisi*. Interestingly these enzymes showed different specificities. A cutinase from *T. fusca* was most active on PTT fibers and films and was able to cleave cyclic PTT oligomers in contrast to a lipase from *T. lanuginosus* which did not hydrolyze the PTT film, and cyclic oligomers. In contrast to alkaline hydrolysis of PPT which leads to crater-like structures, the enzyme hydrolysis seemed to be more uniform (Figure 15.1).

15.3.3
Surface Analytical Tools

Some recent studies mechanistically investigated the mode of action of PET-hydrolases comparing different enzymes and enzyme and alkaline hydrolysis. Soon it was clear that imparting a given surface hydrophilicity to PET was at the expense of considerable weight losses in the case of the alkaline treatment (>6% for 1 M NaOH) in contrast to the enzyme treatment (<<1%) [24]. In addition, alkaline treatment released only monomers from PET and lead to a crater-like surface of PET fibers (Figure 15.1). On the other hand, at the same final hydrophilicity enzymes had only released small amounts of oligomers but did not significantly alter the surface characteristics. This was the first indication that enzymes obviously acted more *endo*-wise than did alkaline hydrolysis. Consequently, there were several attempts by using sophisticated analytical tools to prove this assumption. MALDI-TOF MS analysis clearly indicated an *endo*-type enzymatic hydrolysis for PET ($M_W = 3500$) although there was a slight preference of enzymes to act repeatedly on the same polymer chain, as indicated by higher amounts of smaller fragments (e.g., m/z 854) compared to larger fragments (m/z 1623) [55]. XPS data showed broader carbon peaks after enzyme treatment of PET in contrast to alkaline hydrolysis [24, 50, 55]. Provided complete removal of adsorbed protein (e.g., no nitrogen peak in XPS), novel carboxyl and hydroxyl groups should clearly indicate *endo*-type hydrolysis. However, extensive enzymatic hydrolysis (high enzyme dosage, long incubation times) will finally lead to release of oligomers and monomers and this is probably the reason why some authors did not obtain a conclusive picture with XPS [57]. In other words, there is an optimum in terms of the extent of enzymatic surface hydrolysis. At this optimum, surface polymer chains are hydrolyzed at all different positions at a similar rate (i.e., *endo*-wise). Consequently, there is no significant concomitant release of short oligomers or weight loss. However, when hydrolysis proceeds, the resulting large fragments will be successively cleaved into smaller and smaller oligomers until the outermost layers are degraded. Interestingly, this mechanism is reflected by findings obtained in a study on enzymatic PVC-coating of PET where prolonged hydrolysis was found to reduce beneficial increases in binding strength obtained initially by enzyme treatment [27].

15.4
Surface Hydrolysis of Polyamides

15.4.1
Enzymes and Processes

Polyamide-6 (Nylon-6, Perlon) and polyamide-6.6 (Nylon-6.6) are the most well known polyamides. Polyamide-based filaments find wide spread applications as yarns for textile or industrial and carpet materials [70]. However, nylon-based textiles are uncomfortable to wear and difficult to finish due to their hydrophobic character. This characteristic also leads to fouling of PA-based ultrafiltration membranes by proteins and other biomolecules which increases the energy demand for filtration and requires cleaning with aggressive chemicals or replacement [71–73]. Consequently the enhancement of the hydrophilicity of nylon is a key requirement for many applications and can be achieved by using plasma treatment [74–76]. As a promising alternative, enzymatic hydrophilization of PA requires less energy and is not restricted to planar surfaces.

Due to the chemical similarities of synthetic polyamides with their natural analogs, the search for polyamide hydrolyzing enzyme activities was first focused on proteases [77, 78]. Using a protease from *B. subtilis*, hydrolysis of PA was shown based on detection of reaction products released. Hydrolysis of PA led to increased hydrophilicity and enhanced binding of reactive dyes [78]. Despite the large number of proteases commercial available, only few representatives were found to hydrolyze PA. Thus, in order to allow fast screening for new polyamidases, water insoluble oligomeric model substrates were developed and it was demonstrated that their activity correlated to activity on PA [56]. Screening experiments with these model substrates led to the discovery of a number of amidases acting on PA but not showing protease activity. A fungal amidase from *Beauveria brongniartii* and a bacterial amidase from *Nocardia farcinica* were purified and characterized in detail related to their activities on polyamides [79, 80]. The 55 kDa amidase from *B. brongniartii* was active on both aliphatic and aromatic substrates with higher activity on longer chain amides up to C6. Upon incubation with this enzyme the hydrophilicity of PA6 was drastically increased based on reduction of the drop dissipation time from 60 s was reduced to 7 s after 60-minute treatment which correlated to rising high measurements. Using tensiometry, the surface tension σ increased upon 3-minute enzyme treatment from 46.1 mNm to 67.4 mNm [79]. Although more hydrophilic, already polyamidase treatment led to a further increase of the hydrophilicity of PA 6.6. Interestingly, after prolonged incubation a decrease of the hydrophilicity was observed. It has been hypothesized that extensive surface hydrolysis might cause solubilization of the upper layer (i.e., oligomers) of the material thus reducing the number of new carboxylic acid groups on the surface [79].

Apart from the *B. brongniartii* fungal amidase enzyme, a bacterial amidase from *N. farcinica* (also without protease activity) was recently shown to hydrolase PA [80]. Based on rising height and tensiometry measurements a large increase of hydrophilicity was measured after only 10-minutes enzyme treatment. To take into account possible artifacts due to protein (= enzyme) adsorption, surface hydrolysis by the polyamidase was compared with mercury chloride inhibited controls. Like with the *B. brongniartii* amidase, a plateau hydrophilicity increase was seen which decreased after prolonged incubation probably for the same reasons as described above. The polyamidase also hydrolyzed various small amides and esters including *p*-nitroacetanilide, *p*-nitrophenylbutyrate which is typical for aryl acylamidases [81]. Also, as a typical amidase the polyamidase of *N. farcinica* catalyzed the transfer of the acyl group of hexanoamide to hydroxylamine [82].

15.4.2
Mechanistic Aspects

As an important issue for future engineering/screening for more efficient polyamidases, the *N. farcinica* amidase has been compared with other homologous enzymes. The polyamidase belongs to the amidase signature family. Within this group of enzymes the Ser-Ser-Lys triad is involved in the catalytic reaction, unlike serine proteases, lipases and esterases which are characterized by the catalytic triad Ser-His-Asp [83, 84]. The corresponding catalytic mechanism has been recently confirmed, based on the availability of the crystal structure of the *Stenotrophomonas maltophilia* peptide amidase (Pam). Apart from the common hydrolysis of amide bonds, individual representatives of the amidase signature family enzymes show very distinct substrate specificities [84]. Most likely this is due to binding of the substrate by residues outside the signature sequence [83]. Interestingly, within the amidase signature family an amidase cleaving cyclic nylon oligomers has been described which, however, did not show activity on

PA [85, 86]. Enzymatic degradation of linear and cyclic nylon oligomers has been described extensively as these molecules are undesirable by-products in nylon production which are released to the environment [87]. Three enzymes have been found to be involved in nylon oligomer degradation by the *Arthrobacter* sp. KI72 and also by *Pseudomonas* sp. NK87 namely a 6-aminohexanoate-cyclic dimer hydrolase (EI), a 6-aminohexanoate-dimer hydrolase (EII) and an endo-type 6-aminohexanoate oligomer hydrolase (EIII) [87]. EIII hydrolyses the cyclic tetramer and dimer as well as linear oligomers *endo*-wise [85]. Interestingly, only the cyclic dimer hydrolases belong to the amidase signature while the linear dimer hydrolase activity (EII) has evolved in an esterase with β-lactamase folds. Surprisingly the endo-acting 6-aminohexanoate oligomer hydrolase (EIII) showed the least homology to the *N. farcinica* polyamidase [80, 87].

Apart from proteases and amidases, a cutinase from *Fusarium solani* has been shown to hydrolyze polyamides and genetic engineering was successfully used to achieve higher activity [88–90]. Interestingly, the same cutinase was also able to hydrolyze PET.

Fungal oxidases have previously been demonstrated to degrade PA [91–94]. Using a laccase-mediator system, researchers have shown hydrophilicity increases of PA 6.6 based on rising height measurements [91] while other authors have demonstrated disintegration of PA membranes. Mechanistic studies revealed that peroxidases attack methylene groups adjacent to the nitrogen atoms while the reaction then proceeds in an auto-oxidative manner [92, 95, 96]. Nevertheless it seems that oxidative enzymatic modification of PA is difficult to control and thus not suitable for targeted surface modification.

15.5
Surface Hydrolysis of Polyacrylonitriles

Like for other synthetic polymers discussed above, a variety of chemical and physical techniques for the functionalization of polyacrylonitriles is available including plasma treatment, oxidation by hydrogen peroxide or hydrolysis with acids and bases. Enzymes such as nitrilases or the nitrile hydratase/amidase system offer an interesting alternative as they specifically can hydrolyze nitrile groups on the surface of PAN [25]. However, polyacrylonitrile is a collective name for polymers that are composed of at least 85% acrylonitrile as monomer, while fiber products typically contain 4–10% of a nonionic co-monomer like vinyl acetate moieties, which are a target for enzymatic hydrolysis [97]. Although PAN-based materials were for a long time believed to be resistant to biodegradation, it has recently been shown that a novel strain of *Micrococcus luteus* can degrade this material, see Figure 15.2. Using ^{13}C-labeled PAN, release of polyacrylic acid was measured with NMR analysis during bacterial degradation [99, 100].

Enzymatic hydrolysis of PAN was first shown by Tauber *et al.* who monitored the formation of ammonia during hydrolysis of nitrile groups [99, 100]. Considering the fact that only a low 1.1% of nitrile groups are displayed on the polymer

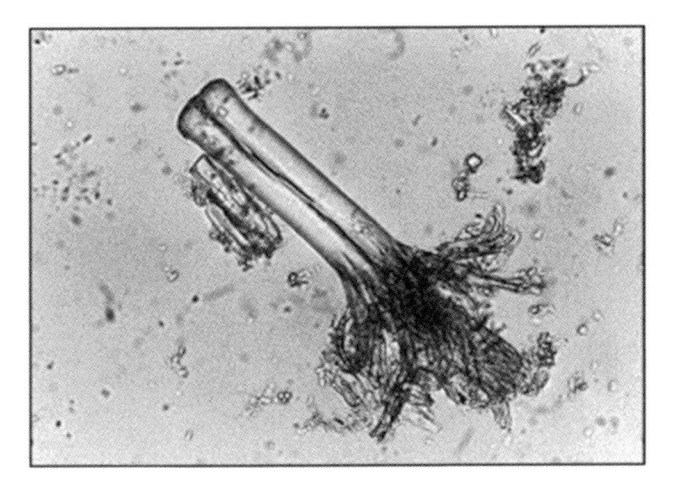

Figure 15.2 Degradation of PAN fibers by *Micrococcus luteus*.(Reproduced with permission from [96]. Copyright © (2007) Elsevier).

surface [98, 101], this conversion corresponds to a much higher degree of surface modification. After two-step surface hydrolysis of PAN by nitrile hydratase and amidase from *Rhodococcus rhodochrous* improved dye up-take was found while hydrolysis was found to be faster for shorter chain polymers. Other authors used a commercial nitrilase and found an increase of color levels by 156% and in the presence of additives by 199%. During hydrolysis the release of ammonia and polyacrylic acid was quantified [102]. Apart from these dye binding assays, XPS analysis and FTIR have been used to demonstrate and quantify chemical changes upon enzymatic hydolysis [98, 101].

Increases of the O/C ratio of 60 to 80% were measured with XPS for PAN treated with nitrilases from *Arthrobacter sp* and *A. tumefaciens*, respectively. This clearly indicates incorporation of oxygen into the polymer surface due to enzymatic hydrolysis [103]. Using FTIR analysis the conversion of nitrile groups into amide groups was demonstrated based on the formation of new bands at 1649 cm^{-1} and 1529 cm^{-1}. The band at 1649 cm^{-1} was assigned to the stretching of the carbonyl group of the amide while the band at 1529 cm^{-1} is due to CN stretching and NH bending in the same amide configuration [101]. Likewise nitrile hydratases from *Rhodococcus rhodochrous*, *Brevibacterium imperiale* and *Corynebacterium nitrilophilus* were able to convert nitrile into amide groups [99, 104]. Interestingly, apart from pure nitrile hydratases, limited hydrolysis with a nitrile hydratase and amidase enzyme system also leads to amide rather than acid groups since further hydrolysis of amide groups by the amidase seems to be slower [99, 101].

The potential of lipases and cutinases for the hydrolysis of vinyl acetate moieties (about 7%) in commercial PAN materials was assessed. Indeed it was shown that the commercial esterase Texazym PES and a cutinase from using *Fusarium solani pisi* were able to release acetate from PAN [97]. Furthermore it was demonstrated that enzymatic hydrolysis can be enhanced both in the presence of organic solvents (e.g., N,N-dimethylacetamide) and by addition of enzyme stabilizers such as glycerol. Generally mechanical agitation was shown to be an important factor in enzymatic hydrolysis of PAN [97]. In contrast to enzymatic hydrolysis of PET [53], no change in crystallinity as determined by X-ray diffraction was found after enzymatic hydrolysis of PAN with lipases and cutinase [97]. On the other hand, in a study on the enzymatic hydrolysis with PAN materials with different comonomer content, lower hydrolysis rates were measured for highly crystalline PAN materials [98, 101].

15.6
Future Developments

Esterases and polyurethanases can hydrolyze polyurethanes resulting in changes on the surface of PU film. *In vivo*, cholesterol esterase can initiate polyurethane degradation while a number of PU-degrading enzymes have been described from micro-organisms, such as *Candida rugosa* lipase or esterases from *Pseudomonas chlororaphis*, for example, polyurethanase [105].

Although a number of attractive applications of PET-hydrolases have been described, enzymatic hydrolysis is still a rather slow process. Thus, more effective enzymes could enhance implementation of existing applications (such as in textiles or detergents) or open up new fields such as enzymatic PET recycling. It has already been shown that additives such as plasticizers or detergents can improve PET hydrolysis by cutinases and lipases, respectively [55, 66]. In addition, the effect of surface active molecules designed by nature on PET-hydrolysis should be investigated in more detail. Hydrolysis of polyesters by *Aspergillus oryzae* cutinases is assisted by proteins called hydrophobins (RolA protein and HsbA) which guide the enzyme along the polymer surface [106, 107]. In addition, genetic enzyme engineering offers a number of tools to make PET-hydrolases more efficient. Site-directed mutagenesis was used to enlarge the active site of a cutinase from *F. solani* leading to five-fold higher activity on PET (Figure 15.3). To better exploit genetic engineering in future approaches a more detailed knowledge about structure function relationships of PET-hydrolases is necessary. Since PET-hydrolases have been reported from both lipases and cutinases, it would be interesting to identify common structural motifs influences this capability of hydrolyzing PET.

Acknowledgment

The work was financed by the SFG, the FFG, the city of Graz and the province of Styria within the MacroFun project and supported by the European COST868 program.

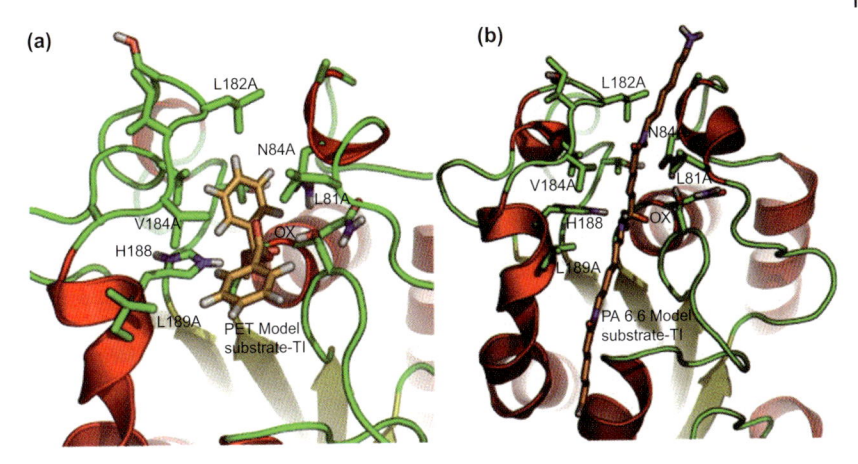

Figure 15.3 Detail of the active site X-ray structure of cutinase with the energy minimized structure of the TI of (a) 1,2-ethanodiol dibenzoate (PET model substrate) and (b) PA 6,6. The catalytic histidine (H188) and oxianion-hole (OX) are shown. Residues mutated in this study are labeled as: L81A, N84A, L182A, V184A and L189A.

References

1 Musculus, F., and von Mering, I. (1879) On the action of diastase, of sativa and of pancreatic juice on starch and glycogen. *J. Am. Chem. Soc.*, **1** (5), 173.

2 Malfitano, G. (1900) Over the protease of the aspergillus niger. [machine translation]. *Ann. Inst. Pasteur*, **14**, 420–447.

3 van Iterson, G., Jr. (1903) The decomposition of cellulose by aerobic microorganisms. *Proc. K. Akad. Wet. Amst.*, **5**, 685–703.

4 Felby, C., Hassingboe, J., and Lund, M. (2002) Pilot scale production of fibreboards made by laccase oxidized wood fibres: broad properties and evidence for cross-liking of lignin. *Enzyme Microb. Technol.*, **31**, 736–741.

5 Selinheimo, E., Autio, K., Krijus, K., and Buchert, J. (2007) Elucidating the mechanism of laccase and tyrosinase in wheat bread making. *J. Agric. Food Chem.*, **55** (15), 6357–6365.

6 Eker, B., Zagorevski, D., Zhu, G., Linhardt, R.J., and Dordick, J.S. (2009) Enzymatic polymerization of phenols in room-temperature ionic liquids. *J. Mol. Catal. B Enzym.*, **59** (1–3), 177–184.

7 Milstein, O., Huttermann, A., Frund, R., and Ludemann, H.D. (1994) Enzymatic co-polymerization of lignin with low-molecular-mass compounds. *Appl. Microbiol. Biotechnol.*, **40** (5), 760–767.

8 Felby, C., Thygesen, L.G., Sanadi, A., and Barsberg, S. (2004) Native lignin for bonding of fiber boards – evaluation of bonding mechanisms in boards made from laccase-treated fibers of beech (Fagus sylvatica). *Ind. Crops Prod.*, **20** (2), 181–189.

9 Kudanga, T., Nugroho Prasetyo, E., Sipilä, J., Eberl, A., Nyanhongo, G.S., and Guebitz, G.M. (2009) Coupling of aromatic amines onto syringylglycerol [beta]-guaiacylether using Bacillus SF spore laccase: a model for

47 Liu, Y.B., Wu, G.F., and Gu, L.H. (2008) Enzymatic treatment of PET fabrics for improved hydrophilicity. *AATCC Rev.*, **8** (2), 44–48.

48 Nimchua, T., Punnapayak, H., and Zimmermann, W. (2007) Comparison of the hydrolysis of polyethylene terephthalate fibers by a hydrolase from *Fusarium oxysporum* LCH I and *Fusarium solani f. sp. pisi. Biotechnol. J.*, **2**, 361–364.

49 Alisch-Mark, M., Herrmann, A., and Zimmermann, W. (2006) Increase of the hydrophilicity of polyethylene terephthalate fibres by hydrolases from Thermomonospora fusca and Fusarium solani f. sp.pisi. *Biotechnol. Lett.*, **28**, 681–685.

50 Vertommen, M.A.M.E., Nierstrasz, V.A., Veer, M., and Warmoeskerken, M.M.C.G. (2005) Enzymatic surface modification of poly(ethylene terephthalate). *J. Biotechnol.*, **120** (4), 376–386.

51 Nimchua, T., Punnapayak, H., and Zimmermann, W. (2006) Comparison of the hydrolysis of polyethylene terephthalate fibre by a hydrolase from Fusarium oxysporum LCH I and Fusarium solani f. sp. pisi. *Biotechnol. J.*, **2**, 361–364.

52 Andersen, B.K., Borch, K., Abo, M., and Damgaard, B. (1999) Method of treating polyester fabrics, US Patent, vol. Nr. 5,997,584, pp. 1–20.

53 Donelli, M., Taddei, P., Smet, P.F., Poelman, D., Nierstrasz, V.A., and Freddi, G. (2009) Enzymatic Surface Modification and Functionalization of PET. A Water Contact Angle, FTIR, and Fluorescence Spectroscopy Study. *COST868 Conference, Istanbul, Turkey, 19.–20.2.2009.*

54 Eberl, A., Heumann, S., Kotek, R., Kaufmann, F., Mitsche, S., Cavaco-Paulo, A., and Guebitz, G.M. (2008) Enzymatic hydrolysis of PTT polymers and oligomers. *J. Biotechnol.*, **135**, 45–51.

55 Eberl, A., Heumann, S., Brückner, T., Araujo, R., Cavaco-Paulo, A., Kaufmann, F., Kroutil, W., and Guebitz, G.M. (2009) Enzymatic surface hydrolysis of poly(ethylene terephthalate) and bis(benzoyloxyethyl) terephthalate by lipase and cutinase in the presence of surface active molecules. *J. Biotechnol.*, **143** (3), 207–212.

56 Heumann, S., Eberl, A., Pobeheim, H., Liebminger, S., Fischer-Colbrie, G., Almansa, E., Cavaco-Paulo, A., and Gubitz, G.M. (2006) New model substrates for enzymes hydrolysing polyethyleneterephthalate and polyamide fibres. *J. Biochem. Biophys. Methods*, **69** (1–2), 89–99.

57 Alisch, M., Feuerhack, A., Mueller, H., Mensak, B., Andreaus, J., and Zimmermann, W. (2004) Biocatalytic modification of polyethylene terephthalate fibres by esterases from actinomycete isolates. *Biocatal. Biotransformation*, **22** (5–6), 347–351.

58 Lee, C.W., and Do Chung, J. (2009) Synthesis and biodegradation behavior of poly(ethylene terephthalate) oligomers. *Polymer-Korea*, **33** (3), 198–202.

59 Nechwatal, A., Blokesch, A., Nicolai, M., Krieg, M., Kolbe, A., Wolf, M., and Gerhardt, M. (2006) A contribution to the investigation of enzyme-catalysed hydrolysis of poly(ethylene terephthalate) oligomers. *Macromol. Mater. Eng.*, **291** (12), 1486–1494.

60 Yoon, M.Y., Kellis, J., and Poulouse, A.J. (2002) Enzymatic modification of polyester. *AATCC Rev.*, **2**, 33–36.

61 Michels, A., Pütz, A., Maurer, K.H., Eggert, T., and Jäger, K.-E. (2007) Use of esterases for separating plastics, WO/2007/017181 (patent).

62 Kellis, J., Poulose, A.J., and Yoon, M.Y. (2001) Enzymatic Modification of the surface of a polyester fiber or article; US 6,254,645 B1, US 6,254,645 B1 (patent).

63 Fojan, P., Jonson, P.H., Petersen, M.T.N., and Petersen, S.B. (2000) What distinguishes an esterase from a lipase: a novel structural approach. *Biochimie*, **82** (11), 1033–1041.

64 Pleiss, J., Fischer, M., and Schmid, R.D. (1998) Anatomy of lipase binding sites: the scissile fatty acid binding site. *Chem. Phys. Lipids*, **93** (1–2), 67–80.

65 Grochulski, P., Li, Y., Schrag, J.D., Bouthillier, F., Smith, P., Harrison, D., and Cygler, M. (1993) Insights into

interacial activation from an open structure fo Candida rugosa lipase. *J. Biol. Chem.*, **286** (17), 12843–12847.

66 Kim, H.R., and Song, W.S. (2006) Lipase treatment of polyester fabrics. *Fibers Polym.*, **7** (4), 339–343.

67 Fischer-Colbrie, G., Heumann, S., Liebminger, S., Almansa, E., Cavaco-Paulo, A., and Gübitz, G.M. (2004) New enzymes with potential for PET surface modification. *Biocatal. Biotransformation*, **22**, 341–346.

68 Herzog, K., Müller, R.J., and Deckwer, W.D. (2006) Mechanism and kinetics of the enzymatic hydrolysis of polyester nanoparticles by lipases. *Polym. Degrad. Stab.*, **91** (10), 2486–2498.

69 Müller, R.J., Schrader, H., Profe, J., Dresler, K., and Deckwer, W.D. (2005) Enzymatic degradation of poly(ethylene terephthalate): rapid hydrolyse using a hydrolase from T. fusca. *Macromol. Rapid Commun.*, **26** (17), 1400–1405.

70 Saurer Management AG. (2007) *The Fibre Year Book 2006/07*.

71 Asatekin, A., Kang, S., Elimelech, M., and Mayes, A.M. (2007) Anti-fouling ultrafiltration membranes containing polyacrylonitrile-graft-poly(ethylene oxide) comb copolymer additives. *J. Membr. Sci.*, **298** (1–2), 136–146.

72 Kim, H.A., Choi, J.H., and Takizawa, S. (2007) Comparison of initial filtration resistance by pretreatment processes in the nanofiltration for drinking water treatment. *Separ. Purif. Technol.*, **56** (3), 354–362.

73 Qiao, X., Zhang, Z., and Ping, Z. (2007) Hydrophilic modification of ultrafiltration membranes and their application in Salvia Miltiorrhiza decoction. *Separ. Purif. Technol.*, **56** (3), 265–269.

74 De Geyter, N., Morent, R., Leys, C., Gengembre, L., and Payen, E. (2007) Treatment of polymer films with a dielectric barrier discharge in air, helium and argon at medium pressure. *Surf. Coat. Tech.*, **201** (16–17), 7066–7075.

75 McCord, M.G., Hwang, Y.J., Hauser, P.J., Qiu, Y., Cuomo, J.J., Hankins, O.E., Bourham, M.A., and Canup, L.K. (2002) Modifying nylon and polypropylene fabrics with atmospheric pressure plasmas. *Text. Res. J.*, **72**, 491–498.

76 Tusek, L., Nitschke, M., Werber, C., Stana-Kleinschek, K., and Ribitsch, V. (2001) Surface characterization of NH3 plasma treated polyamide foils. *Colloids Surf. A Phys. Eng. Asp.*, **195**, 81–95.

77 Miettinen-Oinonen, A., Puolakka, A., and Buchert, J. (2007) Method for modifying polyamide, EP1761670 (patent).

78 Silva, C., Araujo, R., Casal, M., Gubitz, G.M., and Cavaco-Paulo, A. (2007) Influence of mechanical agitation on cutinases and protease activity towards polyamide substrates. *Enzyme Microb. Technol.*, **40** (7), 1678–1685.

79 Almansa, E., Heumann, S., Eberl, A., Kaufmann, F., Cavaco-Paulo, A., and Gubitz, G.M. (2008) Surface hydrolysis of polyamide with a new polyamidase from Beauveria brongniartii. *Biocatal. Biotransformation*, **26** (5), 371–377.

80 Heumann, S., Eberl, A., Fischer-Colbrie, G., Pobeheim, H., Kaufmann, F., Ribitsch, D., Cavaco-Paulo, A., and Guebitz, G.M. (2009) A novel aryl acylamidase from Nocardia farcinica hydrolyses polyamide. *Biotechnol. Bioeng.*, **102** (4), 1003–1011.

81 Yoshioka, H., Nagasawa, T., and Yamada, H. (1991) Purification and characterization of aryl acylamidase from nocardia globerula. *Eur. J. Biochem.*, **199** (1), 17–24.

82 Fournand, D., Bigey, F., and Arnaud, A. (1998) Acyl transfer activity of an amidase from Rhodococcus sp. strain R312: formation of a wide range of hydroxamic acids. *Appl. Environ. Micrbiol.*, **64** (8), 2844–2852.

83 Labahn, J., Neumann, S., Buldt, G., Kula, M.R., and Granzin, J. (2002) An alternative mechanism for amidase signature enzymes. *J. Mol. Biol.*, **322** (5), 1053–1064.

84 Valina, A.L.B., Mazumder-Shivakumar, D., and Bruice, T.C. (2004) Probing the Ser-Ser-Lys catalytic triad mechanism of peptide amidase: computational studies of the ground state, transition state, and intermediate. *Biochemistry*, **43** (50), 15657–15672.

85 Kakudo, S., Negoro, S., Urabe, I., and Okada, H. (2000) Nylon oligomer degradation gene, nylC, on plasmid pOAD2 from a Flavobacterium strain encodes endo-type 6-aminohexanoate oligomer. *Appl. Environ. Microbiol.*, **59** (11), 3978–3980.

86 Negoro, S. (2000) Biodegradation of nylon oligomers. *Appl. Microbiol. Biotechnol.*, **54**, 461–466.

87 Negoro, S., Ohki, T., Shibata, N., Sasa, K., Hayashi, H., Nakano, H., Yasuhira, K., Kato, D.I., Takeo, M., and Higuchi, Y. (2007) Nylon-oligomer degrading enzyme/substrate complex: catalytic mechanism of 6-aminohexanoate-dimer hydrolase. *J. Mol. Biol.*, **370** (1), 142–156.

88 Araujo, R., Silva, C., O'Neill, A., Micaelo, N., Guebitz, G., Soares, C.M., Casal, M., and Cavaco-Paulo, A. (2007) Tailoring cutinase activity towards polyethylene terephthalate and polyamide 6,6 fibers. *J. Biotechnol.*, **128** (4), 849–857.

89 Silva, C., Carneiro, F., O'Neill, A., Fonseca, L.P., Cabral, J.S.M., Gübitz, G.M., and Cavaco-Paulo, A. (2005) Cutinase-A new tool for biomodification of synthetic fibers. *J. Polym. Sci.*, **43**, 2448–2450.

90 Silva, C., Matama, T., Guebitz, G.M., and Cavaco-Paulo, A. (2005) Influence of organic solvents on cutinase stability and accessibility to polyamide fibers. *J. Polym. Sci. [A1]*, **43** (13), 2749–2753.

91 Miettinen-Oinonen, A., Puolakka, A., Reinikainen, T., and Buchert, J. (2004) Enzymatic modification of polyamide and polyacrylic fibres. *3rd Int. Conf. Textile Biotechnology. Graz, Austria, 13–16 June 2004* p. Abstract 6.

92 Deguchi, T., Kitaoka, Y., Kakezawa, M., and Nishida, T. (1998) Purification and characterization of a nylon-degrading enzyme. *Appl. Environ. Microbiol.*, **64** (4), 1366–1371.

93 Friedrich, J., Zalar, P., Mohorcic, M., Klun, U., and Krzan, A. (2007) Ability of fungi to degrade synthetic polymer nylon-6. *Chemosphere*, **67** (10), 2089–2095.

94 Klun, U., Friedrich, J., and Krzan, A. (2003) Polyamide 6 fibre degradation by a lignolytic fungus. *Polym. Degrad. Stab.*, **79**, 99–104.

95 Deguchi, T., Kakezawa, M., and Nishida, T. (1997) Nylon biodegradation by lignin-degrading fungi. *Appl. Environ. Microbiol.*, **63** (1), 329–331.

96 Nomura, N., Deguchi, T., Shigeno-Akutsu, Y., Nakajima-Kambe, T., and Nakahara, T. (2001) Gene structures and catalytic mechanisms of microbial enzymes able to biodegrade the synthetic solid polymers nylon and polyester polyurethane. *Biotechnol. Genet. Eng. Rev.*, **18**, 125–147.

97 Matama, T., Vaz, F., Gubitz, G.M., and Cavaco-Paulo, A. (2006) The effect of additives and mechanical agitation in surface modification of acrylic fibres by cutinase and esterase. *Biotechnol. J.*, **1** (7–8), 842–849.

98 Fischer-Colbrie, G., Matama, T., Heumann, S., Martinkova, L., Cavaco Paulo, A., and Guebitz, G. (2007) Surface hydrolysis of polyacrylonitrile with nitrile hydrolysing enzymes from Micrococcus luteus BST20. *J. Biotechnol.*, **129** (1), 62–68.

99 Tauber, M.M., Cavaco-Paulo, A., Robra, K.-H., and Gübitz, G.M. (2000) Nitrile hydratase and amidase from *Rhodococcus rhodochrous* hydrolyse acrylic fibers and granulates. *Appl. Environ. Microbiol.*, **66** (4), 1634–1638.

100 Tauber, M.M., Cavaco-Paulo, A., and Gübitz, G.M. (2001) Enzymatic treatment of acrylic fibers and granulates. *AATCC Rev.*, **1** (9), 17–19.

101 Fischer-Colbrie, G., Herrmann, M., Heumann, S., Puolakka, A., Wirth, A., Cavaco-Paulo, A., and Guebitz, G. (2006) Surface modification of polyacrylonitrile with nitrile hydratase and amidase from A*grobacterium tumefaciens*. *Biocatal. Biotransformation*, **24** (6), 419–425.

102 Matama, T., Carneiro, F., Caparrós, C., Gübitz, G.M., and Cavaco-Paulo, A. (2007) Using a nitrilase for the surface modification of acrylic fibres. *Biotechnol. J.*, **2** (3), 353–360.

103 Wang, N., Xu, Y., and Lu, D. (2004) Enzymatic surface modification of acrylic fiber. *AATCC Rev.*, **4** (9), 28–30.

104 Battistel, E., Morra, M., and Marinetti, M. (2001) Enzymatic surface modification of acrylonitrile fibers. *Appl. Surf. Sci.*, **177** (1–2), 32–41.

105 Christenson, E.M., Patel, S., Anderson, J.M., and Hiltner, A. (2006) Enzymatic degradation of poly(ether urethane) and poly(carbonate urethane) by cholesterol esterase. *Biomaterials*, **27** (21), 3920–3926.

106 Ohtaki, S., Maeda, H., Takahashi, T., Yamagata, Y., Hasegawa, F., Gomi, K., Nakajima, T., and Abe, K. (2006) Novel hydrophobic surface binding protein, HsbA, produced by Aspergillus oryzae. *Appl. Environ. Microbiol.*, **72** (4), 2407–2413.

107 Takahashi, T., Maeda, H., Yoneda, S., Ohtaki, S., Yamagata, Y., Hasegawa, F., Gomi, K., Nakajima, T., and Abe, K. (2005) The fungal hydrophobin RolA recruits polyesterase and laterally moves on hydrophobic surfaces. *Mol. Microbiol.*, **57** (6), 1780–1796.

16
Enzymatic Polysaccharide Degradation

Maricica Munteanu and Helmut Ritter

16.1
The Features of the Enzymatic Degradation

The increasing strong interest in biodegradable materials for drug delivery systems and other biomedical uses, as well as for environmental applications, prompted us to evaluate their enzymatic degradation.

The biodegradation of polymers was defined as: (i) a biophysical effect, such as mechanical damage on a material by the swelling and bursting effect of growing cells; (ii) a secondary biochemical effect, resulting from the excretion of substances other than cell enzymes, which might act directly on the polymer or by changing the pH or redox conditions of the surroundings; (iii) direct enzymatic actions leading to splitting or oxidative breakdown of the material.

Biodegradable polymers, defined as polymers that are degradable *in vitro* to produce biocompatible or nontoxic by-products are classified into three groups [1], namely:

1) natural polymers: gelatin, alginate, albumin, collagen, starch, dextran, chitosan, chitin;
2) semisynthetic polymers obtained by modification of chitosan, alginate or hyaluronic acid;
3) synthetic polymers: polylactic acid, polygycolic acid, poly(lactide-co-glycolide), poly(orthoester), polyhydroxy butyrate, polyhydroxy valerate, polyanhydride.

Enzymatic degradation of polymers is widely studied. The enzymes facilitate extracellular degradation processes, which progressively reduce the molecular weight and the structural identity of macromolecules. Generally, the high polymer is first reduced to a low polymer by a combination of scission and removal of repeat units from chain ends. This leads to fragmentation or dissolution of the polymer. Further enzymatic action yields oligomeric fragments and simple organic intermediates of the biodegradation process. If biodegradation is allowed to continue, complete mineralization would ensue, with the transformation of the organic polymer into carbon dioxide and water.

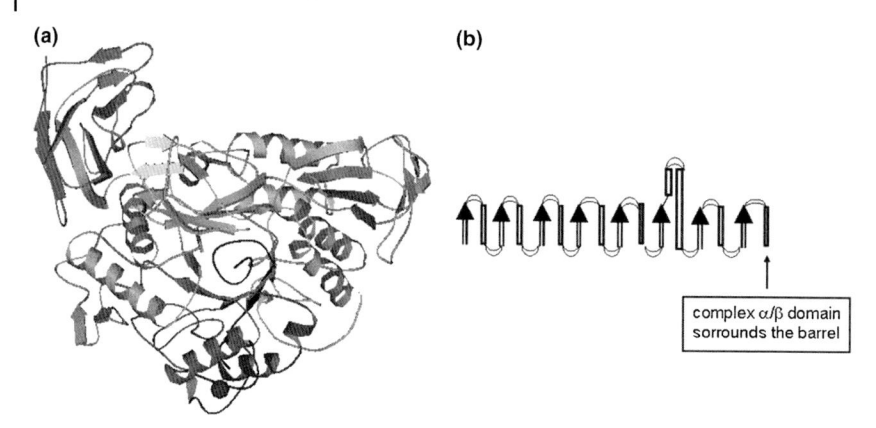

Figure 16.3 (a)The three-dimensional structure of cyclodextrin glycosyltransferase and (b) the catalytic $(\beta/\alpha)_8$ barrel.(Reproduced with permission from [30]. Copyright © (2000) Elsevier).

Domain E contains a raw-starch binding motif forming two maltose-binding sites (MBS) that are responsible for starch binding (MBS1) and for guidance of the substrate into the active site (MBS2). These MBSs also bind CDs, thus they play an important role in the product inhibition of CGTase.

Cyclodextrin glycosyltransferase catalyzes the breakdown of starch and linear maltodextrin substrates to produce CDs, while α-amylase converts starch into linear oligosaccharides, resulting in a rapid decrease in viscosity [37].

The systematic name of CGTase is 1,4-α-D-glucan 4-α-D-(1,4-α-D-glucano)-transferase (cyclizing). It is also called cyclodextrin glycosyltransferase, cyclomaltodextrin glucanotransferase or cyclomaltodextrin glycosyltransferase. Cyclodextrin glycosyltransferase is a hexosyltransferase and belongs to the transferase group of enzymes (EC 2.4.1.19) [38]. The enzyme displays its cyclic action on substrates with α-1,4-glycosyl chain, such as starch, amylose, amylopectin, dextrins, or glycogen [39]. The distinctive feature of this enzyme is the formation of cyclic products, CDs, which have the systematic name of cyclic α-D-(1,4)-linked D-glucose oligosaccharides consisting of 6, 7, 8 glycosyl units, well known as α-, β-, γ-CD (CD6, CD7, CD8). CGTases have been further classified into three subgroups (α-, β-, γ-CGTases), according to their product specificity.

16.2.2.2 Cyclodextrin Glycosyltransferase: Cyclodextrin-Forming Activity

Cyclodextrin glycosyltransferases catalyze the breakdown of starch and linear maltodextrin substrates into CDs via an intramolecular transglycosylation reaction (cyclization). They also catalyze two other transferase reactions: (i) the coupling reaction between CDs and linear dextrins, the opening of a CD ring being followed by transfer to a linear oligosaccharide; and (ii) disproportionation in which two linear oligosaccharides are converted into linear oligosaccharides of different sizes, by transferring a part of a linear oligosaccharide to another

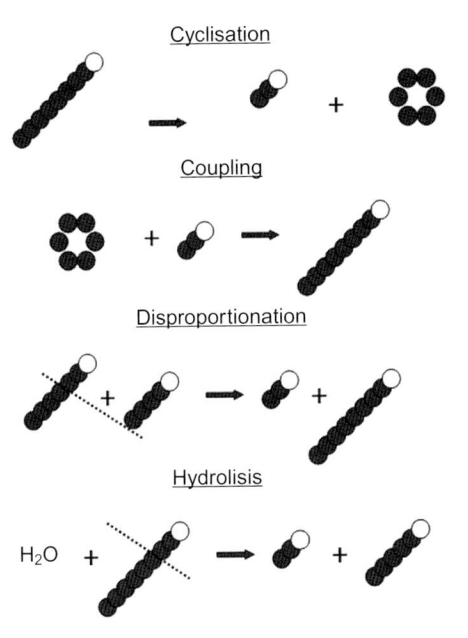

Figure 16.4 Cyclodextrin glycosyltransferase-catalyzed reactions (• represents glucose residues and ○ glucose residues with free reducing ends). (Reproduced with permission from [30]. Copyright © (2000) Elsevier).

oligosaccharide (Figure 16.4). In addition, these enzymes possess a weak hydrolyzing activity in which part of a linear oligosaccharide is transferred to water to produce linear dextrins [35, 40–42].

Since the coupling reaction is the reverse reaction of cyclization, the enzyme is able to degrade CDs in the presence of suitable acceptor molecules. This activity can have a negative effect on conversion of starch into CDs. Reaction conditions are essential for directing the enzyme action to CD production [39].

Cyclodextrin glycosyltransferases can synthesize all forms of CDs, thus the product of the conversion results in a mixture of the three main types of cyclic molecules, in ratios that are strictly dependent on the enzyme used: each CGTase has its own characteristic $\alpha:\beta:\gamma$ synthesis ratio. The amount of CDs and the ratio of α-, β- and γ-forms in the product are determined not only by the CGTase but also by the reaction conditions, including reaction time and temperature. Purification of the three types of CDs takes advantage of the different water solubility of the molecules: β-CD which is very poorly water soluble (18.5 g l^{-1} or 16.3 mM) (at 25 °C) can be easily retrieved through crystallization. The more soluble α- and γ-CDs (145 and 232 g l^{-1} respectively) are usually purified by means of expensive and time consuming chromatography techniques.

The product selectivity of industrially used CGTases is not very high and expensive separation techniques are required to isolate individual CDs. As an alternative a 'complexing agent' can be added during the enzymatic conversion step.

16.2.3.2 Cyclodextrin Enzymatic Degradation

Since the glycosidic bond (bridge) oxygens are buried in the nonpolar ring center, these compounds are resistant to a multitude of glycosyl hydrolases. Therefore, CDs are most suitable for preserving unstable bioactive compounds and for releasing them at a slow, defined rate.

The CDs are relatively stable compared to the corresponding linear malto-oligosaccharides, but are readily degraded to glucose and maltose by specific hydrolases including CDase, maltogenic amylase, and neopullulanase-like TVA amylase.

Some studies have been made in connection with the interaction between carbohydrate-degrading enzymes and CDs [65, 66]. The first member of CD-degrading enzymes was detected four decades ago [67]. Since then, new members have been given rather uncorrelated names [68, 69]. These enzymes were unified to the common name cyclomaltodextrinase (CMD). Five CMD structures are published; namely, those of the enzymes from *Thermoactinomyces vulgaris* (TvCMD), *Thermus sp.* IM6501 (TspCMD), alkalophilic *Bacillus sp.* I-5 (BspCMD), *Flavobacterium sp.* no. 92 (FspCMD) and *Bacillus stearothermophilus* (BsCMD) [70]. All of them consist of the four domains N (N-terminal), A, B and C. Domain A forms a $(\beta/\alpha)8$-barrel that, as commonly observed, accommodates the active center at the carbonyl ends of the β-strands. Domain B is an insertion of about 50 residues in domain A, and domain C forms a β-sheet sandwich. Domains A, B and C correspond to related domains of α-amylases and cyclodextrin glycosyltransferases.

Most reported CDases have an optimum temperature below 50 °C and mainly produce maltose as the final product (Table 16.4) [71]. It has been proposed that a multiple attack mechanism is an inherent property of the depolymerization enzymes.

Table 16.4 Characteristics of cyclomaltodextrinases.

Origin	Molecular mass (Da, monomer)	Optimum Temp. (°C)	Optimum pH
B. sphaericus ATCC7055	91 200–95 000	40	6.0–6.5
B. sphaericus E-244	72 000	45	8.0
B. stearothermophilus K-12481	67 000	60	6.5
T. ethanolicus 39E	66 000	65	5.9
Alkalophilic Bacillus sp.	67 000	50	6.0
Alkalophilic Bacillus I-5	65 000	45	6.5
B. coagulans	62 000	50	6.2
Flavobacterium sp.	62 000		6.0–7.5
Bacteroides ovatus		42	7.0
Klebsiella oxitoca	69 000		
Xanthomonas campestris K-11151	55 000	55	4.5
E. coli	66 000		
Thermotoga maritime	55 000	85	6.5

Amylases can be divided into two categories, endoamylases and exoamylases. Endoamylases catalyze hydrolysis in a random manner in the interior of the starch molecule producing linear and branched oligosaccharides of various chain lengths. Exoamylases act from the nonreducing end successively resulting in short end products. The activity of α-amylase is different from that of β-amylase, which splits off maltose units from the nonreducing end of the glucan in a zipper fashion. The activity and stability of α-amylase depends on the presence of chlorine ions and calcium; the latter probably form intramolecular cross-links similar to disulfide bridges. The enzyme is exceptionally stable at high pH and temperature ranges.

Due to the lack of nonreducing glucose residues in the molecule, CDs are not substrates of the exo-type starch-degrading enzymes. In fact, they are potential inhibitors of some amylolytic enzymes such as β-amylase and pullulanase.

However some α-amylases, such as human salivary α-amylase, human- and porcine pancreatic α-amylase are able to hydrolyze CDs. Moreover, several bacterial and fungal α-amylases were shown to be able to hydrolyze CDs (Table 16.5), although the rate of hydrolysis is much slower than that of amylase [66]. α-Amylase (EC 3.2.1.1) is an important *endo*-type carbohydrate that hydrolyzes α-1,4 glycosidic linkages of D-glucose oligomers and polymers. It is a key enzyme of carbohydrate metabolism in mammals, plants, and bacteria [72]. Since it was first termed as 'diastase' by French in 1833, α-amylases from fungal and bacterial sources were developed and commercialized for a variety of applications such as brewing, textiles, paper, and corn syrup industries. The amylase production and use was extended after the isolation of *Bacillus subtilis* α-amylase and *Aspergillus niger* glucoamylase [73].

It can be proposed that degradation of α-, β-, and γ-CDs follows a parallel series of reactions as shown with α-CD as an example in Figure 16.6.

No marked differences in the K_m values of the three CDs was observed, however, the V_{max} value obtained for the hydrolysis of β- and γ-CD was almost 30 and 500

Table 16.5 Cyclodextrin-degrading α-amylases.

Strain	pH optimum	Temp. optimum (°C)	pH stability	Thermostability
Aspergillus oryzae	4.8–6.6	35–37		
A. oryzae	5.0–5.9		5.8–7.2 10°C over a year 5.0–8.2 37°C 30 min	70%, 60°C, 90 min (+Ca) 70%, 50°C, 30 min (−Ca)
Bacillus coagulans	6.2	50	6.0–7.3 40°C, 2 h	below 45°C, 15 min
Pseudomonas Msl	5.5	50	4.7–7.5 40°C 30 min	95%, 50°C, 30 min 0%, 60°C, 30 min

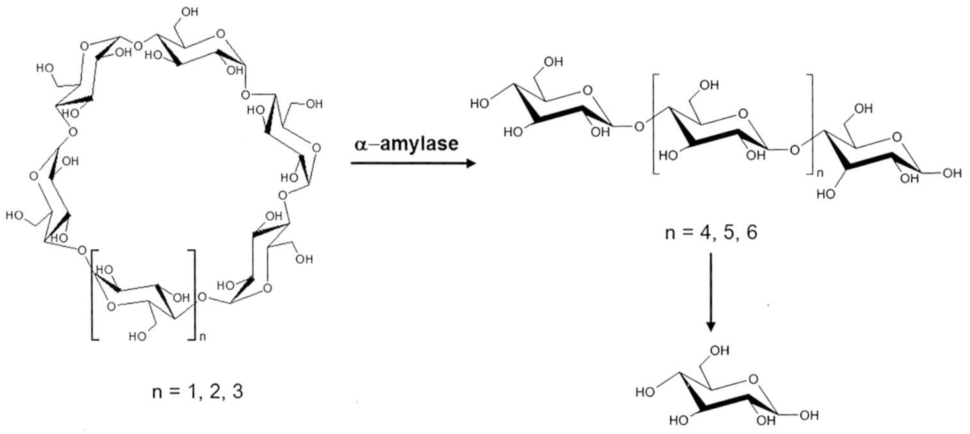

Figure 16.6 Enzymatic ring opening of cyclodextrin and the hydrolysis of linear oligosaccharides to glucose.

Table 16.6 Rate parameters for the hydrolysis of cyclodextrin by *Aspergillus oryzae* α-amylase [a][65].

Cyclodextrin	α-CD	β-CD	γ-CD
K_m (10^{-3} mM)	7.13	4.35	3.12
V_{max} (min^{-1})	5.8	166	2300

a) At pH 5.2 and 37 °C.

Table 16.7 Distribution of malto-oligomers at the hydrolysis of cyclodextrin by *Aspergillus oryzae* α-amylase [a].

Cyclodextrin	α-CD	β-CD	γ-CD
Glucose	0.4	11.96	17.24
Maltose	1.56	1.88	1.5
Triose	—	2.97	71.25
Tetraose	—	—	—
Unmodified CD	98	83.18	—

a) At pH 5.2 and 37 °C, after 24 h.

times, respectively, greater than those measured for α-CD (Table 16.6). This finding pointed to the formation of enzyme–substrate complexes of similar stability in each case, but with markedly different decomposition rates amylase [66].

According to liquid chromatographic analysis, glucose, maltose, maltotriose were identified upon hydrolysis of β-CD for 3 h (Table 16.7). At the beginning of the hydrolysis of γ-CD, HPLC analysis indicated also relatively large quantities of the higher malto-oligomers (maltohexaose, maltopentaose, maltotetraose), beside glucose and maltose [66].

According to these results α-, β- and γ-CD are substrates of *Aspergillus oryzae* amylase. However, the three cyclic substrates are hydrolyzed with rather different rates. γ-Cyclodextrin seems to be the best substrate, while α-CD is hydrolyzed only very slowly. Probably the size and flexibility of the ring of γ-CD is the most similar to the windings of the helical structure of amylose chain. In the early stage of the process all malto-oligomers from glucose to maltohexaose are present in the reaction mixture. Szejtli extended also the investigations concerning the degrading ability of Taka amylase A on α-and β-CD to γ-CD [66]. The enzymatic hydrolysis of γ-CD with *A. oryzae* α-amylase was studied by monitoring of the composition of the reaction mixture and by identifying the hydrolysis products with HPLC method. The main product obtained by the enzymatic hydrolysis of γ-CD was maltotriose. A small amount of maltotetraose was also present in the hydrolysate of both β- and γ-CD but no higher-membered oligomers could be detected.

The concept of multiple attack of α-amylase on γ-CD was investigated in the presence of β-amylase. The γ-CD was treated with crystalline porcine pancreatic α-amylase together with a 60-fold excess of crystalline sweetpotato β-amylase. β-Amylase, requiring a nonreducing end-group for its actions, acts rapidly on linear oligosaccharides in the range G4 and higher, but it has no action on cyclic substrates. However, as soon as the CD ring is open, β-amylase will rapidly degrade the linear oligosaccharides to G1, if the substrate contains an even number of glucose units, or to G2 plus G3 and glucose if the oligosaccharide contains an odd number of units [74]. A similar experiment was performed with crystalline *A. oryzae* α-amylase and sweetpotato β-amylase on γ-CD and crystalline ε-CD. These reactions produced only G2, thus indicating that the *Aspergillus* α-amylase shows little if any multiple attack.

The inhibitory effect of maltose, maltotetraose and maltopentaose was also investigated in the hydrolysis of β-CD and γ-CD with *A. oryzae* α-amylase. According to these data it can be established that maltose, maltotriose and maltotetraose are not good substrates of *A. oryzae* α-amylase, whereas maltopentaose and maltohexaose act as good substrates of the enzyme, inhibiting the degradation of the cyclodextrin 'competitively'. On the other hand, glucose, maltose and maltotriose are not substrates of the enzyme, but by linking to the enzyme-protein they may become 'noncompetitive' inhibitors. The 'noncompetitive' inhibitory effect could be explained by the decreasing of the linear higher malto-oligomers, with the simultaneous increase of concentration of glucose, maltose and maltotriose in the course of CD-hydrolysis. Since 83% of unmodified β-CD was detected even after 24 h, it was supposed that the large amount of β-CD could competitively inhibit the hydrolysis of the linear malto-oligomers.

By interrupting the hydrolysis of γ-CD at the appropriate moment, significant quantities of maltotriose can be obtained. The preparative liquid chromatographic technique makes the isolation of the oligomers possible in pure form and they can be used either as reference material in the synthesis of oligosaccharides or for the substrate-specificity examinations of carbohydrases. There are manifold possibilities and demands on the use of maltotriose in biochemistry and in food industry.

16.2.3.3 Cyclodextrin Degradation by the Intestinal Flora

Cyclodextrins are barely hydrolyzed by amylases in the upper part of the gastrointestinal tract and only slightly absorbed in the stomach and small intestine, but are readily fermented by the colonic microbial flora [75, 76]. After fermentation into small saccharides, they are absorbed in the large intestine. The peculiar hydrolyzing property of CDs makes them useful for colon drug targeting. Cyclodextrin prodrugs which survive passage through stomach and small intestine can release the drug by enzymatic degradation of the CD ring in the colon [77]. Therefore, the CD prodrug approach can provide a versatile means for constructions of colon-specific delivery systems of certain drugs (Figure 16.7).

The release of 5-aminosalicylic acid from tablets coated with β-CD was investigated in a colonic environment using the colonic microflora test [78]. Bisoprolol was also released by colonic bacteria like *Proteus mirabilis* and *Escherichia fergusonii*, after incubating with β-CD coated tablets. Eudragit RS films with incorporated inulin also were described for their successful degradation in human fecal medium [79].

The n-butyric acid–β-CD ester conjugate was also investigated as a prodrug for colon-specific delivery system. The n-butyric acid was rapidly released when the esterase was added after the α-amylase hydrolysis. The ester linkage of the small saccharide conjugates produced after the amylase hydrolysis, was easily cleaved by the esterase, probably because of a relief of steric hindrance of the large β-CD [80].

Biphenyl acetic acid (BPAA) prodrugs for colon-specific delivery were developed by conjugation of the drug onto one of the primary hydroxyl groups of α-, β-, and γ-CDs through an ester or amide linkage. The ester prodrugs, particularly the α- and γ-CD forms, are subject to ring-opening followed by hydrolysis to the maltose and triose conjugates. In the case of ester prodrugs, the maltose and triose conjugates released the free drug after initial hydrolysis of the susceptible

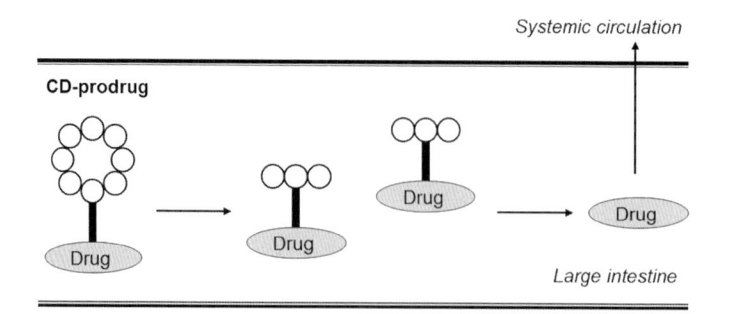

Figure 16.7 Release mechanism of the drug from γ-CyD prodrugs in cecum and colon.

ester linkage; in the case of amide prodrugs, the conjugates provided delayed release due to the resistance of the amide bond to hydrolysis. Therefore, the CD prodrug approach can provide a versatile means for constructions of colon-specific delivery systems of certain drugs [77].

16.2.4
Enzymatic Synthesis of Cyclodextrin-Derivatives

Enzymes can be also used in the synthesis of CD derivatives. β-Cyclodextrin and its methyl and hydroxypropyl derivatives were acylated by means of lipases with a series of carboxylic acids [81], and the synthesis of CD fatty acid esters was enzymatically catalyzed by some proteases [82].

Peracetylated β-CD has a beneficial action on the lipase-catalyzed enantioselective transesterification of 1-(2-furyl)ethanol in organic solvents. The use of CD as a regulator of lipase was assumed to be the result of increasing the conformational flexibility of the enzyme and undergoing host–guest complexation with the product, with an enhancement of the enantiomeric ratio E and the reaction rate [83].

16.2.5
Cyclodextrin-Based Enzyme Mimics

Cyclodextrins have proven to be the most popular enzyme mimics, catalyzing various reactions. Cyclodextrin-based neoglycoenzymes with improved efficiency have also been designed and synthesized. Cyclodextrin-modified enzymes have potential application as biosensors as well as in the formulation of effective and biodegradable drug delivery systems for enzyme replacement therapy [84].

The cleavage of esters in basic medium, mediated by CDs, is the system most studied. In some cases, such as that with *meta*-substituted phenyl acetates, cleavage is strongly accelerated; on the contrary, the reaction of *para*-substituted isomers is accelerated only modestly. The observed effects depend on the chain length of the ester, the CD, and the position of the substituent on the phenyl group. The mechanism typically involves nucleophilic attack by the ionized secondary hydroxyl groups of CD, but in other cases, general base catalysis competes with nucleophile catalysis.

Cyclodextrins can serve as enzyme models because of their stereospecific binding followed by stereospecific reactions [85]. Several modifications of CDs, specifically of their primary and secondary hydroxyls as tosylates, methoxylates, allyl ethers and many others have been described.

Cyclodextrins with two imidazole groups on the primary hydroxyl side can enhance the enolate formation [86] of a simple bound ketone by bifunctional acid–base catalysis and accelerating the intramolecular aldol condensation of bound ketoaldehyde and dialdehyde. The aldolase mimics which catalyzed crossed aldol condensations were obtained by the assembly of β-CD and various amino moieties as Schiff base [87].

16.2.6
Specific-Base-Catalyzed Hydrolysis

Although CD complexation of drug molecules usually results in increased drug stability, there are examples of accelerated degradation. Cyclodextrins can act efficiently as an acid–base multiple catalyst, due to the ability of the hydroxyl group to act as an acid in a wide range of pH and to dissociate to the corresponding anion in a strongly alkaline solution, acting as a base. In these circumstances, CDs themselves satisfy the minimum necessary and sufficient conditions to be models of certain enzymes, which involve acid and/or base catalyses [88]. As a monoanion, CD accelerates the hydrolytic cleavage of certain pyrophosphates [89] or carboxylates [90]. Cyclodextrins facilitated the specific-base-catalyzed hydrolysis of the β-lactam ring by simultaneous hydrogen bonding between two adjacent hydroxyl groups on the glucose residue and the amide carbonyl and β-lactam carbonyl groups of the antibiotic. The (hydroxypropyl)-CDs have a destabilizing effect on cephalothin at pH 9.7, where the specific-base catalysis dominates. Cyclodextrin also accelerated the specific-base-catalyzed degradation of aztreonam, aspirin, and of the antiallergic drug tranilast [91].

16.3
Hyaluronic Acid Enzymatic Degradation

16.3.1
Hyaluronic Acid: Structure, Biological Functions and Clinical Applications

After its discovery in 1934 by Meyer and Palmer [92], hyaluronan (hyaluronic acid or hyaluronate, HA) was found to be an ubiquitous component in vertebrate tissues, particularly in cartilage, skin and vitreous humor [93] (see also Chapter 9). Hyaluronan is a high-molecular-weight, highly anionic glycosaminoglycan, with a low charge density [94]. The negative charge on the repeating disaccharide is the cause of many of the unique properties of this polyacid, which is able to produce a highly viscous solution in water [95–97].

Hyaluronic acid is a linear polysaccharide built from two alternating monosaccharide units, D-N-acetyl glucosamine (GlcNAc) and D-glucuronic acid (GlcA), connected by β-(1→3) and β-(1→4) glycosidic bonds, respectively (Figure 16.8).

Hyaluronan is continuously synthesized and secreted by fibroblasts, keratinocytes, chondrocytes and other specialized cells in the extracellular matrix (ECMs) throughout the body. It is synthesized by HA synthase (see also Chapter 9) at the inner face of the plasma membrane [98]. The level of HA synthesis is very high in skin and cartilage [99]. Hyaluronic acid is not one of the major components of the ECMs of the connective tissues, but it is found in various locations such as synovial fluid, vitreous humor, and umbilical cords [100]. Its biological functions include the maintenance of mechanical properties such as swelling in connective tissues and control of tissue hydration, providing lubricating properties in synovial fluid and heart valves.

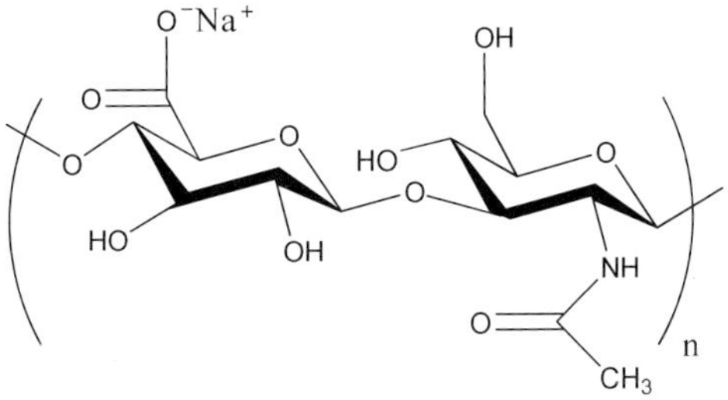

Figure 16.8 Structure of sodium hyaluronate (hyaluronic acid, HA).

The molecular weight of HA strongly depends on its biological origin [101]. When produced by microbial fermentation, the molecular weight ranges from just below 1×10^6 up to $4 \times 10^6\,\mathrm{g\,mol^{-1}}$ [102]. When prepared by extraction, molecular weights up to 6×10^6 have been reported [103]. Today, two main microbial sources are exploited for HA production, *Streptococcus* strains [104] and recombinant *Bacillus subtilis* [105]. In streptococci, HA is produced as part of the capsule bound to the cell membrane, making it challenging to isolate. On the other hand, HA from *Bacillus* is excreted extracellularly and can therefore conveniently be isolated as a high purity product [95].

The basic areas of the clinical applications of HA and its derivatives are classified by Balazs [106] as follows:

1) **viscosurgery:** to protect delicate tissues and provide space during surgical manipulations, as in ophthalmological surgeries,

2) **viscoaugmentation:** to fill and augment tissue spaces, as in skin, sphincter muscles, vocal and pharyngeal tissues,

3) **viscoseparation:** to separate connective tissue surfaces traumatized by surgical procedures or injury, in order to prevent adhesions and excessive scar formation,

4) **viscosupplementation:** to replace or supplement tissue fluids, such as replacement of synovial fluid in painful arthritis, and to relieve pain,

5) **viscoprotection:** to protect healthy, wounded, or injured tissue surfaces from dryness or noxious environmental agents, and to promote the healing of such surfaces [107].

The high-molecular-weight of HA together with its special viscoelastic features and biological functions has made HA as an attractive material to prepare biocompatible devices with applications in drug delivery and tissue engineering [108–111].

The physical and biological properties of HA appear to be affected by many factors including HA concentration and chain length [112]. In addition, viscosity of the HA gel and the ability to hydrate large amounts of water were shown to be dependent on the molecular size of the HA chain.

High- and low-molecular-weight forms of HA exhibit opposite effects on cell behavior [113]. Extracellular high-molecular-weight HA (HMWHA) inhibits endothelial cell growth, and is thereby anti-angiogenic in nature, whereas low-molecular-weight HA (LMWHA) stimulated cell growth leading to induction of angiogenesis. The HMWHA polymers have an increased ability to bind fibrinogen; this is one of the first reactions to occur in clot formation, which are critical in early wound healing [114]. These HA polymers are also anti-inflammatory and immunosuppressive in nature [115]. The degradation of LMWHA results in small oligomers that can induce heat shock proteins and that are anti-apoptotic. The single sugar products, glucuronic acid and a glucosamine derivative are released from lysosomes to the cytoplasm, where they become available for other metabolic cycles.

High-molar-mass forms of HA are reflections of intact normal tissue, while fragmented forms of HA are usually indications of stress. Indeed, various size HA fractions have an enormous repertoire of functions and constitute an information-rich system [116].

16.3.2
Hyaluronidase: Biological and Clinical Significance

Hyaluronic acid is found abundantly in the body and has an extraordinarily high rate of turnover. There are 15 g of hyaluronan in the 70-kg individual, of which 5 g are cycled daily through the catabolic pathways, mostly as the result of the hyaluronidases [97]. Hyaluronan deposition and turnover is even more abundant and more rapid in malignant tissues [116].

Hyaluronidase (HAse, EC 3.2.1.35), is an endoglycosidase which randomly cleaves internal β-N-acetyl-hexosamine glucosidic linkages in HA and chondroitin-4- and -6-sulfate and their desulfated derivatives to liberate oligosaccharides containing equimolar glucuronic acid residues at their reducing terminals and N-acetylglucosamine (NAG) at the nonreducing ends [117–119] (see also Chapter 9). The enzyme has been demonstrated in a wide range of mammalian tissues, such as synovial fluid, serum, alveolar macrophages, brain, skin, kidney, liver, spleen, and lung [117].

The hyaluronidases were divided into three classes [120], based on analysis of the their reaction products:

1) Bacterial hyaluronidases (EC 4.2.99.1) are endo-β-acetyl-hexosaminidases that function as eliminases yielding disaccharides. In marked contrast with their eukaryotic counterparts, they are specific for HA.

2) Endo-β-glucuronidase types of hyaluronidase (EC 3.2.1.36) found in leeches, crustaceans (Karlstam et al., 1991) [121] and some parasites, generate tetra- and hexasaccharide end-products.

3) The mammalian types of hyaluronidase (EC 3.2.1.35) are also endo-β-acetyl-hexosaminidases, but function as hydrolases, with tetrasaccharides as the

predominant end-products. They lack substrate specificity, able to digest chondroitin sulfates (CS), though at a slower rate. In addition, they have transglycosidase activity that generates complex cross-linked chains *in vitro*. This ability has not been documented *in vivo*.

Based on their pH-dependent activity profile, hyaluronidases are classified into two groups. Acid active hyaluronidases are active between pH 3 and 4, and this group includes human liver and serum hyaluronidases. Neutral active hyaluronidases are active between pH 5 to 8 and include PH-20, snake venom, and bee venom hyaluronidases [122].

The key role of hyaluronidases has been recognized in a number of physiological and pathological processes such as embryogenesis, angiogenesis, inflammation, disease progression, wound healing, bacterial pathogenesis and the diffusion of systemic toxins/venoms [118, 119]. Sperm hyaluronidase plays an important role in successful fertilization in most mammals, including human [123–125].

Northup *et al.* postulated that HAse may play a role in cancer invasiveness [126]. Stern *et al.* described elevated levels of HAse in urine of children with Wilm's tumor and suggested that HAse may be used as a tumor marker [127].

The major clinical use of hyaluronidase is in ophthalmology, by improving the local anesthetic infiltration [128, 129]. The anesthetic-hyaluronidase combinations has also found applications in pain relief for high ligations in varicose vein surgery or hernia surgery [130]. Clinically, the enzyme is used in chemotherapy to enhance the antineoplastic activity of cytostatics and in the treatment of acute myocardial infarction [131].

The balance between the anabolism and catabolism of HA is maintained by the inhibitors of hyaluronidases. Heparin is a known and well-characterized inhibitor of hyaluronidase [132]. Chitosan inhibits HA degradation by venom and bovine testicular hyaluronidases [133]. *Cis*-unsaturated fatty acids can also inhibit the hyaluronidase activity [134]. Certain anti-inflammatory drugs including salycylates, indomethacin and dexamethasone are also known to exert anti-hyaluronidase activity [133, 135]. Ascorbic acid [136], as well as plant derived bioactive compounds, such as flavonoids, tannins, pectins, curcumins, coumarins, gylcyrrhizin are used to block hyaluronidase activity [100].

16.4
Alginate Enzymatic Degradation

16.4.1
Alginate as Biocompatible Polysaccharide

Alginate is a gelling polysaccharide found in great abundance as part of the cell wall and intracellular material in the brown seaweeds (Phaeophyceae). Most of the alginate used commercially is obtained from *Macrocystis*, *Laminaria*, and *Ascophyllum* [137]. Recently, *Azotobacter vinelandii*, a nitrogen-fixing bacterium

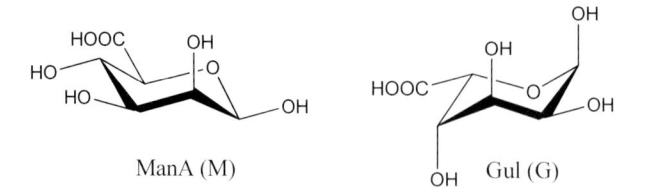

Figure 16.9 The building block of alginate.

from soil, and several species belonging to the related genus *Pseudomonas* have been found to produce the polymer alginate [138]. The function of alginates in algae is primarily skeletal. The gel located in the cell wall and intercellular matrix confers the strength and flexibility which is necessary to withstand the force of water in which the seaweed grows [139].

Alginate is one of the most extensively investigated biopolymers. After it was discovered by Stanford 1883 [140], the commercial production of alginate started in 1927, and has expanded to about 50 000 tonnes per year worldwide. Only 30% of the alginate production is destined for the food industry, the rest being used in industrial, pharmaceutical and dental applications [141].

Alginates are composed of (1→4)-linked β-D-mannuronic acid (M units) and α-L-guluronic acid (G units) monomers (Figure 16.9), with different sequential distribution along the polymer chain [142]. The composition, sequence, and molecular weight generally determine the physical properties of alginate [143].

The relative amount and distribution of these two residues vary according to the species and growth conditions [144]. Some of the M residues in alginates may be O acetylated at C-2, C-3, or both C-2 and C-3 [145]. The acetylated form is produced as an exocellular polymer by certain heterotrophic bacteria, the *Pseudomonadaceae* and the *Azotobacteriaceae* [146]. Acetylation affects the water-binding properties and ion-binding selectivity of the polymer [137]. Divalent cations like Ca^{2+} are cooperatively bound between the G blocks of adjacent alginate chains, creating ionic interchain bridges which cause gelling of aqueous alginate solutions [147]. A mild gelling reaction of alginate hydrogels occurs in the presence of calcium chloride [142, 148–151]. Alginate/cell suspensions may also be gelled *in situ*, providing a means for cell transplantation with minimally invasive surgical procedures [152].

Alginate can also form porous sponges, beads, and microfibers, all of which have been used for many biomedical applications. It has excellent biocompatibility, low toxicity, nonimmunogenicity, relatively low cost, and simple gelation behavior with divalent cations such as Ca^{2+}, Mg^{2+}, Ba^{2+}, and Sr^{2+} [153]. In particular, the sodium alginate hydrogels cross-linked by calcium ions are widely used for the encapsulation of cells, proteins, oligonucleotides or DNA and also as scaffolds for cartilage tissue engineering [154].

As a result of their biocompatibility and their gelling capacity, alginates have been applied in the engineering of biomaterials [155]. Beside their major applications as devices for tissue engineering and drug delivery, alginate-based materials are used as adsorbents for the elimination of heavy metals and organic pollutants from contaminated environments [156].

16.4.2
Alginate Depolymerization by Alginate Lyases

Alginate can be degraded by alginate lyases, which cleave the polysaccharide by a β-elimination reaction [138]. Alginate lyases are also known as alginases or alginate depolymerizes. The susceptibility of alginates to degradation is determined by both the block structure and degree of O-acetylation within the macromolecule [137]. Although all lyases perform essentially the same depolymerization action on alginate, each enzyme is defined by its individual characteristics and its preference for the glycolytic bond connecting M and G monomers. Most organisms produce a single alginate lyase with defined substrate specificity. Some marine animal and bacterial strains can produce two or more enzymes, and at least one bacterial strain, *Alteromonas* sp. H-4, produces a lyase with multiple substrate specificities.

The oligosaccharides prepared from alginate can exhibit a variety of biological activities, including anti-allergy properties, anti-hypertensive activity, an ability to enhance the growth of human endothelial cells and keratinocytes, and an ability to induce cytokine production [157].

16.5
Chitin and Chitosan Enzymatic Degradation

16.5.1
Enzymatic Hydrolysis of Chitin

Chitin is the second most abundant natural polymer in nature after cellulose and it is found in the structure of a wide number of marine invertebrates such as crabs and shrimps and the fungal walls [158].

Three polymorphic forms of chitin are widely spread in nature: α-, β- and γ-chitins. α-Chitin, the most available isomorphous form, is tightly compacted due to its crystalline structure where the chains are in antiparallel fashion favoring strong hydrogen bounding. β-Chitin has an arrangement in parallel with weaker intermolecular forces that leads to a less stable molecule than α-chitin. γ-Chitin is a mixture of both α- and β-chitins. Generally β-chitin is preferred due to its higher solubility and swelling compared with α-chitin [159].

The solubility of chitin is enhanced by partial deacetylation under mild conditions that do not degrade the polymer, but increase the polarity and electrostatic repulsion of the amino groups. Besides, the loss of the crystalline structure is a consequence of the reduction of the hydrogen bounds by the elimination of acetyl groups. It has been reported that chitins with a degree of acetylation (DA) of

β(1→4)-2-acetamido-2-deoxy-D-glucopyranose

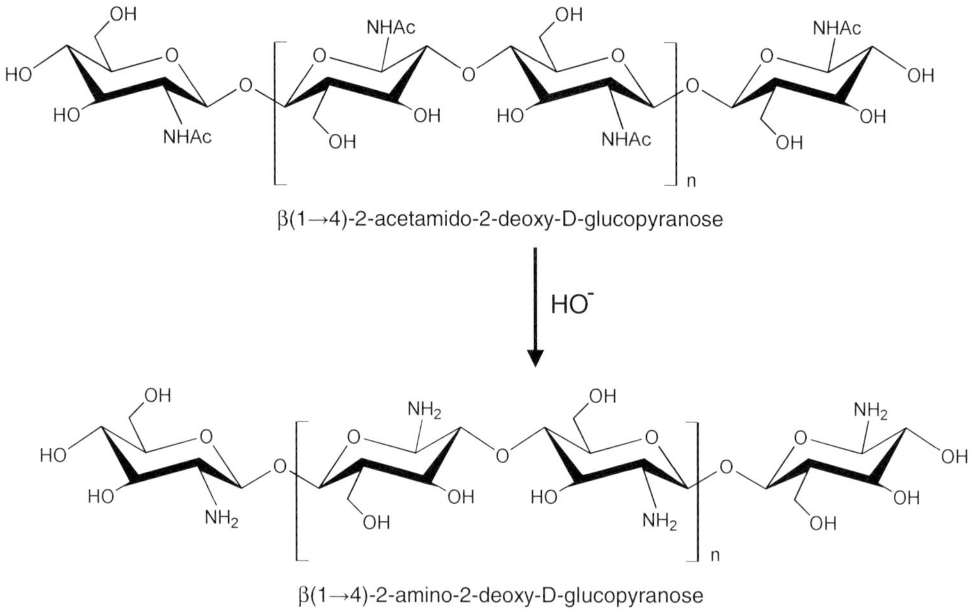

β(1→4)-2-amino-2-deoxy-D-glucopyranose

Figure 16.10 Chemical structure of chitin and chitosan.

0.45–0.55 have good solubility in aqueous media. The polymer obtained by the partially or totally deacetylation of chitin [(β(1→4)-2-acetamido-2-deoxy-D-glucan] is called chitosan [(β(1→4)-2-amino-2-deoxy-D-glucan] (Figure 16.10).

The hydrolysis of chitin is rarely conducted to full completion; hence chitosan polymeric chain is generally described as a copolymeric structure comprised of D-glucosamine along with N-acetyl residues (Figure 16.11). The DA is one of the most important chemical characteristics of chitosan, which can influence the biochemical and biopharmacological properties of chitosan [160–162]. The chitosan-related biological properties are also highly dependent on the average molecular weight [163].

In nature, chitosan is found only in cell wall of *Zygomycetes*, a group of phytophatogenic fungi. Chitosan is much less hydrophobic and is soluble in diluted acids. In contrast to chitin, the presence of free amine groups along the chitosan chain allows this macromolecule to dissolve in diluted aqueous acidic solvents due to the protonation of these groups, rendering the corresponding chitosan salt in solution [158].

Enzymes that hydrolyze chitin – chitinases – have low specific activities and act very slowly.

Enzymatic hydrolysis of chitosan is much easier than that of chitin, because of its significant content of free amino groups, chitosan has a cationic character and has a positive charge at most pH levels.

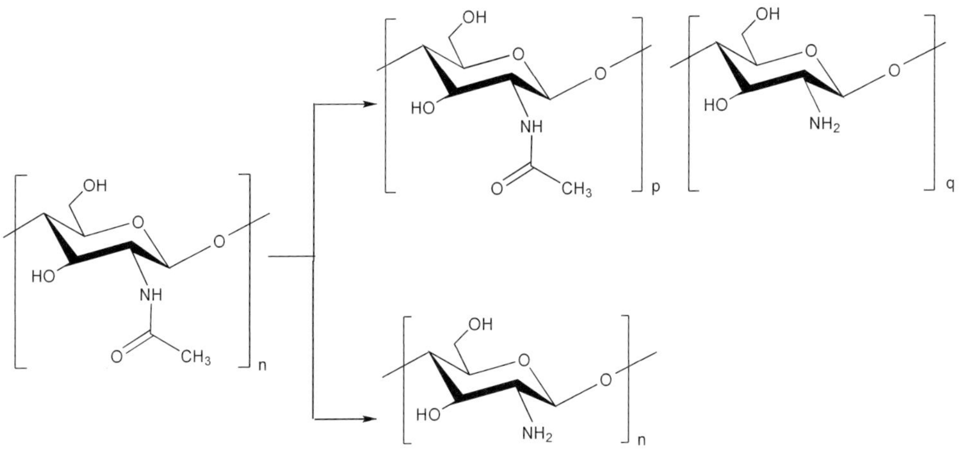

Figure 16.11 Synthesis of chitosan from chitin.

16.5.2
Enzymatic Hydrolysis of Chitosan

The most effective enzymes capable of hydrolyzing chitosan are chitinase and chitosananse, which can be found in fungus, bacteria and plants [164].

Besides the specific chitinase, chitosanase and lysozyme, chitosan could also be hydrolyzed by some nonspecific enzymes such as cellulase, protease, lipase and pepsin. The cellulase also has high activity on chitosan. Almost all the cellulases produced by different kinds of micro-organisms could degrade chitosan to chito-oligomers. A bifunctional cellulase–chitosanase with cellulase and chitosanase activity has also been reported [165].

Chitosan-degrading enzymes have been isolated from diverse bacteria, including *Streptomyces griseus*, *Bacillus circulans*, and members of the actinomycetes [166]; and from vegetable sources. Latex sap from *Carica papaya* contains lysozyme [167], as well as chitinase enzymes [168]. Lysozyme from hens' eggs has been investigated and has been shown to be most efficacious when the chitosan is only partially deacetylated [169]. The enzymatic activity is dependent on the pH and the DA. The low-molecular-weight chitosan (LMWC) with an average M_w in the range of 5000–20000 Da obtained after enzymatic hydrolysis was shown to possess superior biological activities [170].

Since the colonic bacterial enzymes are capable of hydrolyzing various di- and oligosaccharides (glycosidases) and polysaccharides (polysaccharidases), the chitosanolytic activities of rat cecal and colonic enzymes on chitosan was investigated. The results revealed that the rat bacterial enzymes have the ability to degrade chitosan, depending on both the M_w and DA of the chitosan sample [171].

Chitosan is a biocompatible and biodegradable natural biopolymer, with applications in food processing, agriculture, biomedicine, and biotechnology. Chitosan is already used as functional food in some Asian countries, but in the field of medical applications, it has not yet been approved by the FDA [158].

Its uses are limited by its insolubility in water because of its high molecular weight. Chemical or enzymatic degradation of chitosan is able to produce low molecular weight and water-soluble oligochitosan [164]. Chitosan oligosaccharides are widely used as bioactive compounds in the fields of food, health and agriculture [172]. Beside the biological properties, chitin and chitosan have other interesting properties, such as nontoxicity, film- and fiber-forming properties, adsorption of metal ions, coagulation of suspensions or solutes. Chitin and chitosan are thus biofunctional polymers having much higher potential than cellulose in many fields and important not only as abundant resources but also as a novel type of environmentally benign materials [173]. The amount of crustacean shells produced worldwide was estimated in the range of 150×10^3 metric tons per annum [174]. It has been estimated that there are more than 200 potential applications of chitin, chitosan and their derivatives [175].

16.6
Cellulose Enzymatic Degradation

Cellulose is the most abundantly produced biopolymer, found in the protective cell walls of plants, but also in bacteria, fungi, algae and even in animals [176]. It consists of β-D-glucopyranose units, linked by β-(1→4)-glycosidic bonds (Figure 16.12) [177]. The disaccharide cellobiose is regarded as the smallest repetitive unit of cellulose and can be converted into glucose residues. The size of cellulose molecules (degree of polymerization) varies from 7000 to 14 000 glucose moieties per molecule in secondary walls of plants but may be as low as 500 glucose units per molecule in primary walls [178, 179].

An important feature of cellulose is its crystalline structure [180], which shows highly ordered crystalline domains interspersed by amorphous regions [177]. The high-crystallinity of cellulose fibrils renders the internal surface of the biopolymer inaccessible to hydrolyzing enzymes, as well as water.

The degradation of cellulose requires a complex of enzymes, consisting of at least three classes of enzymes, working together [181]. The components of cellulase systems were first classified based on their mode of catalytic action and have more recently been classified based on structural properties [182] (see also

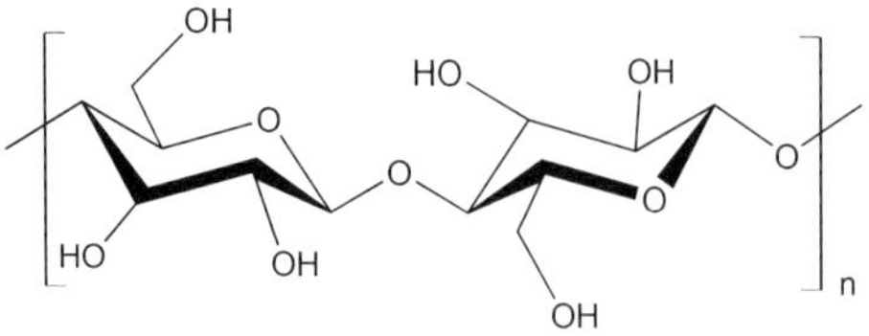

Figure 16.12 Structure of cellulose.

Chapter 9). The cellulose-hydrolyzing enzymes (i.e., cellulases) are divided into three major groups: endoglucanases, cellobiohydrolases (exoglucanases), and β-glucosidases. The endoglucanases catalyze random cleavage of internal bonds of the cellulose chain. They act presumably mainly on the amorphous or disordered regions of cellulose, generating oligosaccharides of various lengths and consequently new chain ends. Cellobiohydrolases attack the chain ends, releasing either glucose (glucanohydrolases) or cellobiose (cellobiohydrolase) as major products [183]. β-Glucosidases are only active on cello-oligosaccharides and cellobiose, and release glucose monomers units from the cellobiose [184].

Cooperative action, often designated *synergy*, of the three cellulolytic enzyme classes is essential for an efficient enzymatic hydrolysis process [181].

The hydrolysis rate of the microcrystalline cellulose can be increased by treatment with ionic liquids (ILs) (see also Chapter 13). The essentially amorphous or a mixture of amorphous and partially crystalline cellulose were recovered with an antisolvent. With appropriate selection of IL treatment conditions and enzymes, the initial hydrolysis rates for IL-treated cellulose were up to 90 times greater than those of untreated cellulose [185].

16.7
Conclusion

The interest in enzymatic degradation of polysaccharides is strongly associated with their biomedical applications. The process is dependent on the polysaccharide structures and conformation. Enzymatic degradation of these biopolymers has received attention for many years and is becoming an attractive alternative to chemical and mechanical processes.

Since the degree of polymerization of the polysaccharides is very important for their biological activity, strategies based on their enzymatic degradation can find application in developing new polysaccharide-based biomaterials.

References

1 Ranade, V.V., and Hollniger, M.H. (2003) *Drug Delivery Systems*, 2nd edn, Taylor & Francis, Inc..
2 Hamid, S.H. (2000) *Handbook of Polymer Degradation*, 2nd edn, Marcel Dekker, New York.
3 French, D. (1957) *Adv. Carbohydr. Chem.*, **12**, 189–260.
4 French, D., Pulley, A.O., Effenberger, J.A., Rougvie, M.A., and Abdullah, M. (1965) *Arch. Biochem. Biophys.*, **111**, 153–160.
5 Entzeroth, M. (1986) *J. Org. Chem.*, **51**, 5307.
6 Szejtli, S. (2004) *Pure Appl. Chem.*, 1825–1845.
7 Villiers, A. (1891) *Compt. Rend. Acad. Sci. Paris*, **112**, 435–437.
8 Schradinger, A. (1903) *Z. Untersuch. Nahrungsm. Genussm.*, **6**, 865–880.
9 Freudenberg, K., Blomqvist, G., and Ewald, L. (1936) *Chem. Ber.*, **69**, 1258–1266.
10 Freudenberg, K., Engler, K., Flickinger, E., Sobeck, A., and Klink, F. (1938) *Ber. Deutsch. Chem. Ges.*, **71**, 1821.

11 Freudenberg, K., Schaaf, E., Dumpert, G., and Ploetz, T. (1939) *Naturwissenschaften*, **27**, 850–853.
12 Freudenberg, K., Cramer, F., and Plieninger, H. (1953) Ger. Patent 895,769.
13 Jansook, P., Kurkov, S.V., and Loftsson, T. (2009) *J. Pharm. Sci.*, **9999**, 1–11.
14 Wenz, G. (1994) *Angew. Chem. Int. Ed.*, **33**, 803–822.
15 Wenz, G., Han, B.-H., and Müller, A. (2006) *Chem. Rev.*, **106**, 782–817.
16 Ritter, H., and Tabatabai, M. (2002) *Prog. Polym. Sci.*, **27**, 1713.
17 Harada, A., Furue, M., and Nozakura, S. (1976) *Macromolecules*, **9**, 701–704.
18 Kretschmann, O., Steffens, C., and Ritter, H. (2007) *Angew. Chem. Int. Ed.*, **46**, 2708–2711.
19 Koopmans, C., and Ritter, H. (2007) *J. Am. Chem. Soc.*, **129**, 3502–3504.
20 Kretschmann, O., Choi, S.W., Miyauchi, M., Tomatsu, I., Harada, A., and Ritter, H. (2006) *Angew. Chem. Int. Ed.*, **45**, 4361–4365.
21 Köllisch, H., Barner-Kowollik, C., and Ritter, H. (2006) *Macromol. Rapid Commun.*, **27**, 848–853.
22 Schmitz, S., and Ritter, H. (2005) *Angew. Chem. Int. Ed.*, **44**, 5658–5661.
23 Amajjahe, S., Choi, S.W., Munteanu, M., and Ritter, H. (2008) *Angew. Chem. Int. Ed.*, **47**, 3435. Angew. Chem. Int. Ed., 2008, 120, 3484.
24 Munteanu, M., Choi, S.W., and Ritter, H. (2008) *Macromolecules*, **41**, 9619–9623.
25 Amajjahe, S., Munteanu, M., and Ritter, H. (2009) *Macromol. Rapid Commun.*, **30**, 904–908.
26 Munteanu, M., Choi, S.W., and Ritter, H. (2009) *Macromolecules*, **42**, 3887–3891.
27 Slomihska, L., and Sobkowiak, B. (1997) *Starch/Stärke*, 301–305.
28 Janecek, S., and Baláz, S. (1993) *J. Protein Chem.*, **12**, 509–514.
29 Del-Rio, G., Morett, E., and Soberon, X. (1997) *FEBS Lett.*, **416**, 221–224.
30 van der Veen, B., Uitdehaag, J., Dijkstra, B., and Dijkhuizen, L. (2000) *Biochimet. Biophys. Acta*, **1543**, 336–360.
31 Stam, M., Danchin, E., Rancurel, C., Coutinho, P., and Henrissat, B. (2006) *Protein Eng. Des. Sel.*, **19**, 555–562.
32 Endo, T., Zheng, M., and Zimmermann, W. (2002) *Aust. J. Chem.*, **55**, 39–48.
33 Takano, T., Fukuda, M., Monma, M., Kobayashi, S., Kainuma, K., and Yamane, K. (1986) *J. Bacteriol.*, **166**, 1118–1122.
34 Svenson, B. (1994) *Plant Mol. Biol.*, **25**, 141–157.
35 Wind, R.D., Buitelaar, R.M., and Dijkhuizen, L. (1998) *Eur. J. Biochem.*, **253**, 598–605.
36 Costa, H., del Canto, S., Ferrarotti, S., and Biscoglio de Jiménez, M. (2009) *Carbohydr. Res.*, **344**, 74–79.
37 Vihinen, M., and Mäntsälä, P. (1989) *Crit. Rev. Biochem. Mol. Biol.*, **24**, 329–410.
38 Qi, Q., and Zimmermann, W. (2005) *Appl. Microbiol. Biotechnol.*, **66**, 475–485.
39 Tonkova, A. (1998) *Enzyme Microb. Technol.*, **22**, 678–686.
40 Blackwood, A.D., and Bucke, C. (2000) *Enzyme Microb. Technol.*, **27**, 704–708.
41 Penninga, D., van der Veen, B.A., and Knegtel, R.M.A. (1996) *J. Biol. Chem.*, **271**, 32777–33278.
42 Jemli, S., Messaoud, E.B., Ayadi-Zouari, D., Naili, B., Khemakhem, B., and Bejar, S. (2007) *Biochem. Eng. J.*, **34**, 44–50.
43 Lee, Y.D., and Kim, H.S. (1991) *Enzyme Microb. Technol.*, **13**, 499–503.
44 Lee, Y.D., and Kim, H.S. (1992) *Biotechnol. Bioeng.*, **39**, 977–983.
45 Vikmon, M. (1982) *First International Symposium on Cyclodextrins Budapest* (ed. J. Szejtli), D. Reidel Publishing, Dordrecht, The Netherlands, pp. 69–74.
46 Frömming, K.H., and Szejtli, J. (1994) *Cyclodextrins in Pharmacy*, Kluwer Academic, Dordrecht, pp. 13–15.
47 Sin, K.A., Nakamura, A., Masaki, H., Matsuura, Y., and Uozumi, T. (1994) *J. Biotechnol.*, **32**, 283–2884.
48 Vieille, C., and Zeikus, G.J. (2001) *Microbiol. Mol. Biol. Rev.*, **65**, 1–43.
49 Starnes, R.L., (1990) *Cereal Foods World*, **35**, 1094–1099.

50 Wind, R.D., Liebl, W., Buitelaar, R.M., Penninga, D., Spreinat, A., Dijkhuizen, L., and Bahl, H. (1995) *Appl. Environ. Microbiol.*, **61**, 1257–1265.

51 Zamost, B.L., Nielsen, H.K., Starnes, R.L. (1991) *J. Ind. Microbiol. Biol.*, **8**, 71–82.

52 Pedersen, S., Dijkhuizen, L., Dijkstra, B.W., Jensen, B.F., and Jorgensen, S.T. (1995) *Chemtech*, **25**, 19–25.

53 Kometani, T., Terada, Y., Nishimura, T., Takii, H., and Okada, S. (1994) *Biosci. Biotechnol. Biochem.*, **58**, 1990–1994.

54 Kometani, T., Nishimura, T., Nakae, T., Takii, H., and Okada, S. (1996) *Biosci. Biotechnol. Biochem.*, **60**, 645–649.

55 Uchida, R., Nasu, A., Tobe, K., Oguma, T., and Yamaji, N. (1996) *Carbohydr. Res.*, **287**, 271–274.

56 Bruinenberg, P.M., Hulst, A.C., Faber, A., and Voogd, R.H. (1996) Eur. Patent P1995000201751.

57 van Eijk, J.H., and Mutsaers, J.H.G.M. (1995) Eur. Patent 1995000201378.

58 Loftsson, T., and Duchene, D. (2007) *Int. J. Pharm.*, **329**, 1–11.

59 French, D., Knapp, D.W., and Pazur, H. (1950) *J. Am. Chem. Soc.*, **72**, 5120.

60 Freudenberg, K. (1975) *J. Polym. Sci.*, **23**, 791.

61 Vaitkus, R., Grinciene, G., and Norkus, E. (2008) *Chemija*, **19**, 48–51.

62 Swanson, M., and Cori, C. (1948) *J. Biol. Chem.*, **172**, 2.

63 Myrback, K. (1948) *Arkiv. Chem.*, **1**, 161.

64 Szejtli, J., and Budai, Z. (1978) *Acta Chem. Acad. Sci. Hung.*, **91**, 73.

65 Jodal, I., Kandra, L., Harangi, J., Nanasi, P., and Szejtli, J. (1984) *J. I. Phenom.*, **2**, 877–884.

66 Jodal, I., Kandra, L., Harangi, J., Nanasi, P., and Szejtli, J. (1984) *Starch/Starke*, **36**, 140–143.

67 DePinto, J.A., and Campbell, L.L. (1968) *Biochemistry*, **7**, 121–125.

68 Fritzsche, H.B., Schwede, T., and Schulz, G.E. (2003) *Eur. J. Biochem.*, **270**, 2332–2341.

69 Hondoh, H., Kuriki, T., and Matsuura, Y. (2003) *J. Mol. Biol.*, **326**, 177–188.

70 Buedenbender, S., and Schulz, G. (2009) *J. Mol. Biol.*, **385**, 606–617.

71 Park, K.-H., Kim, T.-P., Cheong, T.-K., Kim, J.-W., Oh, B.-H., and Svensson, B. (2000) *Biochimet Biophys. Acta*, **1478**, 165–185.

72 Murayama, T., Tanabe, T., Ikeda, H., and Ueno, A. (2006) *Bioorg. Med. Chem.*, **14**, 3691–3696.

73 Muralikrishna, G., and Nirmalab, M. (2005) *Carbohydr. Polym.*, **60**, 163–173.

74 Lis, A., and Passarge, W. (1966) *Arch. Biochem. Biophys.*, **114**, 593.

75 Antenucci, R.N., and Palmer, J.K. (1984) *J. Agric. Food Chem.*, **32**, 1316–1321.

76 Flourié, B., Molis, C., Achour, L., Dupas, H., Hatat, C., and Rambaud, J.C. (1993) *J. Nutr.*, **123**, 676–680.

77 Minami, K., Hirayama, F., and Uekama, K. (1998) *J. Pharm. Sci.*, **87**, 715–720.

78 Siefke, V., Weckenmann, H.P., and Bauer, K.H. (1993) *Proc. Int. Symp. Control Rel. Bioact. Mater.*, **20**, 182–183.

79 Vervoort, L., and Kinget, R. (1996) *Int. J. Pharm.*, **129**, 185–190.

80 Hirayama, F., Ogata, T., Yano, H., Arima, H., Udo, K., Takano, M., and Uekama, K. (2000) *J. Pharm. Sci.*, **89**, 1486–1495.

81 Pattekan, H.H., and Divagar, S. (2002) *Indian J. Chem. B*, **41**, 1025.

82 Pedersen, N.R., Kristensen, J.B., Bauw, G., Ravoo, B.J., Darcy, R., Larsen, K.L., and Pedersen, L.H. (2005) *Tetrahedron Asymmetry*, **16**, 615–622.

83 Ghanem, A., and Schurig, V. (2001) *Tetrahedron Asymmetry*, **12**, 2761–2766.

84 Villalonga, R., Cao, R., and Fragoso, A. (2007) *Chem. Rev.*, **107**, 3088–3116.

85 Breslow, R. (1980) *Acc. Chem. Res.*, **13**, 170–117.

86 Breslow, R., and Graft, A. (1993) *J. Am. Chem. Soc.*, **115**, 10988.

87 Yuan, D.Q., Dong, S., and Breslow, R. (1998) *Tetrahedron Lett.*, **39**, 7673–7676.

88 Tabushi, I. (1982) *Acc. Chem. Res.*, **15**, 66–72.

89 Van Etten, R.L., Sebastian, J.F., Clowes, G.A., and Bender, M.L. (1967) *J. Am. Chem. Soc.*, **9**, 3242.

90 Carmer, F. (1961) *Angew. Chem. Int. Ed. Engl.*, **73**, 49.
91 Loftsson, T., and Brewster, M.E. (1996) *J. Pharm. Sci.*, **85**, 1017–1025.
92 Meyer, K., and Palmer, J.W. (1934) *J. Biol. Chem.*, **107**, 629–634.
93 Abatangelo, G., and Weigel, P.H. (eds) (2000) *New Frontiers in Medical Sciences: Redefining Hyaluronan*, Elsevier, Amsterdam.
94 He, D., Zhou, A., Wie, W., Nie, L., and Yao, S. (2001) *Talanta*, **53**, 1021–1029.
95 Tømmeraas, K., and Melander, C. (2008) *Biomacromolecules*, **9**, 1535–1540.
96 Laurent, T.C., and Fraser, J.R.E. (1992) *FASEB J.*, **6**, 2397.
97 Stern, R. (2004) *Eur. J. Cell Biol.*, **83**, 317–325.
98 Philipson, L.H., and Schwartz, N.B. (1984) *J. Biol. Chem.*, **259**, 5017–5023.
99 Weigel, P. (2004) The Hyaluronan Synthases, in *Chemistry and Biology of Hyaluronan* (eds H.G. Garg and C.A. Hales), Elsevier, Amsterdam, pp. 553–567.
100 Maleki, A., Kjøniksen, A.-L., and Nysrtöm, B. (2007) *Carbohydr. Res.*, **342**, 2776–2792.
101 Lapcik, L., De Smedt, S., and Demeester, J. (1998) *Chem. Rev.*, **98**, 2663–2684.
102 Armstrong, D.C., and Johns, M.R. (1997) *Appl. Environ. Microbiol.*, **63**, 2759–2764.
103 Lee, H.G., and Cowman, M.K. (1994) *Anal. Biochem.*, **219**, 278–287.
104 Sutherland, I.W. (1990) *Biotechnology of Exopolysaccharides*, Cambridge University Press, Cambridge.
105 Widner, B., Behr, R., Von Dollen, S., Tang, M., Heu, T., Sloma, A., Sternberg, D., DeAngelis, P.L., Weigel, P.H., and Brown, S. (2005) *Appl. Environ. Microbiol.*, **71**, 3747–3752.
106 Balazs, E.A. (2004) Viscoelastic Properties of Hyaluronan and its Therapeutic Use, in *Chemistry and Biology of Hyaluronan* (eds H.G. Garg and C.A. Hales), Elsevier, Amsterdam, pp. 415–443.
107 Kogan, G., Soltés, L., Stern, R., and Gemeiner, P. (2007) *Biotechnol. Lett.*, **29**, 17–25.
108 Yui, N., Okano, T., and Sakurai, Y.J. (1992) *Control. Release*, **22**, 105–116.
109 Pouyani, T., and Prestwich, G.D. (1994) *Bioconjug. Chem.*, **5**, 339–347.
110 Vercruysse, K.P., and Prestwich, G.D. (1998) *Crit. Rev. Ther. Drug Carrier Syst.*, **15**, 513–555.
111 Etienne, O., Schneider, A., Taddei, C., Richert, L., Schaaf, P., Voegel, J.-C., Egles, C., and Picart, C. (2005) *Biomacromolecules*, **6**, 726–733.
112 West, D.C., and Kumar, S. (1989) *Ciba Found. Symp.*, **143**, 187–207.
113 Stern, R., and Asari, A.A. (2006) *Eur. J. Cell. Biol.*, **85**, 699–715.
114 Chen, W.J.Y., and Abatangelo, G. (1999) *Wound Repair Regen.*, **7**, 79–89.
115 Milner, C.M., Higman, V.A., and Day, A.J. (2006) *Biochem. Soc. Trans.*, **34**, 446–450.
116 Stern, R. (2008) *Semin. Cancer Biol.*, **18**, 275–280.
117 Chen, S.S., Hsu, D.S., and Hoffman, P. (1979) *Clin. Chim. Acta*, **95**, 277.
118 Belsky, E., and Toole, B.P. (1983) *Cell Differ.*, **12**, 61.
119 Kulyk, W.M., and Kosher, R.A. (1987) *Dev. Biol.*, **120**, 535.
120 Meyer, K. (1971) Hyaluronidases, in *The Enzymes*, vol. 5 (ed. P.D. Boyer), Academic Press, New York, pp. 307–320.
121 Karlstam, B., Ljungloef, A. (1991) *Polar Biol.*, **11**, 501–507.
122 Girish, K.S., and Kemparaju, K. (2007) *Life Sci.*, **80**, 1921–1943.
123 Manzel, E.J., and Farr, C. (1988) *Cancer Lett.*, **131**, 3–11.
124 Frost, G.I., Csoka, T., and Stern, R. (1996) *Trends Glycosci. Glycotechnol.*, **8**, 419–434.
125 Borrelli, F., Antonetti, F., Martelli, F., and Caprino, L. (1986) *Thromb. Res.*, **42**, 153–164.
126 Northup, S.N., Stasiw, R.O., and Brown, H.D. (1973) *Clin. Biochem.*, **6**, 220.
127 Stern, M., Longaker, M.T., Adzick, N.S., Harrison, M.R., and Stern, R. (1991) *J. Natl. Cancer Inst.*, **83**, 1569–1574.
128 Eccarius, S.G., and Gordon, E.G. (1990) *Ophthalmology*, **97**, 1499–1501.

129 Mindel, J.S. (1978) *Am. J. Ophthalmol.*, **85**, 643–646.
130 Watson, D. (1993) *Br. J. Anaesth.*, **71**, 422–425.
131 Pillwein, K., Fuiko, R., Slavc, I., Czech, T., Hawliczek, G., Bernhardt, G., Nirnberger, G., and Koller, U. (1998) *Cancer Lett.*, **131**, 101–108.
132 Wolf, R., Glogar, D., Chaung, L.Y., Garrett, P.E., Ertl, G., Tumas, J., Braunwald, E., Kloner, R.A., Feldstein, M.L., and Muller, J.E. (1984) *Am. J. Cardiol.*, **53**, 941–944.
133 Girish, K.S., and Kemparaju, K. (2005) *Biochemistry (Mosc.)*, **70**, 708–712.
134 Suzuki, K., Terasaki, Y., and Uyeda, M. (2002) *J. Enz. Inh. Med. Chem.*, **17**, 183–186.
135 Mio, K., and Stern, R. (2002) *Matrix Biol.*, **21**, 31–37.
136 Botzki, A., Rigden, D.J., Braun, S., Nukui, M., Salmen, S., Hoechstetter, J., Bernhardt, G., Dove, S., Jedrzejas, M.J., and Buschauer, A. (2004) *J. Biol. Chem.*, **279**, 45990–45997.
137 Wong, T.Y., Preston, L., and Schiller, N. (2000) *Annu. Rev. Microbiol.*, **54**, 289–340.
138 Gimmestad, M., Ertesvåg, H., Bjerkan Heggeset, T.M., Aarstad, O., Glærum Svanem, B.I., and Valla, S. (2009) *J. Bacteriol.*, **191**, 4845–4853.
139 Ertesvåg, H., Valla, S., and Skjåk-Bræk, G. (1996) *Carbohydr. Eur.*, **14**, 14–18.
140 Stanford, E.C. (1883) *Chem. News*, **47**, 254–257.
141 Gomez d'Ayala, G., Malinconico, M., and Laurienzo, P. (2008) *Molecules*, **13**, 2069–2106.
142 Rowley, J., Madlambayan, G., and Mooney, D. (1999) *Biomaterials*, **20**, 45–53.
143 George, M., and Abraham, T.E. (2006) *J. Control. Release*, **114**, 1–14.
144 Smidsrød, O., and Draget, K.I. (1996) *Carbohydr. Eur.*, **14**, 6–13.
145 Skjåk-Bræk, G., Grasdalen, H., and Larsen, B. (1986) *Carbohydr. Res.*, **154**, 239–250.
146 Goycoolea, F., Lollo, G., Remuñán-Loópez, C., Quaglia, F., and Alonso, M. (2009) *Biomacromolecules*, **10**, 1736–1743.
147 Shoichet, M.S., Li, R.H., White, M.L., and Winn, S.R. (1996) *Biotechnol. Bioeng.*, **50**, 374–381.
148 Sun, A.M.F., Goosen, M.F.A., and Oshea, G. (1987) *Crit. Rev. Ther. Drug Carrier Syst.*, **4**, 1–12.
149 Smith, A.M., Harris, J.J., Shelton, R.M., and Perrie, Y. (2007) *J. Control. Release*, **119**, 94–101.
150 West, E.R., Xu, M., Woodruff, T.K., and Shea, L.D. (2007) *Biomaterials*, **28**, 4439–4448.
151 Alsberg, E., Anderson, K.W., Albeiruti, A., Franceschi, R.T., and Mooney, D.J. (2001) *J. Dent. Res.*, **80**, 2025–2029.
152 Atala, A., Kim, W., Paige, K.T., Vacanti, C.A., and Retik, A.B. (1994) *J. Urol.*, **152**, 641–643.
153 Draget, K.I., Ostgaard, K., and Smidsrød, O. (1990) *Carbohydr. Polym.*, **14**, 159–178.
154 Yao, B., Ni, C., Xiong, C., Zhu, C., and Huang, B. (2009) *Bioprocess Biosyst. Eng.*.
155 Augst, A.D., Kong, H.J., and Mooney, D. (2006) *J. Macromol. Biosci.*, **6**, 623–633.
156 Barbetta, A., Barigelli, E., and Dentini, M. (2009) *Biomacromolecules*, **10**, 2328–2337.
157 Burana-Osot, J., Hosoyama, S., Nagamoto, Y., Suzuki, S., Linhardt, R., and Toida, T. (2009) *Carbohydr. Res.*. doi: 10.1016/j.carres.2009.06.027
158 Aranaz, I., Mengíbar, M., Harris, R., Paños, I., Miralles, B., Acosta, N., Galed, G., and Heras, A. (2009) *Curr. Chem. Biol.*, **3**, 203–230.
159 Ramírez-Coutiño, L., Marín-Cervantes, M., Huerta, S., Revah, S., and Shirai, K. (2006) *Process Biochem.*, **41**, 1106–1110.
160 Li, Q.D., and Dunn, E.T. (1992) *J. Bioact. Compat. Polym.*, **7**, 370–397.
161 Baxter, A., Dillon, M., Anthony Taylor, K.D., and Roberts, G.A.F. (1992) *Int. J. Biol. Macromol.*, **14**, 166–169.
162 Tharanathan, R.N., and Kittur, F.S. (2003) *Crit. Rev. Food Sci. Nutr.*, **43**, 61–87.
163 Brzezinski, B. (1996) US Patent US005482843A.
164 Ilyina, A.V., Tikhonov, V.E., Abulov, A.I., and Varlamov, V.P. (2000) *Proc. Biochem.*, **35**, 563–568.

165 Xia, W., Liu, P., and Li, J. (2008) *Bioresour. Technol.*, **99**, 6751–6762.

166 Ohtakara, A., Matsunaga, H., and Mitsutomi, M. (1990) *Agric. Biol. Chem.*, **54**, 3191–3199.

167 Kendra, D.F., and Hadwiger, L.A. (1984) *Exp. Mycol.*, **8**, 276–281.

168 Azarkan, M., Amrani, A., Nijs, M., Vandermeers, A., Zerhouni, S., Smoulders, N., and Looze, Y. (1997) *Phytochemistry*, **46**, 1319–1325.

169 Nordtveit, R.J., Vårum, K.M., and Smidsrød, O. (1996) *Carbohydr. Polym.*, **29**, 163–167.

170 Abdel-Aziz, S., and Moafi, F. (2008) *J. Appl. Sci. Res.*, **4**, 1755–1761.

171 Zhang, H., and Neau, S. (2002) *Biomaterials*, **23**, 2761–2766.

172 Roncal, T., Oviedo, A., López de Armentia, I., Fernández, L., and Villarán, M.-C. (2007) *Carbohydr. Res.*, **342**, 2750–2756.

173 Kurita, K. (2006) *Mar. Biotechnol.*, **8**, 203–226.

174 Pelletier, A., and Sygusch, J. (1990) *Appl. Environ. Microbiol.*, **56**, 844–848.

175 Kumar, M.N.V.R. (2000) *React. Funct. Polym.*, **46**, 1–27.

176 O'Sullivan, A. (1997) *Cellulose*, **4**, 173–207.

177 Leschine, S. (1995) *Annu. Rev. Microbiol.*, **49**, 399–426.

178 Ljungdahl, L.G., and Eriksson, K.E. (1985) *Adv. Microb. Ecol.*, **8**, 237–299.

179 Robson, L.M., and Chambliss, G.H. (1989) *Enzyme Microb. Technol.*, **11**, 626–642.

180 Lynd, L., Weimer, P., van Zyl, W., and Pretorius, I. (2002) *Microbiol. Mol. Biol. Rev.*, **66**, 506–577.

181 Andersen, N., Johansen, K., Michelsen, M., Stenby, E., Krogh, K., and Olsson, L. (2008) *Enzyme Microb. Technol.*, **42**, 362–370.

182 Henrissat, B., Teeri, T.T., and Warren, R.A.J. (1998) *FEBS Lett.*, **425**, 352–354.

183 Teeri, T.T. (1997) *Trends Biotechnol.*, **15**, 160–167.

184 Zhang, Y.-H.P., Himmel, M.E., and Mielenz, J.R. (2006) *Biotechnol. Adv.*, **24**, 452–481.

185 Dadi, P., Schall, C., and Varanasi, S. (2007) *Appl. Biochem. Biotechnol.*, **137**, 407–422.

Index

Page numbers in *italics* refer to figures or tables.

a

AA-BB monomers
- dynamic kinetic resolution 290
- polyesterification 92–97, *98*, *99*

ab initio process 349, 352

AB monomers
- dynamic kinetic resolution 290, 292, 293
- polyesterification 86, 87, *88–90*, 91, 92

Accurel 74, 79, 80, 327
acetone powder 178
acetylacetone 148, *152*, 153, *156*, *157*
acid active hyaluronidases 409
acidolysis 86
acrylamides *151*, *152*, *158–160*
acrylonitrile 267
acyl-enzyme intermediate
- CALB catalysis 278, *279*
- molecular modeling 359, 360, *361*, *362*, *363*, 364
- ring-opening of lactones 102, *103*, 104

acylation
- enantioselective 298, 299, *300*
- lipases 278, *279*

adenosine diphosphate (ADP) 262
adenosine triphosphate (ATP) 262, 263, *262*, 265

adenylation 247, 248
adipic acid 92, *283*
- esterification 357, *358*
adsorption, clays 37, 38, 42
Agrobacterium sp. 237
AK lipase 119
b-alanine 359, 364, 365
albumin 42, *51*
alcohol dehydrogenase 39, 47
- enantioselectivity 298
- nanoparticles 54

alcohol oxidase 145, 159, 160
alcohols, dynamic kinetic resolution 288, 289
alcoholysis 86
- esters 354–357
alditols 96, 284
- anhydroalditols 6, *7*, *8*
aldonic acid lactones 220, 221
aldopentoses *11*
alginate 409–411
alginate lyases 411
alkalophilic *Bacillus* 400
aluminosilicates 36, 37
Amberzyme oxiranes 87, 93
amidases 376, 377
amides, *see also* acrylamides; polyamides
- secondary 288, 289
amines
- diamines 134, *135*
- secondary 288, 289, 293
aminolysis, esters 354–357
a-amylase 176, *330*, 393, 394, 401, *402*, 403
b-amylase 401, 403
amyloglucosidase 51
amylomaltase 228, 229
amylopectin 214, *215*, 227, 390
amylose 214, *215*, 390
- amylosucrase 227
- comb-like copolymers with 222, 223
- copolymers with 221, 222
- hybrids with short alkyl chains 220
- linear block copolymers with 223, 224
- polymerization with glycogen phosphorylase 215–219
amylose brushes 220, 221
amylosucrase 227
Anabaena cylindrica 260

Biocatalysis in Polymer Chemistry. Edited by Katja Loos
Copyright © 2011 WILEY-VCH Verlag GmbH & Co. KGaA, Weinheim
ISBN: 978-3-527-32618-1

anhydroalditols 6, 7, 8
aniline, *see* polyaniline
antimicrobial functionalization 370
antioxidants 181
arginine 258–260, *262*, 265–267
Arthrobacter sp. 379
aspartate *262*, 263, 265
Aspergillus 68
Aspergillus niger 39, 88, *107*, 330, 373
Aspergillus nomius 373
Aspergillus oryzae 373, 380, *401*, *402*, 403
atom transfer radical polymerization (ATRP) 296, *297*
– bifunctional initiation 330, 331, *332*
– block copolymer synthesis 310–314, 317, *318*
ATP, *see* adenosine triphosphate (ATP)
2,2′-azino-bis(3-ethylbenzothiazoline-6-sulfonate) diammonium salt 204, *205*
aziridine-2-carboxylates 355

b

b-lactam polymerization 357, 359–367
Bacillus coagulans 401
Bacillus licheniformis 237, 249, *330*
Bacillus megaterium 254
Bacillus sphaericus 400
Bacillus stearothermophilus 400
Bacillus subtilis 376, 407
Bäckvall system 293
bacterial cutinases 373
bacterial esterases 373, 380
bacterial hyaluronidases 408
bacterial hydrogenase 39
bacterial laccases 373
bacterial lipases 373
bacterial storage
– cyanophycins 257–268
– PHAs 249–257
Bacteroides ovatus 400
Beauveria brongniartii 377
Betula pendula 26
bifunctional initiators 329–332
bilirubin oxidase 200
biodegradable polymers 389
– PHAs 256, 257
– polyesters 83, 117
biodegradation
– alginate 409–411
– cellulose 414, 415
– chitin and chitosan 411–414
– cyclodextrin 390–406
– features of 389, 390
– hyaluronic acid 406–409
– PANI fibers 378, *379*

biofuel cells 46, 47
biomass
– degradation 328, 329, *330*
– oxpropylation 25
biomimetic synthesis, PANI 194, 195
biorefinery 1
biosensors
– CNTs 45, 46, *47*, 48
– nanoparticles 54, 55
– nanowires 201, *202*
biphasic polymerization 339–342
– fluoruos 342
– ionic liquid-supported catalyst 340, *341*
– polyphenols 342
bisphenols 168, 173, 174
block copolymers
– amylose, linear 223, 224
– from bifunctional initiators 329–331
– chiral, from enzymatic catalysis 296–298
– enzymatic synthesis 305, 306–318
– – macroinitiation 306, 307–309
– – macroinitiators followed by chemical polymerization 310–318
boronic acid-containing PANIs 188
bovine serum albumin 42
BPEC pathway 255, 256
branching enzymes 224–227
Burkholderia cepacia lipase 282, 373
1,4-butanediamine 267
1,4-butanediol 92, 93
1-butyl-3-methylimidazolium bistriflamide 335, 337, 338, 339, 340
1-butyl-3-methylimidazolium dicyanamide 335, 338
1-butyl-3-methylimidazolium hexafluorophosphate 335, 337, 338, 340
1-butyl-3-methylimidazolium tetrachloroferrate 335, 338
1-butyl-3-methylimidazolium tetrafluoroborate 335, 336, 337, 338
1-butyl-1-methylpyrrolidinium dicyanamide 335, 338
1-butyl-1-methylpyrrolidinium tetrafluoroborate 335, 336
b-butyrolactone *339*
g-butyrolactone *339*

c

cadmium sulfide 175
CALB, *see Candida antarctica* lipase B
calcium alginate hydrogel beads *219*
Caldariomyces fumago 147, 198
Cambridge structural database 349
Candida antarctica lipase 373, 374

Candida antarctica lipase A   119
Candida antarctica lipase B   65–68
– b-lactam polymerization   357, 359–367
– binding sites   *351*
– catalyzed polymer synthesis   71–80
– on clay   40, *41*, 42
– in exotic solvents   324
– in ionic liquids   338, 339
– molecular modeling   353–356, *357*
– Novozym 435   69–71
– polyester synthesis   88, 89, *91*, 357, 358
– polyesterification   93–97, 99, 100
– reaction mechanism   278, *279*
– ring-opening copolymerization   115, 119
– ring-opening polymerization   104, 105
Candida cylindracea lipase   87, 89, 105
Candida rugosa lipase   40, 47, 90, 105
– nanoparticles   54
– optically pure monomers from   280
e-caprolactone
– in bifunctional initiation   330, 331
– chiral block copolymers from   296, *297*, 298
– dual initiator   311, *312*
– dynamic kinetic resolution   289, 290
– in ionic liquids   339
– iterative tandem catalysis   294, *295*
– kinetic resolution   286, 287
– macroinitiation   307, 308, *309*
– polymerization   72
– ring-opening copolymerization   113, *114*, 296, *297*
– ring-opening polymerization   104, *107*, 108, *108*, 327
– synthesis time   74, 76
carbamate formation   325
Carbohydrate-Active enZYmes Database   214
carbohydrates, *see* polysaccharides; starch; sugars
carbon dioxide, *see* supercritical carbon dioxide
carbon layered materials   43, *43*, 44
carbon nanotube–nanoparticle conjugates   55
carbon nanotubes
– applications   45, 46, 47, 48
– immobilization approaches   46, 48–50
– immobilized enzymes, structure and catalytic behaviour   50, *51*, 52
– introduction   44, 45
carboxylated self-doped PANI/PDADMAC complex   *191*
cardanol   178
cashew nut shell liquid   178

cast films   203
catalytic triad   278, 353
catalytical chemical vapor deposition   44
catechins   179, *180*, 181
catechols   177
cell labeling/separation   53
cellobiohydrolases   415
b-D-cellobiosyl fluoride   232, *233*
cellulase   232–234, *330*, 415
cellulose
– biodegradation   414, 415
– degradation   *330*
– industrial uses   212
– synthesis   233, 234
Cerrena unicolor laccase   43, 47
chain elongation   366
chiral affinity chromatography   221
chiral polymers
– conclusions and outlook   301
– introduction   277, 278
– from optically pure monomers   280–284
– from racemic monomers   284, 295
– tuning polymer properties with chirality   296–301
chitin   411–414
a-chitin   411
b-chitin   411
g-chitin   411
chitinase   413
chitosan   193, 411–414
chitosanase   413
cholesterol biosensors   46
cholesterol oxidase   54
cholic acid   91
Chromobacterium viscosum lipase   280
a-chymotrypsin   39, 47, 51
– polypeptide synthesis   133
circular dichroism   42, *51*
citric acid   28
clays   36–38, *39*, 40, 40–42
CNTs, *see* carbon nanotubes
coatings   177–179
coenzyme A   251, 252, 255
colloids, PANI   192, *193*
colophony   4–6
comb-like copolymers   222, 223
condensed tannins (polyflavonoids)   21, 22
coniferyl alcohol   174, 177
conventional docking   351
copolymerizations, ring-opening   113–119, *119*, 120
– ε-caprolactone   113, *114*, 296, *297*
copolymers, amylose   221–224
Coprinus cinereus   173, 196
Coriolus hirsutus   197

cork   26, 27
corn starch   *330*
cotton   *212*
cotton boll fibers   *330*
covalent docking   351, 360, 362
crosslinking   68–71, 80
– polypeptides   175, 176
crystallinity   390
cutin   371, 372
cutinases   371, 372, *373, 374, 375, 381*
cyanobacteria   258, 267
cyanophycin   136, 137, 249
– biotechnological relevance   267, 268
– chemical structure   258, 259
– variants   259
cyanophycin synthetase   249
– catalytic cycle   263, *264*
– embedding in general metabolism   267
– enzyme activity assay   260, 261
– enzyme granule structure   261
– occurrence   257, 258
– primary structures and essential motifs   262, 263
– reaction catalyzed by   260
– wild type enzyme, kinetic data   261, 265
cyanophycin synthetase (CphA)
– catalytic cycle   265
– mutant variants   265, 266
– *in vitro* synthesis   266
cyanophycinase   267
cyclic esters, ring-opening polymerizations   111–113
α-cyclodextrin   *391*, 392, 395, *396*, *402*, 403
β-cyclodextrin   *391*, 392, 395–397, 399, *402*, 403
γ-cyclodextrin   *391*, 392, 395, *396*, *402*, 403
cyclodextrin glycosyltransferase   392, *393*
– cyclodextrin-forming activity   394–397
– other industrial applications   397, 398
– structure and catalytic activity   393, 394
cyclodextrins
– hydrolysis   398–405
– – acidic   399
– – degradation by intestinal flora   404, 405
– – enzymatic degradation   400–404
– – enzyme mimics   405
– – synthesis of derivatives   405
– structure and physicochemical properties   390–392
– synthesis via biodegradation of starch   392–398
cyclomaltodextrinases   392, *400*
cytochrome c   *47, 48, 51, 55*

d
deacylation
– chitin   411, 412
– enantioselective   298, 299, *309*
– lipases   278, *279*
degradation, *see* biodegradation; depolymerization
Deinococcus geothermalis   226, 227
deoxyribonucleic acid (DNA)   188, 201, *202*
depolymerization
– alginate   411
– lignin   26
– lipase-catalysed   328, *329, 330*
– suberin   27
dextran   *212*
dextrins   398
diacyl donors, enantioselective polymerization   284, *285*
dialkyl esters, polyamides from   134, *135*
diamines, polyamides from   134, *135*
dicarboxylic acids, polyesterification   92–97, *98*
dicarboxylic anhydrides   *117*
Diels–Alder reaction   15, *16, 17*
diesters, dynamic kinetic resolution   290–293
diethylene octane-1,8-dicarboxylate   *337*
diethylene triamine   *135*
differential scanning calorimetry   299
difuran monomers   *13, 15, 16*
β-diketones   150, *151, 152,* 157
dimethylsulfoxide   70, *396*
diols
– dynamic kinetic resolution   290–293
– in polyesterification   92, 93, 98, 99, 100, 102
β-dipeptides   268
divinyl esters   99, *100,* 102
docking   351, 360, 362, *363*
ω-dodecanolactone   109
double-wall carbon nanotubes   44
drug delivery   54
dual initiator approach   310–316
dynamic kinetic resolution   278, 287–295

e
electrical conductivity
– PANIs   191, 194, 195, 197, 198
– polypyrroles   205
electrodes, CNT modified   45, *47*
electrostatics, CNTs   46, 48
enantioselective polymerization
– acylation/deacylation   298, 299, *300*
– diacyl donors   284, *285*

enantioselectivite polymerization, lipases 278–280
endo-β-acetyl-hexosaminidases 408, 409
endo-β-glucuronidases 408
endoamylases 401
endoglucanases 415
enol esters, *in vitro* polyesterification 97–100, *101*, 102
enzyme activated chain segment 306
Enzyme Commission classification scheme 214
enzyme degradation, *see* biodegradation
enzyme immobilization 35, 36, 68–71
– carbon layered materials 43, *43*, 44
– clays 38, *39*, *40*, 40–42
– CNTs 45, 46, *47*, 48, 48–50, *51*, 52
– horseradish peroxidase on surfaces 200–202
– nanoparticles 52–57
enzyme leakage 48, 65, 68, 70, 71
enzyme production 66, 67
epigallocatechin gallate 180, 181
epoxidation, vegetable oils 18, *20*
epoxy-containing polyesters 101
equilibrium-controlled synthesis 231
Escherichia coli 225, 226, 229, 250, *400*
esterase *373*, 380
esters, *see also* polyester synthesis
– alcoholysis and aminolysis 354–357
– cleavage 405
– cyclic 111–113
– dialkyl 134, *135*
– diesters 290–293
– kinetic resolution 288
1-ethyl-3-(3-dimethylaminopropyl) carbodiimide 48
ethyl esters 99
ethyl valerate 338
3,4-ethylendioxythiophene 205, 206
exoamylases 401
exotic solvents
– biphasic polymerization 339–342
– enzymes employed in *324*
– introduction 323, 324
– ionic liquids 334–339
– other 342, 343
– supercritical CO_2 313, 320, 324–334

f

fatty acids 18, *19*, *20*, 27
fermentation 66
ferriprotoporphyrin IX *147*, 148, *166*, 194, 195
ferritin 49

fibers, PANI *197*, 202, 378, *379*
films, PANI 199–203
Flavobacterium sp. *400*
flavonoids 179–181
fluoruos biphasic polymerization 342
formaldehyde 165
formulation 67
Fourier transform infrared (FTIR) spectroscopy 41, 42, *51*, 379
free radical polymerization
– from bifunctional initiators 329, 330
– in ionic liquids 334–336
– living 300–301
– using initiators 333, 334
functionalization 50, 56
fungal cutinases 371, 372, *373*
fungal lipases *373*
fungal oxidases 378
furandialdehyde 12, *14*, *15*
furans 11–16, *17*
furfural 11, *12*, *13*
furfuryl alcohol 12
Fusarium oxysporum *373*
Fusarium solani *373*, 374, *375*, 378–380

g

gel permeation chromatography 73, 75, 76, 77
glass-transition temperature 390
glassy carbon electrode *47*
glucansucrase 228
glucansucrase glycosyl-transferase R 228
glucoamylase 397
α-D-glucopyranose 390, 391, 414
glucose 214–215, 226, *402*, 403, 414
– biosensor *47*, 55
– priming activity 218
glucose-1-phosphate 216, *217*, *218*, 219, *226*
glucose oxidase *39*, 43, 44, *47*, 55
– in PANI synthesis 198
β-glucosidases 42, *47*, 415
glutaraldehyde 56, 68–70
glycerol 8–10, *9*, 96
glycidyl methacrylate *311*, 314, 317
glycogen 215, 225, *226*
glycogen phosphorylase 215
– amylose polymerization with 215–219
– hybrid structures with amylose blocks 220–224
glycosaminoglycans 236
glycosidases 212, 213, 231–237
– cellulase 232–234
– glycosynthases 236, 237
– hyaluronidase 234–236

glycosidic linkages/bonds 213, 215, 216, 393
– branching enzymes 225, *226*
– cellulose 414
glycosyl fluorides 232
glycosylation 213
glycosyltransferases 213–230
– amylomaltase 228, 229
– branching enzyme 224–227
– hyaluronan synthase 229, 230
– phosphorylase 214–224
– sucrase 227, 228
glycosynthases 236, 237
gold nanoparticles 53, *54*, *55*, 56
graft copolymers 319, 320, 332
'grafting to/from' approaches 220
granules
– cyanophycin 261
– PHA 252, 253, 257
graphene *43*, 44
graphite *43*, 44
green solvents, *see* exotic solvents
green tea 179, 180
GTF180 enzyme 228

h

H-shape block copolymers 318
hematin 194, 195
heme-containing proteins 195
hemoglobin *51*, 195
1,6-hexanediol 357, 358
high-molecular weight hyaluronan 408
horseradish peroxidase 44, 46, *47*, 51
– biphasic polymerization 340
– exotic solvents 324
– free radical production 333, 334
– nanoparticles *51*, 55
– PANI synthesis 188, 189, 195, 196
– – immobilization on surfaces 200–202
– phenolic polymerization 165, *166*, 167, 171, 174, 175
– polythiophene synthesis 206
– vinyl polymerization 149, 153, 154, *155*, 156
Humicola insolens cutinase 87, *90*, 93, *105*, *107*, *330*, 373
hyaluronan/hyaluronic acid 175, 176, 229, 230, *234*, 235
– biological and clinical significance 408, 409
– structure, biological functions, clinical applications 406–408

hyaluronan synthase 229, 230, 406
hyaluronidase 234–236, 408, 409
hydrogel *175*, 176
hydrogen peroxide *147*, 153
– concentration 148–150
– in polypyrrole synthesis 203, 204, *205*
hydrolases 213
hydrolyzable tannins 21, *22*
hydrophilicity
– PET 371, 372, 374, 375
– polyamides 376, 377
hydrophobicity 68
– CNTs 46
hydrophobins 380
hydroxyacids/esters, polycondensation 85, 86, *88*
hydroxyalkanoic acids *250*, 251
3-hydroxybutyric acid 339
10-hydroxydecanoic acid *311*, 314
hydroxymethylfuraldehyde 12, 13, *14*
hyperthermia 55

i

immobilization, *see* enzyme immobilization
immobilization supports, *see also* Lewatit; Novozym 435; NS81018
– CALB 73, 74
– polycaprolactone synthesis 79, 80
– properties 68
initiation *144*, *145*, 148
– macroinitiation *306*, 307–309
initiators
– bifunctional 329–332
– free radical polymerization using 333, 334
– macroinitiators 310–318
intercalation 36–38
intermolecular esterification 86
intestinal flora 404, 405
intrinsically conducting polymers 187
ionic liquid-supported catalyst 340, *341*
ionic liquids 334–339
– cellulose hydrolysis 415
– free radical polymerization 334–336
– lipase-catalyzed polymerization 337–339
iron protoporphyrin, *see* ferriprotoporphyrin IX
isoelectric point 38
isoidide 6, *7*
isomannide 6, *7*
isoprene (2-methyl-1,4-butadiene) 2
isosorbide 6, *7*, *15*
iterative tandem catalysis 294, 295

k

Kazlauskas rule 354, 355
kinetic resolution
- dynamic, racemic monomers 287–295
- racemic monomers 284–287
kinetically controlled synthesis 132, 231
Klebsiella 90, 400

l

laccase, *see also individual laccases*
- bacterial 373
- functional molecules from 369, 370
- in PANI synthesis 197, 198
- in phenolic polymerization 176, 177, 179, 180
- in polypyrrole synthesis 204, *205*
- in vinyl polymerization 144, *145*, 156–158
laccase-mediator-system 157
laccol 177
lactate oxidase 44
lactic acid derived O-carboxy anhydrides 282
D,D-lactide 280
D-lactide 282
L-lactide 282
Lactobacillus reuteri 228
Lactococcus lactis 249
lactones, *see also* ε-caprolactone; polycaprolactone
- aldonic acid 220, 221
- iterative tandem catalysis 294, 295
- kinetic resolution 286, 287
- ring-opening copolymerizations 113, *114*, 116
- ring-opening polymerizations 85, 286, 287
- – substituted 109, 110
- – unsubstituted 102, *103*, 106, *105–107*, 107, 109, *109*
lag phase 149
layer-by-layer films 202, 203
Leloir glycosyltransferases 213
Lewatit 65, 69, 70, *74*, 79
lid 67
lignin
- acrylamide grafting 158
- fragments 23, *24*, 25, 26
- *in vitro* synthesis 174
lignocelluloses 369, 370
linear block copolymers, amylose 223, 224
lipases 67, 68, 277, 278, *see also Candida antarctica* lipase B (CALB); *individual lipases*
- AA-BB polyesterification 92–96, 97
- AB polyesterification 87, 88, 90, 91
- block and graft copolymer synthesis 305
- depolymerization in supercritical CO_2 328, 329, *330*
- enol ester polyesterification 97, 99, 100, *101*, 102
- fungal and bacterial 373
- homopolymerizations in supercritical CO_2 326–328
- kinetic resolution of racemic monomers 284, 295
- molecular modeling 349, *350*, 354–357
- PET hydrolysis 372
- polyamide synthesis 134–136
- polyester synthesis 84, 357, *358*
- polymerization in ionic liquids 337–339
- reaction mechanism and enantioselectivity 278–280
- ring-opening copolymerizations 119, *120*
- ring-opening polymerizations 102, *103*, 104, *104–107*
- synthesis and polymerization of optically pure monomers 280–284
- tuning polymer properties with chirality 296–301
lipolase 330
lipoxygenase 145
liquid chromatography under critical conditions *312*
living free radical polymerization 300, 301
low-molecular weight hyaluronan 408
lysozyme *51*, 413

m

macroinitiation *306*, 307–309
macroinitiators
- dual initiator approach 310–316
- enzymatic blocks forming 316–318
magnetic nanoparticles 52–56
magnetite nanoparticles *54*, 55
magnetofection 55
magnetosomes 53
bis-maleimide monomers 15, *16*, 17
tris-maleimide monomers 16
malic acid 97, 283
maltooligosaccharides 218, 220, 224, 229
maltopentaose *218*, 223
maltose 402, 403
maltose-binding sites 394
maltotetraose 218, *222*, 402, 403
maltotriose 218, 219, 402, 403

D-mannitol   96
maximal enzymatic conversion rate   145
medium   66
11-mercaptoundecanoic acid   91
4-methyl-ε-caprolactone   296–301, *311*, 312
6-methyl-ε-caprolactone   294, 295
methyl ε-hydroxyhexanoate   87
methyl methacrylate   296, 297
– bifunctional initiation   330, 331
– dual initiator   *311*, 313
methyl valerate   338
ω-methylated lactones   294, 295
5-methylfurfural   11
micelles, anionic, PANI synthesis   193, 194
Michaelis constant   145, 146
Michaelis–Menten kinetics   145, 146
Micrococcus luteus   378, 379
molecular dynamics   351
molecular mechanics   352, 362, 364, *365*
molecular modeling
– enzymatic polymerization   352–367
– – alcoholysis and aminolysis of esters 354–357
– – CALB   353–356, *357*
– – – β-lactam   357, 359–367
– – polyester formation   357, *358*
– introduction   349–352
Mucor   90, 92, 93, *105*, 324
multi-angle light scattering (MALS) detector   73, 75, 77
'multi-chain' polymerization   218, 219
multi-wall carbon nanotubes   44, 47, 48, 51
multiarm heteroblock star-type copolymers   116
murein ligases   263, *264*
mutans streptococci   181
mutant variants, CphA   265, 266
myoglobin   51

n

nanoentrapment   56
nanomaterials   35, 36
– carbon layered materials   43, *43*, 44
– clays   36–38, *39*, 40, 40–42
– CNTs   44–46, *47*, 48, 48–50, *51*, 52
nanoparticles
– applications   53, *54*, 55
– immobilization approaches   55, 56
– immobilized enzymes, structure and catalytic behavior   57
– introduction   52, 53
nanowires, PANI   201, 202
naringin   398
neohesperidin   398

neutral active hyaluronidases   409
nitroxide mediated living free radical polymerization (NMP)   297, 310, *311*, 315
Nocardia farcinica   377
non-Leloir glycosyltransferases   213
Novozym 435   65, 69–71, 326
– kinetic resolution   287
– NS81018 comparison   75, 76
– optically pure monomers from   280, *281*, 283
– and PCL molecular weight   74, 75
– polyamide synthesis   135
– in polyesterification   93–95, 96, *97*, 100
– in ring-opening polymerization   109
– synthesis parameters   78
– vacuum drying   78, 79
Noyori type catalyst   294
NS81018   66, 71
– crosslinking effects   80
– Novozym 435 comparison   75, 76
– synthesis parameters   78
– vacuum drying   78, 79
nucleophilic elbow   278
nylon   376, 378

o

1-octyl-3-methylimidazolium dicyanamide 335, 338
oligoglycerols   9
optically pure monomers, polymerization 280–284
organo-modified clays   40, *41*, 42
ornithine   259, 267
orthophosphate   216, *217*, 218
oxazolinium ion intermediate   234, 235
oxidoreductases   165, 370
oximes   102
oxiranes   87, 93, 117
oxpropylation   25
oxyanion hole   278, 353, 355, 360
oxyphenylene   167

p

P-loop motif   263
palm tree peroxidase   196
palygorskites   40
PANI, *see* polyaniline
papain   133
Pasturella multicoda   230
PCL, *see* polycaprolactone
PEG, *see* polyethylene glycol
Penicillium citrinum   373, 375
Penicillium roruefortil lipase   105
Penicillium vitale   203

pentablock copolymers *317*
pentadecalactone
– ring-opening copolymerization 113, *114*, 115, 119
– ring-opening polymerization 104, *104*
2-pentylpropanoic acid 356
peptide synthesis 248
perfluorooctyl methacrylate *311*, 314
peroxidase, *see also* horseradish peroxidase; soybean peroxidase
– in functional phenolic synthesis 170–176
– nanoparticles 51
– in PANI synthesis 196, *197*
– in phenolic polymerization 165–170
– in vinyl polymerization 144, *145*, 146–156
persistence length 221
PET, *see* polyethylene terephthalate
PHAs, *see* polyhydroxyalkonates
phasins 253, 257
phenolics
– enzymatic preparation of coatings 177–179
– flavonoid polymerization 179–181
– functional, peroxidase-catalyzed synthesis 170–176
– introduction 165
– laccase-catalyzed polymerization 176, 177
– peroxidase-catalyzed polymerization 165–170
phenol(s) 166, 167
– bisphenols 168, 173, 174
– free radical polymerization 335, *336*
phenylene 167
1-phenylethanol 289, 290
phosphoglucomutase 215
phosphohydrolase 39
phosphonate ligand *353*, 354, 357
phosphorolysis 216, *217*
phosphorylase 214–224, 226, 227
– amylose polymerization with glycogen phosphorylase 215–219
– hybrid structures with amylose blocks 220–224
phosphorylation 247, 248
photoresists 173, 174
physical adsorption 68
β-pinene 2, 3, 4
ping-pong bi-bi mechanism 357
polyacetylenes 298
polyacrylonitriles, surface hydrolysis 378–380
poly(β-alanine) 359, 360

polyamides
– surface hydrolysis 376–378
– synthesis
– – from lipases 134–136
– – modes 247, 248
– – molecular modeling 352
– – from other enzymes 136, 137
– – from proteases 133, 134
– – from sugars 7, 8
poly(amino acids) 248
polyaniline
– biomimetic synthesis 194, 195
– films and nanowires prepared from 199–203
– introduction 187
– synthesis in template-free, dispersed and micellar media 192–194
– synthesis using enzymes different from HRP 195–199
– synthesis using templates 188–191
polyanion-assisted polymerization 188–190
poly(aspartic acid) 268
polybutadiene macroinitiator 307
polycaprolactone 108, *109*
– depolymerization 328
– in phenolic polymerization 176, 177
polycaprolactone synthesis
– block copolymers 308, 311, 312, 314
– graft copolymers 319
– immobilization supports 79, 80
– molecular weight and synthesis time 74, *75*, 76
– molecular weight determination by GPC 75, *76*, 77
– procedures 72
– in 1,1,1,2-tetrafluoroethane 342, *343*
– solvent effects 78
– in supercritical CO_2 327
– termination 77, 78
– water effects 78, 79
polycarbonates 99, 111, *112*, 282, 283
poly(catechin) 179–181
polycation-assisted polymerization 190, 191
polycondensations 85–102
– AA-BB type polyesterification 92–97, *98*
– AB type polyesterification 86, 87, 88–90, 91, 92
– block copolymers 308, 309
– enol esters for *in vitro* polyesterification 97–99, *101*, 102
– four basic modes 86
– and ring-opening combined 119–121

poly(depsipeptides)   283
polyester synthesis   84–121
– AA-BB type   92–97, 98
– AB type   86, 87, 88–90, 91, 92
– from anhydroalditols   7
– from citric acid   28
– combined condensation and ring-opening polymerization   119–121
– from enol esters   97–99, 101, 102
– from furans   15
– introduction   83, 84
– lipases   84, 357, 358
– polycondensations   85–87, 88–90, 91–102
– ring-opening copolymerizations   113–119, 119, 120
– ring-opening polymerizations   102, 113
poly(2,5-ethylene furancarboxylate) (PEF)   14
polyethylene glycol   49, 169, 194, 195
– di- and tri-block copolymers   307, 308
poly(ethylene oxide)   223
polyethyleneimine (PEI)   69, 71
poly-(γ-glutamate)   249
polyglycerols, hyperbranched   10
polyglycidols   319
polyhydroxyalkonate synthases   249–257
– catalytic mechanism   254
– embedding in general mechanism   255, 256
– enzyme activity assay   252
– enzyme location and granule structure   252, 253
– enzyme primary structures   253, 254
– reaction catalyzed by   251
– special motifs and essential residues   254
– in vitro synthesis   255
polyhydroxyalkonates
– biotechnological relevance   256, 257
– chemical structures   250, 251
– occurrence   249, 250
poly(3-hydroxybutyrate)   249, 296, 308
poly-L-lactide   277, 281, 282
polylactones   338, 339
poly-(ε-lysine) synthetase   249
polymer brushes   220, 221, 371
polymer modification
– future developments   380
– introduction   369
– from natural to synthetic materials   369, 370
– surface hydrolysis of polyacrylonitriles   378–380
– surface hydrolysis of polyamides   376–378
poly(methyl methacrylate)   65, 74

polypeptides
– crosslinking   175, 176
– from other enzymes   136, 137
– from proteases   132–134
polyphenol oxidase   39
polyphenols   179, 342
poly(phenylene oxide)   170, 171, 176
polypropylene   74
polypyrrole   187, 203–205
polyricinoleate   91
polysaccharides, see also cellulose; hyaluronan/hyaluronic acid
– alginate   409–411
– chitin and chitosan   193, 411–414
– conclusion   237, 238
– cyclodextrins   390–406
– glycosidases   213, 231–237
– glycosyltransferases   212, 213–230
– industrial uses   212
– introduction   211–213
polystyrene(s)   77, 155
– sulfonated   188, 189, 195, 200, 206
polythioesters   251, 252
polythiophenes   187, 205, 206
polytransesterification   14
poly(trimethylene terephthalate)   10, 374, 375
polyurethanes   7, 8, 380
porcine pancreatic lipase   90, 94, 105, 285
– in exotic solvents   324
– in ring-opening copolymerization   120
potato phosphorylase   218, 220, 221, 227
priming activity   218, 219
prodrugs   404, 405
production strains   66
Proleather   95
propagation reaction   144
β-propiolactone   339
propyl laurate units (PLU) assay   71, 79, 80
proteases
– classes   132
– in polypeptide and polyamide synthesis   132–134
– surface hydrolysis   376, 377
protein database (PDB)   349, 350, 354
protoporphyrin   147, 148, 166, 194, 195
Pseudomonas   40, 373
– in ring-opening copolymerization   120
– in ring-opening polymerization   106
Pseudomonas aeruginosa   90, 105, 253
Pseudomonas cepacia lipase   40, 87, 90, 104, 105, 106, 106, 373
– in exotic solvents   324
– optically pure monomers from   280, 281

Pseudomonas fluorescens lipase   99, 106, *106*, 110
– optically pure monomers from   280
– ring-opening copolymerization   *118, 120*
Pseudomonas Msl   401
Pseudomonas putida   253
purification   66, 67, 72, 73
pyridoxal-5′-phosphate   216, 217

q
Q-enzyme   225
QDE 2-3-4   327
quantum mechanics   351, 352, 362, 364, 365
Quecus suber   26, 27

r
R-134a   342, *343*
racemic monomers
– dynamic kinetic resolution   287–295
– kinetic resolution   284–287
radical chain polymerization   144, *152*
radical coupling   *170*
radical formation   *170*
radical transfer reaction   *170*
Ralstonia eutropha   253
recovery step   66, 67
regioselectivity, enzyme   95, 97
renewable resources   1–29
resin acids   4, 5
reverse micellar polymerization   175
reversible addition fragmentation chain transfer (RAFT)   4, 310, *311*, 315, 331
Rhizopus delemer lipase   106
Rhizopus japonicus lipase   106
Rhodococcus rhodochrous   379
Rhodococcus ruber   253
ring-opening copolymerizations   113–119, *119, 120*
– ε-caprolactone   113, *114*, 296, *297*
ring-opening polymerizations
– and condensation combined   120–122
– cyclic ester related monomers   110–113
– lactones   85, 286, *287*
– and living free radical polymerization   299, 300
– substituted lactones   109, 111
– in supercritical CO$_2$ 327
– unsubstituted lactones   102, *103*, 104, *105–107*, 107, 108, *109*
RNAase   48, *51*
rod–coil systems   221
rosin   4–6
rutin   181

s
scanning electron microscopy (SEM)   42, 69
sebacic acid   92, 93
self-doped PANIs   190, 191, 194
Ser$_{105}$ 357, *358*, 364, 365
Shvo's catalyst   295
silica supports   74, *79*, 80
silicon substrates   199, 200, *201*
sinapyl alcohol   174
single-wall carbon nanotubes   44, *47*, 48, 49, *51*
size exclusion chromatography   *312*
smectite clays   36, *37*, 40
sodium dodecylsulfate   195
sodium dodecylbenzensulfonate   193, 194
sorbitol   96
soybean peroxidase   *51*, 167, 168, 174
– in exotic solvents   *324*
– in PANI synthesis   192, 196
starch   214, 390
– biodegradation   392–398
– biosynthesis   225, *226*
– industrial uses   *212*
– metabolism   216
starch branching enzyme   225
starch-urea phosphate   178
stevioside   397
Streptococcus   407
Streptococcus equi   229
Streptococcus oralis   228
Streptococcus pyogenes   230
Streptococcus zooepidemicus   229
Streptomyces coelicolor   373
styrene, dual initiator   *311*
styrene polymerization   151
– enantioselective   298, *299, 300*
suberin fragments   26, 27
suboxide dismutase   159
succinic acid   29
sucrase   227, 228
sucrose phosphorylase   226
sugars   6–8, *see also* glucose; maltose; polysaccharides
– UDP-sugars   230
sulfonated polystyrene   188, 189, 195, 200, 206
sunflower oil   *330*
supercritical carbon dioxide   313, 320, 324–334
– in biphasic polymerization   340, *341*
– free radical polymerization   333, 334
– lipase-catalyzed depolymerization   328, *329, 330*

– lipase-catalyzed homopolymerizations 326–328
– polymerization from bifunctional initiators 329–332
– supercritical region 325
supercritical region 325
superoxide anion scavenging 179, 180, 181
surface confinement 200, 201
– nanowires and thin films from 201, 202
surface hydrolysis
– polyacrylonitriles 378–380
– poly(alkyleneterephthalate)s 370–376
– polyamides 376–378
surface-initiated polymerization 220, 221
surfactants
– CNT immobilization 49
– lipase homopolymerizations 326
– PANI synthesis 193–195
– peroxidase polymerization 152, 153
– from polysaccharides 220
synchrotron infrared microspectroscopy 69, 70

t

tannins 21, 22
tartaric acid 28, 282
template-free PANI synthesis 192
templates, PANI synthesis using 188–191
terephthalates 7
termination reaction 144
terpenes 2, 3, 4
terpolyesters 96
terthiophene 206
tetrahedral intermediate TI1 360, 361, 362
tetrahedral intermediate TI2 361, 363, 364, 365
tetrahedral intermediate TI3 361, 364, 365
tetrahydrofuran 72, 77
Thermoanaerobacter 396, 397
Thermobifida alba 373
Thermobifida fusca 372, 373, 374, 375, 375
Thermococcus 397
Thermomyces lanuginosus 372, 373, 374, 375
Thermotoga maritime 400
Thermus ethanolicus 400
thiamin 398
thitsiol 177
tissue repair 54
toluene 396
Trametes hirsuta laccase 47, 197, 198

Trametes versicolor laccase 39, 43, 48, 204
transesterification
– enzymatic, on polymer backbone 299, 300
– polytransesterification 14
transferases 212
transglutaminases 370
transglycosylation 231, 232, 237, 394
Trichoderma cirude 330
triglycerides 18, 19, 84
trimethylcarbonate 107, 108, 111, 117, 118
trimethylolpropane 95
Triticum aestivum 373, 374
turpentine 2, 3
Tween 349, 350, 354
tyrosinases 44, 55, 370

u

UDP-sugars 230
unconventional solvents, *see* exotic solvents
urease 39, 44
uric acid 180
urishi/urishiols 177, 178

v

δ-valerolactone 104, 105, 339
vegetable oils 16–18, 19, 20
vine-twining polymerization 224
vinyl polymerization
– general mechanism and enzyme kinetics 143–146
– introduction 143
– laccase-initiated 144, 145, 156–158
– miscellaneous enzyme systems 159, 160
– peroxidase-initiated 144, 145, 146–156
– – enzyme concentration 148, 149
– – hydrogen peroxide concentration 148–150
– – mechanism 147, 148
– – mediator/mediator concentration 150, 151, 152
– – miscellaneous 152, 153
– – selected examples 153, 154, 155, 156
– state of the art and future developments 160, 161
viscoaugmentation 407
viscoprotection 407
viscoseparation 407
viscosupplementation 407
viscosurgery 407
vitamin B_6 enzymes 216
vitamin P 181
volatile organic compounds 323

w

water-in-oil microemulsion 56
water solubility 390
white-rot fungi 198

x

X-ray crystallography 349
X-ray diffraction 40, 41

X-ray photoelectron spectroscopy (XPS) 41, 374–376, 379
xanthine oxidase 145, *159*, 180, 181
Xanthomonas campestris 400
xylanase 47

z

Zygomycetes 412